Physical Activity and Health

SECOND EDITION

Claude Bouchard, PhD
Pennington Biomedical Research Center

Steven N. Blair, PED
University of South Carolina

William L. Haskell, PhD
Stanford University

Editors

Human
Kinetics

Library of Congress Cataloging-in-Publication Data

Physical activity and health / Claude Bouchard, Steven N. Blair, and William L. Haskell, editors. -- 2nd ed.
    p. ; cm.
  Includes bibliographical references and index.
  ISBN-13: 978-0-7360-9541-9 (print)
  ISBN-10: 0-7360-9541-1 (print)
  I. Bouchard, Claude. II. Blair, Steven N. III. Haskell, William L.
  [DNLM: 1. Exercise--physiology. 2. Physical Fitness--physiology. 3. Health.
QT 255]
  613.71--dc23

                                        2011040336

ISBN-10: 0-7360-9541-1 (print)
ISBN-13: 978-0-7360-9541-9 (print)

The web addresses cited in this text were current as of July 2011, unless otherwise noted.

Acquisitions Editor: Amy N. Tocco; **Developmental Editor:** Melissa J. Zavala; **Assistant Editors:** Kali Cox and Katherine Maurer; **Copyeditor:** Joyce Sexton; **Indexer:** Betty Frizzell; **Permissions Manager:** Dalene Reeder; **Graphic Designer:** Nancy Rasmus, **Graphic Artist:** Tara Welsch; **Cover Designer:** Bob Reuther; **Photographer (cover):** Arne Trautmann/Panther Media/ age fotostock; **Photographer (interior):** © Human Kinetics, unless otherwise noted; **Photo Asset Manager:** Laura Fitch; **Visual Production Assistant:** Joyce Brumfield; **Photo Production Manager:** Jason Allen; **Art Manager:** Kelly Hendren; **Associate Art Manager:** Alan L. Wilborn; **Art Style Development:** Joanne Brummett; **Printer:** Thomson-Shore, Inc.

Printed in the United States of America   10 9 8 7 6 5 4 3

The paper in this book is certified under a sustainable forestry program.

Human Kinetics
Website: www.HumanKinetics.com

United States: Human Kinetics
P.O. Box 5076
Champaign, IL 61825-5076
800-747-4457
e-mail: humank@hkusa.com

Canada: Human Kinetics
475 Devonshire Road Unit 100
Windsor, ON N8Y 2L5
800-465-7301 (in Canada only)
e-mail: info@hkcanada.com

Europe: Human Kinetics
107 Bradford Road
Stanningley
Leeds LS28 6AT, United Kingdom
+44 (0) 113 255 5665
e-mail: hk@hkeurope.com

Australia: Human Kinetics
57A Price Avenue
Lower Mitcham, South Australia 5062
08 8372 0999
e-mail: info@hkaustralia.com

New Zealand: Human Kinetics
P.O. Box 80
Torrens Park, South Australia 5062
0800 222 062
e-mail: info@hknewzealand.com

E5182

We dedicate this book to two remarkable physical activity epidemiologists, Ralph S. Paffenbarger, Jr., MD, and Jeremy Morris, MD, both of whom recently passed away.

Contents

PART III   Physical Activity, Fitness, and Health

Contents

PART IV Physical Activity, Fitness, Aging, and Brain Functions

PART V   How Much Is Required and How Do We Get There?

PART VI   New Challenges and Opportunities

Preface

The second edition of *Physical Activity and Health* is designed for upper-level undergraduate and graduate students studying the health benefits associated with a physically active lifestyle and a moderate level of fitness versus the potential deleterious consequences of physical inactivity. The book is intended for students in kinesiology, exercise science, physical education, public health, health promotion, preventive medicine, and human biology programs.

This textbook provides an integrated treatise on the relationship between physical activity or sedentarism and health outcomes. It also provides a conceptual framework to help the student relate results from single studies or collections of studies to the overall paradigm linking physical activity and physical fitness to health. The book focuses on the prevention of diseases and enhancement of quality of life and well-being. It does not deal extensively with the role of physical activity in the treatment of diseases and rehabilitation and is not intended to be an encyclopedic review of the field. Rather, each chapter provides an overview of the key concepts and the most important findings emerging from a collection of studies, discusses the limitations of the current knowledge base, and identifies research needs.

The need for this book became evident to us when reviewing this literature in the context of a variety of consensus conferences. A single book has not been available to students that brings together the results of key studies and presents the relevant concepts in a detailed yet concise manner. The book was written with the collaboration of the finest scientists in the field from the United States, Canada, Europe, and Australia.

The book is organized into six parts containing 25 chapters. In the first part, four chapters define basic concepts; trace the history of the field; summarize evidence accumulated on different levels of physical activity and fitness and their variations with age, between women and men, and among ethnic groups; and discuss sedentary time and its impact on physiology and health outcomes. Part II includes five

chapters laying out our current understanding of the effects of acute and chronic exposures to physical activity. The nine chapters in

part III review the relationship between regular physical activity and health outcomes as well as between the level of fitness and the same health outcomes. These health outcomes range from cardiovascular morbidities to all-cause mortality. Part IV focuses on aging, brain functions, and mental health. Part V deals, in two chapters, with dose–response issues and the development of physical activity guidelines. Finally, part VI explores the challenges posed by advances in genetics in understanding the complex relationships among sedentary time, physical activity, fitness, and health resulting from interindividual variability. The last chapter provides an integrated view of the field and discusses new opportunities for research and implementation of these concepts in a public health perspective. Chapter 4 on physical inactivity, chapter 20 on brain functions, and chapter 23 on the development of physical activity guidelines are additions to the volume.

The student will notice that the chapters are concise and the reference lists are not extensive. The intent was to emphasize the key concepts and the most important studies. Many of these references are review papers that the interested student can examine for more in-depth discussion of selected issues and for further references. The student will also recognize common features throughout the chapters, such as untitled special elements that summarize major points; titled special elements that present related topics of interest; key terms and concepts; and study questions. In addition, the chapters are well illustrated with tables and figures.

We hope that the second edition of *Physical Activity and Health* will be as well received as the first one and will generate increased interest in the role that regular physical activity can play as part of a lifelong, comprehensive preventive medicine plan.

Acknowledgments

The impetus for the first edition of *Physical Activity and Health* began with a conversation between Dr. Rainer Martens, president of Human Kinetics, and one of us (CB) at the American College of Sports Medicine annual meeting held in St. Louis in 2002. After much discussion about a proposed publication and its content, we reached a consensus that the most pressing need was for a textbook for advanced undergraduate and graduate students that would succinctly cover a broad spectrum of topics in the area of physical activity and health. This is the goal that the three editors tried to achieve with *Physical Activity and Health*. We hope that we have come even closer to meeting our goal with this second edition of the book.

We have been able to assemble a distinguished panel of collaborators to ensure that this textbook is a state-of-the-art publication. We are grateful to our colleagues from the United States, Canada, Europe, and Australia who accepted our invitation. We hope that they will be as pleased as we are with the quality of the publication.

We all are personally indebted to the pioneers in this field who continue to influence our research and that of others around the world. In this regard, we are dedicating this book to Professors Ralph Paffenbarger Jr. and Jeremy Morris, who have had an enormous impact on us personally and on the field as a whole. They provided direction, inspiration, counsel, and leadership to us and to the overall discipline of physical activity epidemiology. They were marvelous examples not only for their scientific achievement, but also for their kindness and warm human spirit. They are sorely missed.

The three of us have enjoyed the generous and sustained support of the National Institutes of Health for our research over the years. One of us (CB) has also been the beneficiary of numerous grants from the Medical Research Council of Canada (as it was known before his relocation to the Pennington Biomedical Research Center in Louisiana) and other agencies from the government of Canada and the province of Quebec. We thank these agencies for their support over the years.

We also owe a great deal of gratitude to the colleagues, postdoctoral fellows, and students who have contributed so much to our research and productivity over the last few decades. It is impossible to recognize them individually here, but their names can be retrieved from our publications. They have greatly influenced our vision and thinking about issues of critical importance in preventing diseases and preserving health. For their contributions and their patience with us, we are very grateful. Also, we thank our respective institutions, the Pennington Biomedical Research Center, the University of South Carolina, and Stanford University for providing environments conducive to exploring our academic interests.

The staff at Human Kinetics have been very supportive of this endeavor. They also have been very patient with us despite the fact that we missed some early deadlines. In particular, we would like to recognize and thank Amy Tocco and Mike Bahrke, acquisitions editors; Melissa Zavala, developmental editor; and Kali Cox, assistant editor. Amy, Mike, Melissa, and Kali nurtured this project as if it were their own and were very patient with the authors and editors. We would also like to thank John Laskowski, knowledge management coordinator; Dalene Reeder, permissions manager; and Tara Welsch, graphic artist at Human Kinetics.

Finally, this second edition of the book would not have been produced and would not have been edited, with extraordinary attention to countless details, if it had not been for the dedication and competent contribution of Allison Templet from the Pennington Biomedical Research Center. Nina Rumler was involved in the planning of the second edition and in the development of the whole manuscript for the first edition but was replaced by Allison early in the preparation of this edition. Allison made our responsibility as editors much more enjoyable. Her incredible attention to detail in all aspects of the book, including interactions with the network of collaborators and the staff at Human Kinetics, is in the end responsible for much of its quality.

Even though Allison and the staff at Human Kinetics took great care in correcting errors and inconsistencies that may have existed in the manuscript, we take full responsibility for any omissions or errors that may remain in the publication.

PART I

History and Current Status of the Study of Physical Activity and Health

Part I includes an overview of the evolution and emergence of physical activity and health as an area of scientific investigation. This is a young field; systematic research on physical activity and health has been under way only since the middle of the 20th century. The field is maturing rapidly, the research database is extensive, and sedentary habits are now identified as a major public health problem in many countries of the world.

Four chapters in part I set the stage for what is to follow. These four chapters provide the organizational and conceptual framework for the book, a historical review of key developments, and the current status of physical activity and of sedentary time in the general population. Chapter 1, by the editors, provides an overall view of the purposes of the book. Chapter 2 includes an overview of key events in the development of physical activity and health as a scientific discipline in biomedical science and the ways in which this accumulated research helped make physical activity and health an important public health issue. Chapter 3 discusses variations in physical fitness and physical activity level with age and among sex and ethnic groups. Chapter 4 emphasizes the importance of the time spent in a sedentary state, particularly sitting, its biological implications, and how it affects the risk of disease and premature death. We hope that these introductory chapters will show you how the subdisciplines in exercise science and sports medicine are interrelated and are moving into an integrated field.

Why Study Physical Activity and Health?

Claude Bouchard, PhD; Steven N. Blair, PED; and William L. Haskell, PhD

CHAPTER OUTLINE

Human Evolution, History, and Physical Activity

Homo Sapiens Is the Product of an Active Mode of Life

Physical Activity: From the Advent of Agriculture to This Millennium

Burden of Chronic Diseases

Health and Its Determinants

Health and Morbidity

Genetic, Behavioral, and Environmental Determinants of Health

Aging and Health

Defining Physical Activity and Physical Fitness

Defining Physical Activity

Defining Physical Fitness

Physical Inactivity Versus Physical Activity

Physical Inactivity as a Risk Factor

Benefits of Regular Physical Activity

Summary

Review Materials

©Art Explosion

> "We in the West are the first generation in human history in which the mass of the population has to deliberately exercise to be healthy. How can society's collective adaptations match?"
>
> *Morris 2009*

Because of the dramatic changes in the lives of people in industrialized countries over the past century, the necessity for most people to engage in challenging **physical activity** has disappeared. As physical activity has diminished, a host of physical ills related to inactivity have become manifest. Thus, intentional physical activity has become an important component of a healthy lifestyle. As you explore this issue, you will discover that *Homo sapiens* won the war against physical work of all kinds but in the process has become afflicted by diseases brought about by a physically inactive lifestyle.

Your journey begins with this introductory chapter. In it you will learn about how health is defined in the 21st century compared with past periods; the concepts of health, quality of life, and longevity; the global burden of chronic diseases related to inactivity; some of the challenges posed by the aging of the population; the definition of the physical activity–fitness–health paradigm; and why **sedentary time**, physical activity, and fitness are poised to occupy a central place in preventive medicine and the public health agenda.

It is often said that the human body is designed for activity. Even though it is difficult to test this hypothesis in a formal experimental setting, at least three lines of evidence support this view.

- First, the human organism can adapt to a wide range of physical demands imposed by work and exercise. A young adult can easily increase his metabolic rate by 10-fold when exercising and can sustain this rate of energy expenditure for a few minutes. It is not uncommon to find people who increase their energy output 100-fold above resting in maximal performance of very short durations. So, the human body architecture and physiology appear to be well organized to perform muscular work over a wide range of metabolic rates.

- Second, spending excessive amounts of time in a sedentary state and a low level of physical activity have been associated with a poor risk profile for common diseases, loss of functional capacity, and premature death.

- Third, the early humans could not have survived in life-threatening environments without having both adequate motor skills and the ability to perform demanding physical work.

Human Evolution, History, and Physical Activity

Evolution teaches us that those carrying genetic alleles favoring motor skills, strength, speed, stamina, and other physical attributes at relevant genes were more likely to have enjoyed greater reproductive fitness because of their greater probability of securing food, attracting mates, and staying alive long enough to have children. The current chapter develops this line of reasoning. Other aspects of genetics are dealt with in greater detail throughout the book, particularly in chapter 24.

Homo Sapiens Is the Product of an Active Mode of Life

Although debate persists about the time and circumstances of the emergence of *Homo sapiens*, enough pieces of our evolutionary history are available to allow us to conclude that the basic theorem of the Darwinian theory of evolution of species is correct. The emergence of human beings was intimately related to progressive molecular changes in genes affecting posture, bipedal locomotion, and brain functions of closely related nonhuman primates. The evolution of the brain meant not only greater brain capacity, progressive mastery of language, and more refined intelligence, but also growing control over an expanded movement repertoire. From the research of paleontologists, anthropologists, anatomists, archaeologists, and molecular biologists, the main events in the evolution of our species can be briefly outlined with an emphasis on those that have implications for physical activity.

Physical activity and **physical fitness** have been major factors in the evolutionary history of *Homo sapiens*. The most important events in the evolution of modern *Homo sapiens* occurred within the past 10 million years. The exact details of the molecular events that fueled this evolution are still a matter of debate. However, the end results of this complex journey and the molecular distances between *Homo sapiens* and closely related species are faithfully registered within the human genome and the genome of these closely related nonhuman primate species. In brief, molecular alterations in the deoxyribonucleic acid (DNA) of germ cells of our primate ancestors, combined with the effects of natural selection, led over millions of

The Human Body Is Designed for Activity

- The human organism can adapt to a wide range of metabolic demands imposed by work or exercise.
- A low level of physical activity is associated with risk for common diseases and premature death.
- Evolutionary history teaches us that early humans could not have survived without the ability to perform very demanding physical work.

years to small creatures, clearly hominid in appearance, that are collectively referred to as *Australopithecus.* Several types of *Australopithecus* were uncovered on the African continent and dated as far back as 6 million years ago. In general, *Australopithecus* had a stature of about 1.5 m (5 ft) with a brain size about 40% of that of modern *Homo sapiens.* Darwinian selection, new genetic mutations, variation in gene copy numbers, chromosomal rearrangements, and undoubtedly other random genomic events progressively shaped human-like forms of life until *Homo habilis,* then *Homo erectus,* and finally modern *Homo sapiens* emerged more than 100,000 years ago.

It is difficult to establish with precision the role played by motor ability and physical performance capacity in the evolutionary journey of our species. Comparative studies at the DNA and protein level between *Homo sapiens* and the most closely related nonhuman primates indicate that the genomic differences are generally on the order of 2%, which is actually less than the genomic distance observed among individuals in a large heterogeneous population. However, these relatively small molecular differences between nonhuman primates and *Homo sapiens* carry major functional and behavioral implications, as is obvious to anyone who has observed today's primates in their natural habitat or in zoos around the world. Thus, human beings have optimized several traits that carried evolutionary advantages such as upright posture, bipedal locomotion, well-articulated thumbs

for better hand prehension, vertical head position facilitating visual scanning, and refined language capacity, to name but a few. All of these characteristics had enormous selective advantages in the hostile environment prevailing during the evolutionary journey. They conferred an improved capacity to walk and run, grasp, carry, catch, throw, and perform well in activities requiring quick responses, precision, speed, strength, and endurance.

The emergence and spread of *Homo sapiens* involved conditions that required a high level of habitual physical activity, especially relative to today's standard. Furthermore, performance capacity and motor skill played a major role in survival. The best performers had a clear advantage in the quest for food and in defense against animal predators and during times of conflict. Thus, the best male performers are likely to have contributed more genes to the next generations than the other males. Likewise, fitness and performance capacities played an important role in the success of females, who were called on to bear children as well as to help in gathering food, firewood, and other necessities while caring for their offspring.

Physical activity, then, has been a major force in the evolution of *Homo sapiens.* Hunting, gathering, escaping, and fighting were essential actions for the survival of our ancestors. They had to throw, lift, carry, climb, walk, run, and perform all kinds of basic motor skills throughout their lives. Thus, it is hard to

Traits With Selective Advantages in the Hostile Evolutionary Environment

- Upright posture
- Bipedal locomotion
- Well-articulated thumbs for better hand prehension
- Vertical head position to facilitate visual scanning
- Refined language capacity

imagine that physical activity and performance capacity were not important features during the evolutionary history of our species, conferring mating advantages to the carriers of the genetic alleles associated with these traits. Our ancestors would not have reached the age of reproduction if they had poor endurance, lacked speed and power, or had been clumsy. In other words, survival and reproductive success over tens of thousands of years required our ancestors to be physically active and good performers. Darwinian fitness was closely associated with physical fitness in the early ages of our species. Human reproductive capacity today, however, is less likely to depend on the level of physical activity and fitness than it did in the past (Malina 1991).

Physical Activity: From the Advent of Agriculture to This Millennium

With the advent of agriculture and animal domestication, when humans began to live in larger settlements, muscular work remained important. Strength, endurance, and skill must have been associated with economic success and survival in those times. Our ancestors learned to use various metals; the wheel was conceived; and tools of all kinds were developed to ease the burden of hunting, agriculture, and various domestic chores. Soon, some people began to have more leisure time. Indeed, archaeological records have provided us with ample evidence of leisure activities in communities 5,000 to 8,000 years ago in several parts of the world. It is evident from museum artifacts that among these leisure activities, physical activities were quite popular. Images of foot races, throwing contests, wrestling, dances, and hunting are well represented in the archaeological findings of this era.

By about three millennia ago, physical activities had become popular, in part because people believed that these activities influenced normal development and health. This was such a strong idea that elite performers were deified. In ancient Greece, the Olympic Games were started about 776 B.C. and were intimately associated with the civilization of the time.

During the next millennium, men and women remained interested in physical activities, as shown by the various types of games, tournaments, dances, and hunting expeditions that captured the attention of the nobles and the richest people. For most, however, physical activity meant long, hard days of physical labor that was necessary to subsist and satisfy a demanding master. Living conditions obviously had improved, but muscular work remained absolutely essential for the survival of the majority of people. Numerous wars also tested the fitness levels and performance capacities of the soldiers and of the people caught between rival factions.

The Renaissance period, with its taste for beauty and knowledge, changed the Western world. Physical activities remained quite popular, and not only among the rich and noble. Games evolved but remained largely influenced by preoccupations with war and hunting. Large tournaments were regular occurrences on the aristocrats' agenda. Dances were also highly popular with the nobility. Peasants continued working hard, but they also enjoyed wrestling matches, horse racing, archery competitions, and dances. Jean-Jacques Rousseau introduced proposals to reform the education of children; these were quite compatible with the teachings of those who thought that physical activity should be part of the educational system. Widespread interest in sport soon appeared on the scene.

Throughout this journey, the struggle to free human beings from muscular work and physical exertion was a constant feature. It made impressive gains during the industrial revolution and even more in the past century through the technological progress achieved in industrialized countries. Lately, however, the alarm bell has begun to sound, and it is proposed that the reduction in the amount of physical work may have gone too far (figure 1.1). The benefits of a physically active lifestyle have been compared with those of an inactive mode of life; and although additional evidence is needed, it seems that human beings are better off when they keep sedentary time (such as in sitting) at a minimum and maintain a physically active lifestyle. This is certainly one of the most striking paradoxes of the evolutionary and historical journey of *Homo sapiens*.

Burden of Chronic Diseases

Chronic diseases are the most serious public health burden that the world faces today. Among these chronic diseases, cardiovascular diseases and cancer are the most important—and they are, as we shall see throughout this text, related to physical activity or lack of it. Moreover, the burden of chronic diseases is rapidly increasing around the world. It has been estimated that chronic diseases contributed approximately 60% of the 59 million reported deaths in the world in 2004 (World Health Organization 2009). Almost half of the total deaths related to chronic disease are attributable to cardiovascular diseases. Table 1.1 lists the 10 leading risk factors for death in the world. It is important to note that high blood pressure and tobacco use are the two major causes of death. They are closely followed by high blood glucose, physical inactivity, and excess body weight.

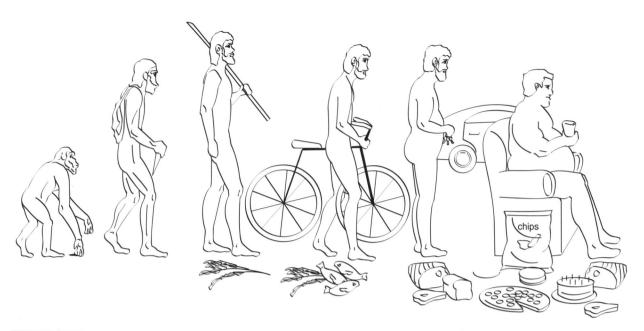

Is evolution leading *Homo sapiens* to *Homo sedens*, with a panoply of undesirable behavioral and biological features?

A Paradox From the Evolutionary Journey of *Homo Sapiens*

Now that the industrialized world has eliminated much of the need for hard physical labor to survive, the benefits of a physically active lifestyle have been compared with those of an inactive mode of life. Although not all the evidence is in, it seems that human beings are better off when they maintain a physically active lifestyle. This is certainly one of the most striking paradoxes emanating from the evolutionary and historical journey of *Homo sapiens*.

TABLE 1.1 Ten Leading Risk Factors for the Causes of Death in the World, 2004 (59 Million Deaths Total)

Risk factor	Deaths (millions)	% of total
1. High blood pressure	7.5	12.8
2. Tobacco use	5.1	8.7
3. High blood glucose	3.4	5.8
4. Physical inactivity	3.2	5.5
5. Overweight and obesity	2.8	4.8
6. High cholesterol	2.6	4.5
7. Unsafe sex	2.4	4.0
8. Alcohol use	2.3	3.8
9. Childhood underweight	2.2	3.8
10. Indoor smoke from solid fuels	2.0	3.3

Adapted, by permission, from World Health Organization, 2009, *Global health risks: Mortality and burden of disease attributable to selected major risks* (Geneva, Switzerland: World Health Organization), 11.

Obesity and diabetes are also on the rise. This trend is worrisome because these are both strong **risk factors** for vascular diseases and have started to appear earlier in life—even before puberty. The number of people with diabetes in the developing world will have increased from 84 million in 1995 to 228 million in 2025 (Aboderin et al. 2001). As for overweight and obesity, they are already epidemic in the developed nations and continue to increase in prevalence around the world, with more than 1 billion adults affected. The public health implications of these trends are staggering. It has been projected that by 2020, chronic diseases will account for almost three-fourths of all deaths worldwide (World Health Organization 2009). These projections may turn out to be underestimates, as people of low- and moderate-income countries are likely to become more sedentary as they experience growing economic development and urbanization.

The situation for chronic diseases as leading causes of death in the United States is also quite striking. There are approximately 2.4 million deaths per year in the United States, according to 2005 data from the Centers for Disease Control and Prevention, as summarized recently (Kung et al. 2008). Table 1.2 lists the 10 leading causes of deaths in 2005. Together, these causes account for almost 1.9 million of the deaths registered in the United States. Among these leading causes, heart disease (26.6%), cancer (22.8%), and cerebrovascular disease (5.9%) are the dominant causes of premature death. They are the three diseases with the highest rates of occurrence per 100,000 persons. Physical activity and diet play an important role in at least four of these leading causes of death: heart disease, malignant neoplasm, cerebrovascular disease, and diabetes mellitus. These four diseases were responsible for more than 1.4 million or about 58% of the deaths in 2005.

The burden of mortality from cardiovascular causes is predicted to continue to increase in developed countries, as illustrated in figure 1.2. However, the projected increase in the number of deaths attributable to the same causes is much more dramatic in the developing world. For instance, there were 9 million deaths per year from cardiovascular causes in the developing countries around 1990. In 2020, this number is projected to reach approximately 19 million deaths per year. These numbers, taken together with the already high prevalence rates of high sedentary time, lack of physical activity, and obesity and the predicted increases in cases of type 2 diabetes mellitus, suggest that a devastating epidemic of common chronic diseases is currently in the making.

TABLE 1.2   Leading Disease-Related Causes of Death in the United States in 2005

Causes of death	Number of deaths	% of all deaths	Rate/100,000
Heart disease	652,091	26.6	220.0
Malignant neoplasm	559,312	22.8	188.7
Cerebrovascular disease	143,579	5.9	48.4
Chronic lower respiratory disease	130,933	5.3	44.2
Unintentional injuries	117,809	4.8	39.7
Diabetes mellitus	75,119	3.1	25.3
Alzheimer's disease	71,599	2.9	24.2
Influenza and pneumonia	63,001	2.6	21.3
Nephritis, nephrotic syndrome, and nephrosis	43,901	1.8	14.8
Septicemia	34,136	1.4	11.5
Other	556,537	22.7	187.8
Total	2,448,017	100	825.9

Kung et al. 2008.

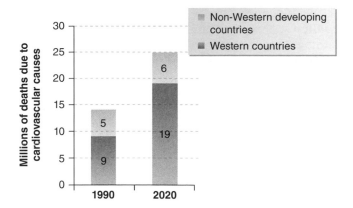

Burden of Common Chronic Diseases

- Currently there are about 59 million deaths per year in the world.
- Chronic diseases account for about 60% of these deaths.
- Cardiovascular disease accounts for about half of the latter.
- Causes of death from chronic diseases include sedentary time, a low physical activity level, poor diet, and obesity.

Health and Its Determinants

To take meaningful action on any social issue, one must first be clear on what that issue is, as well as the factors that affect it. Thus, before discussing the influence of physical activity and physical inactivity on health and morbidity, we will consider the definitions and components of health and morbidity and the factors that affect them. In doing so, we will examine the concepts of active life expectancy, disability-free life expectancy, and wellness; and we will quickly survey the roles that genetic factors, behavioral traits, socioeconomic class, and quality and availability of medical care play in health and morbidity.

Health and Morbidity

Defining health remains a major challenge, despite the progress made in treating diseases and increasing the average life duration in Western societies. The World Health Organization described **health** as "a state of complete physical, mental, and social well-being and not merely the absence of disease or infirmity" (World Health Organization 1948). Health is "a human condition with physical, social

and psychological dimensions, each characterized on a continuum with positive and negative poles. Positive health pertains to the capacity to enjoy life and to withstand challenges; it is not merely the absence of disease. Negative health pertains to morbidity and, in the extreme, with premature mortality" (Bouchard and Shephard 1994, p. 9).

Because health is complex and multifactorial and is not merely the absence of disease, traditional illness and mortality statistics do not provide a full assessment of health. A more comprehensive approach requires that the profile of the individual be established in terms of common health end points; risk factor profile; morbidities; temporary and chronic disabilities; physical and mental functional level; absenteeism; overall productivity; health-related fitness status; objective and perceived level of well-being; and use of all forms of medical services, including prescribed and nonprescribed drugs.

If the **health-related quality of life** is less than optimal, **life expectancy** should be adjusted to reflect a quality-adjusted value. Important concepts are **active life expectancy** and **disability-free life expectancy**. Active life expectancy is simply the age to which a given person is expected to live free of conditions that may restrict her activities. On the other hand, disability-free life expectancy is the number of years of life remaining at a given age with no limitations attributable to physical or mental function impairments. Both concepts can be predicted based on age, gender, education, socioeconomic circumstances, ethnic background, current health status, and other characteristics.

Morbidity can be defined as any departure from a state of physical or psychological well-being, short of death. Morbidity can be measured as

- the number of persons who are ill per unit of population per year,
- the incidence of specific conditions per unit of population per year, and
- the average duration of these conditions.

On the other hand, **wellness** is a holistic concept, describing a state of positive health in the individual and comprising physical, social, and psychological well-being.

Genetic, Behavioral, and Environmental Determinants of Health

Chronic diseases have complex etiologies. They are also heterogeneous in the sense that the paths leading to a disease manifestation vary from disease to disease and are also characterized by considerable individual differences. A large body of evidence indicates that genetic differences, behavior, and the physical and social environment all contribute in varying degrees to the burden of chronic diseases in any country.

Heart disease, stroke, cancer, type 2 diabetes, obesity, and other chronic conditions aggregate in families. The level of familial aggregation varies from condition to condition, with a range from about 30% to 50% of the age- and gender-adjusted variance. This strongly suggests that genetic factors are involved. And indeed, a good number of molecular genetics studies have identified specific genes and mutations contributing to the burden of these common chronic diseases (see chapter 24). However, there is also strong evidence that behavioral factors contribute to the etiology of these diseases. Chronic diseases are largely preventable. Smoking, poor nutritional habits, excessive alcohol consumption, time spent in sedentary pursuits (e.g., sitting time), a low level of physical activity, low physiological fitness, substance abuse, and high-risk sexual behavior are among the behaviors and states typically associated with an increased risk of death or morbidity. The exact contribution of these behavioral traits to the

global burden of disease is not easily quantified but appears to account for a substantial fraction of the risk of disease. Even though much more research is needed before an evidence-based breakdown of the causes of common chronic disease or of premature deaths can be established, the evidence suggests that genetic factors and behavioral traits contribute the most, perhaps as much as 80%, to the level of risk.

Beyond genetic factors and behavioral traits, which together explain most of the predisposition to the major chronic diseases, the social and physical environment plays an important role. For example, people in low socioeconomic classes or with less education are more likely to be economically disadvantaged and are at a greater risk of being affected by chronic diseases and dying prematurely. Additionally, limitations in the health care delivery system, medical errors, and other situations out of personal control contribute to the fact that some affected individuals die prematurely.

Existing reports on the population attributable risk (PAR) for physical inactivity show that it is one of the major causes of death in the United States and in the world as a whole. It also should be mentioned that PAR estimates are likely underestimates of the true effect of physical inactivity. This is the case because they are based on self-report of physical activity, which results in substantial misclassification and thus lower estimates.

The results from the INTERHEART study provide strong support for this notion. A total of 30,000 men and women from 52 countries were enrolled in this study (Yusuf et al. 2004). Half of these participants had experienced a myocardial infarction. Irrespective of ethnic background and country of residence, the risk factor profile associated with a cardiac event was the same. This risk profile included smoking, sedentarism, and poor nutrition.

Poor nutritional habits play a key role in the etiology of several chronic diseases. A high-fat, high-sugar, energy-dense diet, with a substantial content of animal foods, appears to be the origin of many current health problems. However, diet is only one of the risk factors. Sedentary time, a low physical activity level, and poor physiological fitness are also increasingly recognized determinants of health. A sedentary lifestyle and poor nutritional habits, together with other risk factors such as tobacco use and stressful aspects of modern life, are potent enough to accelerate the development of chronic diseases, accentuate their severity, and contribute to the loss of function leading to the frailty that often accompanies aging. More research is clearly needed on the mechanisms linking dietary habits and physical activity level to health. However, the available scientific evidence is already sufficiently strong to justify implementing preventive measures right now. The public health approach is likely to be the most cost-effective approach to coping with the chronic disease epidemic.

Public health is concerned not only with reducing smoking rates, improving the diet, and promoting physical activity. It is also concerned with the fact that as many as 22,000 deaths from alcohol-induced causes occurred in the year 2005; 44,000 deaths were caused by injuries related to motor vehicle traffic; 20,000 persons died as the result of falls; and 31,000 deaths were caused by firearms in the United States alone. A global public health approach needs to take into account all unhealthy behaviors and agents. Understanding what underlies the pervasiveness of such risk factors is the first step in reducing them and consequently diminishing their negative effects.

Aging and Health

The fact that the population is progressively getting older has an enormous impact on the importance of the physical activity and health paradigm (see

chapter 19). The average life expectancy has substantially increased over the past 100 years. On average, women are now expected to live about 80 years in some countries, whereas men will live about 75 years. There are obviously considerable differences in these life expectancy estimates even among developed countries.

Although Americans are living longer because of recent declines in heart disease and strokes, chronic diseases such as high blood pressure and diabetes are becoming more common among older adults. Aging can be defined as a progressive decline in the ability of an organism to resist stress, tissue damage, and disease. It is characterized by an increase in the incidence of degenerative disorders. The United States population over 65 is projected to grow from 40.3 million in 2010 to 88.5 million in 2050. A large increase in the aging population is occurring among those 85 years and older. By the year 2050, there will be 19 million Americans above the age of 85, a more than threefold increase compared to the 2010 figure (figure 1.3).

More than 20% of U.S. adults over the age of 65 live with at least partial disability, defined as some degree of difficulty in performing activities of daily

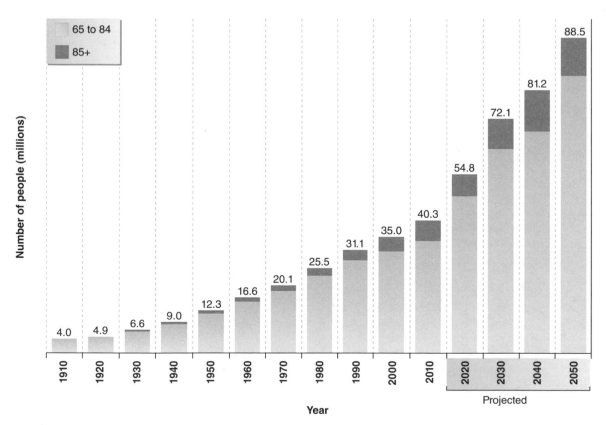

FIGURE 1.3 Changes since 1910 and the projection until 2050 of the number (in millions) of U.S. adults aged 65 to 84 years and those 85 and older.

Data from the U.S. Census Bureau 2010.

living, according to the National Center for Health Statistics. Above the age of 85, 45% of persons need some assistance with one or more basic activities of daily living. This trend has enormous implications for health care systems. It also represents a challenge for all those who believe that regular physical activity is essential to preserve autonomy as we age. The statistics showing that the prevalence of common chronic diseases is on the rise worldwide, together with the fact that people live longer, suggest that societies will likely face a major escalation of health care costs and a public health crisis in the coming decades.

Aging is a very complex process and is only just beginning to be understood. Many factors are involved, including endogenous cellular processes, environmental insults, and interactions among environmental factors—including those associated with nutrients and the demands of physical work—with one's genome. No single factor can explain the variation in the way we look, feel, or behave as we get older. But it remains clear that regular physical activity is one of the most important lifestyle components for preventing the age-related decline in overall physical independence and well-being.

Defining Physical Activity and Physical Fitness

In this section, we consider the definitions of physical activity and fitness. Both are complex concepts that cover a number of components with potential applications to several fields of study and practice.

Defining Physical Activity

Physical activity comprises any bodily movement produced by the skeletal muscles that results in an increase in metabolic rate over resting energy expenditure. Under this broad concept, we need to consider leisure-time physical activity, exercise, sport,

transportation, occupational work, and chores. The energy expenditure associated with physical activity is the only discretionary component of total daily energy expenditure. Energy expenditure of activity is typically only about 25% of daily energy expenditure in a sedentary person, whereas it may be as high as 50% in an endurance athlete on a training day or in persons performing heavy labor for many hours during the day.

Leisure-Time Physical Activity

In most developed societies, after completion of work, traveling, domestic chores, and personal hygiene, the average person has 3 to 4 h of "free," leisure, or discretionary time per day. However, there is wide interindividual variation, depending in part on such personal circumstances as the duration of paid work, the division of labor in the home, the need for self-sufficiency activities, daily commuting time, and the number and age of dependents.

- **Leisure-time physical activity** is an activity undertaken in the individual's discretionary time that increases the total daily energy expenditure. The element of personal choice is inherent to the definition. Activity is selected on the basis of personal needs and interests. When the motivation is to improve health or fitness, the pattern of activity undertaken will be consonant with this objective. But there are many other possible motivations (Dishman 1988), including aesthetic motivations (pursuit of a desired body type or an appreciation of the beauty of movement), ascetic issues (the setting of a personal physical challenge), the thrill of fast movement and physical danger, chance and competition, social contacts, fun, mental arousal, relaxation, and even addiction to endogenous opioids (Bouchard and Shephard 1994).

- **Exercise** is a form of leisure-time physical activity that is usually performed repeatedly over an extended period of time (exercise training) with a specific external objective such as the improvement

Components of Total Daily Energy Expenditure

- Basal and resting metabolic rate account for about 65% of daily energy expenditure.
- Because of their high metabolic rates, cardiac muscle, liver, brain, kidney, pancreas, and other organs account for about 70% of the energy expended at rest.
- The thermic response to food (absorption, digestion, transport, and storage) accounts for about 10% of daily energy expenditure.
- Physical activity and movement of all types account for about 25% of the energy expended in a typical day by a sedentary person.

of fitness, physical performance, or health. When prescribed by a physician or exercise specialist, the regimen typically covers the recommended mode, intensity, frequency, and duration of such activity. For example, figure 1.4 shows six intensity levels of endurance exercise, ranging from very light to maximal. Figure 1.4*a* displays the relationship between intensity of exercise and heart rate expressed as a percentage of the maximal attainable heart rate. Figure 1.4*b* illustrates the power output in **metabolic equivalents (METs)** across the six categories of exercise intensity for two individuals, one with a **maximal oxygen uptake ($\dot{V}O_2$max)** of 10 METs (i.e., equivalent to 10 times the resting energy expenditure) and the other with a $\dot{V}O_2$max of 5 METs, as in some elderly people.

- Sport is a form of physical activity that involves competition. In general, a sport is a competitive activity undertaken in the context of rules defined by an international regulatory agency. However, in some parts of the world, the term *sport* may also embrace exercise and recreation (as in the UNESCO "Sport for All" movement) (McIntosh 1980).

Work

Work is also an important component of daily activities. In the past, energy expenditures required by occupational work and the associated demands of transportation (on foot or on a bicycle) or work around the house accounted for a major fraction of the total daily energy metabolism in a large segment of the labor force.

- Heavy occupational demand has been of considerable epidemiological interest (Paffenbarger, Hyde, and Wing 1990) in the past because it was typically sustained for 30 to 40 h per week over many years. This is still true in some developing societies, and even in the Western world there are still occupational categories with a high energy demand. However, "heavy" employment is commonly accompanied by low levels of leisure-time physical activity. The standards defining a high or a very high intensity of occupational activity differ from those applicable to "exercise" (table 1.3). For instance, heavy work is defined as an energy expenditure of 5 to 7.5 kcal/min. This is so because in industry, the duration of individual activity bouts is usually prolonged; other circumstances are often adverse (such as a high environmental temperature, awkward posture, or a heavy loading of small muscle groups); and normally the pace of working is set by such factors as a machine, a supervisor, or a union contract rather than by the individual (Bouchard and Shephard 1994).

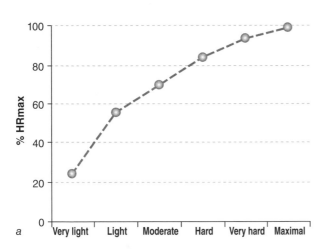

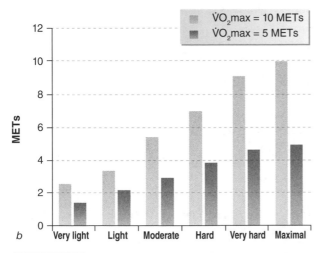

FIGURE 1.4 The definition of exercise intensity varies according to the level of fitness and can be expressed in terms of (*a*) % HRmax or (*b*) METs.

TABLE 1.3 Intensity of Occupational Work

Intensity	Energy expenditure in kcal/min
Sedentary	<2.0
Light	2.0-3.5
Moderate	3.5-5.0
Heavy	5.0-7.5
Very heavy	>7.5

Data from Brown and Crowden 1963.

- Household and other chores also need to be considered. Automation has progressively reduced the energy demands associated with the operation of a household in developed societies. Although some individuals may deliberately seek out heavy activities, most necessary domestic chores now fall into the "light" category on the industrial scale. Possible exceptions are the care of dependents (playing with young children and caring for elderly relatives) and vigorous gardening, which can on occasion involve heavy work and higher energy costs.

Figure 1.5 illustrates trends over a decade in the occupational and domestic physical activity–related energy expenditure for adult men and women in China. The energy expenditure is reported in terms of MET-hours per week, and the data are from the China Health and Nutrition Survey. It is rather striking that significant reductions in occupational and domestic energy expenditures of activity can be so clearly identified within a single decade (Monda et al. 2008).

Defining Physical Fitness

There is no universally agreed-upon definition of fitness and of its components. The World Health Organization defined fitness as "the ability to per-form muscular work satisfactorily" (World Health Organization 1968). Fitness implies that the individual has attained those characteristics that permit an acceptable performance of a given physical task in a specified physical, social, and psychological environment. Fitness is typically defined with a focus on two goals: performance or health.

- **Performance-related fitness** refers to those components of fitness that are necessary for optimal work or sport performance (Bouchard and Shephard 1994; Pate 1988). It is defined in terms of the individual's ability in athletic competition. Performance-related fitness depends heavily on motor skills, cardiorespiratory power and capacity, muscular strength, speed, power or endurance, body size, body composition, motivation, and nutritional status. Performance-related fitness is not considered in detail in this book.

- **Health-related fitness,** in contrast, refers to those components of fitness that are affected favorably or unfavorably by habitual physical activity habits and that relate to health status. Health-related fitness has been defined as a state characterized by an ability to perform daily activities with vigor and by traits and capacities that are associated with a low risk for the development of chronic diseases and premature death (Pate 1988). Important components of health-related fitness include those listed in "Health-Related Fitness Components and Traits" on page 16. These biological traits relate to health outcomes as assessed by the profile of risk factors and by morbidity and mortality statistics. They are grouped under five major components: morphological, muscular, motor, cardiorespiratory, and metabolic fitness. These components and embedded traits are addressed in various chapters of this book.

Physical Inactivity Versus Physical Activity

The main emphasis of this book is on the deleterious effects of physical inactivity, as evidenced by excessive sedentary time, and the benefits of a physically active lifestyle. The central problem is briefly defined in the following paragraphs.

Physical Inactivity as a Risk Factor

Homo sapiens has attempted for millennia to reduce the amount of muscular work and physical activity required in daily life. The war on muscular work has been a remarkable success. Thus, the amount of energy expended by individuals to ensure a sus-

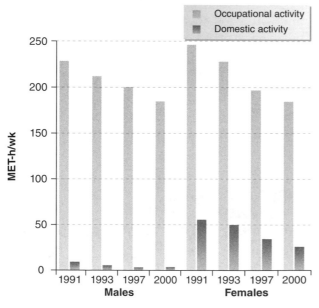

FIGURE 1.5 Trends in average energy expenditure in MET-hours per week from occupational and domestic physical activity for adult men and women in the China Health and Nutrition Survey.

Reprinted by permission from Macmillan Publishers Ltd: *European Journal of Clinical Nutrition*, K.L. Monda, L.S. Adair, F. Zhai, and B.M. Popkin, "Longitudinal relationships between occupational and domestic physical activity patterns and body weight in China," 62: 1318-1325, copyright 2008.

Definition of Fitness and Related Concepts

- Fitness is the ability to perform muscular work satisfactorily.
- Performance-related fitness refers to the components of fitness that are necessary for maximal sport performance.
- Health-related fitness refers to those components of fitness that benefit from a physically active lifestyle and relate to health.

tained food supply, decent housing under a variety of climatic conditions, safe and rapid transportation, personal and collective security, and diversified and abundant leisure activities has decreased substantially. The decline in the amount of physical activity has been so dramatic that a variety of health problems, furthered by a sedentary mode of life, increased considerably in the 20th century. These health problems were referred to as "hypokinetic diseases" 50 years ago (Krauss and Raab 1961).

Many agents reduced our overall amount of muscular work and increased our sedentary time. Motorized transportation is undoubtedly at the top of the list. Labor-saving devices and systems in the work environment also play a major role. Computers and a large number of electrically powered tools

and gadgets have dramatically reduced the need to rely on muscular work. Television, video games, and domestic labor-saving devices have all contributed to the increase in sedentary time. Elevators, escalators, and other convenient modes of moving up and down in the urban environment have made the situation even more serious. Urban design generally favors the use of the automobile and has made it more challenging for people to be physically active.

In industrialized countries, most citizens who want to adopt a physically active lifestyle have to do so using some of their leisure time. However, in some of these countries, a substantial fraction of the people get plenty of physical activity simply because they choose to cycle or walk as their preferred mode of transportation. Table 1.4 lists travel modes in 12

TABLE 1.4 Travel Modes in Europe and North America

Country	Percentage of trips by travel mode				
	Bicycle	Walking	Public transport	Auto	Other
Netherlands	30	18	5	45	2
Denmark	20	21	14	42	3
West Germany	12	22	16	49	1
Switzerland	10	29	20	38	3
Sweden	10	39	11	36	4
Austria	9	31	13	39	8
East Germany	8	29	14	48	1
England and Wales	8	12	14	62	4
France	5	30	12	47	6
Italy	5	28	16	42	9
Canada	1	10	14	74	1
United States	1	9	3	84	3

Adapted with permission of the Eno Transportation Foundation, Washington DC, from "Bicycling Boom in Germany: A Revival Engineered by Public Policy," *Transportation Quarterly* 51: 31-36. Copyright 1997 Eno Transportation Foundation.

countries of Western Europe and North America from more than a decade ago. Note that 40% or more of the adult populations were walking or bicycling as their mode of transportation in four of these countries: Austria, Denmark, Netherlands, and Sweden. In contrast, only about 10% of the adults in Canada and the United States did so. These observations suggest that promoting a physically active lifestyle represents a greater challenge in North America. These issues are discussed in subsequent chapters of this book.

Benefits of Regular Physical Activity

The relationship between physical activity and health is more complex than is apparent from a cursory glance. Figure 1.6 depicts the simplest path linking physical activity and health. The message from this diagram is quite straightforward: Physical activity is associated with health benefits. A low level of physical activity is likely to translate into unfavorable health outcomes, whereas the converse would be true for a high level of physical activity. However, the reality is more complex. For instance, consider the path diagram in figure 1.7. On average, and in most people, regular physical activity increases health-related fitness. This implies an increase in cardiorespiratory endurance and in insulin action in tissues such as skeletal muscle, an increase in high-density lipoprotein cholesterol, a decrease in blood pressure, and a decrease in whole-body adiposity, to name but a few. Such improvements in fitness are likely to have favorable effects on overall health.

However, a body of data shows that some health benefits are derived from being physically active even though there may be no or little associated gain in fitness as it is traditionally measured. Thus, two paths potentially contribute to the relationship between regular physical activity and health, one independent of changes in fitness and the second mediated by the gains in physical fitness (figure 1.8).

The reality is even more complex than suggested by the path diagram in figure 1.8. For instance, those who are healthier are typically more active and are the fittest in a population. Thus, the paths from activity or fitness to health are not necessarily causal paths. Furthermore, habitual physical activity can influence fitness, which in turn may modify the level of habitual physical activity (figure 1.9). For example, with increasing fitness, people tend to become more active, and the fittest become the most active. These potentially confounding relationships need to be taken into account if we are to understand the relationships between regular physical activity and health.

Health-Related Fitness Components and Traits

- Morphological component
  - Body mass for height
  - Body composition
  - Subcutaneous fat distribution
  - Abdominal visceral fat
  - Bone density
  - Flexibility
- Cardiorespiratory component
  - Submaximal exercise capacity
  - Maximal aerobic power
  - Heart functions
  - Lung functions
  - Blood pressure
- Muscular component
  - Power
  - Strength
  - Endurance
- Motor component
  - Agility
  - Balance
  - Coordination
  - Speed of movement
- Metabolic component
  - Glucose tolerance
  - Insulin sensitivity
  - Inflammatory markers
  - Lipid and lipoprotein metabolism
  - Substrate oxidation characteristics

Adapted, by permission, from C. Bouchard, R.J. Shepherd, T. Stephens, 1994, *Physical activity, fitness and health: International proceedings and consensus statement* (Champaign, IL: Human Kinetics), 81.

FIGURE 1.6 The simplest model depicting the relationship between physical activity and health.

FIGURE 1.7 A model in which the effects of physical activity on health are mediated by increases in fitness.

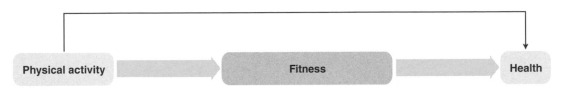

FIGURE 1.8 In this model, physical activity has direct fitness benefits but also improves health.

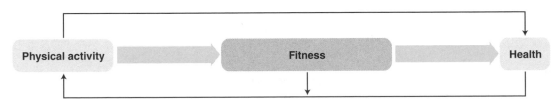

FIGURE 1.9 This model specifies not only that physical activity and fitness are positively associated with health but also that healthier individuals are more inclined to be physically active.

Needless to say, the relationships among levels of physical activity, health-related fitness, and health are even more complex than suggested by the models discussed so far. Figure 1.10 illustrates additional complexities in these relationships. This model shows that habitual physical activity can influence fitness, which in turn may modify the level of habitual physical activity. It not only shows that people tend to become more active with increasing fitness and that the fittest individuals tend to be the most active; it also specifies that fitness is related to health in a reciprocal manner. That is, fitness influences health, and health status also influences both habitual physical activity level and fitness level. Other factors are associated with individual differences in health status. Likewise, the level of fitness is not determined entirely by an individual's level of habitual physical activity. Other lifestyle behaviors, physical and social environmental conditions, personal attributes, and genetic characteristics also affect the major components of the basic model and determine their interrelationships.

Subsequent chapters discuss in greater detail how habitual physical activity and health-related fitness are related to various health outcomes. In this context, understanding the extent and causes of human variation is an important issue that is addressed in chapter 24. However, it is useful to appreciate early in this book that the causes of human variation are legion. For instance, cardiorespiratory fitness, a key component of health-related fitness, results from the contributions of a number of effectors. This is illustrated in figure 1.11. Here we focus on the interindividual differences in cardiorespiratory fitness typically observed in a large population of sedentary adults. If the overall variance in cardiorespiratory fitness adjusted for body mass is set at 100%, a substantial fraction (up to 10%) will be accounted for by measurement errors and other uncontrolled factors. Age, sex, and ethnic differences may account for as

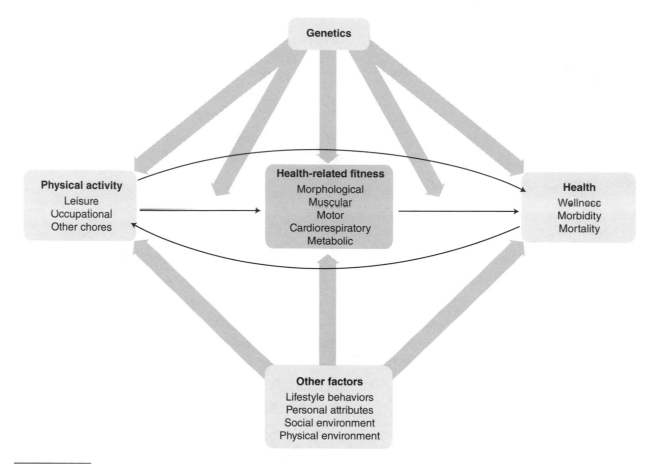

FIGURE 1.10 A more complete model defining the relationships among physical activity, health-related fitness, and health status. This model allows for contributions of inherited factors and other lifestyle behaviors, personal attributes, and social and physical environmental factors.

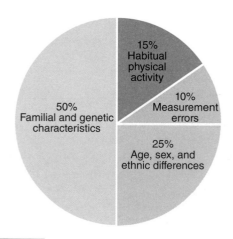

FIGURE 1.11 Partitioning of variance in cardiorespiratory fitness in a heterogeneous population of sedentary individuals.

much as 25% of the variation seen among adults. The remaining 65% can be divided into two major components. About 15% of the individual differences in cardiorespiratory fitness can be explained by the slight variation in habitual physical activity even among sedentary people. Finally, cardiorespiratory fitness is determined in part by familial and genetic characteristics. The heritability level of this attribute is thought to reach almost one-half of the overall variance. This simple partitioning of the population heterogeneity in a key component of health-related fitness has considerable implications for understanding the physical activity–fitness–health paradigm. These issues are considered in more detail in the chapter devoted to genetic differences (chapter 24).

There is no doubt that regular physical activity is accompanied by substantial benefits on health and quality of life indicators. Most chapters of this book are devoted to a review of the systems and attributes, as well as risk factors and morbid conditions, that are favorably influenced by a physically active lifestyle.

Summary

More than half of all yearly deaths in the United States are attributable to common chronic diseases, which are related to lifestyle, social environment, and built environment characteristics. As physical activity has diminished dramatically in our lives and the amount of time spent in a sedentary state has increased, a host of physical ills have become more prevalent. This chapter provides a flexible framework relating physical activity to health. Key concepts such as health, wellness, morbidity, and mortality are defined. Then the global burden of common chronic diseases is described, and the effects of biological, behavioral, and environmental determinants as well as aging are highlighted. To ensure that terms such as *physical activity* and *physical fitness* are interpreted in a consistent fashion in the remainder of this book, their complexities are considered. Physical inactivity as a risk factor is briefly discussed. A practical model linking physical activity, physical fitness, and health is introduced; and this serves as the basic conceptual framework for the organization of the text.

Key Concepts

active life expectancy—Age to which a person is expected to live free of disabling diseases or conditions.

disability-free life expectancy—Average number of years of life remaining to a person at a particular age without limitations in physical or mental functions.

exercise—Planned, structured, and repetitive bodily movement done to improve or maintain one or more components of physical fitness.

health—State of complete physical, mental, and social well-being and not merely the absence of disease or infirmity.

health-related fitness—Components of fitness that are affected favorably by habitual physical activity or unfavorably by a sedentary mode of life. Health-related fitness is a state characterized by (a) an ability to perform daily activities with vigor and (b) traits and capacities that are associated with a low risk of premature development of hypokinetic diseases and conditions.

health-related quality of life—The quality of one's personal and mental health and the ability to react to factors in the physical and social environments; years lived with the full range of functional capacity.

leisure-time physical activity—Activity undertaken in the individual's discretionary time that substantially increases total daily energy expenditure. The element of personal choice is inherent to the definition.

life expectancy—Average number of years of life remaining to a person at a particular age.

maximal oxygen uptake ($\dot{V}O_2max$)—Maximal capacity for oxygen consumption by the body during maximal exertion; product of cardiac output and oxygen extraction. It is also known as aerobic power and is considered a valid measure of cardiorespiratory fitness.

metabolic equivalent (MET)—Unit used to estimate the metabolic cost (oxygen consumption) or intensity of physical activity. One MET equals the sitting metabolic rate of approximately 3.5 ml · kg^{-1} · min^{-1}, or 1 kcal · kg^{-1} · h^{-1}. A collection of activity-specific MET values for use in research and other applications has been published by Ainsworth and colleagues as the Compendium of Physical Activities.

morbidity—Measures pertaining to disease states and conditions that have not achieved a mortal end point. Common measures of morbidity are disease- or condition-specific incidence rates, hospital admissions, bed days, treatment costs, loss of physical function and independence, and lost days of work for specific causes.

performance-related fitness—Components of fitness that are necessary for optimal work or sport performance. This is defined in terms of the individual's ability in athletic competition, a performance test, or occupational work and depends heavily on motor skills; cardiorespiratory power and capacity; muscular strength, power, or endurance; body size; body composition; motivation; and nutritional status.

physical activity—Bodily movement that is produced by the contraction of skeletal muscle and that substantially increases energy expenditure.

physical fitness—A set of attributes that people have or achieve that relates to the ability to perform physical work.

risk factor—An aspect of personal behavior or lifestyle, an environmental exposure, or an inborn or inherited characteristic that is known to be associated with health-related conditions. When present over an extended period of time, a risk factor can significantly either increase the probability of developing a common degenerative disease such as cardiovascular disease, type 2 diabetes mellitus, or osteoporosis or increase the probability of premature death.

sedentary time—Prolonged periods of being physically inactive, typically involving sitting or screen time or both.

wellness—Holistic concept describing a state of positive health in the individual and comprising physical, social, and psychological well-being.

Study Questions

1. Identify three lines of evidence in support of the view that the human body is designed for physical activity.

2. Name five of the traits that distinguish *Homo sapiens* from other primates and that favored the *Homo sapiens* population expansion despite a hostile environment.

3. What are the major causes of death around the world?

4. What is the prevalence of lack or partial lack of autonomy among Americans 85 years of age and older?

5. A fraction of total daily energy expenditure is accounted for by the energy costs of all forms of movement and activity. Describe how variable this fraction is among people who are sedentary and physically active.

6. Compare the classifications of physical activity intensities and occupational work intensities and define the ways in which they differ.

7. Define morbidity and show its relationship to health-related fitness.

8. Compare the use of the automobile between countries of Western Europe and the United States and Canada.

9. Explain the path from physical activity to health versus that from physical activity to fitness to health.

10. Define the major source of variation in cardiorespiratory fitness, as assessed by $\dot{V}O_2$max adjusted for body mass, among a population of sedentary individuals.

References

Aboderin, I., A. Kalache, Y. Ben-Shlomo, J.W. Lynch, C.S. Yajnik, D. Kuh, and D. Yach. 2001. *Life course perspectives on coronary heart disease, stroke, and diabetes: Key issues and implications for policy and research.* Geneva: World Health Organization.

Bouchard, C., and R.J. Shephard. 1994. Physical activity, fitness, and health: The model and key concepts. In *Physical activity, fitness, and health,* ed. C. Bouchard, R.J. Shephard, and T. Stephens (pp. 77-88). Champaign, IL: Human Kinetics.

Brown, J.R., and G.P. Crowden. 1963. Energy expenditure ranges and muscular work grades. *British Journal of Industrial Medicine* 20:227.

Dishman, R.K. 1988. Determinants of participation in physical activity. In *Exercise, fitness and health: A consensus of current knowledge,* ed. C. Bouchard, R.J. Shephard, T. Stephens, J.R. Sutton, and B.D. McPherson (pp. 33-48). Champaign, IL: Human Kinetics.

Krauss, H., and W. Raab. 1961. *Hypokinetic diseases.* Springfield, IL: Charles C Thomas.

Kung H.-C., D.L. Hoyert, J. Xu, and S.L. Murphy. 2008. Deaths: Final data for 2005. In *National vital statistics reports.* Hyattsville, MD: National Center for Health Statistics.

Malina, R.M. 1991. Darwinian fitness, physical fitness and physical activity. In *Applications of biological anthropology,* ed. C.G.N. Mascie-Taylor and G.W. Lasker (pp. 143-184). Cambridge, UK: Cambridge University Press.

McIntosh, P.C. 1980. *"Sport for All" programmes throughout the world.* Paris: UNESCO.

Monda, K.L., L.S. Adair, F. Zhai, and B.M. Popkin. 2008. Longitudinal relationships between occupational and domestic physical activity patterns and body weight in China. *European Journal of Clinical Nutrition* 62:1318-1325.

Morris, J.N. 2009. Foreword. In *Epidemiologic methods in physical activity studies,* ed. I-M. Lee, S.N. Blair, J.E. Manson, R.S. Paffenbarger. (pp. 3-12). New York: Oxford University Press.

Paffenbarger, R., R.T. Hyde, and A.L. Wing. 1990. Physical activity and physical fitness as determinants of health and longevity. In *Exercise, fitness and health: A consensus of current knowledge,* ed. C. Bouchard, R.J. Shephard, T. Stephens, J.R. Sutton, and B.D. McPherson (pp. 33-48). Champaign, IL: Human Kinetics.

Pate, R.R. 1988. The evolving definition of fitness. *Quest* 40:174-179.

Pucher, J. 1997. Bicycling boom in Germany: A revival engineered by public policy. *Transportation Quarterly* 51:31-36.

Reddy, K.S. 2004. Cardiovascular disease in non-Western countries. *New England Journal of Medicine* 350:2438-2440.

U.S. Census Bureau. 2010. 2010 census summary file 1. http://factfinder2.census.gov/bkmk/table/1-0/en/DEC/10_SF1/QTP1. Accessed December 7, 2011.

U.S. Census Bureau. 2008. U.S. population projections. http://www.census.gov/population/www/projections/index.html.Accessed August 25, 2011.

World Health Organization. 1948. *Constitution of the World Health Organization. Basic documents.* Geneva: World Health Organization.

World Health Organization. 1968. *Meeting of investigators on exercise tests in relation to cardiovascular function.* Geneva: World Health Organization.

World Health Organization. 2009. *Global health risks: Mortality and burden of disease attributable to selected major risks.* Geneva: World Health Organization.

Yusuf, S., S. Hawken, S. Ounpuu, T. Dans, A. Avzeum, F. Lanas, M. McQueen, A. Budaj, P. Pais, J. Varigos, and L. Lisheng. 2004. Effect of potentially modifiable risk factors associated with myocardial infarction in 52 countries (the INTERHEART study) case-control study. *Lancet* 364:937-952.

2

Historical Perspectives on Physical Activity, Fitness, and Health

Russell R. Pate, PhD

CHAPTER OUTLINE

©J.Fishkin/CMSP

The body of knowledge in any scientific discipline evolves gradually over time, each new bit of information building on a base established by the results of earlier investigations. So it is with exercise science. Our current understanding of the relationship between physical activity and health is heavily influenced by the findings of studies completed rather recently—mostly in the last 20 years. But the key studies completed in recent decades used methods refined by earlier investigators and tested hypotheses that were suggested by the results of previous investigations. Without a doubt, today's exercise scientists stand on the shoulders of the pioneers who first applied the principles of scientific inquiry to the exercise–health relationship.

The goal of this chapter is to provide a historical context for our current beliefs about the effects of physical activity on health. One specific purpose of this chapter is to overview the broad developmental trends that are evident in the scientific study of physical activity and its effects on human health. Particular attention is given to the gradual expansion of exercise science into several distinct subdisciplines, each with its own unique set of scientific methods. A second purpose is to identify the seminal events that shaped our understanding of the effect of exercise on health. Finally, in recent decades, numerous agencies and expert panels have applied the growing body of knowledge on physical activity and health by advancing guidelines on the types and amounts of physical activity needed to maintain or promote health. Therefore, a third and central purpose of this chapter is to track the evolution of physical activity guidelines that have been issued to promote public health.

Early Beliefs About Physical Activity and Health

Many ancient cultures, scientists, and physicians recognized the role of physical activity in promoting the health of mind and body. In China and India, concepts of health and prevention were developing as early as 3000 B.C. Ancient Chinese writings promoted harmony and prevention as keys to longevity. A medical document written in India between 1000 B.C. and 800 B.C., the *Ayur Veda*, recommended massage and exercise in the treatment of rheumatism. Over thousands of years, both cultures developed philosophies, including Taoism and yoga, that emphasized the importance of a system of exercise to health.

Among Western cultures, the ancient Greeks dominated the study and understanding of the effects of physical activity on health, quality of life, and life span. As early as the fifth century B.C., Greek physicians promoted the "laws of health"—breathe fresh air, eat good foods, drink the proper beverages, participate in exercise, and get adequate sleep (U.S. Department of Health and Human Services [DHHS] 1996). Herodicus (fifth century B.C.) was the first of the Greek gymnasts (a type of medical practitioner) to prescribe therapeutic exercise. **Hippocrates** (fifth century B.C.), the father of preventive medicine, wrote extensively about the benefits of exercise for a variety of ailments, including mental illnesses. Although he criticized Herodicus for prescribing exercise that was too strenuous, Hippocrates recommended walking and other forms of moderate-intensity exercise. Many Greek physicians who practiced medicine and taught in the medical schools during the fourth through second centuries B.C., including Herophilus, Eristratus, Asclepiades, and Celsus, prescribed either moderate or vigorous exercise to maintain health and treat a variety of diseases (figure 2.1).

"Eating alone will not keep a man well; he must also take exercise. For food and exercise, while possessing opposite qualities, yet work together to produce health."

Hippocrates, Regimen

Claudius Galenus (Galen), born about A.D. 131, was a brilliant Greek physician who worked in Rome and whose teachings and views dominated European medicine for 1,000 years. He wrote and lectured extensively on anatomy and epidemiology and was the first scientist to systematically describe the human body and to recognize that contraction is the primary action of muscle. He believed that some form of exercise could be used to treat virtually every form of disease, and he classified exercises according to their purposes. In his medical practice and his writings, most notably *On Hygiene,* Galen promoted his belief that everyone—athletes, healthy adults and children, invalids, and even babies—could benefit from exercise.

The influence of the ancient Greek physicians on European medicine faded during the Middle Ages but reemerged during the Renaissance, when some of the original Greek manuscripts were rediscovered. In the 15th century, Vergerius and Vittorino da Feltre of Italy became early advocates of regular exercise for children. Da Feltre established a school in which children participated in exercises designed to meet their individual needs and in which all children participated in gymnastics and played many sports. In

"The uses of exercise, I think, are twofold; one for the evacuation of the excrements, the other for the production of good condition of the firm parts of the body. For since vigorous motion is exercise, it must needs be that only these three things result from it in the exercising body—hardness of the organs from mutual attrition, increase of the intrinsic warmth, and accelerated movement of respiration. These are followed by all the other individual benefits which accrue to the body from exercise."

Galen, On Hygiene

FIGURE 2.1 The value that the ancient Greeks placed on exercise is clearly illustrated by the fact that many of their surviving vases depict athletes. This amphora dates from the time of Hippocrates (fifth century B.C.) and features three nude athletes running under the watchful eye of Athena, the patron goddess of Athens, who is pictured on the side of the vase that is not shown.

Courtesy of The Spurlock Museum, University of Illinois at Urbana-Champaign.

1553 in Spain, Cristobal Mendez, a physician, published one of the earliest printed books on exercise, in which he prescribed exercise for elderly people and those who were ill, noting that "the easiest way of all to preserve and restore health without diverse peculiarities and with greater profit than all other measures put together is to exercise well." Mercurialis, one of the most influential European physicians of the Renaissance, published the six-volume *Art of Gymnastics* in 1569. Mercurialis classified exercises as preventive or therapeutic and recommended that all sedentary people begin to exercise. His recommendations for exercises for people who were sick or infirm established the foundation for the future development of rehabilitation medicine.

Although scientific support for the belief that exercise could prevent or ameliorate disease did not emerge until the 20th century, investigations of that

"Those who sit at their work and are therefore called 'chair workers,' such as cobblers and tailors, become bent and humpbacked and hold their heads down like people looking for something on the ground. . . . These workers, then, suffer from general ill health caused by their sedentary life . . . they should be advised to take physical exercise, at any rate on holidays. Let them make the best use they can of some one day, and so to some extent counteract the harm done by many days of sedentary life."

Bernardo Ramazzini, Diseases of Workers

possibility began much earlier. Italian physician Bernardo Ramazzini wrote *Diseases of Workers,* the first published work on occupational diseases, in 1713. Ramazzini observed that runners who worked as messengers avoided the health problems suffered by cobblers and tailors and recommended that sedentary workers exercise on their holidays. In the 1840s, **W.A. Guy** compared the morbidity and mortality rates of men in active and sedentary occupations and noted the superior health of those in active occupations. He recommended that those in sedentary jobs perform physical exercise during their leisure time to improve their health (Paffenbarger, Blair, and Lee 2001).

Early Beliefs About Physical Activity and Health

- Scientists and physicians in China and India recognized a link between physical activity and health more than 5,000 years ago.
- Ancient Greek physicians, including Herodicus and Hippocrates, prescribed exercise to prevent and treat a variety of ailments as early as the fifth century B.C.
- During the 1500s, Italian physicians prescribed exercise for the healthy growth and development of children and for the treatment of elderly and ill people.
- In the early 1700s, Ramazzini identified the negative health effects of certain occupations. He noted that runners (messengers) avoided many of the health problems that affected cobblers, tailors, and other sedentary workers.

Scientific Inquiry on Exercise and Health

Although great scholars had recognized the importance of physical activity to health as far back as the ancient cultures, it was not until the early 20th century that scientists began systematic study of the effects of exercise on the human body. The development of "exercise science" began with the work of physiologists who became interested in the body's functional responses to different types of exercise. Although the discipline of exercise physiology has contributed much to our knowledge of the relationship between physical activity and health, many other fields of scientific inquiry have come into play over the past century. In this section, we review the contributions made by several scientific disciplines to our rapidly expanding knowledge of the impact of exercise on human well-being.

Exercise Physiology

Exercise physiology is a scientific discipline dedicated to the study of the body's function during exercise and its physiological adaptations to regular participation in exercise. In the early 20th century, one of the earliest exercise physiologists, **R. Tait McKenzie** of the University of Pennsylvania, began studying the effects of regular physical activity on the body, using a system of regular medical examinations of athletes before and after sport participation. In 1909, A.V. Hill of Cambridge University began studying the physiology of muscle contraction. He also conducted pioneering studies of the thermal changes associated with muscle function. August Krogh and Marie Jorgensen of Denmark studied carbon dioxide transport in the lungs, metabolism, and the role of insulin. Krogh also designed a bicycle ergometer with which he studied exercise intensity.

One of the first exercise laboratories, the **Harvard Fatigue Laboratory,** was founded in 1927. Although the lab conducted research on a number of topics, including nutrition, blood chemistry, and the stresses imposed on the body by altitude and climate, it is perhaps best known for its work in exercise physiology. Scientists at the lab made some of the first measurements of the body's capacity to consume oxygen. The director, D.B. Dill, studied thermoregulation during exercise and the effects of environment on exercise. Researchers who trained at the Harvard Fatigue Laboratory established many of the first exercise physiology programs at universities in the United States.

The landmark work of McKenzie, Hill, Krogh, Dill, and other founders of exercise physiology established certain basic principles and procedures of exercise physiology, though most of the early research of exercise physiologists focused on human performance, not on the health effects of exercise. But as the field developed during the middle to late decades of the 20th century, exercise physiology came to focus on the effects of acute and chronic exercise on factors that were thought to be associated with health. For example, William L. Haskell identified the positive effects of exercise on plasma triglycerides, high-density lipoproteins, and other blood lipids (Haskell 1984). Other investigators identified the potential of exercise to lower blood pressure in both normotensive and hypertensive individuals (Tipton 1984).

Epidemiology

Epidemiologists study the rates at which diseases occur in populations and identify factors that are associated with the incidence of specific diseases. For example, in 1897, a British officer in the Indian Medical Service first demonstrated that mosquitoes transmit malaria to birds. Between 1898 and 1910, epidemiologists and other scientists uncovered the mechanism of transmission in humans and developed eradication and treatment procedures. One of the first large-scale tests of eradication procedures was implemented during construction of the Panama Canal. More recently, epidemiologists established cigarette smoking as a cause of lung cancer and coronary heart disease. The era of modern physical activity epidemiology really began in 1949, when

Jeremy N. Morris and colleagues began to carefully examine the health effects of active occupations on workers' health (Paffenbarger, Blair, and Lee 2001). In the **London Bus Study,** the Morris group found that conductors on London's double-decker buses, who climbed the buses' stairs many times each day and spent 90% of their shifts on their feet, had lower rates of coronary heart disease than the bus drivers, who were almost entirely sedentary (Morris et al. 1953). Morris also led a large necroscopy study of British workers that provided additional evidence of the health benefits of a physically active occupation (Morris and Crawford 1958). From these important beginnings, evidence of the link between physical activity and health continued to build throughout the 20th century.

AMERICAN
Journal of Epidemiology
Formerly AMERICAN JOURNAL OF HYGIENE

© 1978 by The Johns Hopkins University School of Hygiene and Public Health

VOL. 108   SEPTEMBER, 1978   NO. 3

Original Contributions

PHYSICAL ACTIVITY AS AN INDEX OF HEART ATTACK RISK IN COLLEGE ALUMNI[1]

RALPH S. PAFFENBARGER, Jr., ALVIN L. WING, AND ROBERT T. HYDE

Paffenbarger, R. S., Jr. (Stanford University School of Medicine, Stanford, CA 94305), A. L. Wing and R. T. Hyde. Physical activity as an index of heart attack risk in college alumni. *Am J Epidemiol* 108:161–175, 1978.

Risk of first heart attack was found to be related inversely to energy expenditure reported by 16,936 Harvard male alumni, aged 35–74 years, of whom 572 experienced heart attacks in 117,680 person-years of followup. Stairs climbed, blocks walked, strenuous sports played, and a composite physical activity index all opposed risk. Men with index below 2000 kilocalories per week were at 64% higher risk than classmates with higher index. Adult exercise was independent of other influences on heart attack risk, and peak exertion as strenuous sports play enhanced the effect of total energy expenditure. Notably, alumni physical activity supplanted student athleticism assessed in college 16–50 years earlier. If it is postulated that varsity athlete status implies selective cardiovascular fitness, such selection alone is insufficient to explain lower heart attack risk in later adult years. Ex-varsity athletes retained lower risk only if they maintained a high physical activity index as alumni.

coronary disease; hypertension; kilocalories; obesity; physical fitness; smoking; sports

FIGURE 2.2 Abstract of the landmark 1978 study in which Paffenbarger and colleagues first quantified a relationship between health and physical activity. They found that the risk of first heart attack is inversely related to physical activity up to a level corresponding to approximately 2,000 kcal of energy expenditure in physical activity per week.

Reprinted from R.S. Paffenbarger, Jr., A.L. Wing, and R.T. Hyde, "Physical activity as an index of heart attack risk in college alumni," *American Journal of Epidemiology*, 1978, 108(3): 161, by permission of Oxford University Press.

Particularly notable is the work of **Dr. Ralph Paffenbarger,** who conducted long-term epidemiological studies focusing on the relationship between physical activity and outcomes such as death attributable to cardiovascular disease (figure 2.2). His studies included groups as diverse as former longshoremen, who as young men performed heavy physical labor on the docks in San Francisco, and alumni of Harvard College, most of whom were engaged in rather sedentary occupations but some of whom were very physically active in their leisure time. Also central to the development of physical activity epidemiology as a scientific discipline has been the research of Steven N. Blair. Blair and colleagues at the Cooper Institute for Aerobics Research have conducted a series of landmark investigations, collectively known as the Aerobics Center Longitudinal Study, examining the relationship between physical fitness, measured as performance on a treadmill exercise test, and a wide range of chronic disease outcomes. Considered collectively, these studies have demonstrated that achieving and maintaining at least a moderate level of physical fitness provide very important health benefits, including reduced risk of death attributable to cardiovascular disease and increased longevity. The work of Paffenbarger, Morris, Blair, and other epidemiologists has been extremely influential because it established the impact of physical activity on health at the population level. Also, as is discussed in more detail later in this chapter, the work of epidemiologists has been important in establishing the types and amounts of physical activity needed to provide health benefits.

Clinical Science

Clinical research is typically designed to observe the effects of an experimental treatment on patients who have been diagnosed with a disease or clinical condition. For example, before a new drug can be brought to market, it must be subjected to clinical trials in which the effects of the drug are compared with the effects of a placebo on persons who have a condition that the drug is designed to treat. Exercise was first used as a form of treatment for diseased patients in the 1950s, when pioneering cardiologist **Paul Dudley White** began prescribing exercise as part of a rehabilitation program for patients with coronary heart disease. At about the same time, Herman Hellerstein and other physicians and scientists recognized that bed rest, the common prescription following a heart attack, was detrimental to people with heart disease. Hellerstein promoted many ideas that helped form the emerging field of cardiac rehabilitation, including exercise and a multidisciplinary approach to recovery from cardiac events.

By the mid-1970s, clinical research supported the benefits of exercise for cardiac patients; the American Heart Association (AHA) noted in 1975 that "data from several sources suggest that survivors of myocardial infarction who participate in physical training programs have a mortality rate approximately one-third below that of patients who remain inactive" (AHA 1975, p. 22). These studies demonstrated important health benefits of physical activity and paved the way for clinical studies of exercise as a treatment for a wide range of chronic diseases, including type 2 diabetes mellitus, cancer, pulmonary disease, and many other maladies.

Behavioral Science

Physical activity is central to normal human function. Accordingly, psychologists and other behavioral scientists have long been interested in physical activity. However, much of the initial interest of behavioral scientists focused on the learning of motor skills and enhancement of performance in sports, neither of which relates directly to the effects of physical activity on health. But in recent decades, with the growing interest of the scientific community in the physical activity–health relationship, health psychologists have begun to focus on topics that bear directly on physical activity as a behavior that influences the health of both individuals and entire populations.

Psychologists have studied the psychosocial factors that influence participation in physical activity; as a result, it is now known that numerous personal, social, and environmental factors combine to determine a person's habitual physical activity level. In addition, behavioral scientists have conducted studies of interventions to increase physical activity. These studies have been conducted with individuals, small groups, and communities. For example, Marcus and other investigators applied the concept of stages of change to physical activity to better understand the responses of individuals to exercise programs (Marcus et al. 1992; Marcus and Simkin 1993). King and others developed community approaches to physical activity promotion that included environmental, organizational, and policy strategies in addition to individual strategies (King 1991, 1994). Sallis and colleagues applied the principles of health psychology to the study of physical activity and its promotion in children and adolescents (Sallis, Prochaska, and Taylor 2000).

Molecular Biology and Genetics

Essentially all human characteristics are determined, at least in part, by the individual's genetic background. That is, much of what we are and can become is programmed in the genetic material, the DNA that we inherit from our parents. This principle has been developing as a scientific tenet since Gregor Mendel, an Augustinian monk born in Moravia, began exploring the science of genetics in the 19th century. In 1953, Watson and Crick published their landmark description of DNA, and since then, molecular biology has exploded into countless lines of scientific inquiry. Exercise biochemistry is one of the fields that has been profoundly influenced by the rapidly developing technology for studying the genetic basis of human characteristics.

Claude Bouchard has been a pioneer in the application of genetics to the study of human responses and adaptations to exercise. In the 1980s, Bouchard and his colleagues began studying inheritance of exercise characteristics by observing identical and fraternal twins. These studies demonstrated that the responses to exercise were much more similar in identical twins than in fraternal twins. For example, in a study of 31 pairs of identical twins and 22 pairs of fraternal twins, Bouchard found significant genetic effects on the energy cost of submaximal exercise and the respiratory exchange ratio at low power outputs. Other studies showed genetic influences on resting metabolic rate (RMR) and relative rate of carbohydrate to lipid oxidation (Bouchard 1991). This line of investigation has continued, and the large-scale **HERITAGE Family Study** has shown that the health-related physiological adaptations to exercise training are heavily dependent on genetic background. For example, the HERITAGE Study found that changes in $\dot{V}O_2$max in response to an exercise training program are highly variable and that the variability across families is much greater than within families (Bouchard et al. 1999). It seems certain that molecular biology will play a central role in the future of exercise science. Many exercise scientists believe that elucidating the role of genetics in explaining the individual variability in exercise characteristics will be a key direction for research for the next several decades. For example, we now have evidence that a fit and active way of life prevents or delays the development of hypertension. A next phase of investigation on this topic involving genetics will be to determine whether individuals with a certain genotype are more susceptible than individuals with other genotypes to developing hypertension as a result of a sedentary lifestyle.

Evolution of Physical Activity Guidelines

Scientific knowledge about physical activity and health is of little value if people cannot understand it and apply it to their lives. For the past three decades, there has been a gradual but steady development in the effort to present information on physical activity and health to the general public. This has come through public health messages known as physical activity guidelines.

Groundwork

Health scientists and practitioners have long believed that regular physical activity is essential to maintain good health. So it is not surprising that for a very long time, individual health professionals and health organizations have been making recommendations regarding the types and amounts of physical activity needed for health and fitness. As emphasized in the preceding section, scientific support for the impact of physical activity on health has developed rapidly in recent decades. As the relevant knowledge base has grown, physical activity recommendations for the public have been modified to maintain consistency with the existing research evidence. In this section, we track the evolution of physical activity guidelines as they have been presented to the public since the 1950s.

In 1957, Finnish researcher **Marti Karvonen** and his colleagues published the findings of a study that has become a classic in exercise science. Karvonen observed the effects of exercise training via treadmill running on endurance fitness in a small number of male medical students. He reported that training intensity corresponding to a heart rate of at least

Scientific Inquiry on Exercise and Health

- The modern field of exercise science began to develop in the early 20th century.
- Physiologists were the first scientists to study systematically the effects of exercise on the human body.
- Other disciplines that contributed to understanding the relationship between physical activity and health included epidemiology, clinical science, behavioral science, and molecular biology and genetics.

60% of the heart rate range (maximum heart rate minus resting heart rate) was required to produce significant gains in cardiorespiratory fitness. Although Karvonen's study was very small and quite limited in its research design, his findings became the platform for exercise guidelines for the ensuing three decades. Karvonen's program was presented in terms of minimums for frequency, duration, and intensity of training. A half century later, it seems remarkable that such a small and limited study could have had such a powerful influence on health practice.

> "Heart rate during training has to be more than 60% of the available range from rest to the maximum attainable by running . . . in order to produce a change in the WR [working heart rate]. . . . A decrease of the WR is understood to indicate an increase of the maximum oxygen uptake."
>
> *Karvonen, Kentala, and Mustala 1957, p. 314*

Initial Attempts to Speak to the Public

During the 1960s, two American men, one a track coach and the other a physician, published books that brought practical physical activity guidelines to the masses. In 1963, Bill Bowerman, coach of the University of Oregon's track team, visited a coaching colleague in New Zealand, where he witnessed many middle-aged adults running for health and fitness. He was so impressed by what he had seen that on his return to the United States, he wrote *Jogging,* a small paperback volume that has often been credited with launching a fitness revolution (Bowerman and Harris 1967). Bowerman described a slow running program that emphasized gradual, progressive increases in distance and frequency of exercise. His basic recommendation was that almost everyone can benefit from "an exercise program of relaxed walking and running" and that jogging is something that almost everyone can do.

In 1968, only a year after Bowerman popularized jogging as a specific form of exercise, Dr. Kenneth Cooper, then an Air Force physician, published *Aerobics,* a book in which he laid out a simple point system for determining how much exercise should be accumulated on a weekly basis. With his Aerobics Point System he recommended that adults accumulate a minimum of 30 points per week. Cooper recommended that sedentary adults begin an exercise program by starting at a level compatible with their current fitness (perhaps earning as few as 10 points per week for the first few weeks, for those at the lowest levels of fitness), choose an activity they enjoy, and exercise with others when at all possible. Table 2.1 provides examples of point values assigned to exercises by Cooper. Although neither Bowerman nor Cooper was able to base these recommendations on extensive bodies of directly relevant scientific evidence, both were talented practitioners and gifted communicators who were able to draw on their extensive experience in educating the public about how much physical activity is needed for health and fitness.

TABLE 2.1 Examples of Point Values in Cooper's Original Point System

Activity	Points
Run 1 mi in <8 min	5
Walk 3 mi in <43 min	6
Cycle 5 m in <20 min	5
Swim 600 yd in <15 min	5

Data from Cooper 1968.

Exercise Prescription for the Public

Concurrent with the popularization of the exercise-for-fitness movement and the so-called running boom of the late 1960s and 1970s, exercise scientists began to systematically explore the effects of various types, intensities, durations, and frequencies of endurance exercise on cardiorespiratory fitness. A leader of this extensive scientific effort was Dr. Michael Pollock. During the 1970s, Pollock and his colleagues undertook a series of experimental exercise training studies that, when considered collectively along with the work of other researchers, produced the knowledge needed to recommend exercise in a precise, detailed, and individualized manner. This method became the central dogma in efforts to communicate to the public the types and amounts of exercise needed to promote health and fitness. The American College of Sports Medicine (ACSM) first formally endorsed this detailed approach to recommending exercise in its exercise guidelines book in 1975 (ACSM 1975) and in a position statement issued in 1978 (ACSM 1978). The key recommendations presented in the **ACSM Position Statement** are summarized in table 2.2.

During the same period in which exercise scientists were systematically studying endurance exercise training and its impact on cardiorespiratory fitness in healthy adults, cardiologists and clinical exercise physiologists were studying the effects of exercise training in patients with cardiovascular disease. This research demonstrated the critical and now well-accepted role that exercise can play in rehabilitation of patients with compromised cardiovascular function. But furthermore, this research and the clinical guidelines that it spawned established a medical approach to recommending exercise that came to be referred to as "exercise prescription." This technique drew on the research on normal healthy adults as well as research performed on heart patients. In 1975, the AHA published guidelines on exercise prescription for patients with cardiovascular disease. This document helped to establish a place for exercise in the practice of medicine and was influential in communicating to the public the significant health benefits that accrue to physically active persons, even those with already established cardiovascular disease. Table 2.3 summarizes the AHA's first guidelines for physical activity in people with heart disease or at risk for heart disease (AHA 1975).

Importance of Moderate-Intensity Physical Activity

ACSM's *Guidelines for Exercise Testing and Prescription* has undergone revision approximately every five years since its initial publication in 1975 (ACSM 1975). Note that the first version of this was called *Guidelines for Graded Exercise Testing and Prescription*, but subsequent revisions dropped the word *graded*. Each volume included a primary recommendation on prescription of exercise that reflected the current body of knowledge regarding the types of exercise needed to provide health and fitness benefits to initially sedentary adults. Between the first edition published in 1975 and the eighth edition released in 2009, an interesting trend is evident. As shown by table 2.4, most elements in the exercise prescription guideline remained unchanged. The exception is the recommended range for exercise intensity, the lower end of which decreased from 60% $\dot{V}O_2$max to 40% $\dot{V}O_2$max. The earlier editions of this influential book indicated that rather vigorous exercise was needed to provide benefits, and this concept was widely communicated to the public during the 1970s and 1980s.

Recognition of the importance of moderate-intensity physical activity, as reflected by the changing exercise prescription guidelines of ACSM, evolved gradually during the 1980s and early 1990s as the result of a growing and changing body of research evidence. Two lines of research led to the conclusion that moderate-intensity physical activity (the equivalent of brisk walking) provided important benefits to health and fitness. First and most importantly, the science of physical activity epidemiology matured during the 1980s and produced a series of important investigations. These studies strongly suggested that regular performance of moderate-intensity physical activity provided important health benefits. Not only did these studies show that regularly active persons were less likely than sedentary persons to develop or die from

TABLE 2.2 1978 ACSM Position Statement: The Recommended Quantity and Quality of Exercise for Developing and Maintaining Fitness in Healthy Adults

Quality	Quantity
Frequency	Three to five days per week
Intensity	50-85% $\dot{V}O_2$max (60-90% maximum heart rate)
Duration	15-60 min

Data from ACSM 1978.

TABLE 2.3 The American Heart Association's First Guidelines

Quality	Quantity
Frequency	Three or four times per week
Intensity	70-85% maximum heart rate
Duration	20-60 min

Data from American Heart Association 1975.

TABLE 2.4 ACSM Guidelines for Exercise Prescription

Quality	1975	1995
Frequency	Three times per week	Three to five times per week
Intensity	60-90% $\dot{V}O_2$max	40-85% $\dot{V}O_2$max
Duration	20-30 min	20-30 min

Data from Department of Health and Human Services 1996.

cardiovascular disease, but they also demonstrated that much of the active population's physical activity came from walking and other forms of moderate-intensity physical activity. For example, results of the Third National Health and Nutrition Examination Survey (Crespo et al. 1996) showed that most of the physical activities preferred by American adults were moderate-intensity lifestyle activities, such as walking, gardening, and cycling (table 2.5).

As discussed previously, the exercise prescription method for recommending physical activity to the public was based primarily on the findings of a large number of experimental exercise training studies. The results of these studies had generally been interpreted as indicating that vigorous physical activity (6 METs or 60% or more individual functional capacity) was required to produce benefits. That moderate-intensity physical activity did not provide those benefits became the assumption. However, the epidemiological studies published in the 1980s and early 1990s forced a reexamination of the experimental studies. A closer look revealed that, in studies that compared moderate- and vigorous-intensity physical activity, the moderate level produced increased fitness, although often not to the same extent as the vigorous level. Also, it was seen that moderate-intensity physical activity often provided comparable or even

greater beneficial effects on health outcomes such as blood pressure and high-density lipoprotein (HDL) cholesterol. For example, Duncan and colleagues (1991) found that both women who participated in a vigorous exercise program and those who participated in a moderate exercise program had significant improvements in their lipoprotein profiles (figure 2.3). Although women in the vigorous program had significantly greater gains in fitness, as measured by $\dot{V}O_2$max, increases in HDL were similar in the two groups.

Physical Inactivity: An Independent Risk Factor for Cardiovascular Disease

The AHA is a large and influential organization, and many of its members are physicians and others who provide medical services related to cardiovascular disease. The official positions of the AHA are important because they influence medical practice, health care policy, and public health investments. Before the 1990s, the AHA, although supportive of physical activity through some of its programmatic initiatives, had not officially recognized physical inactivity as a risk factor for cardiovascular disease. That changed in 1992

TABLE 2.5 Physical Activities Preferred by American Adults and Preference Rankings

Activity	Preference rating	
	Men	Women
Walking	2	1
Gardening and yard work	1	2
Calisthenics	3	3
Cycling	4	4
Jogging and running	5	8
Weightlifting	6	9
Swimming	6	6
Dancing	8	5
Aerobics and aerobic dance		6
Basketball		10
Tennis		10
Golf		9

Data from Crespo et al. 1996.

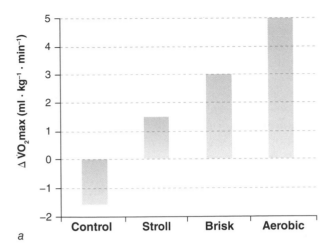

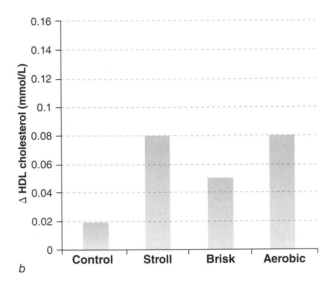

FIGURE 2.3 Effect of walking intensity on changes from baseline in (a) maximal oxygen uptake ($\Delta \dot{V}O_2$max) and (b) high-density lipoprotein cholesterol (Δ HDL cholesterol) after 24 weeks of exercise training.

Data from Duncan, Gordon, and Scott 1991.

"Regular aerobic physical activity increases exercise capacity and plays a role in both primary and secondary prevention of cardiovascular disease. . . . Inactivity is recognized in the AHA Statement on Exercise as a risk factor for coronary artery disease."

Fletcher et al. 1992

tennis, and basketball, when performed at 50% or more of a person's work capacity, were especially beneficial. The statement also noted that "even low-intensity activities performed daily can have some long-term health benefits and lower the risk of cardiovascular disease." However, the document did not provide a specific recommendation regarding the types and amounts of physical activity that should be performed to minimize risk of cardiovascular disease.

Public Health Messages on Physical Activity

The AHA's endorsement of physical inactivity as a risk factor for cardiovascular disease had many important effects. One of these was to heighten the awareness of public health leaders that there was a need to launch large-scale programs to promote physical activity in the American population. But in considering how such programs should be designed, these public health leaders noted a problem. Although many organizations had recommended that Americans become more physically active, there was a lack of clarity regarding the types and amounts of physical activity that should be recommended to promote health in the population. Furthermore, there was a concern that traditional exercise guidelines were not the best ones for broad public health purposes. The conclusion was that a new "public health message" on physical activity should be developed.

In 1993, the U.S. Centers for Disease Control and Prevention (CDC) partnered with the ACSM in drafting a statement on physical activity that would be useful in communicating to the American public how much and which types of physical activity are needed to maintain good health. An expert panel was formed and charged with developing a clear, concise statement on physical activity for enhancement of public health. This panel included exercise scientists,

with the publication of AHA's "Statement on Exercise: Benefits and Recommendations for Physical Activity Programs for All Americans" (Fletcher et al. 1992). This position statement declared that a sedentary lifestyle is a major and independent risk factor for premature development of atherosclerotic cardiovascular disease, the number-one cause of death in the United States and other developed countries. This statement has been updated several times over the past decade to reflect new findings on the role of physical activity in the prevention and treatment of heart disease.

The AHA position stand made several important recommendations. It stated that people of all ages could benefit from a regular exercise program and that activities such as walking, hiking, swimming, epidemiologists, physicians, and health psychologists. In pursuing its task, the panel applied several criteria. First, the panel required that any recommendation be extensively supported by the scientific literature. Second, there was a need to craft a statement that would communicate clearly to the public. And third, there was a desire to recommend a model for physical activity that, while providing important health benefits, would be readily attainable by most people. The following is the recommendation that was issued by the panel.

"Every U.S. adult should accumulate 30 minutes or more of moderate-intensity physical activity on most, preferably all, days of the week."

Pate et al. 1995, p. 402

The release of this statement created considerable controversy, in part because several elements of the recommendation were novel and were perceived by some as inconsistent with previous guidelines. Before issuance of the **CDC–ACSM Recommendation on Physical Activity and Public Health,** there was a broad perception that exercise had to be continuous and rather vigorous to provide important health benefits. The new statement, by endorsing accumulation of moderate-intensity physical activity in bouts as short as 8 to 10 min, was intended to present the public with an approach to a physically active lifestyle that would include a weekly dose of physical activity offering important health benefits—and that would be potentially attainable by and attractive to the average person. A key principle underlying the recommendation was that the greatest public health gain would result from moving the large sedentary segment of the population into a regularly physically active pattern. Hence it was seen as important that the recommendation be viewed as attainable by persons who were currently sedentary.

U.S. Surgeon General's Report on Physical Activity and Health

Although the CDC–ACSM recommendation on physical activity was seen as controversial in some quarters, its core elements were quickly endorsed by several highly credible organizations. The National Institutes of Health (NIH) convened a consensus conference on the health effects of physical activity, and the conclusions of that conference supported the position taken by the CDC and ACSM, that the public health would be greatly enhanced if most American adults performed at least 30 min of moderate-intensity physical activity daily (NIH Consensus Development Panel on Physical Activity and Cardiovascular Health 1996). The World Health Organization, the public health arm of the United Nations, also issued a statement supporting the health benefits of regularly performing 30 min or more of moderate-intensity physical activity (World Health Organization [WHO]–International Federation of Sports Medicine [FIMS] Committee on Physical Activity for Health 1995).

These important statements were followed by a landmark event, the release of *Physical Activity and Health: A Report of the Surgeon General* (DHHS 1996), for which Steven Blair served as senior scientific editor. This extensive document summarized the evidence supporting the health benefits of physical activity and drew several major conclusions (see "Key Conclusions of the Surgeon General's Report"). These recommendations, coming from a prestigious government agency, constituted the strongest support to date for the public health benefits of physical activity.

"Significant health benefits can be obtained by including a moderate amount of physical activity (e.g., 30 minutes of brisk walking or raking leaves, 15 minutes of running, or 45 minutes of playing volleyball) on most, if not all, days of the week. Through a modest increase in daily activity, most Americans can improve their health and quality of life."

U.S. Department of Health and Human Services 1996, p. 4

Daily Physical Activity to Prevent Excessive Weight Gain

The fundamental recommendation that adults should participate in 30 min of physical activity daily, as recommended in the CDC–ACSM guideline and the **Surgeon General's report on physical activity and health**, was widely supported by public health authorities around the globe. Also, during the decade following the initial release of that recommendation, its validity was further substantiated by a significant amount of new research. Nonetheless, some in the scientific community questioned whether 30 min of daily physical activity was enough to provide all the desired benefits of an active lifestyle. In particular, in 2003 a panel of the Institute of Medicine (IOM) concluded that 60 min of daily physical activity is needed to prevent excessive weight gain and obesity. That panel's conclusion was based on an examination of a much narrower body of scientific evidence than that considered for the Surgeon General's report or by the committee that issued the CDC–ACSM recommendation. The IOM panel considered only the relationship between weight status and physical activity, whereas the other panels had taken a much more comprehensive view of the impact of physical activity on health.

A reexamination of these issues was undertaken by the U.S. Dietary Guidelines Advisory Committee in 2004, and that panel concluded that 30 min of daily physical activity provides important public health benefits by substantially reducing the risk of chronic diseases such as coronary heart disease, type 2 diabetes mellitus, and osteoporosis (U.S. Dietary Guidelines Advisory Committee 2004). However, the panel also noted that many people may require more than 30 min of daily physical activity to prevent excessive weight gain and that more physical activity is needed to induce and maintain weight loss in formerly obese individuals than is needed to prevent excessive weight gain in the first place.

2008 Physical Activity Guidelines for Americans and National Physical Activity Plan

The status of physical activity as a public health priority in the United States advanced significantly in 2008 with the release of *Physical Activity Guidelines for Americans* (DHHS 2008). This document provided, for the first time, a comprehensive set of physical activity recommendations that carried the authority and weight of the U.S. federal government. *Guidelines* was produced as the result of a directive from Michael Leavitt, then Secretary of Health and Human Services, who proscribed employment of a developmental protocol similar to that already established for the *Dietary Guidelines for Americans*. These were the key steps in that process:

- The IOM conducted a workshop to advise Secretary Leavitt regarding the need for comprehensive physical activity guidelines. The report on the workshop concluded that the available data "indicate that the development of new or comprehensive guidelines based on such evidence would be feasible. The evidence presented at the workshop clearly indicates that the large volume of high-quality data—much of it recent—could inform new physical activity guidelines" (IOM 2007).

- Secretary Leavitt impaneled the Physical Activity Guidelines Advisory Committee (PAGAC) to conduct an in-depth review of the scientific literature on the effects of physical activity on health and disease outcomes. The PAGAC gave special emphasis to dose–response relationships between physical activity and chronic disease outcomes, including morbidity and mortality due to cardiovascular

Key Conclusions of the Surgeon General's Report

- People of all ages benefit from regular physical activity.
- Moderate physical activity (equivalent to 30 min of brisk walking on most days of the week) can provide significant health benefits.
- Greater amounts of physical activity can provide additional health benefits.
- Physical activity reduces the risk of premature mortality and of coronary heart disease, hypertension, colon cancer, and diabetes.
- More than 60% of American adults are not regularly active, and 25% are not active at all.
- Nearly half of American youth (ages 12-21) are not vigorously active on a regular basis, and physical activity declines significantly during adolescence.

disease, selected cancers, type 2 diabetes mellitus, and osteoporosis and skeletal fractures, as well as all-cause mortality. In addition, committee members reviewed the scientific literature on the effects of physical activity on health for special population groups, including children and adolescents, persons with disabilities, and racial and ethnic minority groups. The PAGAC report constituted a very detailed, comprehensive, and up-to-date summary of the scientific literature on the relationship between physical activity and health (PAGAC 2008).

- Using the PAGAC report as the source of information, DHHS personnel developed a detailed set of recommendations regarding the types and amounts of physical activity needed for health benefits. See a summary of the key elements of *Physical Activity Guidelines for Americans* (DHHS 2008) in "2008 Physical Activity Guidelines for Americans."

To complement and extend the impact of *Physical Activity Guidelines for Americans,* a coalition of national nonprofit organizations developed a strategic plan for increasing physical activity in the U.S. population. Drawing on guidance from WHO and the experience gained in other countries that had developed national physical activity plans (CDC, WHO 2007), the U.S. National Physical Activity Plan was released in 2010. The plan is organized around eight societal sectors, including public health, health

2008 Physical Activity Guidelines for Americans

- Adults should participate in at least 150 min per week of moderate-intensity physical activity or 75 min per week of vigorous-intensity physical activity (or a combination of the two). Additional aerobic activity will provide additional health benefits.
- Adults should do muscle strengthening activities on two days or more per week.
- Children and adolescents should participate in 60 min or more of moderate-intensity or vigorous-intensity physical activity daily. They should participate in vigorous-intensity activity at least three days per week.
- Children and adolescents should participate in muscle strengthening and bone strengthening activities on three days per week.

Evolution of Physical Activity Guidelines

- In 1957, Karvonen's investigation of the heart rate level needed to produce an increase in cardiorespiratory fitness set the stage for the development of physical activity guidelines.
- In 1978, the American College of Sports Medicine issued its first position statement on the quantity and quality of exercise required for developing and maintaining fitness.
- In 1992, the American Heart Association stated officially that a sedentary lifestyle is a major and independent risk factor for the development of cardiovascular disease.
- In 1995, the Centers for Disease Control and Prevention and the American College of Sports Medicine issued the first public health–oriented statement on physical activity and health. The statement said that every American adult should accumulate 30 min or more of physical activity on most, and preferably all, days of the week.
- The Surgeon General's report on physical activity and health, published in 1996, clearly established that regular, moderate physical activity provides health benefits to people of all ages and that additional activity can provide additional benefits.
- In 2005, the U.S. Dietary Guidelines Advisory Committee concluded that 30 min of moderate or vigorous physical activity on most days provides substantial health benefits for adults.
- In 2008, the U.S. Department of Health and Human Services issued *Physical Activity Guidelines for Americans*, which provided specific advice about types and amounts of activity for Americans of all ages.

care, education, mass media, transportation/community planning, nonprofit organizations, sports/recreation/fitness, and business/industry. It includes more than 200 recommended policies, practices, and initiatives, all aimed at changing communities in ways that would result in more people meeting the physical activity guidelines.

Summary

The work of scientists, medical practitioners, and philosophers over several thousand years forms the foundation on which modern exercise science stands. Modern scientists from a number of disciplines, including exercise physiology, epidemiology, the clinical sciences, the behavioral sciences, and molecular biology and genetics, have contributed to our current understanding of the relationship between physical activity and health. Their work has led to breakthroughs in the measurement of physical activity and physical fitness; exercise testing and prescription for people of all ages and abilities; development of interventions that help people adopt and maintain active lifestyles; and a better understanding of the ways in which family, community, and environment influence physical activity. Exercise scientists and their colleagues in related fields have also stepped outside their laboratories and clinical settings and worked to translate their findings into messages that the general public can understand. The result is physical activity guidelines, a series of public health messages designed to help people be active at levels that will improve their health and quality of life. Leading scientists and science organizations have participated in developing and publicizing these guidelines and in revising them to keep up with the pace of modern exercise science. The ideas and hypotheses of ancient Chinese and Greek scholars and European Renaissance physicians—that an active life promotes good health and prevents disease and disability—have been supported by almost a century of modern scientific studies. It seems certain that the scientific disciplines that study physical activity and exercise will continue to grow and change and to apply new techniques to understand the effects of physical activity on the human body.

Key Concepts

ACSM Position Statement, 1978—First statement by the ACSM on the quantity and quality of exercise required for developing and maintaining fitness in healthy adults. ACSM recommended that adults exercise three to five days per week for 15 to 60 min at 60% to 90% of $\dot{V}O_2$max.

CDC–ACSM Recommendation on Physical Activity and Public Health—Issued in 1995, a recommendation that differed from previous recommendations. It focused on moderate physical activity and on accumulation of activity throughout the day on most, if not all, days of the week.

Diseases of Workers—First published work to identify the health risks and benefits of certain occupations (Ramizzini 1713).

Claudius Galenus (Galen)—Brilliant Greek physician, born circa A.D. 131, whose teachings dominated European medicine for 1,000 years. He believed that some form of exercise could be used to treat every form of disease and that all people, healthy and sick, could benefit from regular exercise.

W.A. Guy—Researcher in the 1840s who noted the superior health of men in active occupations. He recommended that men in sedentary occupations perform physical exercise in their leisure time to improve their health.

Harvard Fatigue Laboratory—Lab established in 1927 that conducted research on a variety of subjects, including exercise, nutrition, blood chemistry, and the effects of altitude and climate on the body. Pioneering exercise research included the first measures of the body's capacity to consume oxygen, thermoregulation during exercise, and the effects of environmental factors on exercise.

HERITAGE Family Study—Large-scale study of the role of genetics in cardiovascular and metabolic responses to aerobic training and regular exercise. The study was funded in 1992, with Claude Bouchard as the principal investigator, and included researchers from five field centers in the United States and Canada. One of the earliest findings of the HERITAGE Family Study was that changes in $\dot{V}O_2$max in response to exercise training are highly variable and that variability across families is much greater than within families.

Hippocrates—"Father" of preventive medicine, who wrote extensively in the fifth century B.C. about the benefits of exercise for a variety of illnesses.

Marti Karvonen—Finnish researcher who identified the intensity of exercise training required to produce gains in cardiorespiratory fitness.

London Bus Study—First physical activity epidemiology study, conducted by Jeremy Morris and colleagues in the 1940s. The study showed that London's double-decker bus conductors, who climbed stairs and were on their feet throughout their shifts, had lower rates of coronary heart disease than bus drivers, who were almost entirely sedentary during their shifts.

R. Tait McKenzie—One of the first physiologists to study exercise. McKenzie investigated the effects of exercise on athletes before and after sport participation.

Ralph Paffenbarger—Pioneering physical activity epidemiologist. Paffenbarger conducted large-scale studies on the relationship of physical activity to health outcomes. Perhaps the best known of his studies, the Harvard Alumni Health Study, clearly established that physical activity reduces the rate and risk of developing coronary heart disease in men.

Surgeon General's report on physical activity and health—Landmark document, published in 1996, that summarized the science on physical activity and health to date and recommended that all adults accumulate 30 min or more of moderate physical activity on all or most days of the week.

Paul Dudley White—Cardiologist who in the 1950s began prescribing exercise as part of a treatment program for people with coronary heart disease.

Study Questions

1. Discuss the views of the ancient Greek physicians on the relationship between physical activity and health.

2. What contributions did the Greek physician Galen make to understanding the relationship between physical activity and health?

3. What information did observational studies of sedentary workers in the 18th and 19th centuries reveal?

4. Name five scientific disciplines that have contributed to the development of the field of exercise science.

5. Describe the work of three scientists who contributed to the development of exercise science in the 20th century.

6. What were the recommendations of the 1978 ACSM position statement on exercise for developing and maintaining fitness? What change in the ACSM position on exercise is reflected in recent versions of the position statement?

7. Why was the 1992 AHA "Statement on Exercise" an important milestone in establishing the physical activity–health connection?

8. Why was the 1995 CDC–ACSM recommendation controversial?

9. List four key conclusions of *Physical Activity and Health: A Report of the Surgeon General.*

10. What did the 2004 U.S. Dietary Guidelines Advisory Committee determine about the relationship of physical activity to weight loss and maintenance of a weight loss?

11. What types and amounts of physical activity did the 2008 *Physical Activity Guidelines for Americans* recommend for adults? For children and adolescents?

References

American College of Sports Medicine. 1975. *Guidelines for graded exercise testing and prescription.* Philadelphia: Lea & Febiger.

American College of Sports Medicine. 1978. Position statement—the recommended quantity and quality of exercise for developing and maintaining fitness in healthy adults. *Medicine and Science in Sports and Exercise* 10:vii-x.

American College of Sports Medicine. 2009. *ACSM's guidelines for exercise testing and prescription.* 8th ed. Philadelphia: Lippincott Williams & Wilkins.

American Heart Association. 1975. *Exercise testing and training of individuals with heart disease or at high risk for its development.* Dallas: American Heart Association.

Bouchard, C. 1991. Heredity and the path to overweight and obesity. *Medicine and Science in Sports and Exercise* 23:285-291.

Bouchard, C., P. An, T. Rice, J.S. Skinner, J.H. Wilmore, J. Gagnon, L. Perusse, A.S. Leon, and D.C. Rao. 1999. Familial aggregation of VO₂max response to exercise training: Results from the HERITAGE Family Study. *Journal of Applied Physiology* 87:1003-1008.

Bowerman, W.J., and W.E. Harris. 1967. *Jogging.* New York: Grosset and Dunlap.

Centers for Disease Control and Prevention, World Health Organization. 2007. *CDC/WHO Collaborating Center workshop on global advocacy for national physical activity plans. Final workshop report.* Atlanta: Centers for Disease Control and Prevention.

Cooper, K.H. *Aerobics.* 1968. New York: Bantam Books.

Crespo, C.J., S.J. Keteyian, G.W. Heath, and C.T. Sempos. 1996. Leisure-time physical activity among US adults. Results from the Third National Health and Nutrition Examination Survey. *Archives of Internal Medicine* 156:93-98.

Duncan, J.J., N.F. Gordon, and C.B. Scott. 1991. Women walking for health and fitness. How much is enough? *Journal of the American Medical Association* 266:3295-3299.

Fletcher, G.F., S.N. Blair, J. Blumenthal, C. Caspersen, B. Chaitman, S. Epstein, H. Falls, E.S. Froelicher, V.F. Froelicher, and I.L. Pina. 1992. Statement on exercise. Benefits and recommendations for physical activity programs for all Americans. A statement for health professionals by the Committee on Exercise and Cardiac Rehabilitation of the Council on Clinical Cardiology, American Heart Association. *Circulation* 86:340-344.

Haskell, W.L. 1984. The influence of exercise on the concentrations of triglyceride and cholesterol in human plasma. *Exercise and Sport Sciences Reviews* 12:205-244.

Institute of Medicine. 2007. *Adequacy of evidence for physical activity guidelines development: Workshop summary.* Washington, DC: National Academies Press.

Karvonen, M.J., E. Kentala, and O. Mustala. 1957. The effects of training on heart rate; a longitudinal study. *Annals of Medicine Experimental Biology Fennica* 35:307-315.

King, A.C. 1991. Community intervention for promotion of physical activity and fitness. *Exercise and Sport Sciences Reviews* 19:211-259.

King, A.C. 1994. Community and public health approaches to the promotion of physical activity. *Medicine and Science in Sports and Exercise* 26:1405-1412.

Marcus, B.H., V.C. Selby, R.S. Niaura, and J.S. Rossi. 1992. Self-efficacy and the stages of exercise behavior change. *Research Quarterly for Exercise and Sport* 63:60-66.

Marcus, B.H., and L.R. Simkin. 1993. The stages of exercise behavior. *Journal of Sports Medicine and Physical Fitness* 33:83-88.

Morris, J.N., and M.D. Crawford. 1958. Coronary heart disease and physical activity of work: Evidence of a national necropsy survey. *British Medical Journal* 30:1485-1496.

Morris, J.N., J.A. Heady, P.A. Raffle, C.G. Roberts, and J.N. Parks. 1953. Coronary heart disease and physical activity at work. *Lancet* 265:1053-1057, 1111-1120.

National physical activity plan. 2010. www.physicalactivityplan.org/NationalPhysicalActivityPlan.pdf. Accessed July 28, 2010.

NIH Consensus Development Panel on Physical Activity and Cardiovascular Health. 1996. Physical activity and cardiovascular health. *Journal of the American Medical Association* 276:241-246.

Paffenbarger, R.S. Jr., S.N. Blair, and I.M. Lee. 2001. A history of physical activity, cardiovascular health and longevity: The scientific contributions of Jeremy N Morris, DSc, DPH, FRCP. *International Journal of Epidemiology* 30:1184-1192.

Paffenbarger, R.S. Jr., A.L. Wing, and R.T. Hyde. 1978. Physical activity as an index of heart attack risk in college alumni. *Journal of Epidemiology* 108:161.

Pate, R.R., M. Pratt, S.N. Blair, W.L. Haskell, C.A. Macera, C. Bouchard, D. Buchner, W. Ettinger, G.W. Heath, A.C. King, A. Kriska, A.S. Leon, S.E. Marcus, J. Morris, R.S. Paffenbarger Jr., K. Patrick, M.L. Pollock, J.M. Rippe, J.F. Sallis, and J.H. Wilmore. 1995. A recommendation from the Centers for Disease Control and Prevention and the American College of Sports Medicine. *Journal of the American Medical Association* 273:402-407.

Physical Activity Guidelines Advisory Committee. 2008. *Physical Activity Guidelines Advisory Committee report, 2008.* Washington, DC: U.S. Department of Health and Human Services.

Sallis, J.F., J.J. Prochaska, and W.C. Taylor. 2000. A review of correlates of physical activity of children and adolescents. *Medicine and Science in Sports and Exercise* 32:963-975.

Tipton, C.M. 1984. Exercise, training, and hypertension. *Exercise and Sport Sciences Reviews* 12:245-306.

U.S. Department of Health and Human Services. 1996. *Physical activity and health: A report of the Surgeon General.* Atlanta: U.S. Department of Health and Human Services, Centers for Disease Control and Prevention.

U.S. Department of Health and Human Services. 2008. *Physical activity guidelines for Americans.* Washington, DC: U.S. Department of Health and Human Services.

U.S. Dietary Guidelines Advisory Committee. 2004. *The report of the Dietary Guidelines Advisory Committee on Dietary Guidelines for Americans, 2005.* Washington, DC: U.S. Department of Health and Human Services and U.S. Department of Agriculture.

WHO-FIMS Committee on Physical Activity for Health. 1995. Exercise for health. *Bulletin of the World Health Organization* 73:135-136.

3

Physical Activity and Fitness With Age, Sex, and Ethnic Differences

Peter T. Katzmarzyk, PhD, FACSM

CHAPTER OUTLINE

This chapter explores age, sex, and ethnic differences in **physical activity** and **physical fitness**. The chapter examines data collected from several countries; and, where available, studies using representative population samples are highlighted. The limitations of the existing databases and studies are also described, and important areas for future research are noted throughout the chapter. By the end of the chapter, you should be able to describe differences in physical activity and fitness between males and females and across different age and ethnic groups. Given the relationships among physical activity, fitness, and health described in part III of this book, we discuss specific population groups that are considered to be at a high risk of morbidity and mortality because of low levels of physical activity or fitness.

It is clear that individuals differ in physical activity levels and physical capabilities. For example, many people have difficulty climbing a flight of stairs without getting winded, whereas others have the ability to complete Ironman-distance triathlons, which involve swimming 4 km (2.4 mi), cycling 180 km (112 mi), and running 42 km (26 mi). There are many reasons for this incredible range of human variation in physical activity levels and physical abilities; social and behavioral influences as well as biological factors play a role. Although physical activity and fitness levels vary considerably from person to person, there are systematic differences according to sex, age, and ethnicity. The field of **epidemiology** is concerned with studying the distribution and determinants of disease, injury, and risk factors in society. As risk factors for chronic disease, physical inactivity and inadequate fitness are important variables in this regard. The purpose of this chapter is to present the **descriptive epidemiology** of physical activity and physical fitness levels across age, sex, and ethnic groups.

Although sex and age are self-explanatory concepts, race and ethnicity are more difficult to define. There are no widely accepted definitions of race or ethnicity; however, there are definite distinctions between the two. For example, the notion of race implies that biological traits can be used to categorize people into subgroups or races. On the other hand, ethnicity implies cultural similarities among individuals rather than strictly biological relatedness. The terms *race* and *ethnicity* are often used interchangeably in the scientific literature or combined into a single dimension such as *race/ethnicity*. Throughout this chapter, the term *ethnicity* is used consistently to refer to racial and ethnic differences in physical activity and physical fitness; however, the authors of the studies described in the chapter may have used either one term or the other in their original reports.

The studies described in this chapter all used either cross-sectional or longitudinal research designs. A **cross-sectional study** is one in which a sample of the population is examined at a given point in time and data from specific groups can be compared. This type of design is commonly used in national population surveys; with this design it is difficult to infer changes in traits or behaviors within individuals, but differences between individuals or groups can be determined. A **longitudinal research design** is one in which participants are measured initially and then followed over time so that changes can be directly measured. Longitudinal studies are generally more difficult and time-consuming to conduct; however, they allow researchers to measure changes within individuals and establish temporal sequences of events.

Physical Activity

As defined in chapter 1, physical activity is a behavior. Like most behaviors, it displays a wide range of variability between males and females and across age and ethnic groups. This section presents the descriptive epidemiology of physical activity levels using the most recently available representative population data from several countries. Figure 3.1 presents the percentage of the U.S. population that is meeting

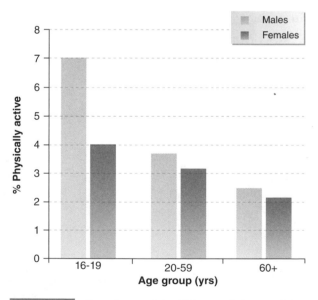

FIGURE 3.1 Prevalence of the U.S. population attaining sufficient physical activity levels. Sufficient physical activity was defined as accumulating 30 min of moderate- or high-intensity physical activity in at least 10-min bouts on at least five of seven days.

Data from National Health and Nutrition Examination Survey; Troiano et al. 2008.

the public health recommendation to accumulate 30 min of moderate-intensity physical activity on most days of the week (Troiano et al. 2008). These data were obtained from the nationally representative National Health and Nutrition Examination Survey (NHANES) and were collected using objective physical activity monitors (accelerometry). Overall, less than 5% of the population is meeting the physical activity recommendations.

Age

This chapter provides a general overview of age-related differences in physical activity and fitness. A more detailed treatment of the topics of physical activity and fitness in childhood and in the aged is provided in chapters 17 and 19.

Several international studies using a cross-sectional research design have documented clear age-related differences in physical activity levels. In general, physical activity levels are highest in childhood and decrease throughout adolescence and into adulthood. Figure 3.2 demonstrates the age-related differences in physical activity levels among adults in Canada (figure 3.2a) and the United States (figure 3.2b). Compared with younger adults, older adults have lower overall levels of **leisure-time physical activity** and typically engage in less vigorous physical activities. Although the trends are similar, it is difficult to compare across the countries because the measures of physical activity are different. These age differences in physical activity levels were also evident

in the Baltimore Longitudinal Study of Aging, in which age-related differences in leisure-time physical activity levels were observed; the data showed a shift from high-intensity to moderate- and low-intensity activities with advancing age (Talbot, Metter, and Fleg 2000).

It is hypothesized that the observed age-related differences in physical activity levels in humans are the result of complex changes in psychological and social factors that occur with age; however, biological mechanisms may also play a role. Age-related declines in physical activity in laboratory and zoo animals, including invertebrates, rodents, dogs, and nonhuman primates, have been well documented. For example, the number of episodes of walking and jumping in three species of female monkeys observed in a zoological park showed a clear age-related trend, as demonstrated in figure 3.3 (Janicke, Coper, and Janicke 1986). Biological mechanisms that have been proposed to explain the age-related declines in physical activity include the general physiological declines that occur with aging, declines in motor abilities, and neurobiological changes that may be involved in the motivation to be physically active.

> Physical activity levels decline across the life span, and evidence from non-human studies suggests that in addition to social and psychological factors, biological mechanisms may play a role.

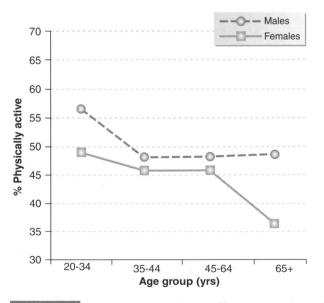
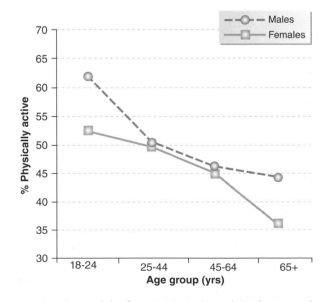

FIGURE 3.2 Age- and sex-related differences in physical activity levels in adults from (a) Canada and (b) the United States. In Canada, physically active was defined as expending at least 1.5 kcal · kg^{-1} · day^{-1}; in the United States, it was defined as participating in at least some leisure-time physical activity in the past month.

Data from the 2007-08 Canadian Community Health Survey and the 2005 U.S. Behavioral Risk Factor Surveillance System.

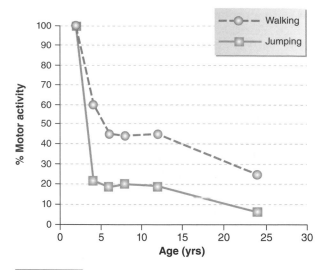

FIGURE 3.3 Age-related differences in physical activity in three species of female monkeys observed in a zoological park.

Adapted, by permission, from B. Janicke, D. Coper, and U.-A. Janicke, 1986, "Motor activity of different-aged Cercopithecidae: Silver-leafed Monkey (Presbytis cristatus Esch.), Lion-tailed monkey (Macaca silenus L.), Moor Macaque (Macaca maura Cuv.) as observed in the zoological garden, Berlin," *Gerontology* 32:133-140, courtesy of S. Karger AG, Basel.

In addition to trends in physical activity levels *across* age groups, an important consideration is **tracking**, or stability of physical activity with advancing age. Public health efforts aimed at increasing physical activity levels of the population assume that physical activity will become a habitual part of an individual's lifestyle. Thus, the degree to which individuals maintain their relative rank or position over time is of interest. In other words, how well does physical activity track from one time to the next? It is often assumed that physical activity levels during childhood track into adulthood, such that active children turn into active adults. However, the available evidence suggests that the relationship between childhood and adulthood physical activity levels is poor. On the other hand, some aspects of childhood physical activity, such as sport participation, may be more strongly related to the persistence of physical activity levels into adulthood than general leisure-time physical activity levels are. Future research should be directed at identifying those aspects and contexts of childhood and adolescent physical activity that best predict adult levels of participation. This information will be important to identify those children and youth who are at an elevated risk of becoming physically inactive adults and who in turn will be at an elevated risk of chronic diseases later in life, and it would allow for targeted early interventions aimed at changing behavior through physical activity and health promotion.

Physical activity levels do not track very well between childhood and adulthood. More research is required to determine those components of physical activity in childhood that best predict continued participation in physical activity in adulthood.

Sex

Clear sex differences in physical activity levels arise during childhood and persist throughout the life span. On average, males are more physically active than females, and they also tend to engage in more vigorous physical activities. Data from the 2005-2006 Health Behaviour in School-Aged Children survey, a cross-national survey of children and youth 11 to 15 years of age from 41 different countries, documented sex differences in physical activity levels that persisted across all countries surveyed (Currie et al. 2008). In each country, more males than females reported regularly participating in physical activity on at least two occasions per week or for 2 h or more per week. There were also lower physical activity levels across incremental age groups in adolescence, and the declines were more apparent in females than males.

Sex differences in physical activity are also evident in figures 3.1 and 3.2. At each age, males engage in higher levels of total leisure-time physical activity than females. These sex differences in physical activity levels were also evident in the Baltimore Longitudinal Study of Aging, and the higher levels of physical activity overall in males were partially explained by the fact that males engaged in higher levels of high-intensity physical activities and lower levels of low-intensity physical activities than females (Talbot, Metter, and Fleg 2000).

Several hypotheses exist to explain the observed sex differences in physical activity levels. For example, a greater socialization toward sport participation in boys versus girls may explain some of the observed differences in level and intensity of physical activity. Other social factors such as peer pressure, family and child care commitments, and access to opportunities for physical activity may also be involved.

Across the life span, males are more physically active than females, and they are more likely to participate in more vigorous forms of physical activity.

Ethnicity

Ethnicity may play a role in explaining variation in physical activity levels. For example, cultural practices and religious beliefs may affect one's level of physical activity, and access to recreational resources may differ by ethnicity, attributable to potential differences in socioeconomic status across ethnic groups. Information on ethnic differences in physical activity levels in most countries is limited, and the validity of the information that is available may be tainted by cultural differences in reporting physical activities or the use of questionnaires that are not culturally specific. Figure 3.4 presents the prevalence of physical activity (reporting leisure-time physical activity in the past month) across various ethnic groups in the United States from the Behavioral Risk Factor Surveillance System. African American and Hispanic groups tend to have lower levels of leisure-time physical activity than white males and females; and within each ethnic group, males are more physically active than females. These ethnic differences in leisure-time physical activity appear to begin in adolescence.

With the incorporation of accelerometry to objectively monitor physical activity in NHANES, a clearer understanding of physical activity patterns is beginning to emerge. The ethnic differences in self-reported physical activity levels presented in figure 3.4 can be contrasted with objectively determined physical activity levels presented in figure 3.5. The

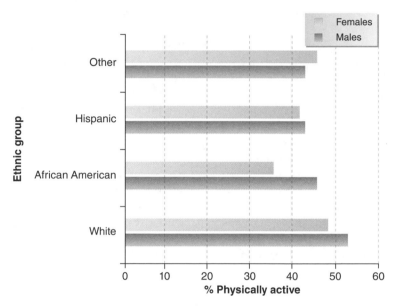

FIGURE 3.4 Prevalence of physical activity among ethnic groups in the United States. Physical activity was defined as engaging in any leisure-time physical activities in the past month.

Data from the 2005 Behavioral Risk Factor Surveillance System.

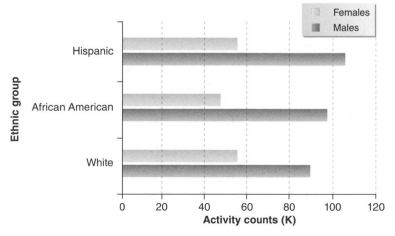

FIGURE 3.5 Daily physical activity counts (thousands) in moderate-to-vigorous physical activity in ethnic groups in the United States.

Data from the 2003-04 NHANES; Hawkins et al. 2009.

accelerometer data show that there are no differences in moderate-to-vigorous physical activity among Hispanic, African American, and white females; however, Hispanic males have higher levels of physical activity than African American and white males (Hawkins et al. 2009). Thus, although Hispanic males report less leisure-time physical activity, they appear to have higher levels of total physical activity than other ethnic groups. These differences can likely be explained by cultural differences in reporting leisure-time physical activities, as well as ethnic differences in nonleisure (i.e., occupational) physical activity levels. Unfortunately, accelerometers cannot determine the context or the mode of physical activity, only the total volume of ambulatory movement that occurs throughout the day.

Recent Trends

Until the end of the 19th century, physical activity was an integral component of people's daily lives because a high level of physical activity was required for many occupations and for the demands of daily living. However, the recent technological revolution has eliminated the requirement for humans to be physically active to make a living or to survive. On the other hand, technological advances have theoretically increased the availability of leisure time, and the potential for people to engage in physically active pursuits during their leisure time has increased. The recent trends in leisure-time physical activity in several countries for which temporal data are available are presented in figure 3.6. Although the

survey questionnaires and sampling designs varied among the countries, the data show a general trend toward increasing levels of leisure-time physical activity among adults within each of the countries. These reported trends for increased physical activity in the face of increased levels of obesity in the same countries are somewhat paradoxical. However, the available data are for leisure-time physical activity, which generally excludes physical activity undertaken at work or in other chores. Thus, it could well be that leisure-time physical activity levels have increased while at the same time total daily energy expenditure levels have decreased because technological advances have made it easier for us to perform our activities of daily living with a minimal expenditure of energy.

Another important factor that could partially explain the increases in reported physical activity levels is the greater public awareness of the health benefits of physical activity and of what constitutes physical activity. As with any health behavior, a certain amount of social desirability is associated with reporting favorable levels of physical activity, and this effect may have increased in recent years. As well, there has been an increasing awareness that physical activities such as walking and gardening "count" as leisure-time physical activities, whereas in the past people may have viewed only traditional activities such as running, cycling, or swimming as physical activities. Given that the most commonly reported leisure-time physical activities in the United States and Canada are indeed walking and gardening or yard work, this may have had an impact. Although these factors may have played a role in the observed

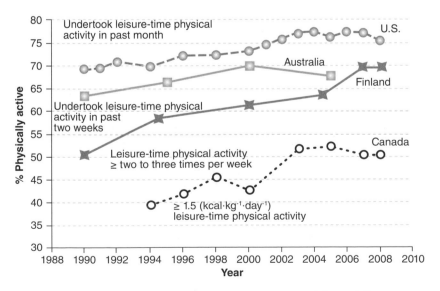

FIGURE 3.6 Recent trends in the prevalence of physical activity among adults in different countries.

Australian Bureau of Statistics 2002; National Population Health Survey and Canadian Community Health Survey; Health Behaviour and Health among the Finnish Adult Population survey; United States, Behavioral Risk Factor Surveillance System.

increases in reported physical activity levels, the extent to which the estimates have been affected is unknown.

Physical Fitness

As outlined in chapter 1, physical fitness has many components and can be conceptualized as either performance related or health related. Because there are so many facets of physical fitness, it is beyond the scope of this chapter to describe age, sex, and ethnic differences of all fitness components. Therefore, this section focuses on the descriptive epidemiology of **cardiorespiratory fitness** and **musculoskeletal fitness,** two of the primary components of physical fitness.

The measurement of physical fitness is more time-consuming and expensive than the measurement of physical activity, which is done using questionnaires in most national surveys. For example, in the Behavioral Risk Factor Surveillance System in the United States, the prevalence of no leisure-time physical activity is assessed by a negative response to the question, "During the past month, other than your regular job, did you participate in any physical activities such as running, calisthenics, golf, gardening, or walking for exercise?" However, objective monitoring of physical activity using accelerometers and measurements of cardiorespiratory fitness have been incorporated into NHANES (Troiano et al. 2008; Wang et al. 2010). Objective measurements of physical fitness are rarely included in nationally representative population surveys because of the greater expense of having personal contact with the respondents and the greater amount of time required to make the measurements. This section presents age, sex, and ethnic differences in physical fitness using population-level data when available.

Age

Among adults, a general age-related physiological decline occurs that affects many organs and organ systems. In turn, this has an effect on physical fitness levels. Both cardiorespiratory fitness and musculoskeletal fitness decrease with advancing age in males and females. Figure 3.7 presents **maximal oxygen uptake ($\dot{V}O_2max$)** values across age among adult males and females from several cross-sectional studies. It is clear that older individuals have lower levels of cardiorespiratory fitness than younger individuals. The absolute levels of fitness should not be compared across the three studies because the studies used different measurements that are not directly

comparable. What is of interest is that the age-related trends within studies are remarkably similar, and although males and females differ in absolute levels of fitness, the observed age differences are similar within a given population. The data from the Baltimore Longitudinal Study of Aging are arguably the best because $\dot{V}O_2max$ was directly measured using a treadmill test; however, the sample was not nationally representative (Talbot, Metter, and Fleg 2000). On the other hand, though the studies from both Canada and England used representative population samples, $\dot{V}O_2max$ was predicted from submaximal exercise tests (Bonen and Shaw 1995; Sports Council and Health Education Authority [HEA] 1992). A further limitation of the Canadian data is that the equation used to predict $\dot{V}O_2max$ included age as a component; thus, some of the observed declines with age could be an artifact of the prediction method. Nevertheless, all of the available data indicate significant age differences in aerobic fitness across the life span. On average, the age-related decrease across incremental age groups in $\dot{V}O_2max$ is approximately 1% per year in adults.

These results have important implications for elderly persons. Any given task, such as climbing a flight of stairs, requires a certain amount of energy expenditure, regardless of age. Thus, a given task for elderly people requires them to work at a higher percentage of their maximal aerobic capacity compared with younger people and requires more physiological effort. Even simple tasks can be very demanding for elderly persons.

The age-related differences in $\dot{V}O_2max$ presented in figure 3.7 are derived from studies using cross-sectional research designs. There is also evidence from studies using a longitudinal research design that aerobic capacity decreases with age. A series of studies was conducted on elite long-distance runners in the late 1960s and early 1970s to determine factors associated with success in this endurance event. Figure 3.8 presents the results of an interesting analysis of $\dot{V}O_2max$ in these same runners 22 years later, based on their current physical activity levels (Trappe et al. 1996). The runners were grouped according to whether they were currently sedentary (untrained), were training f or physical fitness, or were maintaining the highest level of training for competition. Even those runners who were training for competition showed a decline in $\dot{V}O_2max$ with advancing age. However, the decreases in $\dot{V}O_2max$ in those who were training for fitness and those who were completely untrained were progressively greater. These results highlight the importance of maintaining physical activity levels with advancing age to attenuate the decline in cardiorespiratory fitness levels.

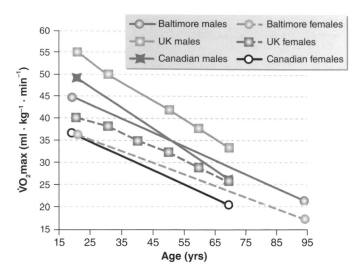

FIGURE 3.7 Age-related trends in V̇O₂max in males and females from the United States (Baltimore Longitudinal Study of Aging, Talbot, Metter, and Fleg 2000); England (Allied Dunbar National Fitness Survey, Sports Council and HEA 1992); and Canada (Canada Fitness Survey, Bonen and Shaw 1995).

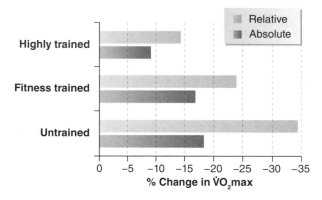

FIGURE 3.8 Longitudinal changes in absolute (L/min) and relative (ml · kg⁻¹ · min⁻¹) V̇O₂max in elite long-distance runners. Runners were currently untrained, were training for physical fitness, or were highly trained for competition after 22 years.

Data from Trappe et al. 1996.

As part of the general physiological decline with advancing age, cardiorespiratory fitness levels decrease throughout adulthood. There is some evidence from longitudinal studies that keeping physically active may attenuate this observed decline in fitness.

Musculoskeletal fitness generally declines with age among adults, although the pattern of decline varies depending on the measurement or component of fitness studied. Figure 3.9 presents cross-sectional age-related differences in **muscular strength** (handgrip, figure 3.9a) and **trunk flexibility** (sit and reach, figure 3.9b) among Canadians (Fitness Canada 1983).

Dramatic increases in muscular strength occur during normal growth and maturation in childhood and adolescence and reach a peak among adults 20 to 29 years of age, after which strength decreases gradually through the rest of the life span. These cross-sectional observations regarding age-related declines in muscular strength have been confirmed in studies using a longitudinal research design. Figure 3.10 shows changes in handgrip strength in Japanese American males participating in the Honolulu Heart Program over 27 years (Rantanen et al. 1998). The results are presented as average changes per year over the entire follow-up period. Although there are reductions in grip strength in all age groups, there is a general trend for greater changes in males who were older rather than younger at the outset of the study. In other words, the rate of decline in muscular strength tends to accelerate with advancing age. These decreases in muscular strength are in large measure related to the loss of muscle mass associated with the aging process.

Flexibility is an important component of musculoskeletal fitness and tends to be joint specific. In other words, if you are quite flexible at the hip, there is no guarantee that you will be flexible at the knee, shoulder, or any other particular joint. Enhanced flexibility is associated with maintaining the range of motion in a given joint and with a reduced occurrence of back pain in the case of hip or trunk flexibility. Trunk flexibility, measured using a sit-and-reach test, is a composite measure of low back and hamstring flexibility, which tends to show age-related variation (figure 3.9b). Performance on the sit-and-reach test tends to decrease across age groups during adolescence. However, this decrease

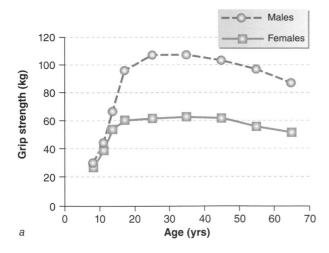

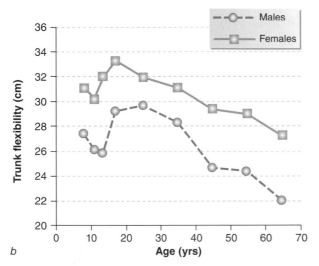

FIGURE 3.9 Age- and sex-related differences in musculoskeletal fitness represented by (a) handgrip strength (combined left and right) and (b) trunk flexibility in Canadian males and females from 7 to 69 years of age.

Data from the 1981 Canada Fitness Survey.

may be more associated with differential changes in the lengths of the arms and legs relative to the trunk as youth go through this period of rapid growth, rather than with actual reductions in the range of motion around the hip joint per se. This observed decline in sit-and reach performance is followed by an increase into adulthood, after which it decreases throughout the rest of the life span.

Sex

Males and females differ in physical fitness levels, but the extent of the difference depends on the specific component of fitness. On average, males have higher levels of cardiorespiratory fitness at a given age. This is quite evident in figure 3.7, which shows that within a particular study, males have greater $\dot{V}O_2$max than

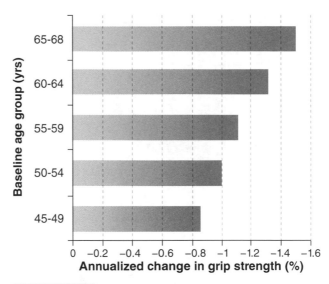

FIGURE 3.10 Longitudinal changes (per year) in maximal handgrip strength over 27 years in Japanese American males.

Data from Rantanen et al. 1998.

females across the life span. The sex differences in cardiorespiratory fitness have generally been attributed to differences in lean body mass, blood hemoglobin levels, and levels of physical activity participation between males and females.

Males are generally stronger than females at any given age (figure 3.9a). This sex difference is mainly attributable to differences in the absolute amount of muscle mass between males and females. When muscle strength is corrected for muscle size, the difference between males and females becomes smaller and sometimes disappears completely. Although males may be stronger than females, females perform better on the sit-and-reach test and tend to be more flexible than males (figure 3.9b). The sex differences in flexibility are likely due to morphological differences in the architecture of the hip joint.

■ Throughout the life span, males are generally stronger than females, whereas females are more flexible than males.

Ethnicity

Data on ethnic differences in physical fitness from population surveys are very limited. NHANES (1999-2004) measured cardiorespiratory fitness in 3,250 people from 20 to 49 years of age (Wang et al. 2010). Figure 3.11 presents the mean $\dot{V}O_2$max values across three ethnic groups in the United States. There were no consistent differences in males, but African American females had lower fitness levels than white and Mexican American females. Note also that in all

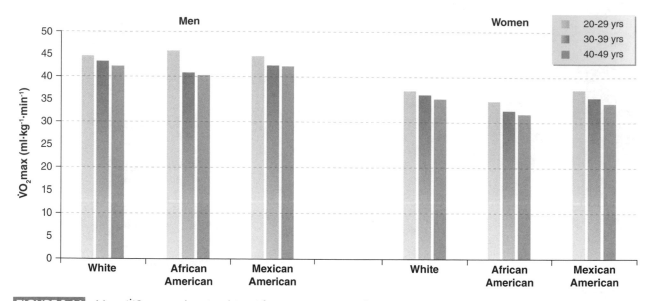

FIGURE 3.11 Mean V̇O$_2$max values in white, African American, and Mexican American adults from 20 to 49 years of age.

Adapted from C.Y. Wang, W.L. Haskell et al., 2010, "Cardiorespiratory fitness levels among US adults 20-49 years of age: findings from the 1999-2004 National Health and Nutrition Examination Survey," *American Journal of Epidemiology* 171:426-435.

ethnic groups, lower fitness levels were observed with advancing age. It should also be noted that although these data come from a national survey, for safety reasons there were many exclusion criteria, and the group that was not eligible for the test had a higher proportion of African Americans and a lower proportion of Mexican Americans than the group that was eligible for the test. The authors of this paper corrected for these differences statistically to account for any bias that may have been introduced by these differences.

Further evidence of ethnic differences in cardiorespiratory fitness comes from a large multiple-center cohort study in the United States that was designed to investigate the development of coronary heart disease risk factors in young adults. The Coronary Artery Risk Development in Young Adults (CARDIA) Study enrolled African American and white participants from 18 to 30 years of age (Carnethon et al. 2003). Males and females performed a maximal treadmill exercise test and then were divided into low (lower 20%), moderate (middle 40%), and high (upper 40%) fitness categories (figure 3.12). Clear trends toward a decreasing proportion of African American participants across low, moderate, and high cardiorespiratory fitness groups were observed in both males and females. This suggests that the overall level of cardiorespiratory fitness was lower in young African American adults compared with young white adults.

The information described in this section provides evidence that there may be systematic differences in cardiorespiratory fitness between African American

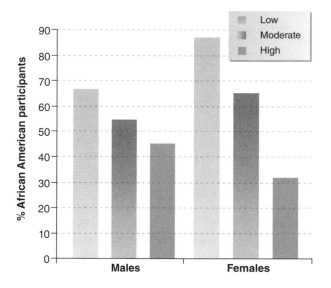

FIGURE 3.12 Distribution of African American participants across cardiorespiratory fitness categories in the Coronary Artery Risk Development in Young Adults (CARDIA) Study.

Data from Carnethon et al. 2003.

and white adults in the United States, particularly among females. More research is required to determine the underlying causes of ethnic differences in cardiorespiratory fitness and whether they are mainly related to differences in behavior or attributable to inherent biological differences. Furthermore, currently we have little information about differences in cardiorespiratory fitness in minority populations living in other countries, which limits our understanding of the range of variability in fitness in the general population.

Unfortunately, very little information on ethnic differences in musculoskeletal fitness in the population is available. This is largely because the national population surveys that have included representation from various ethnic groups have not included measures of musculoskeletal fitness in their battery of tests. This is an important area for future research because musculoskeletal fitness is related to health and independent living among the elderly and because ethnic differences may have implications for identifying high-risk groups.

■ **The available evidence suggests that cardiorespiratory fitness levels are lower among African American persons in North America than among other groups. However, population-level data with which to further explore this issue are lacking.**

Recent Trends

Given the paucity of physical fitness data on nationally representative samples, little information is available on temporal trends. However, there are now data from Canada and the United Kingdom to address this issue. In the 1981 Canada Fitness Survey, a variety of fitness tests were performed in the participants' homes, including submaximal exercise step testing on a portable set of stairs, handgrip dynamometry testing, and sit-and-reach testing (Fitness Canada 1983). Much of this information was updated in the 2007-2009 Canadian Health Measures Survey (Shields et al. 2010), which allows for temporal comparisons to be made. The results of these comparisons indicate that most markers of fitness, including muscular strength, flexibility, and body composition, deteriorated over the past few decades. Figure 3.13 presents the mean grip strength values from 1981 and 2007-2009 in males and females of different ages. In both males and females, the 2007-2009 values were significantly lower than the 1981 values in the youngest two age groups, while the difference was not as great among 60- to 69-year-olds. In addition to showing the observed decrease in strength over time in Canada, the figure shows two trends that were presented earlier in this chapter: (1) Males are stronger than females at every age, and (2) strength declines with advancing age in both genders.

Cardiorespiratory fitness was measured in a nationally representative sample of adults in

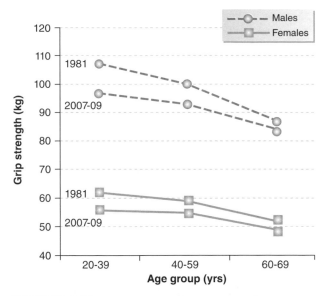

FIGURE 3.13 Changes in handgrip strength in Canadian males and females between 1981 and 2007-2009.

Data from Fitness Canada 1983; Shields et al. 2010.

England in 1990 (Allied Dunbar National Fitness Survey, Sports Council and HEA 1992), and this information was recently updated in the 2008 Health Survey for England (Craig, Mindell, and Hirani 2009). Figure 3.14 presents the observed changes over time in maximal aerobic fitness ($\dot{V}O_2$max) in adults in England between 1990 and 2008. There were decreases in $\dot{V}O_2$max in both males and females over time, and the differences are

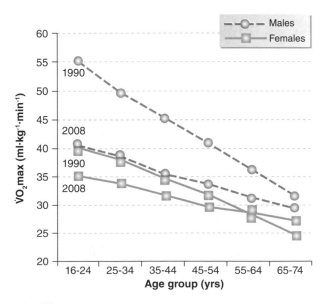

FIGURE 3.14 Changes in $\dot{V}O_2$max in men and women in the United Kingdom between 1990 and 2008.

Data from Sports Council and HEA 1992; Craig et al. 2009.

more noticeable at younger ages compared to older ages. The 1990 data were collected using a treadmill test, whereas the 2008 data were collected using a step test. Thus, minor differences between the protocols are expected; however, the equations for estimating $\dot{V}O_2$max from the step test were obtained through calibration with a treadmill test, so these differences are expected to be small (Craig, Mindell, and Hirani 2009).

■ The results of studies from Canada and the United Kingdom indicate significant declines in musculoskeletal fitness and cardiorespiratory fitness over the past 20 to 30 years.

Summary

This chapter explores the descriptive epidemiology of physical activity and fitness, with a particular emphasis on age, sex, and ethnic differences. The discussion shows systematic differences in physical activity and fitness across various demographic groups. More research is required to further elucidate these differences and the impact they may have on health, particularly among minority populations. The mechanisms behind the observed differences remain largely unknown; researchers should investigate the role of biological, social, and psychological factors in explaining age, sex, and ethnic variation in physical activity and fitness.

Key Concepts

cardiorespiratory fitness—Component of physical fitness that reflects the integrated function of the heart, lungs, vasculature, and skeletal muscles to deliver and use oxygen during dynamic physical activities. Also referred to as aerobic power or cardiorespiratory endurance, it is measured under laboratory conditions as maximal oxygen uptake ($\dot{V}O_2$max) based on ventilatory gas exchange and indirect calorimetry during maximal exercise testing. It can also be estimated from maximal exercise time on a treadmill or bike test, the final speed and grade of a maximal treadmill test, the maximal workload achieved during a bike test, or heart rate responses to multiple submaximal work stages on a bike or treadmill ergometer.

cross-sectional study—A study measuring different attributes at the same time and examining whether they are significantly related to each other. Such studies cannot identify causation, but they nevertheless generate hypotheses that can be tested more rigorously in prospective epidemiological or experimental studies.

descriptive epidemiology—Assessment of variation in the prevalence of a trait or behavior by age, sex, ethnicity, and other geographic or demographic factors.

epidemiology—A scientific discipline that involves study of the distribution and determinants of disease in human populations.

leisure-time physical activity—For definition, see page 19.

longitudinal research design—Research design in which participants are measured initially and then followed over time to directly measure changes over time.

maximal oxygen uptake ($\dot{V}O_2$max)—For definition, see page 19.

muscular strength—Ability of a muscle to produce force.

musculoskeletal fitness—Aspects of physical fitness that are related to muscular strength, muscular endurance, flexibility, and bone health.

physical activity—For definition, see page 19.

physical fitness—For definition, see page 19.

tracking—Stability, or the maintenance of relative rank or position in a group over time.

trunk flexibility—Suppleness and mobility of the hip joint, generally measured using a sit-and-reach test.

Study Questions

1. Differentiate between cross-sectional and longitudinal research designs.

2. Describe the age- and sex-related differences in physical activity observed in many population surveys.

3. Distinguish between age-related differences in physical activity and the tracking or stability of physical activity with age.

4. What evidence is there that the observed age-associated declines in physical activity may have a biological basis?

5. Describe the age-related changes that occur in cardiorespiratory fitness among adults. How do the changes in trained individuals compare with those in sedentary people? What are some potential limitations of the population data that are used to determine these trends?

6. What areas in the descriptive epidemiology of physical activity are still not well understood? Suggest some important areas for future research.

References

Allman-Farinelli, M.A., T. Chey, D. Merom, H. Bowles, and A.E. Bauman. 2009. The effects of age, birth cohort and survey period on leisure-time physical activity by Australian adults: 1990-2005. *British Journal of Nutrition* 101:609-617.

Bonen, A., and S.M. Shaw. 1995. Recreational exercise participation and aerobic fitness in men and women: Analysis of data from a national survey. *Journal of Sports Sciences* 13:297-303.

Carnethon, M.R., S.S. Gidding, R. Nehgme, S. Sidney, D.R. Jacobs, and K. Liu. 2003. Cardiorespiratory fitness in young adulthood and the development of cardiovascular disease risk factors. *Journal of the American Medical Association* 290:3092-3100.

Craig, R., J. Mindell, and V. Hirani (eds.). 2009. *Health Survey for England 2008: Vol. 1. Physical activity and fitness.* London: National Health Survey Information Centre.

Currie, C., S.N. Gabhainn, E. Godeau, C. Roberts, R. Smith, D. Currie, W. Pickett, M. Richter, A. Morgan, and V. Barnekow. 2008. *Inequalities in young people's health: HBSC international report from the 2005/06 survey.* Geneva: World Health Organization.

Fitness Canada. 1983. *Fitness and lifestyle in Canada.* Ottawa: Government of Canada.

Hawkins, M.S., K.L. Storti, C.R. Richardson, W.C. King, S.J. Strath, R.G. Holleman, and A.M. Kriska. 2009. Objectively measured physical activity of USA adults by sex, age, and racial/ethnic groups: A cross-sectional study. *International Journal of Behavioural Nutrition and Physical Activity* 6:31.

Helakorpi, S., M. Paavola, R. Prattala, and A. Uutela. 2009. *Health behaviour and health among the Finnish adult population, spring, 2008.* Helsinki: National Institute for Health and Welfare.

Janicke, B., D. Coper, and U.-A. Janicke. 1986. Motor activity of different-aged Cercopithecidae: Silver-leafed monkey (Presbytis cristatus Esch.), lion-tailed monkey (Macaca silenus L.), moor macaque (Macaca maura Cuv.) as observed in the zoological garden, Berlin. *Gerontology* 32:133-140.

Rantanen, T., K. Masaki, D. Foley, G. Izmirlian, L. White, and J.M. Guralnik. 1998. Grip strength changes over 27 yr in Japanese-American men. *Journal of Applied Physiology* 85:2047-2053.

Shields, M., M.S. Tremblay, M. Laviolette, C.I. Craig, I. Janssen, and S. Connor Gorber. 2010. Fitness of Canadian adults: Results from the 2007-2009 Canadian Health Measures Survey. *Health Reports* 21:1-15.

Sports Council and Health Education Authority. 1992. *Allied Dunbar National Fitness Survey.* London: Sports Council and Health Education Authority.

Talbot, L.A., E.J. Metter, and J.L. Fleg. 2000. Leisure-time physical activities and their relationship to cardiorespiratory fitness in healthy men and women 18-95 years old. *Medicine and Science in Sports and Exercise* 32:417-425.

Trappe, S.W., D.L. Costill, M.D. Vukovich, J. Jones, and T. Melham. 1996. Aging among elite distance runners: A 22-yr longitudinal study. *Journal of Applied Physiology* 80:285-290.

Troiano, R.P., D. Berrigan, K.W. Dodd, L.C. Masse, T. Tilert, and M. McDowell. 2008. Physical activity in the United States measured by accelerometer. *Medicine and Science in Sports and Exercise* 40:181-188.

Wang, C.Y., W.L. Haskell, S.W. Farrell, M.J. Lamonte, S.N. Blair, L.R. Curtin, J.P. Hughes, and V.L. Burt. 2010. Cardiorespiratory fitness levels among US adults 20-49 years of age: Findings from the 1999-2004 National Health and Nutrition Examination Survey. *American Journal of Epidemiology* 171:426-435.

4

Sedentary Behavior and Inactivity Physiology

Marc Hamilton, PhD; and Neville Owen, PhD

CHAPTER OUTLINE

©Yuri Arcurs/fotolia

This chapter addresses **sedentary behavior** (typically, prolonged sitting time at work, in automobiles, and with screen-based entertainment at home) as a newly emerging population health problem. We highlight evidence indicating a unique biology that underlies the deleterious health consequences of too much sedentary time—**inactivity physiology**. Together with recent findings from **cross-sectional studies** and **prospective epidemiological studies**, this evidence on the underlying mechanisms makes a strong case for focusing distinctly on the time that people spend sitting as a behavioral **risk factor** for chronic disease, which is distinct from a lack of **physical activity** or exercise.

Although sitting is now ubiquitous in workplace, transport, and domestic contexts, a noteworthy historical fact is that this is the first time a textbook chapter has been written to focus on this insidious health hazard. Inactivity physiology is an emerging area of medical research and represents a novel way of thinking about the health consequences of a "sedentary lifestyle." Decades of exercise physiology research have provided convincing evidence to recommend that people should perform sustained moderate- to vigorous-intensity physical activity (aerobic exercise) because exercise has an established preventive role with regard to premature mortality, cardiovascular disease, type 2 diabetes, obesity, and some cancers. It is ironic that we know much more about the physiological responses and health effects of only 1 to 3 h per week of sustained endurance exercise than we know about the more elusive responses to ubiquitous sedentary behaviors and normal activities of daily living performed for dozens of hours per week (Hamilton, Hamilton, and Zderic 2007).

The paradigm of inactivity physiology (Hamilton, Hamilton, and Zderic 2004) has raised the testable hypothesis that too much sitting and other behaviors requiring physical inactivity throughout the day have deleterious effects on health, independent of whether people meet the current exercise guidelines, and that physical inactivity is not the biological equivalent of too little exercise (see Hamilton, Hamilton, and Zderic 2007 for the four fundamental tenets and existing evidence).

We conclude our chapter by highlighting the research domains in which further evidence is required, particularly in population-based studies and intervention trials, to demonstrate the feasibility and health outcomes of reducing and breaking up sedentary time. A dire concern for the future may rest with growing numbers of people unaware of the potential insidious dangers of sitting inactive too much and the lack of evidence-based advice promoting effective recommendations (Hamilton, Hamilton, and Zderic 2007). It is especially concerning that dose–response evidence is not yet available from epidemiological studies and experimental trials to inform guidelines for the public on how much physical inactivity is too much.

At this point, the most viable lifestyle alternatives to replace vast amounts of sedentary time in the multiple contexts of peoples' lives remain to be identified through behavioral studies and intervention trials. Thus, this is an exciting time for preventive medicine and public health, with prolonged sitting emerging as a potent preventable risk factor. Here, we demonstrate how a more in-depth and broader understanding of sedentary behavior and inactivity physiology has the potential to be a watershed for novel public health initiatives and innovative lifestyle modifications.

Sedentary Behavior, Physical Activity, and Public Health

The evidence for the negative effects of sedentary behavior on health is increasing, and a new research paradigm is needed.

Modern Humans' Daily Lifestyles: Too Much Sitting

Modern humans are frequently "sitters"—at work, in the car, and at home. Sitting may be the most common human behavior and for many adults exceeds the time spent sleeping! Changes in transportation, communication, workplace, and entertainment technologies have reduced human activity level and energy requirements. Physical, economic, and social environments of populations living in affluent, industrialized countries have been changing in unprecedented ways since the middle of the last century; this is also the case for many urban populations of developing countries.

In this context, prolonged sitting promotes extremely low levels of physical activity, and it is supported by many policies, entrenched practices, and environmental factors that can mandate or encourage sitting. Therefore, a robust research effort is needed to strengthen the relevant evidence on the health risks of sedentary behaviors (from the Latin *sedere*, "to sit") and to identify acceptable strategies to shift ingrained habits within the contexts of people's everyday lives. This requires an interdisciplinary approach and a new physical activity and health research paradigm (Hamilton, Hamilton, and Zderic 2007; Hamilton et al. 2008; Owen et al. 2000; Owen, Bauman, and Brown 2009; Owen et al. 2010) to address a newly identified public health issue of major import: *too much sitting* is a distinct chronic disease risk factor, additional to and independent of *too little exercise*.

Current Recommendations for Healthy Physical Activity

National physical activity and health recommendations for the adult population, such as those in *Physical Activity and Public Health* from the American College of Sports Medicine and the American Heart Association (Haskell et al. 2007), typically identify 30 min or more of moderate-intensity physical activity on most days of the week as the minimum required for health benefits. These recommendations were disseminated broadly in the 2008 *Physical Activity Guidelines for Americans* (U.S. Department of Health and Human Services [DHHS] 2008), which is derived from an evidence-based summary on the benefits of physical activity (Physical Activity Guidelines Advisory Committee [PAGAC] 2008). The U.S. recommendations also state that the "amount of aerobic activity (whether of moderate or vigorous intensity) is in addition to routine activities of daily living which are of light intensity, such as self care, casual walking or grocery shopping, or less than 10 min of duration such as walking to the parking lot or taking out the trash" (Haskell et al. 2007, p. 1084). Doing more such routine activities would lead to reductions in the amount of time spent sitting, but inactivity (*sitting*) is not addressed explicitly in the recommendations.

In the next sections, we outline new scientific evidence suggesting that public health imperatives to reduce the risk of major chronic diseases will require an explicit focus on reducing time spent sitting inactive, which would be in addition to the current recommendations for increasing levels of participation in moderate to vigorous physical activity.

Balancing Sedentary Time With Low-Intensity, Nonexercise Physical Activity

In people who spend about one-third of each 24-h period sleeping, the 16 h remaining for the waking day allow for a large amount of time to be spent sedentary, or alternatively a large amount of time engaged in **low-intensity physical activity** (LIPA). Many factors can influence how much time is spent in these behaviors, and there is tremendous inter-individual (between different people) and intra-individual (changes from day to day) variability. On some days, people can spend almost all of their waking hours sitting (e.g., using computers, driving, watching TV), in addition to other types of inactivity such as reclining to watch TV or lying down to read in bed. Depending on the individual lifestyle, much of the 16-h waking day might involve LIPA such as standing to cook meals, performing household chores, and engaging in many other activities of daily living that are outside the boundaries of recommended exercise. Although the relatively small amount of recommended moderate to vigorous exercise training is known to be beneficial, there are too many hours in the waking day (~16 h) for exercise to significantly displace sedentary time, except in athletes who train for several hours per day or persons who perform manual labor as their regular job. The incidental bouts of daily nonexercise physical activities are almost always too intermittent or too low in intensity (or both) to meet the level historically receiving attention in exercise science research.

Direct measurements of postural allocation using inclinometers have not yet been reported in large population studies or in long-term longitudinal studies. However, the daily upright time (mostly standing and some slow ambulation) and sedentary time (sitting and lying down) were measured in Levine's studies of nonexercise activity thermogenesis (Levine 2007) and earlier studies of lean and obese women in the 1960s (Hamilton et al. 2008). Workdays in the modern office, obesity, advanced age, and even restrictive classrooms for children are directly associated with greater amounts of sitting time.

On average, most people with sedentary occupations are probably spending approximately 9 to 11 h per day sitting or reclining. In leisure time, sedentary time can include 4 h or more stretched out on the couch watching TV or reading in bed. Sitting time also includes the time spent eating, commuting, and socializing. People with desk jobs can sit for about 75% of the workday (~6-7 h), and students can sit for more than 95% of each class period. Thus, it is not difficult to understand why able-bodied people with a functional neuromuscular system and without orthopedic limitations sometimes sit or lie down for all but 2 h of a 24-h period!

There are too many hours (16 h) in the waking day for a significant amount of sedentary time to be replaced by the recommended moderate to vigorous physical activity (sustained aerobic endurance exercise).

Characterizing Adults' Sedentary and Physically Active Time

The pattern in sitting time is highly variable between days (Hamilton, Hamilton, and Zderic 2007) and between individuals. This is illustrated in figure 4.1, in which cluster heat maps (see Owen et al. 2010) represent sedentary and varying levels of physically active time. **Accelerometers**—small electronic devices worn on the hip—provide an objective record of the volume, intensity, and frequency of movement, which may be downloaded to computer databases and used to derive scientifically meaningful activity variables. Accelerometer counts of less than 100 per minute, even though they do not measure posture directly, may be used

to characterize sedentary time (Healy et al. 2007; Matthews et al. 2008).

Figure 4.1 shows seven days of accelerometer-measurement data (starting Monday), with morning to evening hours running from the top to the bottom of each column. The black-to-grayscale-to-white spectrum shows the range of movement detected by accelerometer counts, from totally sedentary (black) through to highly active (white).

Clockwise from top left, four contrasting patterns of sedentary and physically active time are shown:

The high-sedentary, high-exercise pattern (top left) shows accelerometer data for an individual who has long, unbroken blocks of sedentary time but who meets or exceeds the physical activity and health guidelines through regular, continuous bouts of moderate to

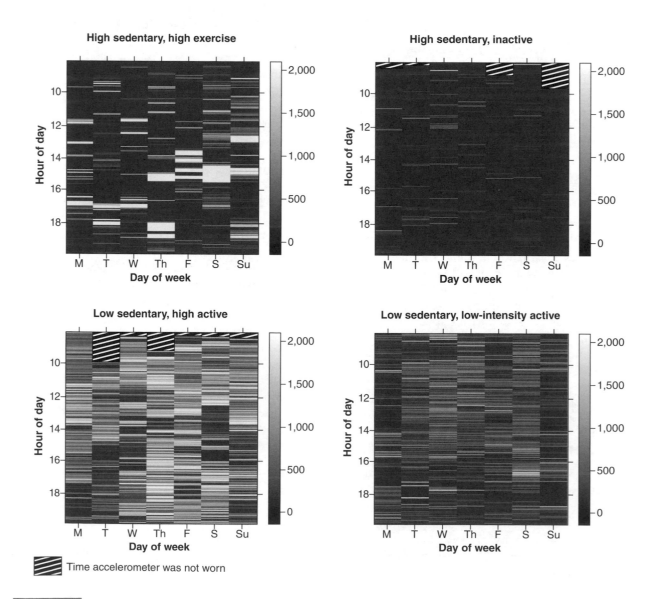

FIGURE 4.1 Cluster heat maps showing seven days of movement data, with four contrasting patterns of sedentary and physically active time.

vigorous exercise. Such a person might, for example, be a young adult who has a desk job and enjoys a large amount of TV viewing but also is in training for a sporting competition.

The *high-sedentary, inactive* pattern (top right) shows an individual who has long unbroken blocks of sedentary time and who engages in very little moderate- to vigorous- or even light-intensity physical activity. Such a person might be an overweight or older adult in a sedentary job who has a long commute by car, can park close to work, and spends evenings watching TV; this person would also take very few breaks from sitting at work.

The *low-sedentary, low-intensity active* pattern (bottom right) shows an individual with extremely frequent low-intensity breaks in sedentary time but with very little moderate to vigorous physical activity. Such a person might work in a service industry, with a job that involves sitting at a desk but with requirements to stand frequently in order to assist the public. As we argue later (see figure 4.2), there may be significant and insufficiently recognized health benefits associated with such a pattern.

The *low-sedentary, high-active* pattern (bottom left) shows an individual whose sedentary time is broken up frequently and who engages in a large amount of moderate to vigorous physical activity, but more intermittently so than with the high-sedentary, high-exercise pattern. Such a person might work in a skilled trade or manual occupation involving large amounts of standing up and moving about and might also have active hobby time in the evening; this person engages in many short- to medium-length bouts of moderate to vigorous physical activity.

Findings from the U.S. National Health and Nutrition Examination Survey (usually referred to by its acronym, NHANES), in which many thousands of people wore accelerometers for up to a week, suggest that the most common patterns in the adult population are those involving large amounts of sedentary time (Matthews et al. 2008).

Focusing on Activities Routinely Performed When Not Sitting

Studies examining sedentary time and its relationships with mortality, disease outcomes, and clinical biomarkers have consistently suggested that limiting the vast amount of daily time spent sitting or reclining could potentially provide healthful benefits, independent of exercise participation. Thus, it is critical to examine what types of activity people routinely perform when they are not sitting (table 4.1). Analogous to the dose–response parameters determining the specific responses to exercise training, the dose–response parameters during daily nonsedentary time are important for understanding how best to limit sedentary time.

When not sitting, people spend most of the waking day engaged in LIPA (Levine 2007). Thus, there is a strong and inverse correlation between sedentary time and LIPA (Healy et al. 2008c). It is not unreasonable to expect that most able-bodied people who are not exercise trained are capable of spending many waking hours per week engaged in LIPA, because this is already what many people do in the incidental activities of daily living. For example, time-use surveys and recent accelerometer data show that some nonexercising adults (e.g., active housewives with children) engage in LIPA for the majority of the waking day.

Thus, one of the major themes for inactivity physiology research will be the development of practical lifestyle recommendations through focusing on the unhealthy effects of different patterns of sitting and determining the dose–response effects for different amounts of daily LIPA.

TABLE 4.1 Inactivity Physiology Dose–Response Parameters

Frequency	Hundreds of bouts of light movements per day; always seven days per week
Intensity	Predominately low-intensity physical activity (LIPA) with modest $\dot{V}O_2$ (<3 METs or 30% $\dot{V}O_2$max)
Duration	Accumulation of a large volume of nonexercise bouts in the LIPA range, often exceeding 8 h per day
Modality	Innumerable but sedentary behaviors (such as TV viewing or screen time) counterbalanced by light activities of daily living (LIPA) when *not sitting*, such as standing to perform incidental tasks or slow walking

The behavioral focus has been on sedentary behaviors involving sitting or lounging (sitting to drive cars, sitting to work, TV viewing and other types of screen time). The frequency, intensity, duration, and modality of physical activity most commonly counterbalancing sedentary time are outlined. It is important to note that the quantitative values of each of these parameters are very unlike the current exercise recommendation in the physical activity guidelines calling for the public to engage in moderate- to vigorous-intensity activity for 150 min per week.

Based on Hamilton et al. 2007.

■ Large amounts of sedentary time can coexist with sufficient moderate to vigorous physical activity that meets current public health guidelines.

Inactivity Physiology: The Underlying Biology of Acute and Chronic Muscular Inactivity

In this section, we review components and characteristics of sedentary behavior, with an emphasis on the physiology of inactivity and sedentary postures.

Energy Expenditure During Sedentary Behaviors

Resting metabolic rate (RMR) is defined as the energy expenditure at rest when no physical activity is occurring. The energy expenditure when sitting comfortably or lying down quietly at rest is approximately 1 kcal · h^{-1} · kg^{-1} (1 MET, a unit expressing multiples of the RMR), or 70 kcal/h in a 70 kg (154 lb) person. Sitting or lying down to watch TV, read a book, type on a computer, or talk on the phone all require about the same low level of energy expenditure. These periods of physical inactivity demand only about 1.0 to 1.2 kcal · h^{-1} · kg^{-1}. To put these levels of energy expenditure into perspective, moderate to vigorous physical activity (walking, bicycling, swimming, or running) requires an energy expenditure of 3 to 10 METs, and even significantly greater expenditures in more highly fit or trained persons.

Most of the energy expenditure during sleep or wakeful rest is required to sustain the basic physiological functions that take place all of the time. Tissues other than skeletal muscle (e.g., the brain, kidneys, diaphragm, heart) are always working, even when persons are physically inactive and skeletal muscles are not contracting. However, because skeletal muscle energy demand is a function of contractile activity, the energy expenditure in specific skeletal muscles and in the whole body declines precipitously during the periods of physical inactivity (prolonged sitting) within the day.

A major distinction between exercise physiology and inactivity physiology dose–response parameters is related to the intensity of the physical activity. By definition in the exercise physiology literature and public health recommendations, moderate intensity is 3 to 5.9 METs, and vigorous intensity is 6 or more METs (table 4.2). Also, the scientific evidence sup-

TABLE 4.2   Classification of the Relative Physical Activity Intensity as a Percentage of $\dot{V}O_2$max, Heart Rate, and Ratings of Perceived Exertion (based on the Borg scale)

	Relative intensity			$\dot{V}O_2$max = 12 METs	
Intensity	%$\dot{V}O_2$R %HRR	%HRmax	RPE	METs	% $\dot{V}O_2$max
Very light	<20	<50	<10	<3.2	<27
Light	20-39	50-63	10-11	3.2-5.3	27-44
Moderate	40-59	64-76	12-13	5.4-7.5	45-62
Hard	60-84	77-93	14-16	7.6-10.2	63-85
Very hard	<85	<94	17-19	<10.3	<86
Maximal	100	100	20	12	100

The best absolute cut points in public health recommendations are ≥3 METs for moderate intensity and ≥6 METs for vigorous intensity. One MET is equivalent to a $\dot{V}O_2$ of 3.5 ml · kg^{-1} · min^{-1} or ~1 kcal · kg^{-1} · h^{-1}. Moderate intensity in an unfit person is often perceived as the intensity when a noticeable increase in ventilation and sweating occurs. The first column lists approximate percentages of $\dot{V}O_2$max (in units approximated by $\dot{V}O_2$ reserve or heart rate reserve) corresponding to relative intensity. Columns are organized to indicate the relative intensity by fitness, such that a moderate intensity for a more fit person would be at a greater MET level than for a less fit person. Notice that the threshold for reaching a moderate intensity is 2.6 METs in the most unfit category and progressively greater as people are more fit.

%$\dot{V}O_2$R = percentage of oxygen uptake reserve; %HRR = percentage of heart rate reserve; RPE = rate of perceived exertion.

Adapted from ACSM 1998; Howley 2001.

porting the recommendations for healthy effects of physical activity has focused on sustained periods of activity (>10 min per bout, preferably longer). In contrast, routine activities of daily living are generally in the very light-intensity category, and much less is known about the physiological and cellular responses in this category or the impact on disease prevention. The low perceived exertion (and physiological responses such as heart rate) at low intensities has supported the perception that reducing prolonged sitting time with light activity would not provide sufficient stimulus for disease prevention. In contrast, the inactivity physiology paradigm has built the case that potent cellular signals are induced by physical inactivity and that these can be short-circuited by LIPA (Hamilton, Hamilton, and Zderic 2004, 2007).

Energy Demand and Metabolism of Muscle Contractions

Skeletal muscle has a far more dynamic range of energy demand than other tissues. All of the muscle fibers within each individual motor unit function in unison. Thus, fibers within a motor unit are either completely resting or are recruited to contract. Electromyography of individual motor units in the legs of rodents shows that many motor units are almost never recruited (less than 1 min per day) during standing and activities of daily living. In contrast, some motor units are utilized for many hours each day, even in the complete absence of exercise. The adenosine triphosphate (ATP) demand of muscle fibers is primarily a function of whether the motor unit is resting or is being recruited to contract. Thus, some motor units contribute significantly to whole-body energy expenditure while others play a minor role (unless more intense exercise training is part of the weekly routine). Given that the muscle mass in the whole body is about 45% of body weight in normal-weight individuals, it is not hard to envision that the sum aggregate of 1 kg of muscle fibers can be working at the same time during LIPA. When muscle fibers are recruited and are contracting, their ATP utilization and energy expenditure increase immediately and by about 100-fold above the resting state (from 0.01 to 1 kcal $\cdot$ kg muscle^{-1} $\cdot$ min^{-1}). The *whole-body* RMR is approximately 1 MET, or 1 kcal/min in a 60 kg (132 lb) woman in the completely sedentary state. Thus, only a *small percentage* of the number of motor units within the total body muscle mass (~3%, or 1 kg muscle) spread throughout the legs or other postural muscles needs to be activated in order to double metabolic rate of the whole body!

$\dot{V}O_2$max = 10 METs		$\dot{V}O_2$max = 8 METs		$\dot{V}O_2$max = 5 METs	
Relative intensity (METs and %$\dot{V}O_2$max) in healthy adults differing in $\dot{V}O_2$max					
METs	% $\dot{V}O_2$max	METs	% $\dot{V}O_2$max	METs	% $\dot{V}O_2$max
<2.8	<28	<2.4	<30	<1.8	<36
2.8-4.5	28-45	2.4-3.7	30-47	1.8-2.5	36-51
4.6-6.3	46-63	3.8-5.1	48-64	2.6-3.3	52-67
6.4-8.6	64-86	5.2-6.9	65-86	3.4-4.3	68-87
<8.7	<87	<7.0	<87	<4.4	<88
10	100	8	100	5	100

Muscle Fiber Type Specificity

Several fundamental principles about muscle physiology are important for understanding inactivity physiology in general, as well as for properly interpreting the existing studies about how physical inactivity affects clinically relevant metabolic markers such as plasma triglyceride and cholesterol levels (Hamilton, Hamilton, and Zderic 2007). Skeletal muscles are composed of phenotypically very diverse types of muscle fibers (i.e., muscle cells). Some muscle fibers resist fatigue; rely heavily on oxidative metabolism for energy; contract inherently at a slow rate; and have a darker red color due to the iron within mitochondria, myoglobin, and blood. These types of fatigue-resistant oxidative muscle fibers are within the motor units that are first recruited to contract at the lowest intensities (figure 4.2). In contrast, the fatigue-sensitive, glycolytic white muscles are in the motor units with the highest threshold for recruitment and thus are not active during LIPA such as standing and slow walking.

Studies examining lipoprotein lipase (LPL) regulation have shown that the red oxidative muscle regions contain about 10 times more functional LPL activity (heparin-releasable LPL in muscle capillaries) than the white glycolytic regions of muscles in the same limb. Importantly, the LPL activity has been shown to decrease 10-fold locally in the oxidative muscle regions within hours after becoming inactive. Low-intensity physical activity (LIPA) produces a potent biochemical signal for activation of LPL activity and subsequently LPL-dependent plasma triglyceride

uptake and maintenance of high levels of plasma high-density lipoprotein (HDL) cholesterol (Hamilton, Hamilton, and Zderic 2004, 2007; Hamilton et al. 2008). Regulation of LPL is important for disease prevention because it has a central role in the plasma lipoprotein profile and regional fat deposition and is associated with clinical outcomes for cardiovascular disease and related chronic metabolic diseases. In contrast to these effects related to physical inactivity and LIPA, the more fatigable white glycolytic types of muscle are not responsive to LIPA (e.g., standing, slow ambulation). Studies are also showing direct associations between LIPA and glucose metabolism (Healy et al. 2007). This is in general agreement with the understanding that impaired metabolism of triglycerides is frequently physiologically related to impaired metabolism of glucose. In summary, light activity does not affect all skeletal muscles to the same degree, but instead affects the recruitment and metabolic responses locally in fatigue-resistant muscle fibers endowed for oxidative metabolism.

Inactivity physiology studies conducted to date in rats indicate that this type of slow red oxidative muscle is responsible for the rapid impairment of plasma lipoprotein and glucose metabolism after physical activity. In addition, this muscle component explains the cellular basis for how maintaining a sufficient amount of LIPA in the course of a day can have positive health benefits, including reducing risk factors associated with cardiovascular disease, metabolic syndrome, type 2 diabetes, and obesity.

> **Inactivity physiology studies indicate that the unloading of slow red oxidative muscle—fatigue-resistant muscle fibers spread throughout the large mass in the legs and trunk—underpins the role of physical inactivity in impaired plasma lipoprotein and glucose metabolism.**

To integrate the points just discussed, activation of only a small percentage of fatigue-resistant muscle fibers, spread throughout the large mass of muscle in the legs and back during light standing activity supporting the weight of the body, requires sustaining very little perceived exertion over prolonged periods. The reason is largely that, as explained earlier, the muscle fibers recruited into action for the lowest intensity of physical activity are of the highly specialized and fatigue-resistant slow red oxidative type (figure 4.2). Controlled animal studies show that fatigue-resistant motor units can be electrically stimulated continuously without rest and can resist fatigue for hours, unlike the far more fatigue-sensitive

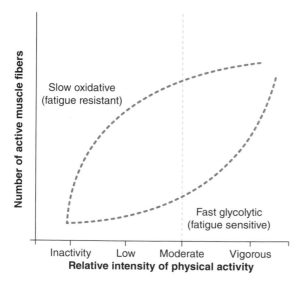

FIGURE 4.2 Differing responses of slow oxidative (fatigue resistant) and fast glycolytic (fatigue sensitive) muscle fibers to varying intensity levels of physical activity.

Inactive leisure time consists of activities such as reading, watching television, and using a computer. Some of this sedentary time could be displaced by more active activities.

fast glycolytic white fibers that typically reach fatigue within minutes of contraction. This in part explains why perceived exertion is very low in even the most unfit of persons performing routine activities of daily living (table 4.2).

A practical implication of this aspect of muscle contraction is that it explains why simple acts such as talking to friends on the phone while reclining on a sofa can reduce total energy expenditure significantly compared to standing upright and engaging in LIPA.

Thus, replacing sedentary time with more LIPA is possible even in people who never exercise or are unfit, although this idea is still largely untested. The average person is already accustomed to spending many hours each week engaged in LIPA during routine activities of daily living (table 4.1). While a doubling of the whole-body metabolic rate may seem small compared to the increase with exercise, it is noteworthy that LIPA can replace far more sedentary time than sustained moderate and vigorous activity because LIPA may be less fatiguing and can be integrated into the entire day more feasibly.

■ No known hormone or safe pharmacological treatment comes close to switching on skeletal muscle and whole-body metabolic rate as much as low-intensity contractile activity does.

The intensity of physical activity that counterbalances the time spent being sedentary is an important consideration for understanding specific health benefits of limiting sitting time. Consider, for example, someone who sleeps 8 h per day, sits 10 h per day, and engages in upright activity for the remaining 6 h per day. Other than a Tour de France cyclist or other highly trained elite endurance athlete, modern humans are not physiologically adapted to engage in 6 h per day of intense physical activity. Instead, the majority of sedentary time is counterbalanced with relatively light activity. Because people can sometimes spend 6 or even 12 h of the waking day engaged in light activity, the *energetic* responses to nonexercise time or nonexercise activity thermogenesis (Levine 2007) are largely, but not totally, driven by the balance between the time spent sitting and the time spent in LIPA. It is also unclear to what extent the *health consequences* of variable amounts of inactivity and LIPA are explained by differences in total activity energy expenditure or, instead, by more direct effects of local contractile inactivity inside of specific skeletal muscles. Arguing for the latter, an experimental model that prevented rats from standing on the hind legs indicated that local contractile inactivity was responsible for rapid reductions in the local LPL activity in the hind legs, plasma triglyceride uptake, plasma HDL cholesterol, and altered gene expression profiles (Hamilton, Hamilton, and Zderic 2007).

Studies measuring energy expenditure of people performing common types of LIPA at work or around the home indicate that many common tasks such as washing dishes, filing papers, or ironing have an energy demand about twofold above the resting metabolic rate. Walking is an interesting behavior because expenditure does not scale perfectly with walking speed as many would suspect. That is, walking at only 1.6 km per hour demands about a 2.5-fold increase above the resting metabolic rate while walking at 3.2 km per hour requires about a 3.1-fold increase.

Sedentary Behavior and Metabolic Health: Emerging Epidemiological Evidence

The evidence for the negative effects of sedentary behavior on health is rapidly accumulating. Here we review the available research from observational studies, from studies that have incorporated objective measurements of sedentary behavior, and from prospective studies.

Cross-Sectional Observational Studies

Laboratory findings on inactivity physiology increasingly are being complemented by evidence from cross-sectional and prospective epidemiological studies, which initially identified TV viewing time as a potential health risk. Previous reviews (Hamilton et al. 2008; Owen et al. 2010) describe the evidence on cross-sectional relationships between sedentary behavior (particularly TV viewing time) and overweight and obesity, the metabolic syndrome, and specific biomarkers of cardiometabolic risk. Findings from studies using objectively measured sedentary time have corroborated the initial TV time findings and have identified breaking up sedentary time as a potential preventive strategy. Type 2 diabetes, the metabolic syndrome, and established clinical risk factors including plasma glucose concentration, HDL cholesterol, plasma triglycerides, and various indices of excess body fat have all been associated with correlates of sedentary time.

An example of the existing evidence comes from a study called AusDiab1 (the baseline measures for the Australian Diabetes, Obesity and Lifestyle study). This national population-based study, which was conducted in 1999 and 2000 with some 11,000 participants, has shown high levels of TV viewing time to be related significantly with the metabolic syn-drome (Dunstan et al. 2005) and with undiagnosed abnormal glucose metabolism (Dunstan et al. 2004). Furthermore, dose–response relationships between TV viewing time and 2-h plasma glucose and fasting insulin were found in women (Dunstan et al. 2007). In all of these studies, the deleterious associations with TV viewing time remained significant after the effects of leisure-time physical activity and waist circumference were controlled for.

TV time and elevated metabolic risk were examined among the physically active (those who reported at least 150 min per week of moderate- to vigorous-intensity physical activity). For both men and women, there were significant detrimental dose–response associations of TV time with waist circumference, systolic blood pressure, and 2-h plasma glucose; for women there were also associations with fasting plasma glucose, triglycerides, and HDL cholesterol (Healy et al. 2008b). Thus, the metabolic correlates of prolonged TV time are adverse and can be observed even among those who are considered sufficiently physically active to reduce their risk of chronic disease.

The Australian TV time study findings described earlier suggest that the adverse role of sedentary behavior for metabolic health may be stronger in women than in men. However, when the relationships of risk biomarkers with both TV time and overall sitting time were examined concurrently (Thorp et al. 2010), the TV time relationships were stronger among women, but the relationships with overall sitting time were similar for men and women. It is possible that there may be behavioral differences in TV-related snacking behaviors for men and women. Or, might the typically smaller muscle mass and higher fat mass of women increase their susceptibility to the metabolic consequences of prolonged sitting following the typically large Australian evening meal?

Sedentary behaviors characterized by self-report measures of TV viewing time and overall sitting time are associated with adverse patterns of biomarkers for cardiometabolic health, particularly for blood glucose and triglycerides, and also for the presence of the metabolic syndrome, a cluster of attributes known to increase risk of diabetes and other chronic diseases.

While there is a need for scientific caution about the use of self-report measures in these sedentary behavior studies, the findings of a recent systematic review (Clark et al. 2009) suggest that the TV time measures used in these studies in general are reliable and valid. However, TV time is one of many sedentary behaviors, and scientific risk will be involved with the use of not yet validated self-report measures to characterize other sedentary behaviors, for which the measurement error potentially could be significant.

Objective-Measurement Studies

Measurement technology advances are reducing concerns about measurement error in the study findings that have relied on self-reported TV time and overall sitting time. Accelerometers were employed to assess sedentary time objectively in a sample of Australian adults. Accelerometer counts of less than 100 per minute were used to characterize sedentary time (Healy et al. 2007; Matthews et al. 2008; Owen et al. 2010). Objective measurement showed that sedentary time was associated with a larger waist circumference and more adverse 2-h plasma glucose and triglyceride levels (Healy et al. 2008c). The most intriguing finding of this study was that the extent to which sedentary time was broken up had beneficial associations with metabolic biomarkers (Healy et al. 2008a). A break was defined as a time when accelerometer output rose to 100 counts per minute or more for at least 1 min. It was assumed that these accelerometer-measured breaks in sedentary time involved either transitions from sitting to standing up or transitions from standing quite still to being ambulatory. Figure

4.3 shows two AusDiab accelerometer-study participants who have the same overall sedentary time but whose free time is broken up quite differently. The "Breaker" graphic shows frequent interruptions to sedentary time, whereas the "Prolonger" shows long unbroken sedentary blocks of time.

Figure 4.4 shows waist circumference across quartiles of breaks in sedentary time among participants in the AusDiab2 accelerometer substudy (Healy et al. 2008a). Those in the lowest 25% of the distribution of breaks in sedentary time had a 6 cm (2.4 in.) greater average waist circumference compared to those in the top 25%. This relationship of waist circumference with the extent to which sedentary time is broken up was found to be independent of total sedentary time, moderate- to vigorous-intensity activity time, and the average intensity of activity. Beneficial associations were also found with triglycerides and 2-h plasma glucose (Healy et al. 2008a). Thus, there may be metabolic health benefits of regular interruptions to prolonged sitting time, which may be in addition to the likely benefits of reducing overall sedentary time.

Findings from the AusDiab study showing adverse associations of cardiometabolic risk with self-reported TV viewing time and overall sitting time have been confirmed using accelerometers to objectively assess sedentary and physically active time.

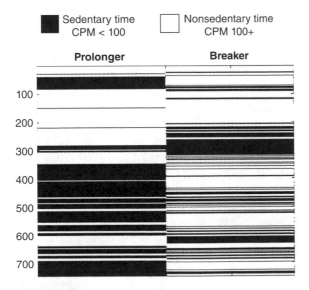

FIGURE 4.3 Accelerometer data showing sedentary time (dark areas) and nonsedentary time (light areas). These two study participants had the same amount of sedentary time but different patterns of accumulation. CPM, counts per minute.

Source: Dunstan, D.W., Healy, G.N., Sugiyama, T., Owen, N., "Too Much Sitting and Metabolic Risk—Has Modern Technology Caught Up with Us?" *US Endocrinology*, 2009; 5:29-33. Reprinted with permission.

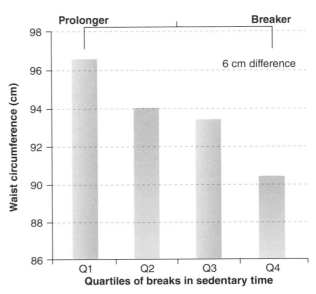

FIGURE 4.4 Associations of breaks in sedentary time with waist circumference in a cross-sectional study.

Adapted from Healy et al. 2008.

Prospective Studies

Hypotheses generated by the findings of the cross-sectional studies described in the preceding section have begun to be tested in prospective investigations. For example, TV viewing time changes predicted significant adverse changes in waist circumference for men and women and in diastolic blood pressure and a clustered cardiometabolic risk score for women. These associations were independent of baseline TV viewing time, baseline physical activity and physical activity change, and other potential confounders (Wijndaele et al. 2010b).

A long-term follow-up of participants in the Canada Fitness Surveys has shown that those who reported spending the majority of their day sitting had adverse mortality outcomes. In this study, relationships with mortality were shown across all levels of a self-report measure of overall sitting time. Relationships between sitting time and mortality were also found for those who were physically active, and the relationships were stronger among the overweight and obese (Katzmarzyk et al. 2009). A recent study with a large European cohort showed TV viewing time to be significantly associated with all-cause and cardiovascular mortality (Wijndaele et al. 2010a). In a 6.5-year follow-up of the AusDiab study participants, each 1-h increment in TV time was found to be associated with an 11% and an 18% increased risk of all-cause and cardiovascular disease mortality, respectively; these relationships were present after controlling for physical activity and waist circumference, as well as for the influence of known risk factors such as smoking, blood pressure, cholesterol, and diet (Dunstan et al. 2010).

Time spent sitting in automobiles is also emerging as a concern for the sedentary behavior and health field. A U.S. study that followed 7,744 men over 21 years (Warren et al. 2010) showed that those who reported spending more than 10 h per week sitting in automobiles (compared to less than 4 h per week), and more than 23 h in combined TV time and automobile time (compared to less than 11 h per week), had an 82% and 64% greater risk of dying from cardiovascular disease, respectively; however, TV time alone was not a significant predictor (Warren et al. 2010).

Prospective observational studies, in which sedentary behaviors such as TV viewing time and time spent sitting in cars have been assessed with subsequent follow-up of participants over several years, have shown sedentary time to predict an increased likelihood of death from cardiovascular disease.

Humans May Not Have Reached the Pinnacle of Physical Inactivity

Several types of evidence indicate that modern humans living in the midst of a sedentary-promoting environment are still remarkably physically active as a species. This perspective is essential for understanding the inactivity physiology studies to date and for understanding why people and laboratory animals are so highly responsive to physical inactivity (Hamilton, Hamilton, and Zderic 2007). Consider, for example, a person who sits for as much as half of the waking day. This leaves 8 h per day, or 56 h per week, available for weight-bearing skeletal muscle activity that meets the definition of LIPA. Actually, 56 h per week of light activity is the mean level for some cohorts of self-professed couch potatoes. In fact, such a high level of LIPA is necessary to explain the well-documented levels of total energy expenditure in the generally sedentary population. On average, nonexercising adults expend approximately 10 kcal/kg per day in physical activity.

Doubly labeled water studies have been able to quantify the total energy expenditure accumulated over the course of one to two weeks. In tandem with indirect calorimetry for measuring RMR, the doubly labeled water studies have described in many persons the energy expended during the abundant nonexercise physical activity, mostly lower-intensity activities of daily living. This is not a trivial amount of physical activity and can be explained only by the fact that humans who do not exercise are nevertheless spending a relatively large amount of time engaged in LIPA when not sitting. Some energy is associated with fidgeting and with use of the hands during sitting (e.g., writing or typing), but it is relatively small. Instead, the better explanation for how the average person who does not exercise still expends so much energy beyond rest is the activation of the posture-dependent musculature to sustain weight-bearing activities for the 40 to 50 h per week during neither sitting nor sleeping. This analysis has led to the key concept that humans have not come close to reaching the extreme levels of physical inactivity of a person in complete bed rest (Hamilton, Hamilton, and Zderic 2007).

A Comprehensive Sedentary Behavior Research Agenda

Building on what we have described in this chapter, further evidence is required to clarify the relationships of sedentary time with cardiometabolic risk biomarkers and adverse health outcomes. Further work in basic science is required to clarify the relevant mechanisms and pathways through which sedentary behavior exerts its adverse influences.

A priority for the sedentary behavior and health research agenda is to gather new evidence from human experimental studies and intervention trials. There is potential, for example, to conduct human experimental studies manipulating different "sedentary break" conditions and examining associated changes in cardiometabolic risk biomarkers. Field studies are also needed to determine the feasibility and acceptability of reducing and breaking up occupational, transit, and domestic sedentary time.

There are also multiple research opportunities to be explored through integrating sedentary behavior interventions into physical activity trials. When accelerometer data are gathered from such studies, sedentary time measures can be derived (as we have described), and unique hypotheses may readily be tested. It is imperative that the field now move to obtain such evidence through intervention trials, which will take the science beyond the inherent logical limitations of observational studies.

Population-based, descriptive epidemiology studies of sedentary behavior are needed if evidence-based public health strategies are to be pursued. Only limited descriptive evidence is as yet available (Owen et al. 2010). In a large sample of Australian women, being in the higher categories of TV time was found to be associated with a broader pattern of leisure-time sedentary behavior and a lower likelihood of meeting physical activity recommendations (Sugiyama et al. 2008). Another Australian study showed that low neighborhood "walkability" (poorly connected streets, low levels of residential density, and limited access to destinations) was associated with higher levels of TV time for women (Sugiyama et al. 2007).

Descriptive studies of sedentary behavior need to be extended beyond these initial studies focused on TV time to understand the correlates and the potential to intervene to influence other common sedentary behaviors. Evidence on the correlates of prolonged sitting in motor vehicles would be particularly informative in light of recent evidence on relationships with premature mortality (Warren et al. 2010). The social and environmental characteristics associated with high levels of time spent sitting in automobiles need to be identified.

The scientific and public health logic required for a comprehensive sedentary behavior research strategy (Owen et al. 2010) calls for the following research phases:

1. The nature of the relationships of sedentary behavior with risk biomarkers and related health outcomes needs to be identified.

2. Prolonged periods of sitting in people's lives need to be measured precisely.

3. The contextual determinants of sedentary behavior need to be identified in domestic, workplace, transportation, and recreation contexts.

4. The feasibility and efficacy of interventions to change certain sedentary behaviors need to be tested rigorously.

5. Public health policy responses need to be informed by evidence from all of these phases.

The research that we have emphasized in this chapter identifies a rapidly developing initial body of evidence in phase 1. Evidence in the subsequent phases is more limited.

Findings from trials of interventions to change sitting time in a different context particularly are required, not only to demonstrate the feasibility of changes in behavior, but also to show changes in biomarkers of chronic disease risk and reductions in subsequent adverse health outcomes. Future public health initiatives to reduce prolonged sitting time will also require novel evidence on the role of built-environment attributes in promoting sedentary behavior, which can be developed within the now well-established field of built environment–physical activity research (see Owen et al. 2010).

The wide range of new research opportunities on sedentary behavior and health includes human experimental studies that manipulate different sedentary break conditions; field studies to determine the feasibility of reducing and breaking up sitting time at work, during transport, and at home; and the integration of sedentary behavior measurement and interventions into physical activity trials.

Public Health Implications

Research on sedentary behavior and inactivity physiology will be an important influence on future public health initiatives in chronic disease prevention. If the findings outlined in this chapter continue to be confirmed by subsequent research studies, we will see a further expansion and differentiation of the physical activity and health research field, with an increase in the diversity of interdisciplinary research focusing explicitly on sedentary behaviors.

Furthermore, it might also be expected that future physical activity guidelines for adults would include explicit recommendations on reducing prolonged sitting time and that public health, transport sector, and urban planning policies would identify reduced sitting time as a major objective, justified by new evidence on environment–sedentary behavior relationships.

■ The potential public health actions that might be taken in relation to sedentary behavior include the following:

- Public information campaigns to emphasize reducing sitting time, as well as increasing physical activity
- Innovative technologies that can provide more opportunities to reduce sitting time (e.g., height-adjustable desks, telephone headsets)
- Regulations in workplaces for reducing or breaking up extended periods of job-related sitting
- Promotion of active transport modes as alternatives to prolonged periods of time spent sitting in automobiles

Summary

This chapter addresses sedentary behavior, highlighting the evidence from the emerging field of inactivity physiology. We summarize recent findings from cross-sectional and prospective epidemiological studies that make a strong case for focusing on the time that people spend sitting as a novel behavioral risk factor for chronic disease in addition to a lack of physical activity. We highlight research require-

ments, arguing that further evidence is needed from population-based studies and intervention trials to demonstrate the feasibility and health outcomes of reducing and breaking up sedentary time. Dose–response evidence from epidemiological studies and experimental trials is particularly required to inform guidelines for the public on how much sitting is too much. A more in-depth and broader understanding of sedentary behavior and inactivity physiology will provide a strong basis for pursuing novel public health initiatives and innovative lifestyle modifications.

Key Concepts

accelerometer—Small electronic device typically worn on the hip that measures the frequency and intensity of movement ("acceleration"); data from accelerometers may be analyzed using criteria for the number of counts in a given period of time (minutes or seconds) that can identify sedentary, low-intensity, and moderate- to vigorous-intensity activity time.

cross-sectional study—For definition, see page 50.

inactivity physiology—A paradigm making the case that too much sitting and other behaviors requiring physical inactivity throughout the day have deleterious effects on health, independent of whether people meet the current exercise guidelines, and that physical inactivity is not the biological equivalent of too little exercise. The paradigm also holds that physical inactivity (e.g., sitting) can rapidly evoke potent cellular and molecular signals that can cause clinically relevant physiological responses. The possible types of activity to offset these signals may require only low-intensity and highly intermittent types of activity that are below the traditional guidelines based on exercise physiology studies.

low-intensity physical activity (LIPA)—Incidental bouts of daily nonexercise physical activities that are almost always too intermittent or too modest in their intensity (or both) to meet the recommended level historically receiving attention in exercise science research.

physical activity—For definition, see page 19.

prospective epidemiological study—A study in which typically large numbers of people are initially assessed and a range of biological, behavioral, and social attributes are measured; then participants are likely to be followed up for repeat measures, and health outcomes or causes of death in the group are ascertained.

risk factor—For definition, see page 19.

sedentary behavior—Prolonged periods of being physically inactive, typically involving sitting in different contexts and for different purposes (derived from the Latin *sedere* "to sit"). The term *sedentary* has previously been used to describe persons who do not meet physical activity guidelines or those grouped in the lowest categories measured.

Study Questions

1. If you were developing a large-scale population survey for several thousand people, what sedentary behaviors would you want to assess, and what questions would you ask to identify sedentary behavior?

2. Describe a research study you might conduct to determine whether people are overweight because they sit too much or whether they sit too much because they are overweight.

3. What amounts and intensities of aerobic exercise, strength and balance exercise, or other activities do you think might be most protective for people who have large amounts of sitting time?

4. What would you consider the most important modifiable determinants (factors that can be changed in order to bring about a change in behavior) of sedentary behavior in the workplace and the home environment?

5. What practical advice would you give people on ways to reduce their sedentary time in the workplace and at home?

6. Consider how you would evaluate changes in sedentary time that might result from "natural experiments" (e.g., the introduction of height-adjustable workstations, changes in community transport infrastructure). What would be the best measures and the best study designs to use?

7. What do you think would be the ideal study to identify whether the time that people spend sitting at work poses long-term risks to their health?

8. What do you think would be the ideal research study to identify whether the time that people spend sitting in cars poses a long-term risk to their health?

9. Based on the evidence presented in this chapter, do you believe that it is now time for a recommendation on reducing sitting time as part of public health guidelines?

10. Based on the evidence presented in this chapter, what do you believe would be the most important public health initiative to reduce the time that people spend sitting?

References

American College of Sports Medicine Position Stand. 1998. The recommended quantity and quality of exercise for developing and maintaining cardiorespiratory and muscular fitness, and flexibility in healthy adults. *Medicine and Science in Sports and Exercise* 30:975-991.

Clark, B.K., T. Sugiyama, G.N. Healy, J. Salmon, D.W. Dunstan, and N. Owen. 2009. Validity and reliability of measures of television viewing time and other non-occupational sedentary behaviour of adults: A review. *Obesity Reviews* 10:7-16.

Dunstan, D.W., E.L. Barr, G.N. Healy, J. Salmon, J.E. Shaw, B. Balkau, D.J. Magliano, A.J. Cameron, P.Z. Zimmet, and N. Owen. 2010. Television viewing time and mortality: The Australian Diabetes, Obesity and Lifestyle Study (AusDiab). *Circulation* 121:384-391.

Dunstan, D.W., J. Salmon, N. Owen, T. Armstrong, P.Z. Zimmet, T.A. Welborn, A.J. Cameron, T. Dwyer, D. Jolley, and J.E. Shaw. 2004. Physical activity and television viewing in relation to risk of undiagnosed abnormal glucose metabolism in adults. *Diabetes Care* 27:2603-2609.

Dunstan, D.W., J. Salmon, G.N. Healy, J.E. Shaw, D. Jolley, P.Z. Zimmet, and N. Owen. 2007. Association of television viewing with fasting and 2-h postchallenge plasma glucose levels in adults without diagnosed diabetes. *Diabetes Care* 30:516-522.

Dunstan, D.W., J. Salmon, N. Owen, T. Armstrong, P.Z. Zimmet, T.A. Welborn, A.J. Cameron, T. Dwyer, D. Jolley, and J.E. Shaw. 2005. Associations of TV viewing and physical activity with the metabolic syndrome in Australian adults. *Diabetologia* 48:2254-2261.

Hamilton, M.T., D.G. Hamilton, and T.W. Zderic. 2004. Exercise physiology versus inactivity physiology: An essential concept for understanding lipoprotein lipase regulation. *Exercise and Sport Sciences Reviews* 32:161-166.

Hamilton, M.T., D.G. Hamilton, and T.W. Zderic. 2007. Role of low energy expenditure and sitting in obesity, metabolic syndrome, type 2 diabetes, and cardiovascular disease. *Diabetes* 56:2655-2667.

Hamilton, M.T., G.N. Healy, T.W. Zderic, D. Dunstan, and N. Owen. 2008. Too little exercise and too much sitting: Inactivity physiology and the need for new recommendations on sedentary behavior. *Current Cardiovascular Risk Reports* 2:292-298.

Haskell, W.L., I.M. Lee, R.R. Pate, K.E. Powell, S.N. Blair, B.A. Franklin, C.A. Macera, G.W. Heath, P.D. Thompson, and A. Bauman. 2007. Physical activity and public health: Updated recommendation for adults from the American College of Sports Medicine and the American Heart Association. *Circulation* 116:1081-1093.

Healy, G.N., D.W. Dunstan, J. Salmon, E. Cerin, J.E. Shaw, P.Z. Zimmet, and N. Owen. 2007. Objectively measured light-intensity physical activity is independently associated with 2-h plasma glucose. *Diabetes Care* 30:1384-1389.

Healy, G.N., D.W. Dunstan, J. Salmon, E. Cerin, J.E. Shaw, P.Z. Zimmet, and N. Owen. 2008a. Breaks in sedentary time: Beneficial associations with metabolic risk. *Diabetes Care* 31:661-666.

Healy, G.N., D.W. Dunstan, J. Salmon, J.E. Shaw, P.Z. Zimmet, and N. Owen. 2008b. Television time and continuous metabolic risk in physically active adults. *Medicine and Science in Sports and Exercise* 40:639-645.

Healy, G.N., K. Wijndaele, D.W. Dunstan, J.E. Shaw, J. Salmon, P.Z. Zimmet, and N. Owen. 2008c. Objectively measured sedentary time, physical activity, and metabolic risk: The Australian Diabetes, Obesity and Lifestyle Study (AusDiab). *Diabetes Care* 31:369-371.

Howley, E.T. 2001. Type of activity: Resistance, aerobic and leisure versus occupational physical activity. *Medicine and Science in Sports and Exercise* 33:S364-S369.

Katzmarzyk, P.T., T.S. Church, C.L. Craig, and C. Bouchard. 2009. Sitting time and mortality from all causes, cardiovascular disease, and cancer. *Medicine and Science in Sports and Exercise* 41:998-1005.

Levine, J.A. 2007. Nonexercise activity thermogenesis—liberating the life-force. *Journal of Internal Medicine* 262:273-287.

Matthews, C.E., K.Y. Chen, P.S. Freedson, M.S. Buchowski, B.M. Beech, R.R. Pate, and R.P. Troiano. 2008. Amount of time spent in sedentary behaviors in the United States, 2003-2004. *American Journal of Epidemiology* 167:875-881.

Owen, N., A. Bauman, and W. Brown. 2009. Too much sitting: A novel and important predictor of chronic disease risk? *British Journal of Sports Medicine* 43:81-83.

Owen, N., G.N. Healy, C.E. Matthews, and D.W. Dunstan. 2010. Too much sitting: The population health science of sedentary behavior. *Exercise and Sport Sciences Reviews* 38:105-113.

Owen, N., E. Leslie, J. Salmon, and M.J. Fotheringham. 2000. Environmental determinants of physical activity and sedentary behavior. *Exercise and Sports Sciences Reviews* 28:153-158.

Physical Activity Guidelines Advisory Committee. 2008. *Physical Activity Guidelines Advisory Committee report, 2008.* Washington, DC: U.S. Department of Health and Human Services.

Sugiyama, T., G.N. Healy, D.W. Dunstan, J. Salmon, and N. Owen. 2008. Is television viewing time a marker of a broader pattern of sedentary behavior? *Annals of Behavioral Medicine* 35:245-250.

Sugiyama, T., J. Salmon, D.W. Dunstan, A.E. Bauman, and N. Owen. 2007. Neighborhood walkability and TV viewing time among Australian adults. *American Journal of Preventive Medicine* 33:444-449.

Thorp, A.A., G.N. Healy, N. Owen, J. Salmon, K. Ball, J.E. Shaw, P.Z. Zimmet, and D.W. Dunstan. 2010. Deleterious associations of sitting time and television viewing time with cardio-metabolic risk biomarkers: AusDiab 2004-2005. *Diabetes Care* 33:327-334.

U.S. Department of Health and Human Services. 2008. *Physical activity guidelines for Americans.* Washington, DC: U.S. Department of Health and Human Services.

Warren, T.Y., V. Barry, S.P. Hooker, X. Sui, T.S. Church, and S.N. Blair. 2010. Sedentary behaviors increase risk of cardiovascular disease mortality in men. *Medicine and Science in Sports and Exercise* 42:879-885.

Wijndaele, K., S. Brage, H. Besson, K.T. Khaw, S.J. Sharp, R. Luben, N.J. Wareham, and U. Ekelund. 2010a. Television viewing time independently predicts all-cause and cardiovascular mortality: The EPIC Norfolk Study. *International Journal of Epidemiology* 40:150-159.

Wijndaele, K., G.N. Healy, D.W. Dunstan, A.G. Barnett, J. Salmon, J.E. Shaw, P.Z. Zimmet, and N. Owen. 2010b. Increased cardiometabolic risk is associated with increased TV viewing time. *Medicine and Science in Sports and Exercise* 42:1511-1518.

PART II

Effects of Physical Activity on the Human Organism

When a person engages in a sustained bout of physical activity, nearly all the systems or organs in the body come into play to support the contracting skeletal muscle. The cardiovascular and respiratory systems immediately respond to increase the availability of oxygen and substrate (glycogen and fat) for energy release in the muscle and to remove metabolic waste products. Free fatty acids are released from adipose tissue, and the liver produces more glucose in response to the increase in energy production. The nervous system is activated along with various hormones and enzymes to help coordinate and regulate all of these functions, and the force generated by muscle contractions and gravity puts strain on bone, ligaments, and tendons. When such bouts of exercise are repeated over weeks or months (exercise training), many of these systems begin to adapt to the activation by increasing their capacity or efficiency.

Chapters 5 through 9 review the major changes that occur in specific systems or organs as a result of an acute bout of exercise and regular physical activity or exercise training. The emphasis is on responses in generally healthy adults with some important exceptions noted. The value of examining these biological and structural responses to acute and chronic exercise lies in their critical role in the effect of regular activity on health. Part II of this book provides a foundation for issues that are studied by physical activity and health researchers around the globe.

5

Metabolic, Cardiovascular, and Respiratory Responses to Physical Activity

Edward T. Howley, PhD

CHAPTER OUTLINE

Knowledge of the metabolic, cardiovascular, and respiratory responses to physical activity and exercise provides a foundation for an understanding of the relative and absolute intensity of an activity and how many calories are expended—important elements in a conversation about "how much exercise is enough." This chapter describes the responses to submaximal exercise and to an incremental exercise test taken to the point of volitional exhaustion. The latter test is used to measure cardiorespiratory fitness, a key variable linked to health outcomes discussed in other chapters. Finally, the impact of training, age, and gender on these responses is discussed. This chapter can provide only a brief overview of these topics; the reader is directed to one of the many exercise physiology textbooks for a more complete treatment.

Relationship of Energy to Physical Activity

Energy is required for movement. Adenosine triphosphate (ATP) is the energy source used by skeletal muscles to generate tension (force), which is, in turn, transmitted by connective tissue to the skeletal system to cause movement. The ATP must be replaced as quickly as it is used for muscles to continue to generate force. The greater the force or the rate at which force is developed (power), the greater is the need for ATP. Muscles have multiple systems for replacing ATP that allow athletes to run at very high speeds for short periods of time (e.g., 100 m dash) or at slower (but still fast) speeds for longer distances (e.g., marathon), or that allow farmers to work all day long in a field. Energy for muscle force generation is made available when ATP is split to adenosine diphosphate (ADP) and inorganic phosphate; energy from other sources is required to regenerate ATP. Two broad classes of reactions are available to regenerate ATP: **anaerobic** reactions, which can regenerate ATP rapidly and without oxygen, and **aerobic** reactions, which are slower to activate but can sustain ATP regeneration for long periods of time through oxygen utilization. In both cases, an increase in the concentration of ADP in the working muscles (from ATP breakdown) activates the enzymes that ultimately lead to the regeneration of ATP.

Anaerobic Energy Sources

There are one-enzyme reactions and a multiple-enzyme pathway that can replace ATP at a very fast rate. These energy systems allow muscle to continue to generate force during strenuous activities or help make the transition from rest to exercise as the slower aerobic system comes up to speed. In the most important of these one-enzyme reactions, catalyzed by creatine kinase, **phosphocreatine (PCr)** reacts with ADP to form ATP. The PCr concentration in muscle is limited and can provide only about 5 s of support during all-out activity. In the multiple-enzyme pathway called **glycolysis**, glucose is metabolized at a high rate, and ATP is generated without oxygen. Lactic acid is produced in the process, leading to hydrogen ion accumulation in the muscle and blood. Glycolysis provides ATP at a high rate and is a substantial contributor to the ATP needed in all-out activities lasting less than about 2 min. In contrast, when glucose is metabolized aerobically, it represents a long-lasting source of ATP, producing about 18 times more ATP per glucose molecule than when glucose is metabolized anaerobically.

Aerobic Energy Production

Oxidative metabolism of carbohydrates (muscle glycogen and blood glucose) and fats (from both adipose tissue and intramuscular sources) provides the long-term source of ATP for physical activities and exercise that we typically associate with health-related outcomes; it also accounts for the majority of the ATP used in all-out performances lasting more than a few minutes (see figure 5.1). The complete oxidation of these fuels takes place in the mitochondria of the cell, which increase in number with endurance training. The larger number of mitochondria increases the capacity of the muscle to use fat as a fuel because fat can be metabolized only via aerobic pathways.

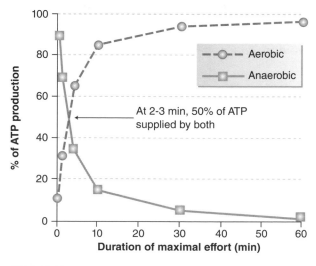

FIGURE 5.1 Percent of energy from aerobic and anaerobic sources for all-out activities of various durations.

Reprinted, by permission, from E.T. Howley and B.D. Franks, 2007, *Fitness professional's handbook*, 5th ed. (Champaign, IL: Human Kinetics), 478.

When fats and carbohydrates are metabolized aerobically, oxygen (O_2) is consumed and carbon dioxide (CO_2) is produced. The ratio of the volume of CO_2 produced ($\dot{V}CO_2$) to the volume of oxygen ($\dot{V}O_2$) consumed by the cell is called the respiratory quotient (RQ). Because we measure $\dot{V}O_2$ and $\dot{V}CO_2$ at the mouth, this ratio is also called the **respiratory exchange ratio (R)**. If fat is the only fuel used for energy during steady-state exercise, the R = 0.70, and 19.6 kJ (4.7 kcal) of energy is produced for each liter of oxygen consumed; when carbohydrates are the only fuel, the R = 1.0 and 20.9 kJ (5.0 kcal) is produced per liter of oxygen. Typically, a mixture of carbohydrates and fats is used to provide ATP for most activities; however, as the intensity of exercise increases, carbohydrates become the primary fuel, as noted by the increase in R (see figure 5.2a). In con-trast, for moderate-intensity activities, the percentage of energy derived from fat increases as the duration of the activity increases, as shown by the decrease in R (see figure 5.2b). The R can go over 1.0 when the hydrogen ion (linked to lactic acid) that is generated in heavy exercise is buffered by plasma bicarbonate and CO_2 is produced; the elevated R is used as an indicator that a person has achieved maximal oxygen uptake ($\dot{V}O_2$max, see later). Oxygen must be delivered to and consumed by the mitochondria to produce ATP from the metabolism of carbohydrates and fats. The roles that the cardiovascular and respiratory systems play in that regard are discussed in the next section.

Oxygen Consumption and Cardiovascular and Respiratory Responses to Exercise

Energy derived from oxygen consumption accounts for the vast majority of energy used in physical activity and exercise, compared with energy derived from anaerobic sources. Consequently, knowledge of how $\dot{V}O_2$ changes during the transition from rest to exercise and then back to rest is helpful for under-standing how the intensity and duration of exercise influence the caloric expenditure associated with exercise. Furthermore, although the cardiovascular and respiratory systems must respond appropri-ately to deliver the correct amount of oxygen to the muscles, the magnitude of these responses is influenced by the individual's **maximal oxygen uptake ($\dot{V}O_2$max)**. The first part of this section briefly describes the $\dot{V}O_2$, **heart rate (HR)**, and ventilation responses to a submaximal exercise task; the second part addresses in more detail the responses to a graded (incremental) exercise test taken to the point of volitional exhaustion. Within the latter context, the effects of physical training are presented.

Rest-to-Exercise Transitions

An individual standing alongside a treadmill set at 200 m/min steps onto the belt and with that one step accelerates to 200 m/min; failure to do so would result in the individual's drifting off the back of the tread-mill. The rate at which ATP is used increases instantly from that needed for standing to that required to run at 200 m/min. However, figure 5.3 shows that $\dot{V}O_2$ does not increase instantaneously to the level required to generate ATP for the task; instead, a gradual

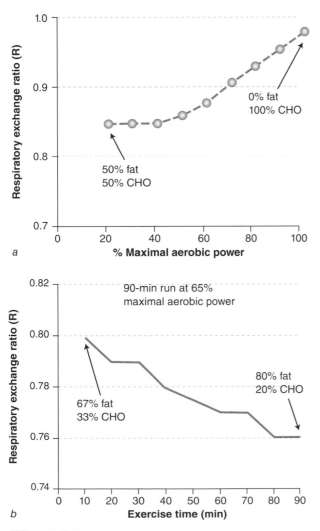

FIGURE 5.2 Effect of exercise (a) intensity and (b) dura-tion on the pattern of carbohydrate and fat use.

Reprinted, by permission, from E.T. Howley and B.D. Franks, 2007, *Fitness professional's handbook*, 5th ed. (Champaign, IL: Human Kinetics), 453.

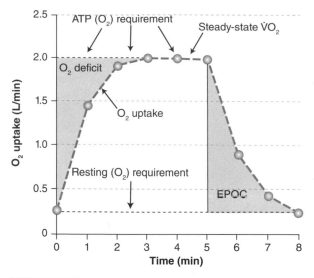

FIGURE 5.3 Oxygen uptake response to submaximal exercise.

Reprinted, by permission, from E.T. Howley and B.D. Franks, 2007, *Fitness professional's handbook*, 5th ed. (Champaign, IL: Human Kinetics), 454.

increase occurs over the first 2 to 3 min until a **steady-state** $\dot{V}O_2$ is achieved. The steady-state $\dot{V}O_2$ is the oxygen requirement for which the oxidative generation of ATP meets the ATP requirement of the task. In fact, the steady-state $\dot{V}O_2$ is used to calculate the caloric cost of the activity. If the steady-state $\dot{V}O_2$ is 2.0 L/min and we use 20.9 kJ (5.0 kcal) as the caloric equivalent of 1 L of oxygen, then 41.8 kJ/min (10 kcal/min) is produced; for 5 min of activity, 209 kJ (50 kcal) of energy is expended. Measurements of oxygen consumption obtained during different physical activities are used to generate tables of the energy cost of those activities.

If one draws a line from the steady-state $\dot{V}O_2$ response back to the start of exercise, it is clear that the $\dot{V}O_2$ does not meet the ATP requirement during the first few minutes of exercise. An **oxygen deficit** is said to exist, and other energy sources have to provide the needed ATP—notably PCr and glycolysis.

The magnitude of the oxygen deficit increases with exercise intensity. When the individual steps off the treadmill at the end of 5 min of exercise, the $\dot{V}O_2$ does not return immediately to the resting level but returns gradually back to rest over several minutes. This has been called the oxygen debt, the recovery oxygen, and, most recently, the **excess postexercise oxygen consumption (EPOC)**. In the first 2 to 3 min of recovery, the $\dot{V}O_2$ falls rapidly as the PCr and oxygen stores in muscle are quickly replenished. After that time, the $\dot{V}O_2$ falls more slowly to the resting baseline value; this portion of the EPOC is associated with the resynthesis of lactic acid to glucose and the elevated HR, ventilation, and hormone levels that have not yet returned to resting values (Gaesser and Brooks 1984).

Why does oxygen uptake increase slowly over the first 2 to 3 min of submaximal exercise when transition from rest occurs? Figure 5.4 shows that

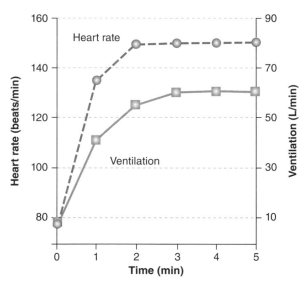

FIGURE 5.4 Heart rate and pulmonary ventilation responses to submaximal exercise.

Data from Ekblom et al. 1968.

Mechanisms for Energy Production During Exercise

Adenosine triphosphate (ATP) is the energy source used by muscles to generate tension or force; it must be replaced as quickly as it is used for exercise to continue. Muscles can produce ATP rapidly using anaerobic mechanisms (i.e., phosphocreatine and glycolysis), but their contribution lasts only 5 s to 2 min during all-out exercise, respectively. Aerobic metabolism of carbohydrates and fats in the mitochondria of the muscle cells is the primary source of ATP in all-out exercise lasting more than 2 to 3 min. Carbohydrates become a more important fuel as exercise intensity increases, and fat provides a greater percentage of the energy as the duration of moderate exercise increases.

Transition From Rest to Exercise

In the transition from rest to submaximal exercise, ATP is delivered to the muscle contractile elements as fast as needed to maintain the intensity of the exercise. Because aerobic systems do not come on line instantly, anaerobic sources of ATP (phosphocreatine and glycolysis) provide the needed ATP at the onset of exercise, and an oxygen deficit is said to exist. By 2 to 3 min into a submaximal exercise task, the oxygen uptake ($\dot{V}O_2$) is adequate to meet the ATP demands, and a steady state (balance between the supply of ATP by aerobic means and the demand for ATP by the muscle) exists. The lag in $\dot{V}O_2$ at the onset of exercise is related to a lag in cardiovascular and respiratory responses and to the time needed for mitochondria to increase their rate of ATP production. When exercise stops, the $\dot{V}O_2$ does not fall immediately back to the resting value. There is at first a very rapid decline, followed by a slower response back to the resting baseline value. The elevated oxygen uptake during recovery from exercise is called the excess postexercise oxygen consumption (EPOC).

HR and ventilation increase in a manner similar to $\dot{V}O_2$, which helps explain, in part, the lag in oxygen delivery to muscle at the onset of exercise. However, the muscle itself also contributes to this lag. It takes some time for the activity of the various mitochondrial enzyme systems involved in oxygen consumption to increase to the level required to meet the ATP demand of the tissue.

Responses to Incremental Exercise Taken to Volitional Exhaustion

A graded exercise test (GXT) is used to evaluate a person's metabolic, cardiovascular, and respiratory responses to systematic increases in exercise intensity. Usually the test begins at a low intensity and progresses in equal-intensity increments. For example, during a treadmill test, one can increase the intensity by changing the speed or the percent grade, or both, depending on the population being studied. During a cycle ergometer test, the work rate might begin at 50 or 100 W and increase in increments of 25 or 50 W over equal time intervals. If athletes are being tested, the initial stage is set at a higher intensity, and the increments between stages are larger. During a GXT, the individual might be monitored for a wide variety of variables including electrocardiogram (ECG), HR, $\dot{V}O_2$, blood lactate concentration, and blood pressure. Measurements are typically made in the last 30 s of each stage of the test to approximate the steady-state response for that stage; data are then plotted against stages of the test. In this section, we begin with metabolic responses to such a test and then consider the associated cardiovascular and respiratory responses.

Oxygen Uptake and Blood Lactate

Figure 5.5*a* shows the oxygen uptake response of an untrained individual to a GXT on a treadmill, in which the speed was set at 80 m/min and the grade increased 3% every 3 min until the participant experienced volitional exhaustion (i.e., participant terminated the test because of the perception of maximal effort). The $\dot{V}O_2$ values measured in the last 30 s of each stage were plotted against the percent grade. The $\dot{V}O_2$ increased with increasing grade except at the end of the test, when the $\dot{V}O_2$ measured at 15% grade did not increase as the grade was increased to 18%. This "plateau" in $\dot{V}O_2$ with an increase in the exercise intensity has been taken, historically, as the criterion for having achieved $\dot{V}O_2$max. However, because only about 50% of individuals experience a plateau in $\dot{V}O_2$ during this type of GXT, investigators have used other measures as indicators that the participant was working maximally when the highest $\dot{V}O_2$ was obtained during a GXT. These indicators include a high blood lactate concentration (>8 mM), a respiratory exchange ratio > 1.10, and an HR response close to an age-predicted maximum value. The latter criterion is not as useful given the inherent variability in any age-predicted estimate of maximal HR (SD = ~10 beats per minute [bpm]) (Powers and Howley 2009). $\dot{V}O_2$max is used as a measure of cardiorespiratory fitness and is also called maximal aerobic power. Figure 5.5 also shows the changes in the $\dot{V}O_2$ response to the GXT following an endurance training program. The $\dot{V}O_2$ at any submaximal stage was the same, but the $\dot{V}O_2$max increased.

There is considerable variability in $\dot{V}O_2$max values across the population. Cardiac patients may have $\dot{V}O_2$max values as low as 20 ml · kg^{-1} · min^{-1}, whereas elite endurance athletes may approach

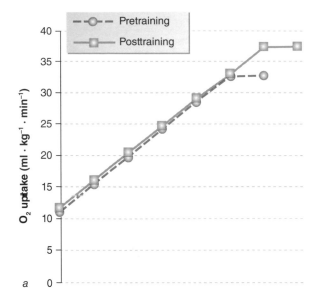

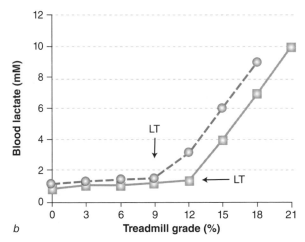

FIGURE 5.5 Oxygen uptake and lactate responses to a graded exercise test.

Part (a) adapted, by permission, from E.T. Howley and B.D. Franks, 2007, *Fitness professional's handbook*, 5th ed. (Champaign, IL: Human Kinetics), 457. Part (b) based on data from the same source.

ences. In the HERITAGE Family Study, genetic factors explained about 50% of the variation that existed in $\dot{V}O_2$max in sedentary adults; however, the investigators believed that the value might be inflated because of nongenetic familial factors (Bouchard et al. 1998).

Figure 5.5b shows that the blood lactate concentration remains close to its resting value until 9% (pretraining measurement) or 12% (posttraining measurement) grade, where it increases systematically. This increase occurs typically between about 50% and 80% of a person's $\dot{V}O_2$max. The work rate (or oxygen consumption) at which the blood lactate concentration increases is called the **lactate threshold (LT)**. The increase in the blood lactate concentration occurs because its rate of appearance is greater than its rate of removal. The rate of appearance is linked to the recruitment of fast-twitch (FT) fibers (see chapter 8) as exercise intensity increases and to accelerated glycolysis that is driven by both intracellular factors (higher ADP levels stimulating key enzymes) and extracellular factors such as the rising concentration of epinephrine (Brooks 1985). The slower rate of removal is associated with a reduced blood flow to the liver (see later discussion on redistribution of blood flow). Regular endurance training delays the LT because of a reduced rate of lactate production, associated with an increase in the number of mitochondria, and because of an enhanced capacity of muscle and other tissues to clear lactate (Brooks 1985). The LT has been used to predict endurance performance.

The $\dot{V}O_2$max can be expressed in absolute terms (L/min), relative to body weight (ml · kg^{-1} · min^{-1}), or as multiples of a standard resting metabolic rate (RMR) (**metabolic equivalents** or **METs**, where one MET is taken, by convention, to be 3.5 ml · kg^{-1} · min^{-1}). In the preceding example (figure 5.5), the participant's pretraining $\dot{V}O_2$max was 31.5 ml · kg^{-1} · min^{-1} or 9 METs. The MET term is used as a simple way to express the more complicated term ml · kg^{-1} · min^{-1} and is used extensively in cardiac rehabilitation and fitness programs, physical activity interventions, and epidemiological investigations to express

80 ml · kg^{-1} · min^{-1}. The differences found in the population are due partly to differences in the quantity and quality of training and partly to genetic influ-

Oxygen Uptake and Blood Lactate Responses During a Graded Exercise Test

Oxygen consumption ($\dot{V}O_2$) increases in a stepwise fashion with increases in exercise intensity during a graded exercise test (GXT), until the final stages during which $\dot{V}O_2$ may not change with an increase in exercise intensity. The highest value achieved in such a test is taken as the individual's maximal oxygen uptake ($\dot{V}O_2$max). The blood lactate concentration remains close to resting values in the early stages of a GXT. The sudden increase in blood lactate concentration at about 50% to 80% of $\dot{V}O_2$max is called the lactate threshold (LT); the LT has been used to predict performance in endurance events.

the intensity (e.g., moderate exercise is 5-9 METs) or volume (MET-hours) of physical activity or exercise (American College of Sports Medicine 2009). Next we look at the cardiovascular and respiratory responses that are linked to $\dot{V}O_2$max.

Heart Rate, Stroke Volume, Cardiac Output, and Oxygen Extraction

Oxygen consumption is the product of the **cardiac output**, expressed in liters of blood pumped from the heart per minute (L/min), and the volume of oxygen extracted from the arterial blood (CaO_2), in ml O_2/L blood. **Oxygen extraction** is the difference between the oxygen content of arterial blood and the oxygen content of mixed venous blood ($C\bar{v}O_2$), the latter of which is measured in the right heart chamber. Cardiac output is the product of HR in beats per minute and **stroke volume (SV)** in liters of blood pumped from the heart per beat (L per beat), so

$$\dot{V}O_2 \text{ (L/min)} = HR \times SV \times (CaO_2 - C\bar{v}O_2).$$

The following sections describe how these variables respond during a GXT and indicate the effect of endurance training on each.

Figure 5.6 shows the changes in cardiac output, HR, SV, and oxygen extraction during a GXT. Cardiac output and HR increase in a linear fashion, but SV does not. In upright exercise, SV increases during the early stages of the test and levels off at about 40% $\dot{V}O_2$max in most individuals. Consequently, further increases in cardiac output (beyond 40% $\dot{V}O_2$max) are attributable entirely to the increase in HR. The fact that HR tracks exercise intensity so well makes it a good indicator of exercise intensity. The increase in HR is attributable to both withdrawal of parasympathetic nerve activity and an increase in sympathetic nerve activity to the sinoatrial node (Rowell 1993).

The SV is the difference between the volume of blood in the heart before contraction (**end-diastolic volume, EDV**) and the volume of blood in the heart after contraction (**end-systolic volume, ESV**). During exercise, an increase in venous return to the heart leads to an increase in EDV, causing a distension of the ventricles and an increase in the force of contraction. The increased venous return is linked to the alternate contraction and relaxation of skeletal muscles acting on the large veins in muscles (muscle pump); the lower intrathoracic pressure caused by the increased depth and rate of ventilation (respiratory pump); and the increase in abdominal pressure, compressing the large intra-abdominal veins (abdominal pump) (Rowell, O'Leary, and Kellogg 1996). The force of contraction of the ventricles is also increased by the increase in sympathetic nerve activity to the ventricles. The ESV decreases with

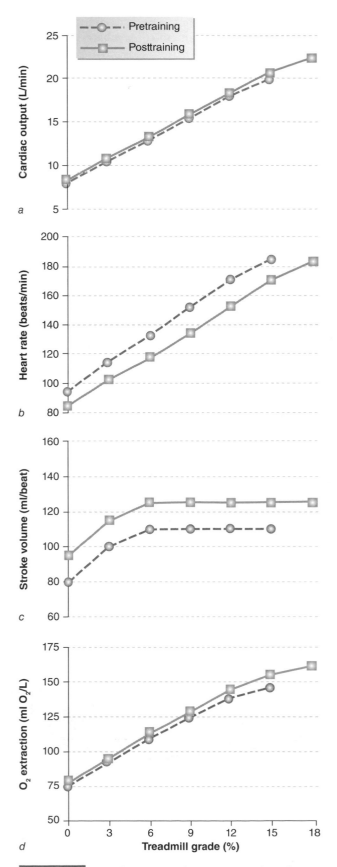

FIGURE 5.6 Cardiac output, heart rate, stroke volume, and oxygen extraction response to a graded exercise test.

Reprinted, by permission, from E.T. Howley and B.D. Franks, 2007, *Fitness professional's handbook*, 5th ed. (Champaign, IL: Human Kinetics), 458-460.

increasing intensity of exercise, and changes in both EDV and ESV contribute to the increased SV during exercise. **Ejection fraction** (SV divided by EDV) is a measure of ventricular function. It is about 0.65 (i.e., 65%) at rest and can increase to as high as 0.85 in peak exercise, attributable to an increase in cardiac contractility (Rowell, O'Leary, and Kellogg 1996).

Figure 5.6, *a* through *c*, shows that following an endurance training program, the cardiac output response to the submaximal stages of the GXT is very similar to the pretraining response, but the manner in which it is realized is different: HR is reduced whereas SV is increased. The increased SV is caused by an increase in EDV. Maximal HR remains the same or is slightly lower as a result of training; however, when coupled with the increased SV, maximal cardiac output is increased and, with it, $\dot{V}O_2$max (Rowell 1993).

The volume of oxygen delivered into the systemic circulation per minute is a product of the cardiac output and CaO_2. The CaO_2 is determined primarily by the hemoglobin concentration and the pressure of oxygen (PO_2) in the arterial blood. The difference in the hemoglobin concentration between men and women (150 g/L vs. 130 g/L, respectively) accounts for some of the difference in $\dot{V}O_2$max between genders. The lower PO_2 at altitude decreases the oxygen saturation of hemoglobin and, consequently, oxygen delivery to muscle, making $\dot{V}O_2$max lower at altitude compared with sea level. Figure 5.6*d* shows the changes in oxygen extraction ($CaO_2 - C\bar{v}O_2$ difference) from light to maximal work.

As exercise intensity increases, more oxygen is extracted by the muscles to support the oxidative generation of ATP, widening the difference between the CaO_2 and $C\bar{v}O_2$. The relative contributions of cardiac output and oxygen extraction to oxygen consumption can be seen in the following calculations for rest and maximal work:

$$\dot{V}O_2 \text{ (L/min)} = \text{Cardiac output} \times (CaO_2 - C\bar{v}O_2)$$

$$\text{Rest: } \dot{V}O_2 \text{ (L/min)} = 5 \text{ L/min} \times (200 \text{ ml } O_2/\text{L blood} - 150 \text{ ml } O_2/\text{L blood})$$

$$\dot{V}O_2 \text{ (L/min)} = 250 \text{ ml/min or } 0.25 \text{ L/min}$$

$$\text{Maximal work: } \dot{V}O_2 \text{ (L/min)} = 20 \text{ L/min} \times (200 \text{ ml } O_2/\text{L blood} - 50 \text{ ml } O_2/\text{L blood})$$

$$\dot{V}O_2 \text{ (L/min)} = 3,000 \text{ ml/min or } 3.00 \text{ L/min}$$

The 12-fold increase in oxygen consumption from rest to maximal exercise is due to a fourfold increase in cardiac output and a threefold increase in oxygen extraction. Relative to $\dot{V}O_2$max, the importance of HR, SV, and oxygen extraction is described clearly in table 5.1. Three different groups are identified: athletes, normally active individuals, and individuals with mitral stenosis (narrow opening at the mitral valve between the left atrium and left ventricle that obstructs blood flow to the ventricle and reduces SV). The $\dot{V}O_2$max values differ dramatically between groups; those of the athletes are almost twice those of the normally active individuals, and those of the normally active individuals are 2.5 times those of the mitral stenosis patients. What explains these large differences? The maximal HR values for all three groups are similar, as are the oxygen extraction values. The major factor accounting for the differences in $\dot{V}O_2$max is the cardiac output, attributable exclusively to the enormous differences in maximal SV between groups (Rowell 1993). The designation of $\dot{V}O_2$max as a measure of cardiorespiratory fitness is attributable to this link between $\dot{V}O_2$max and cardiac output.

The increase in cardiac output is not the only factor responsible for delivery of more oxygen to muscles with increasing exercise intensity. Another is the redistribution of the cardiac output. At rest, only about 20% (~1 L/min) of resting cardiac output (~5 L/min)

TABLE 5.1 Physiologic Basis for Differences in $\dot{V}O_2$max in Three Groups

Participants	$\dot{V}O_2$max (ml/min)	=	Heart rate (bpm)	×	Stroke volume (ml/beat)	×	(a-v̄)O₂ difference (ml/L)
Athletes	6,250	=	190	×	205	×	160
Normally active people	3,500	=	195	×	112	×	160
People with mitral stenosis	1,400	=	190	×	43	×	170

From *Human Cardiovascular Control* by Loring B. Rowell, copyright © 1993 by Oxford University Press, Inc. Used by permission of Oxford University Press, Inc.

is delivered to muscle; however, during maximal work, about 80% (~20 L/min) of maximal cardiac output (~25 L/min) is directed to muscles—a 20-fold increase (Rowell 1993). Blood flow is simultaneously increased to working muscles and decreased to the liver, kidneys, and gastrointestinal tract. This redistribution of blood flow not only is important in directing more blood to working muscles in heavy exercise but also is crucial for maintaining blood pressure.

Blood Pressure

As exercise begins, local factors in the active muscles, such as increases in partial pressure of CO_2 (PCO_2), potassium ion (K^+), hydrogen ion (H^+), nitric oxide, and adenosine, cause a relaxation of arterioles serving those muscles; resistance is decreased and more blood flow is directed to these muscles. This autoregulation of blood flow provides a "supply and demand" arrangement that matches blood flow to the metabolic needs of the muscles. However, this increase in muscle blood flow has to be balanced by reductions in blood flow to other tissues, via sympathetic arteriolar constriction, to prevent blood pressure from falling. This also accounts for the redistribution of blood flow. **Mean arterial blood pressure (MABP)** is the driving force of the blood and is calculated as the sum of the **diastolic blood pressure (DBP)** and one-third of the pulse pressure (difference between **systolic blood pressure [SBP]** and DBP). That is,

$$MABP = DBP + 1/3(SBP - DBP)$$

If MABP were to decrease, so would blood flow to the brain, with dire results. Consequently, MABP is "protected" by control systems that monitor blood pressure and invoke automatic responses to correct discrepancies. The MABP is dependent on both cardiac output and the resistance offered to that blood flow by the whole body **(total peripheral resistance, or TPR)**:

$$MABP = Cardiac\ output \times TPR$$

For example, if blood pressure were to decrease at rest, baroreceptors (pressure or stretch receptors) in the carotid artery and arch of the aorta would sense the fall and signal the cardiovascular control center in the medulla of the brain stem. The control center directs an increased level of sympathetic nerve activity to the heart (to increase cardiac output) and to the arterioles (to increase resistance) to restore blood pressure (Rowell 1993). What happens to blood pressure during exercise?

Figure 5.7 shows the SBP and DBP responses to a GXT. The SBP increases throughout the test, whereas

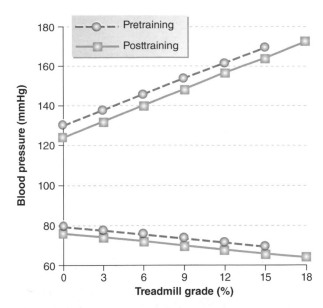

FIGURE 5.7 Changes in systolic and diastolic blood pressures during a graded exercise test.

Reprinted, by permission, from E.T. Howley and B.D. Franks, 2007, *Fitness professional's handbook*, 5th ed. (Champaign, IL: Human Kinetics), 461.

the DBP remains the same or decreases slightly. The MABP increases modestly throughout the GXT, from 90 mmHg at rest to 107 mmHg at maximal exercise, in this example. This small increase in MABP at exercise, in the presence of the large increase in cardiac output, is attributable to the large decrease in TPR (linked to the dilation of arterioles in working muscles). The simultaneous constriction of arterioles in other tissues and organs through the action of sympathetic nerves is sufficient to maintain adequate TPR to preserve the MABP. This appears to be in conflict with the regulation of blood pressure at rest (mentioned previously) where the cardiac output and arteriolar constriction are increased because of a decrease in blood pressure.

The cardiovascular control center in the medulla of the brain stem, however, receives input from a variety of sources during exercise. Clearly, the baroreceptors provide input about blood pressure. However, the combination of efferent input from higher brain centers **(central command)** and afferent input from receptors in the working muscles **(peripheral feedback)** drives the cardiovascular responses mentioned previously. The efferent activity is proportional to the motor unit recruitment and spills over to the medulla as the impulse traffic moves to the muscles. The afferent input comes from both mechanoreceptors and chemoreceptors to help fine-tune the cardiovascular response (Rowell 1993). As with other exercise responses, endurance training reduces the blood pressure responses to exercise

Exercise affects all the major body systems. Chronic exercise contributes to improved health overall.

compared with the pretraining responses at the same exercise intensity.

Respiratory Responses to Exercise

The respiratory system maintains the partial pressures of arterial blood gases (PO_2 and PCO_2) and assists in H^+ regulation. Venous blood, returning from the tissues, enters the right heart chambers and is pumped to the lungs. The pulmonary system is a low-resistance system, and the right ventricle does not have to generate as much force as the left ventricle to have the same SV. As blood flows through the pulmonary capillaries, gas exchange occurs with the alveoli. Carbon dioxide diffuses rapidly down its partial pressure gradient from the blood to the lung and is easily equilibrated with alveolar gas within about 0.1 s; oxygen equilibration takes slightly longer. Obviously, as more oxygen is extracted from the arterial blood with increasing intensities of exercise, **pulmonary ventilation** must increase to bring more oxygen to the alveoli. Figure 5.8 shows the ventilation response to a GXT. Ventilation increases in an almost linear fashion from light through moderate exercise, at which point the rate of ventilation increases dramatically. This sudden break in the pattern of ventilation is called the **ventilatory threshold (VT)**; it is used as a noninvasive estimate of the LT. The connection between the LT and the VT is the buffering of the H^+ by the plasma bicarbonate (HCO_3^-); this buffering results in the generation of CO_2 that drives ventilation to a higher level:

$$H^+ + HCO_3^- \rightarrow H_2CO_3 \rightarrow CO_2 + H_2O$$

The respiratory control center is also located in the medulla of the brain stem and receives input from both central (within the central nervous system) and peripheral (carotid arterial and arch of aorta) chemoreceptors. The PO_2, PCO_2, and H^+ of arterial blood do not change during light to moderate exercise, so what causes the ventilation to increase

Cardiovascular Adjustment During a Graded Exercise Test

Heart rate and cardiac output increase linearly with exercise intensity during a GXT; stroke volume increases until about 40% $\dot{V}O_2$max in upright exercise and remains constant thereafter. Consequently, the increase in cardiac output after that intensity is attributable to heart rate alone. The oxygen extraction (difference between the arterial and mixed venous oxygen contents) increases with exercise intensity. The product of cardiac output and oxygen extraction determines $\dot{V}O_2$. Much of the variation in $\dot{V}O_2$max (L/min) among individuals is attributable to differences in maximal cardiac output, related primarily to differences in maximal stroke volume. $\dot{V}O_2$max is considered a good measure of cardiorespiratory fitness because of its relationship to maximal cardiac output. Systolic blood pressure increases with exercise intensity, whereas diastolic blood pressure remains the same or decreases. Mean arterial blood pressure increases modestly with exercise intensity. The simultaneous dilation of arterioles in working muscles (caused by local factors) and constriction of arterioles in the liver and kidneys (caused by sympathetic nerve activity) results in a redistribution of cardiac output to bring large volumes of blood to muscles during maximal work compared with resting conditions. The small increase in mean arterial blood pressure in the face of a four- to fivefold increase in cardiac output is attributable to the large decrease in total peripheral resistance associated with the dilation of arterioles in working muscles.

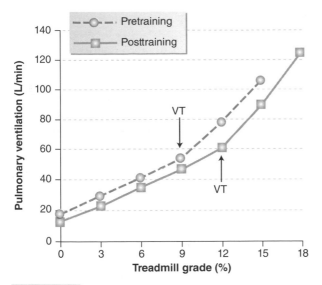

Pretraining
Posttraining

FIGURE 5.8 Changes in pulmonary ventilation during a graded exercise test.

Adapted from Ekblom et al. 1968.

breathing frequency (breaths per minute). Increases in both TV and frequency account for the increase in pulmonary ventilation; however, at very high rates of ventilation, TV levels off, and the remaining increase in pulmonary ventilation is attributable entirely to an increase in breathing frequency (Dempsey et al. 1996).

How effective is pulmonary ventilation in maintaining the blood gases? Even at the maximal cardiac outputs seen in normally active individuals, the cross-sectional area of the pulmonary capillary bed is large enough to slow down blood flow and allow the PO_2 of the red blood cells to equilibrate with the PO_2 of the alveoli; the PO_2 of arterial blood is maintained within narrow limits. In contrast, about 40% to 50% of elite endurance athletes who have very large maximal cardiac outputs (see table 5.1) experience a true desaturation of hemoglobin during heavy exercise. At these very high cardiac outputs, the red blood cells move rapidly through the pulmonary capillaries, not allowing equilibration to occur with the alveolar PO_2. Consequently, the hemoglobin leaves the pulmonary capillaries before it is fully saturated with oxygen. This results in a lower arterial oxygen content and a reduction in oxygen delivery to muscle during heavy exercise in these individuals (Dempsey et al. 1996).

in a linear manner? The same input described previously for the cardiovascular responses appears to be important. Neural input from higher brain centers proportional to motor unit recruitment (central command) and peripheral feedback from receptors in working muscles to the respiratory control center appear to shape the ventilatory response. The sudden change in ventilation that signifies the VT is linked to the appearance of lactate (LT) and the need to buffer H^+ by respiratory compensation—exhaling the CO_2 produced in this reaction. Regular endurance training results in a reduction in the ventilatory response to a given submaximal work rate.

Pulmonary ventilation is the product of tidal volume or volume of air moved per breath (TV) and

Effect of Training, Age, and Gender on Maximal Oxygen Uptake

The $\dot{V}O_2$max and associated cardiovascular responses are affected by training, age, and gender. This section summarizes important observations in regard to these variables.

Respiratory Adjustments During a Graded Exercise Test

Pulmonary ventilation increases in a linear fashion with exercise intensity during light to moderate exercise; after that, it increases at a faster rate. This "break" from linear is called the ventilatory threshold (VT) and is used as a noninvasive estimate of the lactate threshold (LT). The LT and VT are linked to the buffering of the H^+ by the plasma bicarbonate (HCO_3^-), resulting in CO_2 generation that increases ventilation. The increase in ventilation is attributable to an increase in both tidal volume and breathing frequency; the latter is more important at higher rates of ventilation. In normal persons and most endurance-trained athletes, pulmonary ventilation is sufficient to maintain the arterial PO_2 and arterial oxygen content. However, a proportion of endurance athletes with very high $\dot{V}O_2$max values experience a true desaturation of hemoglobin in near-maximal exercise. Part of the reason for this is related to the high rate at which their red blood cells move through the pulmonary capillaries—not allowing enough time for equilibration with the alveolar PO_2.

Training

The average training-induced increase in $\dot{V}O_2$max in formerly sedentary individuals is about 15%; however, there is considerable variability in response. Data from the HERITAGE Family Study showed that although the average increase in $\dot{V}O_2$max attributable to an endurance training program was 16%, some participants did not change at all, and others had an increase of 50%. The variability in response appears to have a strong genetic link, with up to 47% of the gain in $\dot{V}O_2$max being tied to heredity (Bouchard et al. 1999). For more information on heredity and exercise, see chapter 24. The training-induced increase in $\dot{V}O_2$max in previously sedentary individuals is attributable to an increase in maximal cardiac output and maximal oxygen extraction, each contributing equally. The increase in maximal cardiac output is caused exclusively by an increase in maximal SV, because maximal HR changes little with training (Rowell 1993).

Gender and Age

On average, $\dot{V}O_2$max (ml $\cdot$ kg^{-1} $\cdot$ min^{-1}) is about 15% to 30% lower in adult women compared to adult men; average absolute $\dot{V}O_2$max (L/min) differences are greater still. Part of the difference is attributable to a lower maximal cardiac output. The lower maximal cardiac output is linked to a lower maximal SV, secondary to a smaller heart size. Maximal oxygen extraction is also lower in women because of a lower hemoglobin concentration. In addition, when men and women do submaximal exercise at intensities set at the same absolute $\dot{V}O_2$ in L/min (e.g., on a cycle ergometer), considerable differences exist in physiological adjustments to the exercise. At any given submaximal $\dot{V}O_2$, women have a higher HR

response to compensate for the lower SV response (Åstrand et al. 1964).

The $\dot{V}O_2$max decreases with age at the rate of approximately 1% per year in healthy but sedentary men and women. This decrease is attributable to both inactivity and weight gain as well as to any "aging" effect. Trained men experience a decrease of about 0.5% per year, whereas trained women decline at 1% per year. The age-related decrease in maximal HR is the major contributor to the decrease in maximal cardiac output, but maximal oxygen extraction also decreases with age. The latter is probably related more to the effect of inactivity on muscle mitochondria and capillary density than to any specific aging factor. The increase in $\dot{V}O_2$max in older adults in response to an endurance training program is similar to that observed for younger adults (Holloszy and Khort 1995). See chapter 19 for more on physical activity and aging.

Application to Exercise Training and Physical Activity Interventions

An exercise intervention usually specifies the intensity, duration (minutes per session), frequency (sessions per week), and mode of exercise (walk, cycle, swim). Intensity can be prescribed in absolute terms such as walking at a certain speed or MET level or cycling at a specific work rate or absolute $\dot{V}O_2$ in L/min. Exercise intensity can also be set in terms of an individual's maximal oxygen uptake (**percentage of maximal oxygen uptake, %$\dot{V}O_2$max**) or maximal HR (**percentage of maximal HR**). When this is done, the relative effort required is similar among individuals who may differ greatly in terms of $\dot{V}O_2$max. It should

Effects of Training, Gender, and Age on Maximal Oxygen Uptake

In previously sedentary participants, the average training-induced increase in $\dot{V}O_2$max is about 15%, but with large variations in response among participants. Up to 47% of the variation in response can be traced to genetic factors. The increase in $\dot{V}O_2$max is attributable to increases in both maximal cardiac output and oxygen extraction. Women, on average, have a lower $\dot{V}O_2$max than men. In addition, during submaximal exercise set at the same absolute intensity ($\dot{V}O_2$ in L/min), women have a higher heart rate response to compensate for their smaller stroke volume. In adults, $\dot{V}O_2$max decreases about 0.5% to 1% per year because of decreases in maximal heart rate and oxygen extraction. However, the training-induced increase in $\dot{V}O_2$max in older individuals is similar to that of younger adults.

be no surprise that such an approach has been used extensively over the years to specify exercise intensity. The relative intensity can also be expressed as a percentage of the **heart rate reserve (HRR)** (difference between maximal HR and resting HR) or the **oxygen uptake reserve ($\dot{V}O_2R$)** (difference between $\dot{V}O_2$max and resting $\dot{V}O_2$). The HRR has been used extensively as a means to express relative exercise intensity; %$\dot{V}O_2R$ is the newest expression of relative exercise intensity and is used interchangeably with %HRR (American College of Sports Medicine 2009).

Summary

Adenosine triphosphate (ATP) is used by muscles to generate tension or force and is replaced rapidly by anaerobic reactions (PCr and glycolysis) and more slowly by aerobic reactions (metabolism of carbohydrates and fats in the mitochondria). Carbohydrates are more important in heavy exercise, and fats provide a greater percentage of the energy during long-term moderate exercise. In the transition from rest to submaximal exercise, aerobic systems do not come on line instantly, so anaerobic sources (PCr and glycolysis) provide some of the needed ATP. By 2 to 3 min, the oxygen uptake ($\dot{V}O_2$) meets the ATP demands, and a steady state exists. When exercise stops, the $\dot{V}O_2$ declines rapidly, followed by a slower transition back to the resting baseline value. The elevated oxygen uptake during recovery is called the excess postexercise oxygen consumption (EPOC).

Oxygen consumption increases with exercise intensity during a graded exercise test (GXT). The highest $\dot{V}O_2$ value achieved in such a test is taken as the individual's maximal oxygen uptake ($\dot{V}O_2$max). The blood lactate concentration remains close to resting values in the early stages of a GXT but increases suddenly at about 50% to 80% $\dot{V}O_2$max; this is called the lactate threshold (LT). Heart rate and cardiac output increase linearly with exercise intensity, whereas stroke volume levels off at about 40% $\dot{V}O_2$max. Consequently, the increase in cardiac output after that intensity is attributable to heart rate alone. The oxygen extraction (difference between the arterial and mixed venous oxygen contents) increases with exercise intensity. Much of the variation in $\dot{V}O_2$max (L/min) among individuals is attributable to differences in maximal cardiac output, making $\dot{V}O_2$max a good measure of cardiorespiratory fitness.

Systolic blood pressure increases with exercise intensity, whereas diastolic blood pressure remains the same or decreases. The simultaneous dilation of arterioles in working muscles (caused by local factors) and constriction of arterioles in the liver and kidneys (caused by sympathetic nerve activity) result in a redistribution of cardiac output to bring large volumes of blood to muscles during maximal work. The small increase in mean arterial blood pressure in the face of a four- to fivefold increase in cardiac output is attributable to the large decrease in total peripheral resistance associated with the dilation of arterioles in working muscles. Pulmonary ventilation increases in a linear fashion with exercise intensity during light to moderate exercise; after that, it increases at a faster rate. The "break" from linear is called the ventilatory threshold (VT) and is used as a noninvasive estimate of the lactate threshold (LT). The increase in ventilation is attributable to an increase in both tidal volume and breathing frequency. Pulmonary ventilation is sufficient to maintain the arterial PO_2 and arterial oxygen content in most individuals, with the exception of a subgroup of endurance athletes who experience a true desaturation of hemoglobin during near-maximal exercise. Part of the reason for this is related to the high rate at which their red blood cells move through the pulmonary capillaries—not allowing enough time for equilibration with the alveolar PO_2.

In previously sedentary participants, the training-induced increase in $\dot{V}O_2$max averages about 15% and

Application of Physiological Data to Exercise Prescription

The intensity of exercise can be prescribed in numerous ways to reflect energy expenditure on an absolute (L/min) or relative (ml · kg^{-1} · min^{-1}) basis. But in either case, where there are large differences in $\dot{V}O_2$max among individuals participating in exercise programs, the cardiovascular and respiratory responses, as well as the individual's perception of effort, will be markedly different at the same oxygen uptake. However, when the intensity of exercise is set as a percentage of the $\dot{V}O_2$max, percentage of maximal heart rate, or percentage of the heart rate reserve, variability in both physiological and perceptual responses among individuals is markedly reduced.

is attributable to increases in both maximal cardiac output and oxygen extraction. Women, on average, have a lower $\dot{V}O_2$max than men. In addition, during submaximal exercise set at the same absolute intensity ($\dot{V}O_2$ in L/min), women have a higher heart rate response to compensate for their smaller stroke volume. The intensity of exercise can be prescribed in numerous ways to reflect energy expenditure on an absolute (L/min) or relative (ml $\cdot$ kg^{-1} $\cdot$ min^{-1}) basis. When the intensity of exercise is set as a percentage of the $\dot{V}O_2$max, percentage of maximal heart rate, or percentage of the HRR, variability in both physiological and perceptual responses is markedly reduced among individuals who differ in $\dot{V}O_2$max.

Key Concepts

aerobic—ATP-generating reactions that require oxygen and can sustain muscle activity longer; processes occur in the mitochondria of a cell.

anaerobic—ATP-generating reactions that do not require oxygen (e.g., phosphocreatine and glycolysis) and provide short-lived, high-intensity force quickly.

cardiac output—Volume (liters) of blood pumped from the heart per minute; product of heart rate and stroke volume.

central command—Neural activity in higher brain centers associated with the recruitment of motor units; this "spills over" to both the cardiovascular and respiratory control centers in the medulla of the brain stem, driving both cardiovascular and respiratory responses to meet the metabolic demands of the muscles.

diastolic blood pressure (DBP)—Lowest pressure in a cardiac cycle when the ventricles are not contracting (diastole); lower number in blood pressure reading (e.g., 120/80 mmHg).

ejection fraction—Ratio of the volume of blood pumped from the heart per beat (stroke volume) to the volume of blood in the heart prior to contraction (end-diastolic volume).

end-diastolic volume (EDV)—Volume of blood in the heart at the end of diastole, just before ventricular contraction; stretches the ventricle to increase the force of contraction and stroke volume.

end-systolic volume (ESV)—Volume of blood in the heart after ventricular contraction.

excess postexercise oxygen consumption (EPOC)—Also known as oxygen debt or recovery oxygen; oxygen uptake during recovery from exercise that remains above the preexercise resting baseline oxygen uptake.

glycolysis—Metabolic pathway producing ATP from the anaerobic breakdown of glucose; short-term source of ATP that is important in all-out activities lasting less than 2 min and during the transition from rest to steady-state exercise.

heart rate (HR)—Number of times the heart contracts per minute; normally set by the "pacemaker" in the sinoatrial node.

heart rate reserve (HRR)—Difference between maximal HR and resting HR; used as the denominator when exercise intensity is expressed as a percentage of the HRR: (exercise HR – rest HR)/(maximal HR – rest HR) × 100%; used extensively as a means to describe exercise intensity.

lactate threshold (LT)—Exercise intensity (expressed as actual speed, work rate, or oxygen uptake) at which the blood lactate concentration increases in a systematic manner above a baseline established at lower intensities of exercise.

maximal oxygen uptake ($\dot{V}O_2$max)—For definition, see page 19.

mean arterial blood pressure (MABP)—Driving force of the blood relative to blood flow; opposed by the total peripheral resistance; MABP = DBP + 1/3(SBP – DBP).

metabolic equivalents (METs)—For definition, see page 19.

oxygen deficit—Difference between the steady-state oxygen uptake measured during a constant-load exercise test and the actual oxygen uptake measured in the first minutes of the test; a measure of the amount of ATP that had to be provided from anaerobic sources (phosphocreatine and glycolysis).

oxygen extraction—Difference between the oxygen content of arterial blood and the oxygen content of the blood when it returns to the right heart chamber (mixed venous oxygen content).

oxygen uptake reserve ($\dot{V}O_2$R)—Difference between maximal oxygen uptake and resting oxygen uptake; used as the denominator when exercise intensity is expressed as a percentage of the $\dot{V}O_2$R: (exercise $\dot{V}O_2$ – rest $\dot{V}O_2$)/($\dot{V}O_2$max – rest $\dot{V}O_2$) × 100%.

percentage of maximal HR—Ratio of exercise heart rate to maximal heart rate, expressed as a percentage; used extensively as a means to describe exercise intensity.

percentage of maximal oxygen uptake (% $\dot{V}O_2$max)—Ratio of exercise oxygen uptake to $\dot{V}O_2$max, expressed as a percentage; used extensively as a means to describe the relative intensity of exercise.

peripheral feedback—Afferent information from chemoreceptors and mechanoreceptors in working muscles provided to both the cardiovascular and respiratory control centers in the medulla of the brain stem to help shape the cardiovascular and respiratory responses needed to meet the metabolic demands of the tissues.

phosphocreatine (PCr)—Important anaerobic source of ATP; supports the ATP need during the transition from rest to exercise and can provide about 5 s worth of ATP during all-out activity.

pulmonary ventilation—Product of the tidal volume (L per breath) and breathing frequency (breaths/min); increases with exercise intensity to bring sufficient oxygen to the alveoli to saturate the hemoglobin and to clear the CO_2 generated by metabolic reactions.

respiratory exchange ratio (R)—Ratio of CO_2 production to oxygen consumption ($\dot{V}CO_2/\dot{V}O_2$); ratio of 1.0 during steady-state work indicates that 100% of the energy is from carbohydrates; ratio of 0.7 indicates 100% of energy from fat. In heavy exercise in which blood lactic acid accumulates, the R can exceed 1.0 because of buffering of the H^+ by plasma bicarbonate.

steady state—A relatively unchanging response of a physiological measure (e.g., $\dot{V}O_2$, HR) during a submaximal exercise test, usually obtained several minutes into the test. The steady-state $\dot{V}O_2$ response is taken as the oxygen requirement for the task.

stroke volume (SV)—Volume of blood pumped by the heart per beat; difference between the EDV and ESV.

systolic blood pressure (SBP)—Highest pressure measured during a cardiac cycle when the ventricle is contracting (systole); top number in blood pressure reading (e.g., 120/80 mmHg).

total peripheral resistance (TPR)—Whole-body resistance offered to blood flow by the arterioles; decreases dramatically during exercise to allow large blood flow to muscles with only small increases in the mean arterial blood pressure.

ventilatory threshold (VT)—Sharp increase in pulmonary ventilation during a graded exercise test that is linked to the buffering of lactic acid by plasma bicarbonate; used as a noninvasive indicator of the LT.

Study Questions

1. Using figure 5.1, estimate the percent of energy coming from aerobic sources in an all-out 30-min activity. What do you think the value would be if the person performed the activity at a submaximal effort rather than all-out?

2. In the transition from rest to submaximal exercise, why doesn't the oxygen uptake increase immediately to the level required for that activity?

3. What is the link between $\dot{V}O_2$max and cardiovascular function that makes $\dot{V}O_2$max a good measure of cardiorespiratory fitness?

4. What cardiovascular factor explains most of the differences in $\dot{V}O_2$max among individuals?

5. What is the ventilatory threshold, and how is it connected to the lactate threshold?

6. Why do women, on average, have a lower $\dot{V}O_2$max than men?

7. Given the large differences in $\dot{V}O_2$max among individuals, how can exercise intensity be set in a training program to cause most individuals to experience the same relative effort?

References

American College of Sports Medicine. 2009. *ACSM's guidelines for exercise testing and prescription*. Philadelphia: Lippincott Williams & Wilkins.

Åstrand, P-O., T.E. Cuddy, B. Saltin, and J. Stenberg. 1964. Cardiac output during submaximal and maximal work. *Journal of Applied Physiology* 19:268-274.

Bouchard, C., E.W. Daw, T. Rice, L. Pérusse, J. Gagnon, M.A. Province, A.S. Leon, D.C. Rao, J.S. Skinner, and J.H. Wilmore. 1998. Family resemblance for VO$_{2\,max}$ in the sedentary state: The HERITAGE Family Study. *Medicine and Science in Sports and Exercise* 30:252-258.

Bouchard, C., T.P. An, T. Rice, J.S. Skinner, J.H. Wilmore, J. Gagnon, L. Pérusse, A.S. Leon, and D.C. Rao. 1999. Family aggregation of VO$_{2\,max}$ response to exercise: Results from the HERITAGE Family Study. *Journal of Applied Physiology* 87:1003-1008.

Brooks, G.A. 1985. Anaerobic threshold: Review of the concept and directions for future research. *Medicine and Science in Sports and Exercise* 17:22-34.

Dempsey, J.A., L. Adams, D.M. Ainsworth, R.F. Fregosi, C.G. Gallagher, A. Guz, B.D. Johnson, and S.K. Powers. 1996. Airway, lung, and respiratory muscle function during exercise. In *Handbook of physiology: Section 12: Exercise: Regulation and integration of multiple systems*, ed. LB. Rowell and J.T. Shepherd. New York: Oxford University Press.

Ekblom, B., P-O. Åstrand, B. Saltin, J. Stenberg, and B. Wallström. 1968. Effect of training on circulatory response to exercise. *Journal of Applied Physiology* 24:518-528.

Gaesser, G.A., and G.A. Brooks. 1984. Metabolic bases of excess post-exercise oxygen consumption: A review. *Medicine and Science in Sports and Exercise* 19:29-43.

Holloszy, J.O., and W.M. Kohrt. 1995. Exercise. In *Handbook of physiology: Section 11: Aging,* ed. E.J. Masoro. New York: Oxford University Press.

Howley, E.T., and B.D. Franks. 2007. *Fitness professional's handbook.* 5th ed. Champaign, IL: Human Kinetics.

Powers, S.K., and E.T. Howley. 2009. *Exercise physiology: Theory and applications to fitness and performance.* 7th ed. New York: McGraw-Hill.

Rowell, L.B. 1993. *Human cardiovascular control.* New York: Oxford University Press.

Rowell, L.B., D.S. O'Leary, and D.L. Kellogg Jr. 1996. Integration of cardiovascular control systems in dynamic exercise. In *Handbook of physiology: Section 12: Exercise: Regulation and integration of multiple systems,* ed. L.B. Rowell and J.T. Shepherd. New York: Oxford University Press.

6

Acute Responses to Physical Activity and Exercise

Adrianne E. Hardman, MSc, PhD

©Bananastock

Research has repeatedly shown that exposure to regular, frequent bouts of physical activity stimulates physiological and metabolic changes that benefit health. It is helpful to classify these as either (a) **chronic effects**, that is, adaptations to training acquired over weeks or months, or (b) short-term, acute responses to each individual session of activity. Health-related adaptations to training are dealt with in other chapters. This chapter describes selected acute responses that are clearly related to health outcomes and explains their relevance. The extent to which these responses benefit health depends on the type, frequency, and regularity of activity and the extent to which particular acute responses persist into the postactivity period. For people who achieve the minimum recommended amounts of physical activity (Haskell et al. 2007, p. 1423), acute responses should be stimulated on five days per week (for moderate-intensity activity) or three days per week (for vigorous intensity). For reasons often related to statistical power and logistical considerations, the experimental models used to study acute health-related responses have relied mainly on planned, structured exercise, as opposed to physical activity performed during daily living. For this reason, the term *exercise* predominates in this chapter rather than *physical activity*. Studies of short periods of detraining are included because they illustrate that the changes to health outcomes that are rapidly lost when regular training is interrupted are mainly attributable to **acute effects**. Intuitively, unstructured periods of physical activity may be expected to stimulate acute responses that are qualitatively similar to, but less conspicuous than, those arising from planned sessions of exercise.

Lipids and Lipoproteins

Lipoproteins are particles that transport triglycerides and cholesterol in the blood plasma. They have a hydrophobic lipid core and an outer surface layer that allows the particle to mix with the watery plasma. Lipoproteins are classified according to their density, which in turn reflects their composition (see "Characteristics of the Main Classes of Lipoproteins"). There are clear links between markers for disordered lipoprotein metabolism and heart disease: Plasma concentrations of total cholesterol, low-density lipoprotein (LDL) cholesterol, and triglycerides are positively associated with the incidence of coronary heart disease, and there is a clear inverse relationship between high-density lipoprotein (HDL) cholesterol concentration and heart disease incidence. Changes to the concentration of lipoprotein lipids arising from

a session of exercise are therefore of interest because of their implications for cardiovascular risk.

When considerable amounts of energy have been expended, changes to plasma concentrations of triglycerides, total cholesterol, or HDL cholesterol may be observed immediately after an exercise session. For example, men and women who completed the 1994 Hawaii Ironman Triathlon exhibited a nearly 40% decrease in triglycerides and decreases of around 10% in total and LDL cholesterol compared with prerace values. Events such as a marathon can result in an increase in HDL of about 10% after the race. These changes are independent of changes to plasma volume during these endurance events. On the other hand, a session of moderate-intensity exercise of relatively short duration does not lead to clear changes in lipoprotein variables measured immediately afterward. However, these measurements do not reveal the extent of the influence of such an exercise bout on lipoprotein metabolism.

Even a modest session of activity makes important inroads into the body's energy stores, leading to a prolonged period of metabolic "recovery." Thus, when blood samples are obtained hours (rather than minutes) after exercise, effects on lipoprotein metabolism are clear. Decreases in triglycerides and increases in HDL cholesterol measured in the fasted state 24 h after a session of exercise have consistently been observed. The design of the HERITAGE Family Study was such that chronic adaptations to training could be distinguished from acute effects. In samples obtained 24 h following the last exercise session, researchers found that total cholesterol and very low-density lipoprotein (VLDL) triglycerides were lower after 20 weeks of training than at baseline, but there were no changes in samples obtained 72 h posttraining. In other words, these reductions were clearly acute responses to the last bout of exercise. The main factor influencing the extent of acute changes to triglycerides and HDL is the amount of energy expended during the exercise session, irrespective of exercise intensity. It may be that around 4.2 MJ (1,000 kcal or the equivalent of walking more than 16 km [10 mi]) need to be expended to elicit statistically significant changes.

These changes reflect the functional relationship that exists between plasma concentrations of triglycerides and HDL cholesterol. Hydrolysis of triglyceride-rich lipoproteins (chylomicrons and VLDL cholesterol) by the enzyme lipoprotein lipase (LPL) is accompanied by the transfer of cholesterol and other surface materials from triglyceride-rich particles into HDL. Thus, rapid removal of triglyceride-rich lipoproteins is associated with an increase in cholesterol carried in HDL. Prior exercise enhances

Characteristics of the Main Classes of Lipoproteins

There are four main classes of lipoproteins. Density varies with composition; chylomicrons are the least dense, and high-density lipoproteins (HDLs) are the most dense.

1. Chylomicrons consist mainly of dietary triglycerides, with a composition by weight of 90% triglycerides and 5% cholesterol.

2. Very low-density lipoproteins (VLDLs) consist mainly of endogenous triglycerides from the liver. Triglycerides make up 65% of the weight of these particles and cholesterol 13%.

3. Low-density lipoproteins (LDLs) have cholesterol and cholesteryl ester as main lipids. By weight, they are composed of 45% cholesterol and 10% triglycerides, with 20% protein and 23% phospholipids.

4. High-density lipoproteins (HDLs) have cholesteryl ester and phospholipids as main lipids. Cholesterol makes up 18% of HDL and triglycerides 2%; the remainder is protein (50%) and phospholipids (30%).

Note: Different subclasses of lipoproteins have been identified, and the distribution of plasma lipids between these is probably important for cardiovascular health outcomes. For example, small, dense LDLs that are particularly cholesterol rich have been implicated in atherosclerosis.

triglyceride clearance rates, probably by increasing the activity of LPL in muscle—the rate-limiting step in triglyceride clearance. However, there appears to be little cumulative effect of repeated daily sessions of exercise on LPL activity, explaining why the decrease in triglycerides is essentially an acute effect of exercise that dissipates within 24 h.

Effect of Prior Exercise on Postprandial Triglycerides

Changes to lipoprotein metabolism after an exercise session therefore derive mainly from enhanced clearance of triglyceride-rich lipoproteins, a phenomenon most clearly seen when these particles are most numerous, that is, during the **postprandial** state. A rich body of data illustrates that prior exercise markedly decreases the triglyceride response to a subsequent meal. The clinical relevance of this is that an exaggerated postprandial triglyceride response has been linked to the presence of coronary artery disease and to the atherogenic lipoprotein phenotype. Moreover, people spend the majority of their lives in the postprandial state.

Figure 6.1 shows the effect of afternoon exercise (90 min at 60% of $\dot{V}O_2$max) on serum triglyceride concentrations during the 6 h following a standard, high-fat meal consumed the following morning. The subjects were normally active (figure 6.1a) and endurance-trained (figure 6.1b) middle-aged women (Tsetsonis, Hardman, and Mastana 1997). Prior exercise decreased the postprandial response, measured

as the area under the triglyceride concentration versus time curve, by 30% in trained women and by 16% in normally active women compared with values obtained in the control trial (no planned exercise for three days beforehand). Thus, a single session of exercise markedly decreases subsequent postprandial lipemia. This effect has been confirmed in various subject groups, including the obese. However, even in trained athletes, this effect is short-lived—and therefore an acute effect—as shown by a study of detraining. Endurance-trained athletes consumed a test meal on three occasions: 15 h, 60 h, and 6.5 days after their last training session. Compared with the 15-h value, the postprandial triglyceride response was 35% higher after just 60 h without exercise, with little further increase after nearly a week without training. Frequent exercise is therefore needed to maintain the cardiovascular benefits that may be assumed to arise from the triglyceride-lowering effects.

Influence of Intensity and Duration

Prior exercise has such a clear effect on postprandial triglycerides that this model has allowed investigation of the influence of exercise intensity (figure 6.2) and pattern (figure 6.3). Researchers examined the effect of the intensity of prior exercise (controlling for energy expenditure) using a repeated-measures design (Tsetsonis and Hardman 1996). The same participants consumed a high-fat, mixed meal on three occasions: (1) control, no planned exercise

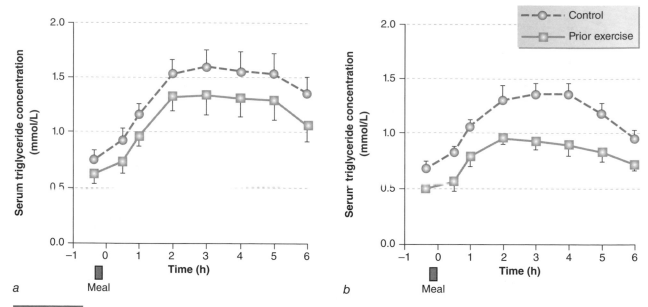

a

b

FIGURE 6.1 Effect of prior exercise on fasting and postprandial concentrations of serum triglycerides in (a) 13 untrained women and (b) 9 endurance-trained women. A high-fat, mixed meal was consumed in the morning after a 12-h fast. Control—participants refrained from exercise for three days; prior exercise—participants walked for 1.5 h at 60% $\dot{V}O_2$max the previous afternoon. Values are mean and standard error (SE).

Reproduced from *American Journal of Clinical Nutrition*. © Am J Clin Nutr. American Society for Clinical Nutrition. Reprinted by permission of Adrianne E. Hardman.

for three days beforehand; (2) 15 h after a 90-min treadmill walk at 60% $\dot{V}O_2$max; and (3) 15 h after a walk that was twice as long (180 min) but at half the intensity (30% $\dot{V}O_2$max). In other words, the researchers asked, "Can intensity be 'traded' for duration?" As figure 6.2 clearly shows, the postprandial triglyceride response to dietary fat was decreased to the same degree (32%) after both exercise sessions, so the answer to this question (for this particular health outcome) is yes.

Different patterns of walking were compared in a study of daylong plasma triglyceride concentrations (figure 6.3). On three occasions, middle-aged participants were followed throughout a day during which they consumed three ordinary meals (Murphy, Nevill, and Hardman 2000). Compared with values from a control trial in which participants sat and worked quietly, plasma triglyceride concentrations were decreased to the same degree (12%) by either one 30-min walk before breakfast or by three 10-min walks taken before breakfast, lunch, and the early-evening meal. A subsequent study that compared the effect of 30 min of continuous walking with that of ten 3-min bouts reported strikingly similar findings (Miyashita, Burns, and Stensel 2008). Thus, the acute decrease in postprandial triglycerides attributable to exercise appears to be determined by the associated energy expenditure rather than by its intensity or pattern. Finally, there is evidence that the triglyceride-lowering effect of a bout of exercise extends into "real-world" settings where there may be a compensatory

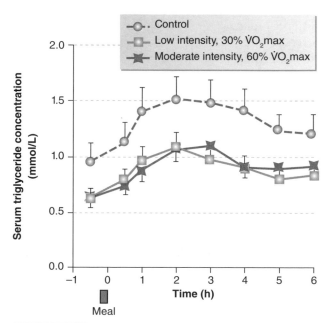

FIGURE 6.2 Influence of intensity of prior exercise on fasting and postprandial serum concentrations of triglycerides. A high-fat, mixed meal was consumed in the morning after a 12-h fast. Control—participants refrained from exercise for three days; prior low-intensity exercise—participants performed treadmill walking at 30% $\dot{V}O_2$max for 3 h the previous afternoon; prior moderate-intensity—participants performed treadmill walking at 60% $\dot{V}O_2$max for 1.5 h the previous afternoon. Values are mean and standard error for nine young adults.

Reprinted, by permission, from N.V. Tsetsonis and A.E. Hardman, 1996, "Reduction in postprandial lipidemia after walking: Influence of exercise intensity," *Medicine and Science in Sports and Exercise* 28: 1235-1242.

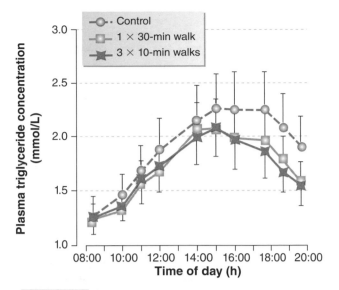

FIGURE 6.3 Influence of the pattern of brisk walking on postprandial plasma concentrations of triglycerides. On different occasions, subjects rested (control), performed one 30-min walk before breakfast, or performed three 10-min walks, one before each meal. Values are mean and standard error for 10 middle-aged participants.

Data from Murphy, Nevill, and Hardman 2000.

increase in food intake that reduces the net energy deficit attributable to the exercise bout (Farah, Malkova, and Gill 2010).

Endothelial Function

As mentioned previously, changes to lipoprotein metabolism during the hours after meal ingestion are likely to play an important role in the atherosclerotic disease process. However, the postprandial state is characterized also by nonlipid disturbances that are central to the progression of atherosclerosis.

Studies using a repeated-measures design have been employed to examine the effect of prior exercise on the postprandial impairment of endothelial function that persists for many hours after a high-fat meal. For example, using laser Doppler imaging of vasodilator responses to acetylcholine, researchers found that prior moderate exercise (90 min at 50% $\dot{V}O_2$max) improved endothelium-dependent vasodilator function by an average of 15% over an 8-h postprandial period (Gill et al. 2004). This technique assesses the cutaneous microcirculation, regarded as a robust surrogate marker of vascular function in other vascular beds more directly involved in the pathogenesis of vascular diseases. Using a different technique, Tyldum and colleagues (2009) found significant attenuation by a single session of moderate exercise of the decrement in flow-mediated vasodilation of the brachial artery associated with a high-fat meal.

Improvements to endothelial function after a single exercise session appear therefore to be clearly discernable in small studies of apparently healthy subjects. Future repeated-measures studies to examine variables of duration, intensity, and pattern are feasible, as well as studies in groups that differ in clinical or training status. Complementary work will no doubt address the mechanisms by which exercise ameliorates postprandial impairment of vascular function: One possibility is a reduction by exercise of postprandial levels of oxidative stress.

Insulin–Glucose Dynamics

Insulin resistance is the main pathology of type 2 diabetes, and skeletal muscle is the body's largest insulin-sensitive tissue (see chapter 13). It is not surprising, therefore, that substrate deficits arising from a session of exercise influence whole-body insulin–glucose dynamics.

It has been known since the 1970s that endurance-trained individuals exhibit normal or improved **glucose tolerance** to a carbohydrate challenge despite a markedly reduced insulin response. More recently, a raft of detraining studies has shown that the improved insulin action underlying these characteristics is rapidly reversed with inactivity, suggesting that much of this benefit arises from acute rather than chronic effects. For example, King and colleagues investigated the effects of a seven-day interruption to training on glucose tolerance and insulin action (King et al. 1995). Participants were middle-aged individuals with a habit of regular moderate exercise, and the measures used were derived from a simple oral glucose tolerance test (OGTT) so as to reflect normal homeostatic mechanisms that pertain to real life. During the five days before these tests, participants performed 45 min of exercise daily at about 70% $\dot{V}O_2$max. Glucose tolerance was poor immediately after the last exercise session, possibly because elevated plasma concentrations of nonesterified fatty acids inhibit glucose uptake into muscle, but had improved markedly 24 h later. This improved insulin action persisted for three days but not for five days, suggesting that the frequency of exercise needed to maintain the exercise-induced improvement in glucose tolerance is once every three days (figure 6.4). These findings were confirmed by the HERITAGE study of more than 500 previously sedentary men and women; 24 h after the last exercise bout of a 20-week training program, fasting insulin concentrations were lower than at pretraining, but this improvement was no longer evident 72 h after the last bout.

As might be expected from these findings on detraining, a single exercise session improves responses to a glucose tolerance test (and **insulin**

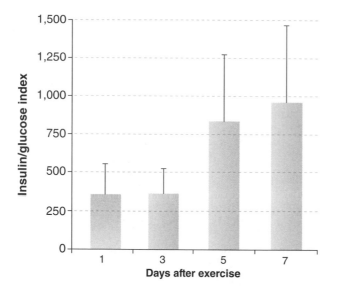

FIGURE 6.4 Time course of changes in insulin/glucose dynamics in moderately trained middle-aged people during a week without planned exercise. Values are mean and standard error for nine participants. Insulin/glucose index (a surrogate measure of insulin sensitivity) was calculated as the product of the areas under the glucose concentration and insulin concentration versus time curves above the baseline level during a 75 g glucose tolerance test.

From D.S. King et al, 1995, "Time course for exercise-induced alterations in insulin action and glucose tolerance in middle-aged people," *Journal of Applied Physiology* 78: 17-22. Used with permission.

sensitivity). For example, Young and colleagues gave untrained, normally active men this test in two conditions: (1) the morning after a day of inactivity (control) and (2) the morning after 40 min of exercise at 40% $\dot{V}O_2$max (Young, Enslin, and Kuca 1989). The plasma insulin response was 40% lower after a single bout of exercise than in the control trial (figure 6.5). Interestingly, the single bout of exercise decreased the insulin response in these untrained men to the same level as that found in participants who were engaged in regular, strenuous endurance training, illustrating the potency of the acute effect.

Studies using the "gold standard" hyperinsulinemic, euglycemic clamp technique have explained the basis of the effects of a single session of exercise on glucose and insulin concentrations. These reflect an increase in whole-body **glucose disposal rate** for a given plasma insulin concentration (in other words, an improvement in the action of insulin) that can persist for as long as two days after a session of aerobic exercise. (Note two exceptions to this finding. First, *immediately* after exercise, insulin action on glucose disposal is impaired. Second, exercise involving predominantly eccentric contractions results in a prolonged decrease in insulin action, possibly because muscle damage is incurred.)

Seventy to ninety percent of the glucose ingested during an OGTT is cleared by skeletal muscle. Therefore, the increased rates of glucose disposal

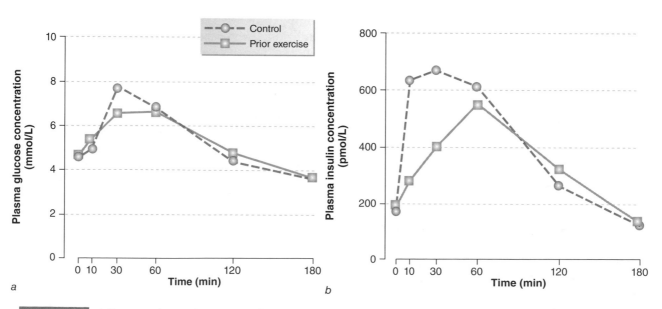

FIGURE 6.5 Influence of a single session of exercise on plasma responses of (*a*) glucose and (*b*) insulin to a glucose tolerance test. Values are means for seven untrained men. Control—at least 40 h after any exercise; prior exercise—40 min on a cycle ergometer at 40% $\dot{V}O_2$max the previous afternoon.

From J.C. Young, J. Enslin, and B. Kuca, 1989, "Exercise intensity and glucose tolerance in trained and untrained subjects," *Journal of Applied Physiology* 67: 39-43. Used with permission.

after exercise mainly reflect increased stimulation by insulin of nonoxidative glucose metabolism, primarily glycogen synthesis, in this tissue. Changes to two regulatory processes are involved: glucose transport across the muscle plasma membrane and the activity of the enzyme glycogen synthase. Glucose transport is increased because of an enhanced recruitment of GLUT4, the insulin-sensitive glucose transporter in skeletal muscle, to the muscle plasma membrane. This facilitates glycogen synthesis by increasing glucose availability. Further stimulus to glycogen synthesis arises from increased activation of glycogen synthase, a rate-limiting enzyme in this process. Both changes demonstrate the adaptive potential of muscle; these changes not only allow muscle to recover to a preexercise state but also can facilitate a degree of glycogen "supercompensation," enabling the muscle to perform better during the next exercise bout. Studies using the one-leg exercise model have shown that these are local responses in the previously exercised muscle.

Few studies have systematically examined the effects of the intensity, duration, and pattern of exercise in relation to acute effects on insulin–glucose dynamics, but limited evidence suggests that the degree of improvement in insulin sensitivity may be independent of the intensity of the exercise bout. The effects of prior low-intensity (50% $\dot{V}O_2$max)

and high-intensity (75% $\dot{V}O_2$max) exercise were compared with a control (no exercise) condition using a repeated-measures design in women with type 2 diabetes (Braun, Zimmerman, and Kretchmer 1995). The duration of the sessions was adjusted so that energy expenditure was the same in both exercise conditions. Participants' plasma glucose (figure 6.6a) and insulin (figure 6.6b) responses to a mixed meal were determined as a test of changes to insulin–glucose dynamics in circumstances in which the normal physiological interrelationship between glucose and insulin was preserved. Postprandially, plasma glucose profiles did not differ among the three conditions, but plasma insulin responses were significantly lower after both low- and high-intensity exercise compared with the control condition. These changes reflected the fact that the rate of glucose disposal per unit of plasma insulin (measured on a separate occasion during infusion at fixed rates of glucose and insulin) was enhanced by almost exactly the same degree after prior low- or high-intensity exercise compared with the control condition.

The paucity of data on the relative importance of intensity, duration, and pattern shows the need for applied studies of these topics. And complementary studies are needed to explore the nature and time course of the cellular mechanisms behind exercise-induced changes in insulin action.

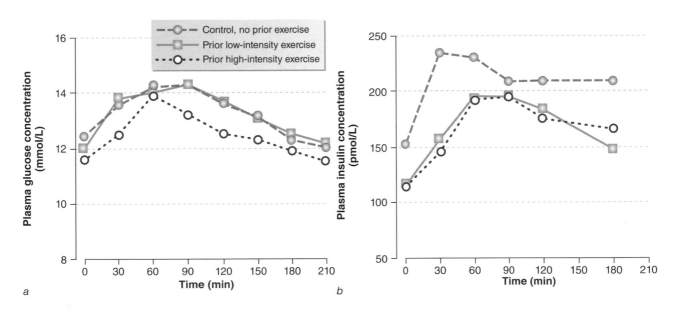

FIGURE 6.6 Influence of the intensity of prior exercise on (a) plasma glucose and (b) insulin responses to a mixed meal in eight women with type 2 diabetes. Control—no prior exercise; low intensity—treadmill walking at 50% $\dot{V}O_2$max on each of the preceding two days; high intensity—treadmill walking at 75% $\dot{V}O_2$max on each of the preceding two days. The duration of walking was adjusted so that the energy expenditure was the same for low- and high-intensity trials. Values are means for eight women.

From B. Braun, M.B. Zimmerman, and N. Kretchmer, 1995, "Effects of exercise intensity on insulin sensitivity in women with non-insulin-dependent diabetes mellitus," *Journal of Applied Physiology* 78: 300-306. Used with permission.

In summary, the nature of regulation of insulin-stimulated glucose disposal and glycogen synthesis in skeletal muscle is not altered by prior exercise; regulation is simply more potently activated by insulin. These improvements—equivalent to those achieved through chronic pharmacological intervention—are restricted to exercised muscle and will therefore be maximized if exercise is undertaken with the body's large muscles. Such improvements have been shown in patients with type 2 diabetes and in individuals at high risk for this disease, as well as in healthy individuals. Frequent exercise involving predominantly concentric, as opposed to eccentric, contraction is indicated. Available but limited evidence indicates that the benefits will be optimal after an exercise session that expends a large amount of energy.

Blood Pressure

Systolic blood pressure rises during dynamic exercise and returns to normal with the cessation of exercise. There is often a transient postexercise pressure "undershoot" caused by the pooling of blood in dilated, previously exercised muscle beds, which can lead to light-headedness, particularly if the individual stands still. Baroreceptor reflexes counter this undershoot to reestablish homeostasis within 10 min. Recent studies of the prolonged postexercise period have shown that resting blood pressure may be decreased for several hours following a single session of exercise. This acute effect may therefore be an important determinant of blood pressure in people with a habit of frequent exercise.

Postexercise hypotension is more consistently reported in individuals with some degree of hypertension than in those with blood pressure in the normal range. In studies that have shown a decline in blood pressure following a session of exercise in people without clinical hypertension, the decrease has typically been around 8/9 mmHg (systolic/diastolic). Greater decreases, in the range of 10-14/7-9 mmHg, have typically been observed in hypertensive individuals. Studies in controlled laboratory settings have shown that postexercise hypotension persists for at least 3 h, but ambulatory blood pressure monitoring provides a much longer observation period. For example, researchers obtained two 24-h recordings, one immediately preceded by an early-morning exercise session (45 min of treadmill walking at about 70% $\dot{V}O_2$max) and another on a control day without exercise (Taylor-Tolbert et al. 2000). The participants were 11 obese, sedentary men aged around 60 with mild to moderate hypertension (140-179/90-109

mmHg). Ambulatory recording of blood pressure started around 9 a.m. and continued until the next morning (figure 6.7). Systolic blood pressure (figure 6.7a) was lower after exercise than on the control day by between 6 mmHg and 13 mmHg for the first 16 h and by an average of 7.4 mmHg over the entire 24-h period. Diastolic blood pressure (figure 6.7b) was lower after exercise than in the control trial by around 5 mmHg for 12 of the first 16 h and by an average of 3.6 mmHg over the 24-h period. Thus, a single session of dynamic exercise leads to substantial and consistent decreases in arterial blood pressure that are sustained over many hours.

Postexercise hypotension also appears to persist for hours when, after the exercise session, participants engage in mild activity designed to simulate activities of daily living. This last finding strengthens the argument for the therapeutic utility of exercise in the management of the large numbers of individuals who have "high-normal" blood pressure.

What features of an exercise session determine the magnitude of the ensuing hypotension? Most but not all direct comparisons have led to the conclusion that postexercise hypotension is independent of exercise intensity. However, few of these studies used 24-h monitoring, and to the author's knowledge, none controlled for the energy expended at different intensities. Few data are available to provide evidence of the influence of exercise duration or the muscle mass engaged in exercise. Moderate-intensity exercise for as little as 10 min has, however, been shown to decrease postexercise blood pressure; it also has been reported that ten 3-min bouts of brisk walking during the day elicited a similar decrease in systolic blood pressure (about 7 mmHg) as one 30-min bout, compared with values on a sedentary control day (Miyashita, Burns, and Stensel 2008).

The etiology of postexercise hypotension is poorly described and probably multifactorial. Postexercise, indexes of vascular resistance are decreased below preexercise values; however, studies of potential mechanisms are contradictory, and no conclusion can be drawn from these. There are racial differences in participants' propensity to exhibit postexercise hypotension; it is not observed in African American women, in whom hypertension is more prevalent and associated with greater comorbidities. The variability in response is probably partly explained by genetic differences in polymorphisms at critical loci involved in systems that affect blood pressure. Future research on these topics may enable clinicians to refine exercise prescriptions for hypertensive persons or those with high-normal blood pressure. More information on aspects of the dose–response relationship is also needed.

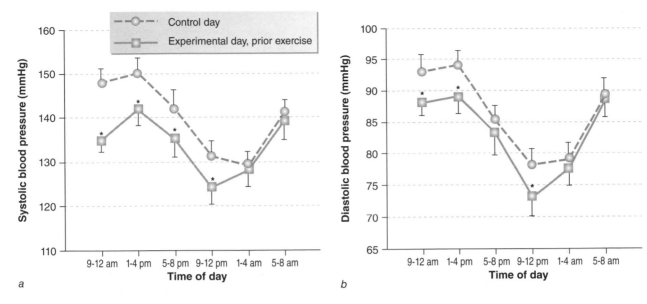

FIGURE 6.7 Influence of prior exercise on ambulatory measurements of (a) systolic and (b) diastolic blood pressures. Experimental day, prior exercise—participants performed 45 min of treadmill walking at 70% $\dot{V}O_2$max immediately before measurements commenced; control—no exercise. Values are mean and standard error for 11 sedentary men with mild to moderate hypertension.

*Difference between the two recordings was statistically significant ($p < .05$).

Reprinted from *American Journal of Hypertension*, Vol. 13, N.S. Taylor-Tolbert et al., "Ambulatory blood pressure after acute exercise in older men with essential hypertension," pp. 44-51, Copyright 2000, with permission from Elsevier.

Hematological Changes

The formation of an intravascular clot is a key event in the pathophysiology of acute cardiovascular events such as heart attack and stroke. The likelihood that this will happen depends on the balance between several processes, namely platelet function, coagulation, and fibrinolysis. Blood flow and the sheer stress it exerts also play an integral role. Acute exercise-induced changes to these processes are therefore of clinical importance. Alterations in the constituents of the coagulation and fibrinolytic pathways, as well as platelet morphology and physiology, have been examined experimentally to determine the hemostatic changes following acute exercise (Thrall et al. 2007). Research on changes in factors influencing

Thrombogenesis

Hemoconcentration arises through fluid shifts between the plasma and interstitial fluid compartments. The concomitant concentrating of the red blood cells in the vasculature leads to unfavorable changes in blood viscosity, promoting thrombogenesis.

Platelets play a key role in the immediate response to vascular injury. A platelet plug is formed when these cell fragments are activated (become "sticky") and start to build up and adhere to the site of damage.

Blood coagulation is the process by which fibrin strands create a mesh that binds together blood components, including platelets, to form a fibrin clot. Two pathways may lead to the formation of a clot; in each, a cascade of reactions occurs by which one activated factor activates another. Fibrinogen, the substrate ultimately converted to a fibrin clot, plays a central role in the final phase of this cascade and is a key determinant of viscosity. **Fibrinolysis** is the dissolution of fibrin (and therefore of a clot). A key stage in this process is the conversion of inactive plasminogen into active plasmin, an enzyme that digests fibrin and thus brings about the breakup of the clot.

blood flow have, in the main, focused on hemoconcentration, determined by changes in hematocrit, plasma volume, or both. An acute bout of exercise leads to an increase in hematocrit of between 1% and 10% (depending on exercise duration and intensity, as well as environmental factors), associated with an increase in blood and plasma viscosity. Robust studies of markers for coagulation and fibrinolysis take hemoconcentration into account when reporting biochemical markers, but many studies have methodological weaknesses in this regard.

What is the evidence concerning factors other than hemoconcentration that lead to hemostatic changes after exercise? There is hardly any evidence specifically for low-intensity exercise. Overall, findings from studies of moderate-intensity exercise have tended to report an improved thrombotic profile with the development of a hyperfibrinolytic state (often in the absence of increases in coagulatory markers), in combination with a reduction in platelet aggregation and adhesion. Most of this research has been conducted on patient populations.

By contrast, much of the evidence relating to high-intensity exercise derives from studies of healthy, fit volunteers. Strenuous exercise has consistently been associated with increases in platelet count in the region of 10% to 40%, but results for platelet function are equivocal. Some studies have shown an increase after exercise in platelet activation or aggregation that is more marked in sedentary than in regularly active individuals. Platelet activation is enhanced by infusion of epinephrine or norepinephrine, so the interaction with training status may be explained by the higher catecholamine response to exercise in sedentary participants.

Most markers of coagulation increase with acute high-intensity exercise. For example, activated partial thromboplastin time (an overall indicator of the intrinsic pathway of blood coagulation) is shortened following exercise. This effect has been demonstrated for different exercise modes and in both men and women. Several other markers of coagulation known to be associated with cardiovascular disease, including fibrinogen, are also increased after exercise. One of these, factor VIII (a component of the final common path of the coagulation cascade), can remain elevated for several hours after a strenuous session. Exercise-induced increases in this and other components of the cascade are largely dependent on exercise intensity, with high-intensity exercise leading to the greatest increase in coagulability. Thrombin formation has been reported to increase following triathlons, high-intensity running, and cycling. Researchers have examined the effect of an exercise bout on the fibrinolytic system by measuring (1) the enzyme that catalyzes the conversion of plasminogen to plasmin and (2) the main circulating inhibitor (plasminogen activator inhibitor-1). Changes in these markers indicate that fibrinolysis increases acutely with exercise and that the magnitude of this response is dependent on the intensity and, to a lesser extent, on the duration of exercise. Values typically return to preexercise levels within 24 h.

Thus, in concert with an increase in blood coagulability, the activity of the fibrinolytic system that opposes coagulability is also enhanced. Yet because the majority of studies have examined only one aspect of thrombogenesis, the overall effect of a session of exercise on this multifactorial process is difficult to evaluate. Thrombogenesis itself has, however, been examined in sedentary men using an experimental model intended to measure the net effect of a session of moderate (50% $\dot{V}O_2$max) or hard (70% $\dot{V}O_2$max) exercise (Cadroy et al. 2002). Platelet thrombus formation and fibrin deposition were determined as arterial blood interacted with collagen (a molecule present in atherosclerotic plaques and primarily responsible for thrombus formation in vivo). Blood flow conditions mimicked those in moderately stenosed small arteries. Moderate exercise did not affect arterial thrombus formation. In contrast, platelet thrombus formation was increased by 20% after 30 min of high-intensity exercise. These findings suggest that the net effect of a session of strenuous—but not moderate—exercise is probably increased risk for arterial thrombogenesis, at least in sedentary men. This proposition is consistent with epidemiological findings that heavy exertion is a potent trigger for myocardial infarction in people unaccustomed to such exertion (discussed in chapter 18).

Immune Function and Inflammation

Since the mid-1980s, researchers have examined the effects of a session of exercise on immune function and related signaling molecules. Stimulated by anecdotal reports that athletes engaged in intense training regimens exhibited enhanced susceptibility to respiratory infections, research initially focused on exercise of high intensity.

Cells of the Immune System

According to studies based on self-reports, there is a 110% to 500% increase in the risk of picking up an upper respiratory tract infection during the weeks following an event such as the marathon. This has been ascribed to the fact that various immune cell functions are temporarily impaired following acute bouts of prolonged, continuous, high-intensity exercise (Gleeson 2007).

A substantial increase in the number of circulating leukocytes occurs following high-intensity exercise, alongside increases in the plasma concentration of various substances that are known to influence leukocyte function, including inflammatory cytokines. Various functions of these immune cells, for example macrophage antiviral function, are impaired for several hours after a prolonged session of high-intensity exercise, and neutrophils may be in an unresponsive state for some time afterward. Acute vigorous exercise temporarily increases the number of circulating natural killer cells that seek out and destroy virus-infected cells. However, large numbers of these cells quickly exit the circulation so that, during the hours after intense exercise of long duration, their concentration declines to 25% to 40% below preexercise values (figure 6.8), as does their **cytolytic** activity. These and other changes during early recovery from intense exercise appear to weaken the potential immune response to pathogens and may provide a "window of opportunity," lasting between 3 and 72 h depending on the parameter measured, during which susceptibility to contracting an infection is increased. Evidence on the effects of exercise of moderate intensity or duration is less consistent than that for prolonged, high-intensity exercise; but in general, it shows that some functions of neutrophils—for instance phagocytosis—may be enhanced whereas others are unaffected.

Inflammatory Response

In response to acute infection, sepsis, or trauma, activated macrophages and other leukocytes release cytokines and cytokine inhibitors at a site of inflammation. This local inflammatory response is accompanied by a systemic response known as the "acute-phase response," mediated largely by so-called proinflammatory cytokines that stimulate the liver to release acute-phase proteins, such as C-reactive protein (CRP). During recent years, it has become clear that inflammatory mechanisms play a key role in the pathogenesis of several chronic diseases, including cardiovascular disease, some cancers, and type 2 diabetes. At least in elderly people, markers in the blood for systemic inflammation such as CRP are strong, consistent, and independent predictors of all-cause mortality and cardiovascular disease morbidity. Effects of exercise on inflammatory markers are therefore of interest.

An acute bout of exercise is accompanied by responses that are remarkably similar in many respects to those that accrue from pathological states. A large number of studies show that exercise induces the release of a cascade of proinflammatory cytokines, particularly interleukin-6 (IL-6). After a marathon race, for example, the plasma concentration of IL-6 has been reported to increase 50-fold. Data from the Copenhagen Marathon between 1995 and 1997 show a correlation between the intensity of exercise and the increase in IL-6 (Febbraio and

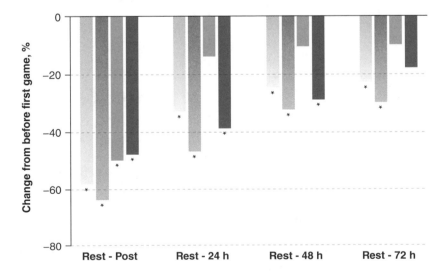

FIGURE 6.8 Decrease in numbers of natural killer cells from before to after two soccer games separated by 20 h. For each time period, bars show values for four different subpopulations of natural killer cells. Values are means for 10 elite Swedish soccer players aged between 16 and 19.

*Difference from before to after the soccer games statistically significant ($p < .01$).

Adapted from Malm, Ekblom, and Ekblom 2004.

Pedersen 2002). During the postexercise period, the level of circulating IL-6 declines. C-reactive protein can increase by as much as 2,000% after a marathon and remain elevated at this level for at least 24 h. Collectively, studies of strenuous endurance events involving cycling, running, and canoeing suggest that the extent of the acute-phase response to exercise is proportional to the amount of activity, the muscle mass involved, and the type of contraction.

Although an acute bout of exercise is accompanied by cytokine responses that are similar in many respects to those induced by infection, there are some important differences; specifically, two classical proinflammatory cytokines, tumor necrosis factor-α (TNF-α) and IL-1β, in general do not increase after exercise. Moreover, although a session of strenuous exercise elicits an acute-phase response, it also produces an acute increase in various anti-inflammatory mediators, a response that develops later than the increase in IL-6. Thus, rather like the opposing changes to coagulation and fibrinolysis described in the previous section, evidence suggests a parallel, "protective" anti-inflammatory counterregulation that opposes the effects of proinflammatory cytokines (Kasapis and Thompson 2005).

It was once thought that muscle injury was the primary stimulus for the IL-6 response after heavy exercise. It is now known, however, that IL-6 is produced locally in skeletal muscle in response to contraction, independently of muscle damage. A subsequent delayed release of IL-6 has a different etiology, reflecting a repair response to muscle damage. As might be expected, therefore, the magnitude by which plasma IL-6 increases is related to exercise duration, intensity, and the muscle mass involved. The time scale of the increase in IL-6, among other features of the evidence, suggests that this cytokine may be a metabolic regulator, perhaps signaling low muscle glycogen. Indeed, the discovery of contracting muscle as a cytokine-producing organ shows that muscle can act as an endocrine organ, producing and releasing "myokines" that may influence metabolism and modify cytokine production in other tissues and organs.

Physical activity, inflammation, and immunity are tightly linked in a complex manner. As already explained, prolonged, high-intensity exercise is associated with a short-term increase in the risk of infection and an increase in markers for systemic inflammation. On the other hand, a session of moderate exercise may stimulate immune function changes that could be beneficial if such exercise is frequent and regular. Convincing epidemiological evidence shows that individuals who regularly do ~2 h of moderate physical activity per day exhibit a 29% lower risk

of picking up an upper respiratory tract infection than their sedentary peers. To study whether acute exercise induces a true anti-inflammatory response, a group from Copenhagen created a model of "low-grade inflammation." They injected two groups of volunteers with the **endotoxin** to bacterium *E. coli* (Mathur and Pedersen 2008). In resting subjects, this led to a threefold increase in TNF-α (the potent proinflammatory cytokine). By contrast, in subjects who received the endotoxin after exercise, the TNF-α response was totally blunted. Moreover, the effect of exercise could be mimicked by infusion of IL-6, suggesting that IL-6 may be involved in mediating the anti-inflammatory effects of exercise.

Much remains to be understood about the way in which physical activity influences inflammatory processes. In particular, there is a dearth of evidence from studies of healthy individuals engaging in moderate-intensity activity. It is possible that the beneficial effects of regular exercise on diseases such as cardiovascular disease and type 2 diabetes may to some extent be ascribed to the anti-inflammatory response elicited by each bout of exercise, but such a link is not established at present.

Responses Related to Energy Balance

An individual's overall level of physical activity exerts a major influence on his energy balance and hence, over time, weight regulation. In addition to the energy expended during activity, acute effects apparent during the hours after each individual session of activity are important in this regard. Findings on two effects—fat oxidation and appetite—are described in this section.

Fat Oxidation

Carbohydrate and protein stores are closely regulated through adjustment of oxidation to intake. It follows that fluctuation in fat balance is the primary determinant of day-to-day fluctuations in energy balance and that fat balance determines energy balance. The evidence summarized next shows that repeated effects of frequent, regular sessions of activity probably help to maintain neutral or negative fat balance.

Fat oxidation is enhanced for some hours after an exercise session compared with a control condition without prior exercise. This is the case for moderate amounts and intensities of exercise, as well as for exhaustive exercise. For example, figure 6.9 shows the influence of just 30 min of brisk walking on the respiratory exchange ratio throughout a normal day.

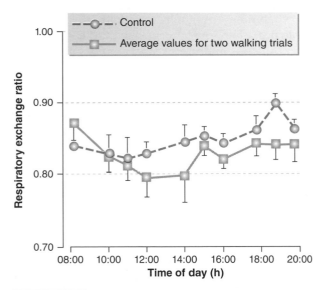

FIGURE 6.9 Influence of 30 min of brisk walking on the respiratory exchange ratio throughout an ordinary day. Values are mean and standard error for 10 middle-aged participants. For details of protocol, see caption for figure 6.3. Average respiratory exchange ratio values for the two walking trials (one 30-min walk or three 10-min walks, treated as a single condition) were significantly lower than control values.

Data from Murphy, Nevill, and Hardman 2000.

(This ratio reflects whole-body substrate oxidation, low values indicating predominantly oxidation of fat.) The protocol for the study (Murphy, Nevill, and Hardman 2000) was described in the earlier section on lipids and lipoproteins. During a daylong observation period, participants consumed breakfast, lunch, and an early-evening meal. In the control condition, they did minimal activity; during the exercise conditions, they walked briskly for 30 min—on one occasion in three 10-min sessions and on the other occasion in one 30-min session. (In figure 6.9, the two exercise conditions are treated as one because the effect on the respiratory exchange ratio was independent of the pattern of walking.) Researchers estimated that, compared with findings for the control trial, an additional 5 g of fat was oxidized over the 11-h observation period, decreasing fat storage by 4% to 5%.

Thus, an exercise session shifts the pattern of substrate utilization at rest toward fat oxidation. The effect is independent of the intensity of exercise, if the energy expenditure is controlled for, and can persist for up to 24 h after an exercise session. Moreover, an increase in postexercise oxidation of fat (compared with a sedentary control situation) is apparent even if the increase in energy expenditure associated with exercise is compensated by an increase in dietary intake (Burton et al. 2008).

Appetite Regulation

What is the effect of a session of exercise on energy intake? Findings of studies that have addressed this question are rather consistent; a session of exercise usually stimulates an increase in energy intake, but this increase only partially compensates for the additional energy expenditure of exercise, leading to a short-term energy deficit.

Methodologically sound studies on this topic over periods of a few days are available. They are discussed here because imprecision reduces the usefulness of studies of the effect of a single exercise session. For example, Stubbs and colleagues observed the effects in women on energy balance and sensations of hunger and appetite of two 40-min sessions of cycle ergometer exercise on each of seven days (Stubbs et al. 2002). Energy intake was higher on the exercise days than on the preceding control days, but the increment in intake was equivalent to only one-third of the additional expenditure, leading to an energy deficit. This finding was consistent with the small increase in hunger reported by the participants.

The effect on energy intake of the opposite intervention, that is, decreasing physical activity to within the sedentary range, has also been studied. In one carefully designed study, in which energy expenditure was measured twice in seven-day protocols in a whole-body indirect calorimeter, energy expenditure was reduced from an average of 12.8 MJ/day (3,050 kcal/day) to 9.7 MJ/day (2,310 kcal/day) (Stubbs et al. 2004). These levels were selected as the high and low ends of the sedentary range of activity in Western societies. The difference was achieved through varying the amount of cycle ergometer exercise. When activity was restricted, there was no compensatory decrease in energy intake so that participants stored more energy than while on the more active regimen. In this study, the greater positive energy balance in the restricted activity condition was largely accounted for by greater fat storage.

This is a difficult field, and available literature is not extensive. Moreover, for methodological reasons, research is largely restricted to cycle ergometry and inevitably confounded by effects on the behavior of human volunteers. Moreover, lean individuals— more commonly studied—may behave differently than those who are overweight or obese.

The finding that a session of exercise creates a short-term energy deficit has, in recent years, stimulated research into a potential role for hormones known to influence food intake and energy expenditure. To date, however, no clear picture has emerged. At least for short-term exercise at a moderate intensity, the consensus is that plasma concentrations of neither leptin (a satiety signal) nor ghrelin (which stimulates appetite) change significantly with acute exercise. One limitation of studies of ghrelin is that they have typically measured only plasma total ghrelin, despite the fact that only the acylated form has a stimulating effect on appetite. In a recent study looking at the impact of 1 h of running, acylated ghrelin was lower during the hours after exercise than in the control condition, alongside a reduction in subjective hunger (Broom et al. 2007). Thus, acylated ghrelin may respond to acute exercise in a different way from total ghrelin, and further studies are needed to clarify this issue.

Augmentation of Acute Effects by Training

In a now classic article, published in 1994, Haskell coined the term "last bout effect," proposing that some of the health benefits of activity—for example, lowering of blood pressure, improved plasma lipo-

protein profiles—might be attributable more to acute biological changes following each bout of activity than to a true training response (Haskell 1994). Does this mean that performing repeated bouts of exercise over weeks or months, progressing by increasing frequency, duration, or intensity—in other words, obtaining a training response—has no synergistic effects on such benefits?

First, if a specific effect lasts for more than 24 h, this effect may be enhanced when exercise is undertaken daily as occasional exercisers progress toward a habit of regular exercise. This is shown schematically in figure 6.10a. As an example, in men with hypertriglyceridemia, fasting plasma triglyceride concentration is progressively decreased over a five-day period when exercise is performed daily. However, the degree of this augmentation of an acute effect must inevitably decrease over time as values for the variable in question approach the end of the physiological range.

As originally suggested by Haskell, the most important way in which training augments acute

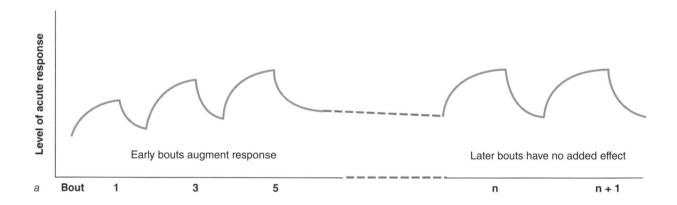

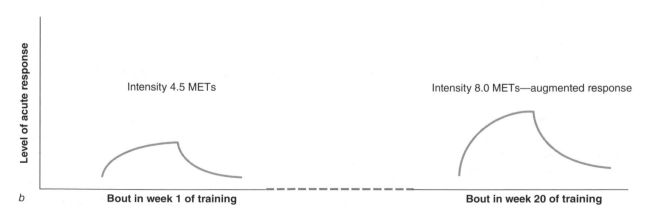

FIGURE 6.10 Schematic representation of potential ways in which training may enhance acute effects of a single exercise session. (a) Repeated bouts of exercise produce increasingly greater acute effects. (b) Acute response is augmented by training because exercise bouts are at a higher intensity and thus higher energy expenditure.

Reprinted, by permission, from W.L. Haskell, 1994, "Health consequences of physical activity: Understanding and challenges regarding dose-response," *Medicine and Science in Sports and Exercise* 26: 649-660.

effects is by enabling performance of more intense, longer, or more frequent exercise. This is depicted schematically in figure 6.10b. The total energy expenditure of an exercise session is an important determinant of the level of the ensuing beneficial effects on several (speculatively many or most) health-related outcomes, as discussed previously. Thus, training enhances acute effects by enabling a greater overall level of exercise energy expenditure.

The data in figure 6.1 illustrate the augmentation by training of an acute effect for one health-related outcome—postprandial lipemia. The decrease in lipemia attributable to a prior 90-min session of treadmill walking was twice as great in trained as in untrained women. Moreover, prior exercise greatly enhanced the difference in lipemic response between these two groups. A likely explanation is that, although all the women walked at the same relative intensity (60% $\dot{V}O_2$max), the trained women expended 50% more energy because of their higher $\dot{V}O_2$max values.

Summary

The acute effects of individual bouts of activity are varied and, if sessions are repeated frequently and regularly, have the potential to make an important contribution to health. Effects on metabolism, cardiovascular functions, and inflammation in particular are relevant to the modern epidemics of coronary heart disease, type 2 diabetes, and obesity. Much remains, however, to be learned about the extent and persistence of acute effects, the mechanisms involved, and their interaction with the intensity and duration of activity.

Key Concepts

acute effects—Short-term responses to an individual session of exercise or activity linked to the body's recovery from that session.

blood coagulation—Process by which fibrin strands create a mesh that binds blood components together to form a fibrin clot.

chronic effects—Adaptations to training (exercise progressing in intensity or frequency) that are acquired over weeks or months and that persist for days or weeks when such a regimen is interrupted.

cytolytic—Referring to destruction or degeneration of cells.

endotoxin—A toxin that forms an integral part of the cell wall of certain bacteria and is released only upon destruction of the bacterial cell.

fibrinolysis—Dissolution of fibrin (and therefore of a clot).

glucose disposal rate or **metabolic clearance rate for glucose**—Rate at which glucose is removed from the circulation.

glucose tolerance—Extent of the increase in blood glucose concentration when glucose is ingested.

insulin sensitivity—Responsiveness to the stimulation of glucose uptake by insulin.

lipoproteins—Particles with a highly hydrophobic lipid core and a relatively hydrophilic outer surface.

postprandial—During the hours after a meal.

Study Questions

1. Why is the effect of exercise on plasma triglycerides particularly clear when subjects are in the postprandial state? Summarize the evidence showing that the energy expenditure of an exercise session mainly determines the extent of the subsequent decrease in postprandial triglycerides.

2. What is meant by the term *glucose tolerance*? Explain why prior exercise improves glucose tolerance.

3. Explain what is meant by the term *postexercise hypotension*. Comment on the extent of this effect and its duration, referring to relevant evidence.

4. Why are the acute effects of exercise on hemostatic factors often described as complex? Under what circumstances is an exercise session likely to cause an overall increase in blood coagulability?

5. Individuals who have participated in strenuous endurance events often experience an increased incidence of upper respiratory tract infections during the weeks that follow. Why might this be?

6. By what means may an exercise session influence energy balance?

7. Explain how training and the acute effects of individual sessions of exercise act synergistically to provide optimal benefits for health.

8. The chapter discusses the argument that exercise benefits health by reducing the systemic inflammation characteristic of several chronic diseases. Explain why you did or did not find this argument convincing. (For further information on this topic, see Bruunsgaard 2005, available at www.jleukbio.org/cgi/content/full/78/4/819)

References

Braun, B., M.B. Zimmermann, and N. Kretchmer. 1995. Effects of exercise intensity on insulin sensitivity in women with non-insulin-dependent diabetes mellitus. *Journal Applied Physiology* 78:300-306.

Broom, D.R., D.J. Stensel, N.C. Bishop, S.F. Burns, and M. Miyashita. 2007. Exercise-induced suppression of acylated ghrelin in humans. *Journal of Applied Physiology* 102:2165-2171.

Bruunsgaard, H. 2005. Physical activity and modulation of systemic low-level inflammation. *Journal of Leukocyte Biology* 78:819-835.

Burton, F.L., D. Malkova, M.J. Caslake, and J.M. Gill. 2008. Energy replacement attenuates the effects of prior moderate exercise on postprandial metabolism in overweight/obese men. *International Journal of Obesity* 32:481-489.

Cadroy, Y., F. Pillard, K.S. Sakariassen, C. Thalamas, B. Boneu, and D. Riviere. 2002. Strenuous but not moderate exercise increases the thrombotic tendency in healthy sedentary male volunteers. *Journal of Applied Physiology* 93:829-833.

Farah, N.M., D. Malkova, and J.M. Gill. 2010. Effects of exercise on postprandial responses to *ad libitum* feeding in overweight men. *Medicine and Science in Sports and Exercise* 42: 2015-2022.

Febbraio, M.A., and B.K. Pedersen. 2002. Muscle-derived interleukin-6: Mechanisms for activation and possible biological roles. *FASEB Journal* 16:1335-1347.

Gill, J.M.R., A. Al-Mamari, W.R. Ferrell, S.J. Cleland, C.J. Packard, N. Sattar, J.R. Petrie, and M.J. Caslake. 2004. Effects of prior moderate exercise on postprandial metabolism and vascular function in lean and centrally obese men. *Journal of the American College of Cardiology* 44:2375-2382.

Gleeson, M. 2007. Immune function in sport and exercise. *Journal of Applied Physiology* 103:693-699.

Haskell, W.L. 1994. Health consequences of physical activity: Understanding and challenges regarding dose-response. *Medicine and Science in Sports and Exercise* 26:649-660.

Haskell, W.L., I.-M. Lee, R.R. Pate, K.E. Powell, S.N. Blair, B.A. Franklin, C.A. Macera, G.W. Heath, P.D. Thompson, and A. Bauman. 2007. Physical activity and public health: Updated recommendation for adults from the American College of Sports Medicine and the American Heart Association. *Medicine and Science in Sports and Exercise* 39:1423-1434.

Kasapis, C., and P.D. Thompson. 2005. The effects of physical activity on serum C-reactive protein and inflammatory markers. *Journal of the American College of Cardiology* 45:1563-1569.

King, D.S., R.J. Baldus, R.L. Sharp, L.D. Kesl, T.L. Feltmeyer, and M.S. Riddle. 1995. Time course for exercise-induced alterations in insulin action and glucose tolerance in middle-aged people. *Journal Applied Physiology* 78:17-22.

Mathur, N., and B.K. Pedersen. 2008. Exercise as a mean to control low-grade systemic inflammation. *Mediators of Inflammation* 2008:109502.

Miyashita, M., S.F. Burns, and D.J. Stensel. 2008. Accumulating short bouts of brisk walking reduces postprandial plasma triacylglycerol concentrations and resting blood pressure in healthy young men. *American Journal of Clinical Nutrition* 88:1225-1231.

Murphy, M.H., A.M. Nevill, and A.E. Hardman. 2000. Different patterns of brisk walking are equally effective in decreasing postprandial lipaemia. *International Journal of Obesity* 24:1303-1309.

Stubbs, J.R., D.A. Hughes, A.M. Johnstone, G.W. Horgan, N. King, and J.E. Blundell. 2004. A decrease in physical activity affects appetite, energy, and nutrient balance in lean men feeding ad libitum. *International Journal of Obesity* 79:62-69.

Stubbs, R.J., A. Sepp, D.A. Hughes, A.M. Johnstone, N. King, G. Horgan, and J.E. Blundell. 2002. The effect of graded levels of exercise on energy intake and balance in free-living women. *International Journal of Obesity* 26:866-869.

Taylor-Tolbert, N., D. Dengel, M.D. Brown, S.D. McCole, R.E. Pratley, M.I. Ferrell, and J.M. Hagberg. 2000. Ambulatory blood pressure after acute exercise in older men with essential hypertension. *American Journal of Hypertension* 13:44-51.

Thrall, G., D. Lane, D. Carroll, and G.Y. Lip. 2007. A systematic review of the effects of acute psychological stress and physical activity on haemorheology, coagulation, fibrinolysis and platelet reactivity: Implications for the pathogenesis of acute coronary syndromes. *Thrombosis Research* 120:819-847.

Tsetsonis, N.V., and A.E. Hardman. 1996. Reduction in postprandial lipidemia after walking: Influence of exercise intensity. *Medicine and Science in Sports and Exercise* 28:1235-1242.

Tsetsonis, N.V., A.E. Hardman, and S.S. Mastana. 1997. Acute effects of exercise on postprandial lipidemia: A comparative study in trained and untrained middle-aged women. *American Journal Clinical Nutrition* 65:525-533.

Tyldum, G.A., I.E. Schjerve, A.E. Tjønna, I. Kirkeby-Garstad, T.O. Stølen, R.S. Richardson, and U. Wisløff. 2009. Endothelial dysfunction induced by post-prandial lipemia: Complete protection afforded by high-intensity aerobic interval exercise. *Journal of the American College of Cardiology* 53:200-206.

Young, J.C., J. Enslin, and B. Kuca. 1989. Exercise intensity and glucose tolerance in trained and untrained subjects. *Journal Applied Physiology* 67:39-43.

7

Hormonal Response to Regular Physical Activity

Peter A. Farrell, PhD

©Photodisc

The endocrine system regulates the physiological and psychological functioning of the human organism. Hormones share this job with the autonomic nervous system, and both hormonal and neural controls are subject to the basic structural and functional constraints that abound in the body. In response to acute physical activity, the concentrations of most hormones in the plasma increase, whereas some do not change and others decrease. This response to physical activity changes as a person progresses from being habitually sedentary to becoming physically active, and those adaptations allow the organism to function at a higher capacity. One limitation to summarizing the literature on endocrine-related health aspects of physical activity is that the predominant objective of most previous investigations in this area has been to describe adaptations to chronic physical activity without reference to the potential positive health benefits of such adaptations. One exception to this limitation relates to the hormone insulin, which has received enormous attention. Regular physical activity results in improved insulin function, and maintaining low concentrations of this hormone has significant health benefits. This chapter provides speculation and comments that should stimulate more research in this area, as the adaptations thus far reported can be interpreted only as potentially positive for human health.

Defining Hormones

Defining what is and what is not a hormone used to be rather simple. Hormones were considered to be endogenous biochemicals, secreted from ductless glands, that traveled through the interstitial fluid to the blood and then to all tissues. Hormones affected only some tissues because those tissues possessed a receptor specific for the given molecule. Hormones were classified as peptides (made up of many amino acids linked by peptide bonds, also classified as water soluble), amines (structures derived from tyrosine, classified as lipid soluble), or steroids (basic biochemical structures similar to cholesterol, classified as lipid soluble). We now find the definition of hormones to be much more complicated. No longer does a ductless gland need to be considered because we know that fat tissue secretes leptin, which circulates and modifies satiety. Fat also secretes adiponectin (which partially regulates both lipid and carbohydrate metabolism and is also found in the brain), tumor necrosis factor, interleukin-6 (IL-6), and resistin. Ghrelin is secreted principally by the stomach, where it stimulates acid secretion and gastric motility, but it also travels to

the brain to stimulate appetite and growth hormone (GH) secretion under basal conditions. Individual cells such as macrophages secrete cytokines, which travel throughout the body to affect many physiological functions such as immune responses. The picture of hormones traveling to other tissues to affect function is further complicated by the fact that many biochemicals that are considered classical hormones are made in one cell and affect either the secreting cell itself (**autocrine** function) or cells in close proximity (**paracrine** function).

A good example of this complexity is insulin-like growth factor 1 (IGF1), which has classic endocrine functions. It is secreted from the liver in response to GH, travels bound to specific binding proteins, and then attaches to IGF1 receptors in muscle and other tissues to stimulate growth. However, IGF1 is also made in muscle cells and affects the cell in which it is made or cells close by. A paracrine function of IGF1 may consist of stimulating adjacent satellite cells to enter into the cell cycle and then to proliferate as new myoblasts and then myofibers. From an autocrine standpoint, IGF1 stimulates transcription of muscle proteins that make up that muscle cell's structure. Proof that these functions, autocrine versus hormonal, can be completely separate lies in the fact that marked increases and decreases in muscle IGF1 after exercise can occur with no change in circulating IGF1. Thus, it is clear that our understanding—and thus our definition—of hormones changes constantly.

Hormone Activity

Another view that must be modified is that circulating concentrations of a hormone accurately reflect "functional activity" of that hormone. Rather, the concentration of a hormone reflects the net result of secretion, the **volume of distribution** (the volume of the body into which the hormone is distributed), the quantity of hormone attached to receptors or binding proteins, hemoconcentration, the rate of blood flow through the endocrine gland, and hormone catabolism. Some have argued that elevations of hormones consequent to physical activity merely reflect decreases in blood volume attributable to large amounts of fluid leaving the extracellular compartment. Yet hemoconcentration cannot account for all of the increase in plasma concentrations observed during physical activity; the usual estimate is that hemoconcentration accounts for no more than 10% of the elevation in a hormone concentration during exercise. As discussed later, some hormone concentrations increase several hundred percent during strenuous prolonged exercise.

In addition to these characteristics, which are external to the target cell, we now know that the status of intracellular pathways can modify the actions of a certain amount of hormone that has attached to a receptor. Intracellular events (biochemical signaling pathways) can make the attachment of a hormone to its receptor either greatly or mildly stimulatory, inhibitory, or ineffective. Furthermore, endocrine responses to acute exercise range widely, with some hormones increasing (catecholamines), some decreasing (insulin), and some not changing, such as luteinizing hormone (LH). Finally, there are marked differences in the temporal pattern of exercise responses. As one example, plasma insulin concentrations begin to decline almost immediately after the start of even mild exercise; yet glucagon, which is made by and secreted from the same organ (pancreas) as insulin and is catabolized by the same organ (liver), normally does not begin to increase until after 30 min of sustained mild to moderate exercise.

Another complicating issue is the intimate relationship (codependence) between the classic endocrine system and the nervous system. It is more appropriate to think in terms of neuroendocrine systems rather than separate endocrine and neural systems because hormones modify neural activity and nerves alter hormone secretion. It is clear that exercise stimulates many hormones, and the final effect of that deluge of hormones resides in the way they function in the aggregate, not individually. Despite the complexity described here, much is known about how regular physical activity alters hormonal status, and this chapter summarizes selected parts of that large body of information. This brief summary relies primarily on data obtained from human studies; however, some experimental manipulations performed in animal models are included.

Measurements of circulating hormone concentrations during exercise reflect a combined influence of hormone secretion, clearance, and volume of distribution of the given hormone. Many other factors must be taken into account before true hormonal activity can be evaluated.

Hormone Functions Related to Physical Activity

Certain general physiological functions are influenced by single bouts of endurance and probably resistance exercise, and many of these functions are at least partially modulated by hormones. They include the following (based on information from Viru 1985a, 1985b; Warren and Constantini 2000; Borer 2003):

- Mobilization of fuels for use by contracting muscles
- Uptake of fuels
- Regulation and distribution of blood flow
- Regulation of electrolyte stability
- Regulation of hydrogen ion status
- Regulation of reproductive function (likely related to energy deficits caused by physical activity)
- Some aspects of thermoregulation (likely related to changes in blood flow)
- Cardiorespiratory function
- Tissue growth stimulated by hormonal changes, which may have an impact on the development of certain cancers (this topic is discussed in chapter 14)

Changes in rate of growth attributable to chronic physical activity also have an endocrine basis (e.g., resistance exercise leads to muscle hypertrophy). Growth hormone, insulin, IGFs, testosterone, estrogens, and other hormones have a role in this adaptation. Additionally, in response to chronic endurance exercise, thyroid hormones are important for the transformation of muscle fiber protein phenotypes.

Each hormone typically responds to a specific major stimulus and also to secondary stimuli. The intimate relationship between insulin and glucose is a good example. As glucose concentrations increase in the arterial blood perfusing the pancreas, the β-cells increase secretion of insulin. As glucose concentrations decrease, insulin secretion subsides. During acute exercise, however, other factors override this relationship; most notably, exercise causes an increase in α-adrenergic stimulation of the pancreatic β-cells, which decreases insulin secretion even when glucose concentrations remain constant during moderate exercise or increase during brief supramaximal exercise. Additionally, glucose ingestion or infusion during exercise does not elevate insulin. Thus, the suppression of insulin secretion during physical activity is difficult to override because the intimate relationship between glucose and insulin is modified during exercise by the sympathetic nervous system. This interruption allows plasma insulin to decline during exercise, which has the following positive effects on fuel mobilization: (1) enhanced hepatic glucose production, (2) enhanced rates of lipolysis (mobilization of stored fat for use during exercise), and (3) enhanced rates of gluconeogenesis. Although it seems counterintuitive that insulin should decrease when more glucose needs to be transported into muscle cells, this seeming aberration is easily explained by the fact that muscle contractions per se augment glucose uptake in the contracting muscle. The need for insulin for glucose uptake during exercise diminishes greatly in active muscle. Thus, regulators that control the hormone–stimulus relationship at rest may not be the principal regulators during exercise.

Another aspect of the endocrine response to physical activity is that, in addition to physical fitness, body composition can influence the hormonal response to physical activity. A classic example of this was seen in Hansen's demonstration that in lean participants, GH increased during and after exercise whereas this response was absent in obese participants (Hansen 1973). Gustafson and colleagues (1990) expanded on this observation by showing that the epinephrine, norepinephrine, and glucagon responses were markedly blunted in massively obese women during submaximal exercise.

Some Control Mechanisms Related to Physical Activity

Significant evidence suggests that the endocrine response to physical activity is regulated by feedforward, feedback, and central command mechanisms, which makes the endocrine system similar to the cardiovascular and respiratory systems (Galbo 1983). Proof of this is the fact that increases in hepatic glucose production and in the hormones needed to accomplish those increases are observed before any decline occurs in circulating glucose during physical activity in rats. Specifically, plasma insulin declines within the first minute of exercise, which leads to a disinhibition of hepatic glucose production. Thus, the endocrine system can respond before metabolic cues, such as reduced glucose, can occur. This suggests that feedforward control, for example from active muscle, is important to the endocrine response. Afferent nerve feedback from muscle fibers also contributes to the regulation of the GH, adrenocorticotropin (ACTH), and β-endorphin (BeP) response to contractions. Moreover, the aldosterone response to acute exercise using two legs is about half the magnitude found when one-leg exercise is performed at the same absolute oxygen consumption. Thus, feedback from the contracting muscle at least partially regulates the endocrine response of some hormones to activity.

The response of several hormones to physical activity is partially regulated by what is referred to as "central command." Proof of this comes from studies that have used neuromuscular blockade, which weakens the muscle and consequently necessitates greater effort (central drive) to produce a given amount of work. Under these conditions, secretion of catecholamines and anterior pituitary hormones (GH, BeP, and ACTH) is greater than during control conditions even though the same absolute work is performed. Thus, some hormones respond to the perceived effort of the task rather than to the actual work performed. Evidence that traditional feedback mechanisms are also important is provided by the glucagon response in that glucagon increases primarily when glucose begins to decline, usually late in exercise.

Therefore, we must consider a variety of regulatory mechanisms as we evaluate endocrine adaptations to regular physical activity. Because the motor cortex partially controls some endocrine responses to physical activity, it is reasonable that adaptations should occur in that part of the brain. Unfortunately, very little is known about adaptations to regular exercise that occur in the brain. Against the background presented thus far, the next section presents several general concepts about how the endocrine system adapts to regular physical activity, along with some comments about possible health implications of those changes.

■ Hormone responses during exercise are regulated by feedforward, feedback, and central command mechanisms. This control is similar to that seen in the cardiovascular and respiratory systems.

Importance of Hormonal Regulation

The human body has developed complex and redundant mechanisms that keep circulating concentrations of all hormones within a very narrow range. For example, during and after exhausting, supramaximal treadmill exercise, circulating concentrations of β-endorphin increase only from 3 to 12 fmol/ml, and ACTH increases from 30 to 40 fmol/ml, in untrained participants. The need for this regulation over a narrow range is evident when we consider that only moderately elevated thyroid hormone concentrations can significantly elevate metabolic rate. Thus, there is a critical need for precise regulation of the balance between secretion and catabolism of all hormones so that the net result (circulating concentration) remains in the physiological range.

Regulation of Hormones at Very Low Concentrations

Other hormones besides insulin could illustrate this concept; however, we consider insulin here because this hormone is at the center of the metabolic syndrome, which is a major health concern. Another chapter in this book (chapter 13) discusses physical activity in diabetic populations; therefore, the following section, "Regular Physical Activity and Hormonal Adaptations," covers some of the same hormones, insulin and glucagon, but concentrates on responses in nondiabetic populations.

A major health-related adaptation of insulin to regular physical activity is a decline in its plasma concentration, which seems to be precisely matched by an increase in insulin sensitivity. Elevated insulin sensitivity means that a given amount of insulin results in a transport of a greater amount of glucose into an insulin-sensitive cell—primarily fat, muscle, and liver. Both acute and regular physical activity increase insulin sensitivity for glucose uptake in both skeletal muscle and adipose tissue in humans. Epidemiological studies show that risk for several diseases increases as basal insulinemia increases. Relative insulinemia, the amount of insulin compared with its effectiveness, is also of concern. For example, an increase of insulin from 30 to 90 pmol may elevate glucose uptake from 2 to 5 mg/kg · body weight^{-1} · min^{-1} in one person while in another person the insulin must increase from 30 to 120 pmol for the same glucose uptake. The second person has a relative hyperinsulinemia, which is characteristic of insulin resistance. Even modest reductions (and elevations) in relative insulinemia can have a significant impact on many physiological phenomena such as fat mobilization. A decline in insulinemia of only 10 to 20 pmol markedly disinhibits fat metabolism, which elevates fat oxidation (Bonadonna et al. 1990).

Numerous studies have shown that insulin concentrations in the plasma are lower either after regular exercise training (longitudinal designs) or when groups that differ in habitual activity levels are compared (cross-sectional designs). Lower basal insulin concentrations are attributable to reduced secretion rather than enhanced insulin clearance. The lower insulin attributable to chronic exercise, coupled with the higher insulin sensitivity, probably allows better glucose control and imposes less demand on the β-cells.

The preceding discussion concentrates on the effects of chronic physical activity on basal insulinemia. However, studies investigating the insulin response to hyperglycemia (nonbasal conditions) also report a reduced insulin secretion in the trained state (Farrell 1992). This reduced insulin response is found at low, moderate, or high levels of glucose stimulation. Training-induced reductions in basal insulinemia and glucose-stimulated insulin secretion are demonstrable in the whole body (humans, rats, dogs), isolated pancreas (dogs and rats), isolated pancreatic islet of Langerhans (rats), and even single isolated pancreatic β-cells (rats) (Farrell 1992). Yet the fact that this adaptation (reduced insulin secretion in response to glucose) is measurable at lower levels of inquiry (cells vs. whole body) does not mean that the entire adaptation can be explained at the cellular or subcellular level. Associated adaptations such as altered total or directional blood flow through the pancreas must also be considered, as well as elevated sensitivity to the inhibitory effects of somatostatin. Somatostatin is a small peptide hormone made in the pancreas (and brain) that has inhibitory effects on insulin as well as many other hormones. Therefore, as the concentration of somatostatin increases in either the blood or pancreas, insulin secretion declines.

> Under nonstressful conditions, hormones are present in the plasma in very low concentrations, and small elevations or reductions in those basal concentrations may affect health significantly. Regular physical activity has the potential to cause such small deviations.

Figure 7.1 provides a compilation of several studies by Mikines and colleagues that demonstrate basic healthful adaptations of insulin in response to both chronic and acute physical activity.

Some information is available on the temporal sequence of adaptations to regular physical activity in endocrine secretion versus adaptations in target tissue. Increases in insulin sensitivity occur with the first bout of exercise. However, an identical acute bout does not change glucose-stimulated insulin secretion (Mikines et al. 1987). Thus, anatomically distinct organs (muscle and the pancreas) adapt to chronic physical activity in a manner that ultimately results in precisely controlled glucose uptake requiring less insulin, but the individual tissues adapt with different temporal patterns. We know very little about the way this coordinated adaptation *between organs*

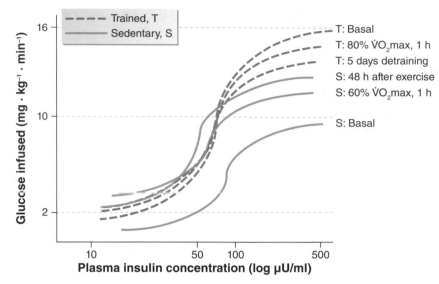

FIGURE 7.1 Positive adaptations of insulin to acute and chronic exercise.

Data from Mikines et al. 1988a; Mikines et al., 1988b; Mikines et al., 1989.

is accomplished. It is unlikely that glucose per se is the common message for insulin–glucose relationships, because resting concentrations of glucose are identical in trained and untrained people, as is the rate of hepatic glucose production.

This same pronounced sensitivity to changes in insulinemia around basal and elevated levels is not evident in some other functions in which insulin has a role. Insulin has a permissive role in allowing rates

of protein synthesis in skeletal muscle to increase after acute resistance exercise. However, insulinemia must be reduced to less than 20% of basal concentrations before rates of protein synthesis do not respond (increase) appropriately after exercise. A general working hypothesis could center on the fact that adaptations to regular physical activity elevate or reduce basal hormone levels, affecting only certain actions of these hormones.

Questions on Figure 7.1*

Fact 1. In the basal state, a trained person has elevated insulin sensitivity (insulin-stimulated glucose uptake at midrange insulin concentrations) and responsiveness (glucose uptake at maximal insulin concentrations).

Question 1: Which two curves (lines) prove this point?

Fact 2. A single bout of moderate-intensity exercise increases both insulin sensitivity and responsiveness in sedentary people.

Question 2: Which two curves (lines) prove this point?

Fact 3. The increase observed in question 2 lasts for at least 48 h in sedentary people.

Question 3: Which two curves (lines) prove this point?

Fact 4. A single bout of exercise in trained people does not alter insulin sensitivity or responsiveness.

Question 4: Which two curves (lines) prove this point?

Fact 5. A period of five days of total inactivity (bed rest) in long-term well-trained athletes does not change insulin responsiveness but decreases insulin sensitivity.

Question 5: Which two curves (lines) prove this point?

*Answers to questions on page 110.

Another important consideration in evaluating endocrine responses to acute exercise is the basic fact that physical activity cannot be performed without creation of an energy deficit at least in the muscles that perform the exercise. Some recent evidence suggests that replacing calories (specific to the carbohydrate and fat used during the exercise) during or immediately after exercise can reduce or nullify improvements in insulin sensitivity. While these studies are not conclusive, they do identify a research area worthy of pursuit. Arguments against energy deficits as the sole reason for elevations in insulin sensitivity reside in studies showing that when comparable calorie deficits are elicited by diet alone or diet plus exercise, the diet plus exercise treatment results in significantly greater elevations in insulin sensitivity. Additionally, short-term fasting (24- to 48-h energy deficit) results in decreased insulin sensitivity, not increased sensitivity (although some reports dispute this). While controversy exists in this area, there is no question that greater attention must be paid to rigorously controlling diet when exercise is used as a treatment.

In contrast to our understanding of exercise-induced changes in glucoregulation by insulin, the effects of changes in insulin sensitivity attributable to chronic physical activity on functions such as protein synthesis, protein catabolism, lipolysis, and vasodilation are much less clear. The effect of chronic physical activity on plasma concentrations of hormones other than insulin is even less clear. Basal plasma epinephrine concentrations may be slightly higher in trained athletes; however, this is not a consistent finding. When higher concentrations are found either at rest or during intense exercise (very high workloads) in trained individuals, sufficient data exist to indicate that such elevations are attributable to an increase in epinephrine secretion rather than a decrease in epinephrine clearance. Resting concentrations (based on single blood samples) of ACTH, BeP, norepinephrine, and glucagon seem to be similar between active and inactive people; but as discussed in the next section, a single measure (one sample) of a hormone concentration is not sufficient to allow solid interpretations about that hormone.

Regulation of the Rhythmic Release of Hormones

Many, if not most, hormones are secreted from glands in a rhythmic manner. Those oscillations can last from minutes to months. The consequence of such secretion is that plasma concentrations of most hormones rise and fall with predictability, and

we can assume that such patterns enhance survivability and function of the organism (Borer 2003). Hormone pulses have many complex characteristics, and only two—pulse frequency (time between pulses) and pulse amplitude (peak amount of hormone secreted during a pulse)—are discussed here. The time between pulse intervals can range from minutes (**ultradian**, e.g., insulin, GH, neuropeptide Y, parathyroid hormone [PTH], ACTH, glucagon, gonadotropin-releasing hormone [GnRH], brain somatostatin) to approximately 24 h (**circadian**, e.g., cortisol, melatonin, leptin) to days (**infradian**, e.g., reproductive hormones, such as follicle-stimulating hormone [FSH], LH, and estrogens). It is also clear that some hormones display rhythms that fall into more than one category, such as LH, which has both ultradian and infradian rhythms. Finally, and adding to the complexity, some hormones change their pattern of secretion because of a change in another hormone.

Many (or most) hormones are released from endocrine glands in a pulsatile manner, and regular physical activity causes adaptations in those pulse profiles that are compatible with better function.

For reasons that are not completely known, a pulsatile release (and consequent presentation to target tissues) of hormones augments target tissue action. Therefore, we would predict that regular exercise should augment the pulsatile release of hormones. In some cases (GH in young women), the literature supports an augmented pulsatility; but in the case of insulin and melatonin, the opposite is seen. Regular physical activity reduces the pulse profile for both insulin and melatonin.

Knowledge of the pattern of a pulse profile is required to correctly interpret exercise data. Reliance on a single blood sample is at best hazardous because any exercise-related response could be masked or accentuated because of a naturally changing baseline. As an example from many years ago, researchers found that if exercise begins during a normal elevation in cortisol, an exercise-induced increase in circulating cortisol concentrations does not occur; but if the same exercise starts when cortisol is at a nadir (lowest level), then circulating cortisol increases significantly. This effect of the constantly changing baseline concentrations and patterns was often neglected in interpretation of research results and probably accounts for a large amount of the confusion in the physical activity–hormone literature.

Answers to Questions on Figure 7.1

Answer 1: Compare T (Basal) versus S (Basal).

Answer 2: Compare S (Basal) versus S (60% $\dot{V}O_2$max, 1 h).

Answer 3: Compare S (Basal) versus S (48 h after exercise).

Answer 4: Compare T (Basal) versus T (80% $\dot{V}O_2$max, 1 h).

Answer 5: Compare T (Basal) versus T (five days detraining).

Insulin

Insulin is secreted from the pancreas in a pulsatile manner and is used here as an example of how regular physical activity can change the pattern of hormone release in a nonstimulated state. Growth hormone and reproductive hormones are also secreted in a pulsatile manner and are reviewed in the next section.

In concert with reduced plasma concentrations of insulin assessed by single-point sampling, minute-by-minute sampling for 90 min shows that the pulse amplitude is markedly lower in exercise-trained men and women whereas pulse frequency is not changed (figure 7.2). This is an interesting observation because it is generally accepted that hormones are more effective (greater target tissue effects) when delivered in a pulsatile manner. Data in figure 7.2 (Engdahl, Veldhuis, and Farrell 1995) were interpreted to suggest that insulin action at target tissues was elevated to the point that delivering insulin in a pulsatile manner was not as necessary to maintain a certain rate of glucose uptake in trained individuals. The logical but unproven conclusion is that because of the greater tissue sensitivity, there is less need for insulin to be secreted (and delivered) in a pulsatile manner.

From an evolutionary standpoint, the pattern for the trained group may, in fact, be the desirable pattern, because there may have been a gradual elevation in burst amplitude over the thousands of years in which we have become less active and probably less insulin sensitive. The coordinated changes in insulin pulsatility and sensitivity with regular physical activity are quite distinct from those observed in people with type 2 diabetes mellitus. In both people with type 2 diabetes mellitus and glucose-intolerant first-degree relatives of those people, the insulin pulse profile is characterized as either absent or erratic, with poorly formed pulse amplitudes and wide disruptions in pulse frequency.

Although meager, the information available on **hormone pulsatility** and chronic exercise suggests that pulse amplitude (especially for insulin, GH, and LH, discussed subsequently) but not pulse frequency is adaptable to chronic physical activity. Interestingly, this is similar to changes found in some diseases in which pulse amplitude changes but pulse frequency may be normal.

Growth Hormone and Insulin-Like Growth Factor

Secreted from the anterior pituitary with an ultradian periodicity, GH plays a critical role in somatic growth. Growth hormone has many isoforms, but the 22-kDa (kilodaltons) isoform is probably the most frequently studied from an exercise perspective. This isoform increases in plasma in an exercise intensity-dependent manner, especially during very heavy physical activity. Growth hormone circulates bound to binding proteins, and the role of these binding proteins in the exercise response is not clear. After puberty, an exercise-induced increase in GH availability may have its greatest impact on maintaining skeletal muscle mass in the elderly.

The complexity of GH regulation becomes quickly evident when we realize that two of the most potent physiological stimuli are sleep (no movement) and exercise (movement). High plasma concentrations

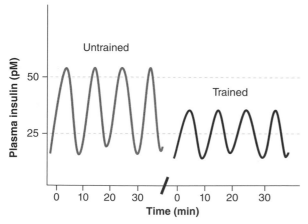

FIGURE 7.2 Effect of regular physical activity on insulin pulsatility.

Data from Engdahl, Veldhuis, and Farrell 1995.

occur 2 to 3 h after sleep begins, but even higher concentrations are recorded after intense acute aerobic and resistance exercise. During nonstimulatory periods, GH output from the pituitary depends on the balance between the stimulatory effects of the hypothalamic peptide GH-releasing hormone and the inhibitory peptide somatostatin. The stimulus for elevated GH during acute exercise is not known. However, cholinergic stimulation, nitric oxide, hydrogen ions, muscle afferent activation (which directly stimulates the anterior pituitary), and muscle vibration have been implicated in various studies (e.g., Gosselink et al. 2004). Cholinergic, dopaminergic, and serotonergic blockades each inhibit the GH response to exercise. Exercise-induced increases in GH are probably not attributable to catecholamines (α- or β-adrenergic) or changes in plasma ghrelin but can be inhibited by opioids (at least those antagonized by naltrexone/naloxone). Plasma ghrelin does not change during exercise when elevations in GH are evident. This finding (no effect of ghrelin on GH), specific to the exercise stimulus, conflicts with evidence from studies in which ghrelin infusion stimulated increases in plasma GH, ACTH, and prolactin. Again this shows that regulators of hormone secretion at rest are not necessarily similar to regulators during exercise, although much overlap obviously occurs.

Chronic physical activity elevates GH pulse amplitude but not frequency in premenopausal women when the training stimulus is above the lactate threshold. Elevations of GH pulse amplitude, as well as the total amount of the hormone secreted over a 24-h period following a single bout of exercise, are found in young women but not young or old men. Data on middle-aged men do not seem to exist. For several hours after heavy resistance exercise, GH pulse amplitude and GH concentrations are significantly lower in young men (22 years). Eventually, however, GH concentrations seem to be higher than under nonexercise conditions. It is obvious that much more work must be done to document whether and how acute exercise alters the pulse profile of GH. Elevations in GH amplitude may have implications for maintaining muscle and bone mass in elderly people, because reductions in GH concentrations are observed in obesity, osteoporosis, and aging.

Growth hormone stimulates production and release of IGF from the liver. Insulin-like growth factor 1 (IGF1) and mechano growth factor (MGF) increase in plasma and in the skeletal muscle, respectively, after resistance exercise, whereas combining exogenous GH with resistance exercise leads to even larger increases in messenger RNA (mRNA) for MGF. Hepatic-derived IGF1 is stimulated by GH;

however, elevations in tissue concentrations of IGF1 in response to contractions can be independent of circulating GH and plasma IGF1. An increase in muscle mass attributable to resistance exercise probably involves both GH-stimulated increases in IGF1 and elevations in intramuscular MGF caused by increased gene expression (MGF mRNA eventually translated into MGF protein). Some lines of evidence suggest differential splicing of the *IGF1* gene in response to chronic muscle contractions. Yet this general finding may be more applicable to rodents than humans, because some reports suggest no training-induced increase in human muscle IGF1 in humans. Acute resistance exercise of moderate to heavy intensity increases serum IGF1, whereas concentrations of IGF binding protein-3 decrease and those of IGF binding protein-1 do not change.

Mounting an appropriate anabolic response to resistance exercise probably involves many hormones, possibly in a redundant fashion. For instance, moderately diabetic rats can increase rates of protein synthesis after resistance exercise, and the reduced insulinemia seems to be compensated for by increased IGF1 concentrations in the exercised muscle in the immediate postexercise period (up to 24 h). This does not occur in nondiabetic rats until two days after exercise. Thus, a relationship seems to exist between IGFs in the muscle and insulin availability during periods of anabolism.

Female Reproduction

Reproductive hormones are also secreted episodically, and chronic, excessive exercise that causes chronic energy deficits also suppresses the pulse profile of several reproductive hormones. It must be noted, however, that normal women can engage in long-term (more than one year) moderate- to high-intensity endurance exercise without affecting reproductive function. Some exercise is probably beneficial to reproductive hormone status, but excessive exercise (very high levels of physical activity that result in an energy deficit) is not compatible with producing offspring (Loucks 2001). Therefore, it is not surprising that a small percentage of female athletes ($\leq 6\%$ depending on the sport) who engage in frequent, high-volume, and high-intensity physical activity temporarily lose reproductive function (cease menstruation) either fully or partially. Although the cause of this decreased function is not completely clear, overlapping lines of evidence support the working hypothesis that it is not the stress of exercise per se that causes the dysfunction, but rather energy deficits caused by the combination of undernutrition and increased energy expenditure from activity.

Exercise requiring significant energy expenditure reduces LH pulse amplitude in women. Such reproductive dysfunction must involve areas in the hypothalamus that determine LH secretion. Luteinizing hormone is under complex control, but the major stimulatory regulator is GnRH, which is made in and secreted from the arcuate nucleus of the hypothalamus. A one-to-one relationship exists between GnRH and LH release, and both are disrupted by energy deficits associated with high levels of physical activity. Consequences of reduced LH pulsatility can include luteal phase defects such as shortening of the luteal phase and lengthening of the follicular phase or insufficient secretion of progesterone by the corpus luteum.

Another anterior pituitary hormone, FSH, is also adversely affected (reduced circulating concentrations) consequent to energy deficits caused by physical activity. Follicle-stimulating hormone attaches to receptors on ovarian follicles and stimulates their maturation. These follicles in turn make and secrete estrogens whose levels are also lower in women who perform extensive exercise. Disruption of the hypothalamic–anterior pituitary–gonadal axis caused by chronic high levels of exercise (or the consequent energy deficit) could occur at any level of the axis. However, gonadotropic receptor sensitivity to GnRH is not altered by chronic exercise. Reproductive dysfunction in women may also be attributable to excessive secretion of corticotropin-releasing factor, ACTH, cortisol, androgens, or BeP. Each has been linked to reduced LH secretion (Borer 2003). At the same time, these women have higher concentrations of GH and androstenedione. Another hormone, leptin, may also be involved in reproductive dysfunction in female athletes because it is also suppressed by low energy status.

Chronic and excessive physical activity early in life can adversely affect sexual maturation. Young girls who perform strenuous, high-volume exercise, especially in sports that require a slim physique, can have delayed menarche of about a year. Characteristics of this delay include a prepubertal pattern of low-frequency LH pulses, increased FSH/LH secretory ratios, and a lack of the enhanced estradiol secretion that is necessary to progress through puberty (Borer 2003).

The ultimate effects on health of disrupted reproductive function in women are not known. If regular exercise chronically lowers circulating estradiol, this may provide protection against heart disease and breast cancer. On the other hand, lack of a normal reproductive cycle causes bone loss and osteoporosis. Again, it must be emphasized that menstrual disturbances are neither a common nor an expected

© Human Kinetics/J. Wiseman, reefpix.org

Extreme exercise causes energy deficits that may suppress normal hormone function. High-performance female athletes may experience amenorrhea.

consequence of regular nonexcessive physical activity participation.

Male Reproduction

Acute exercise results in elevations of free and total testosterone and androstenedione in males; however, heavy, chronic training decreases testosterone. A reasonable working hypothesis is that male reproductive hormones respond negatively to energy deficits in a manner similar to that observed in females. Supporting this hypothesis, chronic, prolonged, high-intensity physical activity decreases testosterone and spermatogenesis. Whether chronic, moderate-intensity physical activity improves reproductive function in either males or females has not been documented.

Regular Physical Activity and Hormonal Adaptations

The suggested relationship between IGFs and insulin availability discussed earlier illustrates the important concept of redundancy of hormone action. This seems to be the case especially in the area of fuel mobilization. Regular physical activity alters hormone availability, and this may alter the pattern of fuel utilization during exercise.

The hormones most involved in the regulation of fuel utilization and mobilization during exercise are epinephrine, glucagon, insulin, and, to a lesser extent, GH, norepinephrine, ACTH, and cortisol. Epinephrine, cortisol, and glucagon are referred to as counterregulatory hormones because they balance the effects of insulin and promote glucose availability for usage as opposed to storage (a major function of insulin). Because of the importance of catecholamines, this section concentrates on two catecholamines (epinephrine and norepinephrine) that alter exercise metabolism and adapt to regular physical activity. Because of the difficulty in differentiating between adrenal-derived catecholamines and those originating from sympathetic nerves, most texts refer to the sympathoadrenal system.

> A significant gap exists in our understanding of how hormonal adaptations to regular physical activity contribute (or not) to improved health.

Epinephrine and Norepinephrine

The catecholamines most studied during exercise are epinephrine, norepinephrine, and dopamine. Epinephrine is made in and secreted from the adrenal medulla; norepinephrine is also made in the adrenal medulla, but the major elevations in plasma found with exercise are attributable to spillover from activated sympathetic nerve terminals. A major neurotransmitter in the central nervous system, dopamine also appears in the peripheral circulation. However, the function and source of peripheral dopamine remain unclear. Plasma epinephrine and norepinephrine concentrations increase exponentially with exercise intensity (figure 7.3), and elevations are most closely associated with the intensity of work relative to maximal aerobic capacity. This increase has been shown to originate both from peripheral factors (local factors in the working muscle) and from central command in the brain. Major physiological functions regulated by catecholamines during exercise are fuel mobilization (epinephrine), partitioning of blood flow (norepinephrine), increased cardiac contractility (norepinephrine), and elevation of blood pressure (norepinephrine). Circulating catecholamines (as opposed to norepinephrine derived from cardiac sympathetic nerves) are probably not critical to the heart rate, systolic blood pressure, and respiratory responses to exercise because these physiological responses are identical when bi-adrenalectomized participants exercise to exhaustion. Circulating catecholamines are rapidly degraded, and their concentrations decline quickly after exercise.

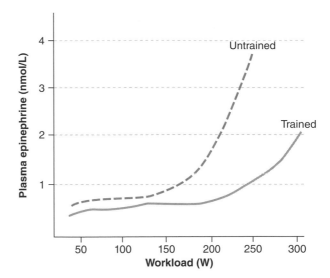

FIGURE 7.3 Plasma epinephrine response to exercise specific to absolute workloads.

Catecholamines account for a major proportion of the exercise-induced increases in adipose tissue lipolysis, but differential increases in blood flow can cause one adipose tissue bed to increase lipolysis to a greater extent than another bed during exercise. For example, sympathetic nerves are not important for exercise-induced increases in lipolysis from umbilical and clavicular beds. Although epinephrine stimulates both hepatic glycogenolysis and adipose tissue lipolysis, the adaptations to training may differ depending on the tissue. Regular physical activity reduces epinephrine-stimulated, hormone-sensitive lipase activation in muscle but elevates it in adipose tissue. Both epinephrine and ACTH stimulate greater lipolysis in isolated adipocytes in animals that have participated in regular physical activity. Catecholamines also stimulate lipolysis indirectly by helping to suppress circulating insulin and by stimulating the release of GH, cortisol, and glucagon, which have lipolytic actions.

Catecholamine response to exercise is a well-established example of the fact that the exercise response of many hormones depends on the relative effort of the exercise and not the amount of work actually being performed, as shown in figures 7.3, 7.4, 7.5, and 7.6 (Kjaer et al. 1986). Most glands that secrete hormones do not sense how much total work is being done but rather how much of the organism's capacity is being used to accomplish that work. This concept has been derived mostly from studies that required large-muscle contractions (running, bicycle exercise), and the applicability of the concept to movements using only a small muscle mass requires documentation.

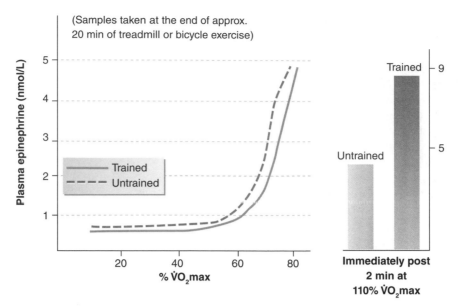

FIGURE 7.4 Changes in plasma epinephrine when exercise workloads are expressed relative to maximal oxygen consumption.

Data from Kjaer and Galbo 1988; Kjaer et al. 1986.

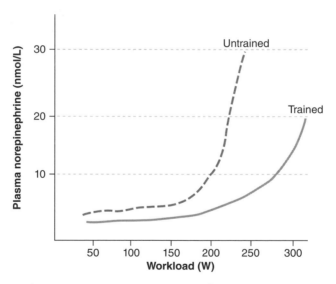

FIGURE 7.5 Plasma norepinephrine response to exercise, specific to absolute workloads.

Regular physical activity probably results in more efficient function of many hormones, which allows lower concentrations of those hormones to meet the requirements of moderate physical activity.

It is difficult to document whether adaptations of the endocrine system result in "better" fuel mobilization and utilization during exercise. Fat mobilization and utilization are appropriate at low work intensi-

ties, but carbohydrate mobilization is appropriate at high intensities. Epinephrine has been shown to be a primary stimulator of adipose tissue lipolysis during exercise (Kjaer and Galbo 1988), and fuel mobilization during exercise depends on these functions to a significant degree. Endurance-trained individuals have an increased lipolytic response to infused epinephrine; however, this adaptation is short-lived (lost in only a few days) once training ceases. This elevated lipolytic response is observed at the whole-body level and when human fat cells are stimulated in vitro.

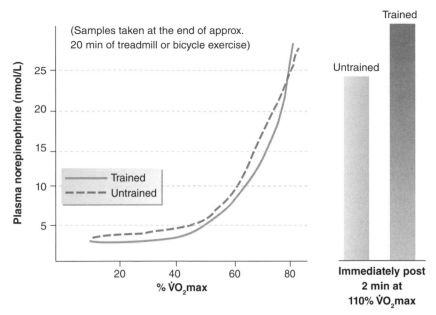

FIGURE 7.6 Changes in plasma norepinephrine when exercise workloads are expressed relative to maximal oxygen consumption.

Data from Kjaer and Galbo 1988; Kjaer et al. 1986.

Some of this adaptation must reside in an enhanced sensitivity of adrenergic receptors (both α and β), as shown in the late 1970s and the 1980s (Krotkiewski et al. 1983). Thus, under several conditions, hormone-stimulated lipolysis from several depots is enhanced by regular physical activity.

> Adequate fuel mobilization is critical to physical activity, and hormones have an important role in this physiological function. It is also clear that redundant endocrine systems affect fuel mobilization during exercise.

Well-trained endurance athletes can secrete greater amounts of epinephrine at high work intensities compared with untrained people (when work is expressed relative to maximal aerobic capacity). This enhanced ability has advantages, for example the mobilization of glucose, as suggested by a linear relationship between hepatic glucose production and plasma epinephrine during exercise when plasma epinephrine increases from 1 to 6 nmol/L. Further support for the necessity of β-adrenergic stimulation during hard exercise comes from evidence that β-adrenergic blockade with propranolol reduces work output at high but not low work intensities. High glucose production would be advantageous during times of high athletic competition or survival. This adaptation

of the adrenal medulla to training may also occur in selected anterior pituitary hormones because BeP, GH, and ACTH all increase in plasma to a greater extent in trained athletes during very hard exercise.

Figure 7.5 shows that plasma norepinephrine also increases with work intensity in an exponential manner, and figure 7.6 demonstrates that well-trained athletes have higher concentrations during very high-intensity exercise. The metabolic or health implications of these changes are not clear.

> Positive hormonal adaptations to regular physical activity occur shortly after the commencement of training and probably disappear as quickly.

Glucagon

A primary function of glucagon is to oppose the glucose-lowering actions of insulin. Glucagon is made in and secreted from the α-cells in the islets of Langerhans of the pancreas. Twenty-four-hour averaged glucagon concentrations are not different between physically active men and inactive men; however, some studies report that resting concentrations of glucagon are lower after chronic physical activity. Although there is a good correlation between plasma concentrations of epinephrine and glucagon during exercise, β-adrenergic blockade does not affect the glucagon response to exercise but increases the epinephrine response. The glucagon response to submaximal exercise is highly dependent on reductions in plasma glucose but is not dependent on either parasympathetic or α-adrenergic blockade. As with epinephrine, the glucagon response to supramaximal exercise is markedly higher in trained than in untrained people, yet the glucagon response to insulin-induced hypoglycemia is lower in trained than in untrained people.

The rate at which hormonal adaptations to regular physical activity disappear during detraining is not clear. The exercise training–induced adaptation of the pancreas is lost quickly (within days), but the adrenal medulla maintains the ability to secrete a high level of epinephrine in response to experimentally induced hypoglycemia even after five weeks of detraining. This is a markedly understudied area of exercise endocrinology.

Hormonal Adaptations to Chronic Physical Activity and Changes in Long-Term Food Intake

Current recommendations for losing weight and maintaining that reduction center on a combination of increased physical activity and reduced calorie intake. Related to fuel utilization during exercise is the issue of fuel storage replenishment after exercise. Leptin is a 146-amino-acid peptide that circulates in both bound and free forms. Early studies clearly demonstrated that leptin availability was closely tied to energy balance; that is, negative energy balance reduced circulating leptin concentrations whereas energy surplus elevated concentrations. Simultaneously, studies showed that obese individuals had higher concentrations of leptin, whereas athletes had reduced levels. Studies progressed from these correlational designs to those using manipulations of energy balance, which advanced and supported the working hypothesis that circulating leptin concentrations do not change with exercise unless a marked energy deficit occurs. Thus, it may be the exercise-related energy deficit rather than the activity per se that reduces circulating leptin concentrations. Such interpretations need to be refined.

Leptin's effects on body weight are mediated through effects on hypothalamic centers that control feeding behavior, hunger, body temperature, and energy expenditure. Daily injections of recombinant mouse or human leptin into *ob/ob* mice (i.e., obese mutants unable to synthesize leptin) led to a dramatic reduction in food intake within a few days and to roughly a 50% reduction in body weight within a month. Weight loss with administration of leptin results from decreased hunger and food consumption and increased energy expenditure. Unfortunately, leptin injection into obese humans has not resulted in consistent weight loss.

Because leptin is acutely sensitive to negative energy balance (fasting or caloric restriction), studies are needed that can distinguish the effects of exercise per se from any attendant change in energy balance or, perhaps more specifically, energy availability. Thus, leptin may act as a sensor of the difference between energy intake and expenditure. The mechanisms by which leptin exerts its effects on metabolism are largely unknown and are likely quite complex.

Related to long-term food intake are questions of whether basal metabolic rate (BMR) changes with chronic physical activity and, germane to this chapter, whether changes in hormones attributable to physical activity could account for such a change. Unfortunately, there is no consensus on whether BMR differs between active and inactive people. Numerous articles either support or refute such changes. Because of the potent effects of thyroid hormones on basal metabolism, physical activity–induced changes in thyroid hormones should also be considered when adaptations that may alter long-range energy balance are evaluated. Hormones that control BMR may also adapt to regular physical activity. Thyroid hormones are secreted from the thyroid gland (located just below the larynx) in response to thyroid-stimulating hormone (TSH), which is secreted by the anterior pituitary gland. Thyroid-stimulating hormone itself is released because of thyrotropin-releasing hormone (TRH) made in and secreted from the hypothalamus. Triiodothyronine (T3), the active thyroid hormone, binds to receptors in the nucleus of specific cells. A key to understanding a role for T3 and physical activity is that thyroid hormones change physiology in general but muscle function in particular by changing gene expression (transcriptional regulation). Most prominently, T3 increases the synthesis and activity of Na^+–K^+–adenosine triphosphatase (ATPase) pumps. If chronic physical activity causes increases in T3 that translate to increased Na^+–K^+–ATPase pumps, then basal metabolism should increase. Increases in BMR have been reported, but other reports on 24-h energy expenditure suggest no difference in trained female athletes compared with sedentary women.

Acute endurance exercise results in increases (in the basal state) in TSH and unbound tetraiodothyronine (T4), whereas the average concentrations for **reverse T3 (rT3)** and T4 are elevated after chronic physical activity. Unbound T3 seems to remain unchanged after chronic physical activity. However, when excessive physical activity is performed, both

T3 and T4 are reduced, as in the case of amenorrheic women athletes or athletes performing very high-intensity training at the peak of a season. This may be a manifestation of chronic energy deficits, because starvation also reduces these thyroid hormones. In terms of a time course for these changes, early training results in initial declines in T3 and T4, which should cause a greater level of TSH; but then as moderate training proceeds, there is an increase in T4 and rT3 but no change in T3. Some reports suggest that the TSH response to TRH is blunted in active people. Uncoupling proteins such as UCP3 are important for stimulating energy metabolism, and T3 stimulates UCP3 mRNA expression. Thyroid hormones are not the only hormonal system that could account for changes in BMR if in fact increases or decreases actually occur with chronic physical activity. Obviously, the endocrine control of individual components of 24-h energy expenditure must be assessed.

Muscle as an Endocrine Organ

As mentioned in the introduction to this chapter, several organs (muscle and adipose tissue) that were not originally thought to be endocrine organs have been shown to secrete biochemicals into the plasma that then reach target tissues. One significant family of these biochemicals is the "myokines." These are cytokines produced in and released from skeletal muscle. Arterial–venous differences in human contracting muscle show that IL-6 is released from muscle after both eccentric and concentric muscle contractions. A mechanism explaining an elevation of muscle-derived cytokines during exercise may reside in catecholamine stimulation. Elevations in muscle-derived IL-6 seem to have anti-inflammatory and metabolic functions; however, the health implications of this response are not clear.

Hormones Affecting Bone Homeostasis

Mechanical strain on bone consequent to many types of physical activity results in greater bone deposition, and this is positive for avoiding bone loss as we age. (This topic is addressed in chapter 15.) Hormonal responses to physical activity may account for some of this positive health adaptation. Parathyroid hormone (PTH) and intact PTH (iPTH, the biologically active sequence of PTH) stimulate osteoblasts, calcium reabsorption from the kidneys, and calcium absorption from the intestine. Therefore, any increase in PTH due to exercise should be beneficial for maintaining calcium and ionized calcium levels in the plasma. A consistent finding is

that acute exercise stimulates the production of PTH. Calcitonin inhibits osteoclasts and thus, if elevated by exercise, should also be beneficial to bone. Regular exercise, however, does not seem to affect basal calcitonin levels. Other hormones such as estrogens, testosterone, adiponectin, GH, IGF1, T3, and leptin have a positive impact on bone, and physical activity stimulates these hormones. A working hypothesis is that the rhythmic pattern of physical activities may amplify the intermittent secretion of anabolic hormones to enhance bone deposition. Increases in GH, IGF1, estrogen, and PTH surges occur during acute exercise, and most of these hormonal responses are amplified by increasing exercise intensity. As with reproductive function, the ability of these hormones to stimulate bone homeostasis after regular exercise is highly dependent on the exerciser's avoiding an energy deficit.

Cross-Adaptations

Generally, the first exposure to a stress results in the greatest response, and the magnitude of the response subsequently declines. Thus, regular exercise lowers the organism's response to subsequent exercise sessions, and this may have positive health benefits.

An underlying assumption concerning health benefits of physical activity is that regular exercise is not performed to an excessive degree but rather on a regular basis and in moderate to high but not excessive volumes. As examples of deleterious effects of excessive training, the menstrual cycle or testicular function can become dysfunctional, and basal catecholamines and cortisol can become chronically elevated. However, when training is not excessive, high levels of chronic physical activity cause adaptations that can be viewed as positive. Teleologically, it is reasonable that the endocrine system in a well-trained person functions at a "better" level than that of a person who is completely sedentary. There are no reports of deleterious hormonal adaptations to regular, moderate-intensity physical activity. From an evolutionary perspective, this makes sense because the current human genome developed during periods that required high levels of physical activity.

Because hormones increase in plasma according to the percentage of the maximal aerobic capacity, a trained person can perform a certain absolute amount of physical work with less of a hormonal response yet still meet the metabolic, cardiovascular, and psychological demands of that work. Therefore, as regular physical activity increases, one's functional capacity to complete normal tasks and challenges should engender less of a stress response. Many animal studies support this concept, but much more work must be done to confirm this phenomenon in humans.

This concept probably crosses several adaptations, because adaptations to one stress usually allow reduced responses to other stresses. As one example, the catecholamine response to mental tasks is lower in trained compared with sedentary people. Although eating is not typically thought of as a stress, many hormones increase or decrease after a meal; and it is clear that the insulin response, for example, to a meal is markedly reduced in trained persons. Considering the frequency of food intake in today's society, this consistently lowered insulin response could have major health benefits. Note, however, that athletes ingest more carbohydrates than nonathletes, and it has been shown that when 24-h calorie intake is considered, integrated 24-h insulin concentrations are not different between trained and sedentary people.

The anterior pituitary–adrenal axis is activated during and after all known stresses. Hormones of both the adrenal cortex (cortisol, aldosterone, and adrenal androgens) and the adrenal medulla (epinephrine and norepinephrine) are activated by stress according to the deviation from homeostasis caused by the stress. Regarding the response to exercise, cortisol is one of the few hormones (along with the catecholamines) in which the increase in plasma concentrations has been proven to be caused by an increase in secretion rather than a decrease in clearance. For cortisol, as exercise intensity increases, the rate of clearance also increases. With epinephrine, the rate of clearance decreases at high work intensities, but during very light work, the rate of clearance increases or does not change. These distinctions are important when we are trying to assess how much the organism is perturbed by stress and how much an endocrine gland increases output to control that perturbation.

Summary

Because the endocrine system is a major regulator of body function, it is reasonable to assume that some of the health adaptations known to occur with regular physical activity are caused by changes in hormone action. Although this is teleologically sound, surprisingly few studies have reported an endocrine adaptation and then offered proof of the health-related benefits of that adaptation. This chapter relies heavily on a review of adaptations of insulin to regular physical activity because of insulin's central role in the metabolic syndrome and the current epidemics of obesity, childhood type 2 diabetes mellitus, and other established chronic diseases that have insulin resistance as a common phenotype. Chronic physical activity reduces basal and glucose-stimulated insulin secretion, and this persistent relative hypoinsulinemia may prove to be the most important beneficial endocrine adaptation in terms of our public health.

Key Concepts

autocrine—Referring to a hormone that is made in a specific cell, is secreted by that cell, and changes the function of that same cell.

circadian—Referring to hormone rhythms that occur with an approximately 24-h cycle.

hormone pulsatility—Regular increase and decrease in the secretion of hormones by an endocrine gland (rather than a constant rate of secretion). Intervals can range from minutes to months.

infradian—Referring to hormone rhythms that are longer than a 24-h cycle.

paracrine—Referring to a hormone that is made by a specific cell, is secreted by that cell, and affects the function of an adjacent cell.

reverse T3 (rT3)—Another form of 3,5,3'-triiodothyronine (T3), with a structure of 3,3',5'-triiodothyronine.

ultradian—Referring to hormone rhythms that occur with a period of less than one day.

volume of distribution—Volume of the body (usually expressed as liters) into which a hormone is distributed once secreted from the gland.

Study Questions

1. Provide an example of a hormone that has classical hormonal actions but also acts in an autocrine and a paracrine manner.

2. Give examples of feedback, feedforward, and central command regulation of the hormone response to physical activity.

3. Name three changes in insulin action or secretion attributable to chronic physical activity that should have positive health outcomes.

4. Name one hormone with a pulse profile that is altered by chronic physical activity, and describe the possible health implications of that alteration.

5. Describe three hormonal adaptations to regular physical activity that result in "better" fuel mobilization during times of stress.

6. Are the hormonal adaptations to regular exercise confined to this particular stress?

References

Bonadonna, R.C., LC. Groop, K. Zych, M. Shank, and R.A. DeFronzo. 1990. Dose dependent effect of insulin on plasma free fatty acid turnover and oxidation in humans. *American Journal of Physiology (Endocrinology Metabolism)* 259:E736-E750.

Borer, K.T. 2003. *Exercise endocrinology.* Champaign, IL: Human Kinetics.

Engdahl, J.H., J.D. Veldhuis, and P.A. Farrell. 1995. Altered pulsatile insulin secretion associated with endurance training. *Journal of Applied Physiology* 79(6):1977-1985.

Farrell, P.A. 1992. Exercise effects on regulation of energy metabolism by pancreatic and gut hormones. In *Perspectives in exercise science and sports medicine,* ed. D.K.R. Lamb, R. Murray, and C. Gisolfi (pp. 383-434). Indianapolis, IN: Brown and Benchmark.

Galbo, H. 1983. *Hormonal and metabolic adaptation to exercise.* Stuttgart: Georg Thieme Verlag.

Gosselink, K., R.R. Roy, H. Zhong, R.E. Grindeland, A.J. Bigbee, and V.R. Edgerton. 2004. Vibration-induced activation of muscle afferents modulates bioassayable growth hormone release. *Journal of Applied Physiology* 96(6):2097-2102.

Gustafson, A.B., P.A. Farrell, and R.K. Kalkhoff. 1990. Impaired plasma catecholamine response to submaximal treadmill exercise in obese women. *Metabolism* 39:410-417.

Hansen, A.P. 1973. Serum growth hormone response to exercise in non-obese and obese normal subjects. *Scandinavian Journal Clinical Laboratory Investigation* 31:175-178.

Kjaer, M., and H. Galbo 1988. Effect of physical training on the capacity to secrete epinephrine. *Journal of Applied Physiology* 64:11-16.

Kjaer M, P.A. Farrell, N.J. Christensen, and H. Galbo. 1986. Increased epinephrine response and inaccurate glucoregulation in exercising athletes. *Journal of Applied Physiology* 61:1693-1700.

Krotkiewski, M., K. Mandroukas, L. Morgan, T. William-Olsson, G.E. Feurle, H. von Schenck, P. Bjorntorp, L. Sjostrom, and U. Smith. 1983. Effect of physical training on adrenergic sensitivity in obesity. *Journal of Applied Physiology* 55:1811-1817.

Loucks, A.B. 2001. Physical health of the female athlete: Observations, effects, and causes of reproductive disorders. *Canadian Journal of Applied Physiology* 26(Suppl):S176-S185.

Mikines, K.J., F. Dela, B. Sonne, P.A. Farrell, E.A. Richter, and H. Galbo. 1987. Insulin action and insulin secretion: Effects of different levels of physical activity. *Canadian Journal of Sports Medicine* 12:113-116.

Mikines, K.J., F. Dela, B. Tronier, and H. Galbo. 1989. Effect of 7 days of bed rest on dose-response relation between plasma glucose and insulin secretion. *American Journal of Physiology (Endocrinology Metabolism)* 257:E43-E48.

Mikines, K.J., B. Sonne, P.A. Farrell, B. Tronier, and H. Galbo. 1988a. Effect of physical exercise on sensitivity and responsiveness to insulin in man. *American Journal of Physiology (Endocrinology Metabolism)* 254:E248-E259.

Mikines, K.J., B. Sonne, P.A. Farrell, B. Tronier, and H. Galbo. 1988b. The effect of training on the dose response relationship for insulin action in man. *Journal of Applied Physiology* 66:695-703.

Viru, A. 1985a. *Hormones in muscular activity: Vol. I. Hormonal ensemble in exercise.* Boca Raton, FL: CRC Press.

Viru, A. 1985b. *Hormones in muscular activity: Vol. II. Adaptive effects of hormones in exercise.* Boca Raton, FL: CRC Press.

Warren, M.P., and N.W. Constantini (eds.). 2000. *Sports endocrinology.* Totowa, NJ: Humana Press.

8

Skeletal Muscle Adaptation to Regular Physical Activity

Howard J. Green, PhD

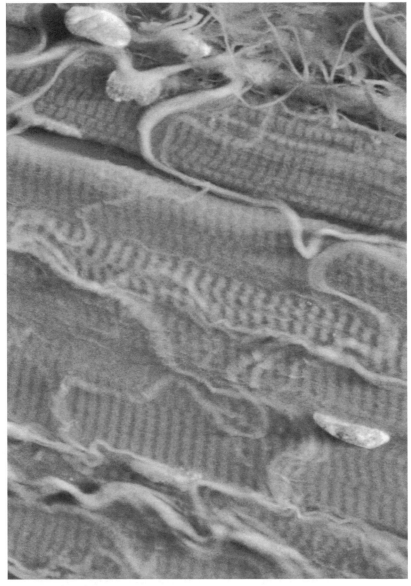

©Robert Becker, Ph.D.

A fundamental property of skeletal muscle is its ability to adapt; skeletal muscle is one of the most malleable tissues of the human body, capable of extensive remodeling if systematically challenged. The remodeling that occurs is to a large degree dependent on the type of contractile activity performed. Skeletal muscle can also rapidly change in response to inactivity and disuse, leading to pronounced deterioration in mechanical capabilities. This can be particularly problematic during aging, when an inevitable decline of muscle mass and function occurs.

Our evolutionary history and the central role of skeletal muscle in survival suggest that our natural state of **acclimatization** is one that includes regular daily activity. Such activity not only promotes optimal muscle function and task performance but also provides health benefits, both directly and indirectly. Research has shown that chronic exercise decreases the risk for both morbidity and all-cause mortality.

In this chapter, we examine the underlying structure and composition of muscle, the factors that allow humans to perform highly specialized tasks, the effects of regular activity on the structure and composition of muscle, and the functional consequences of the **adaptations** that occur. Finally, we examine the effects of aging on muscle and the role of muscular adaptations in improving health and well-being.

Skeletal Muscle and Human Survival

Skeletal muscles are the largest tissue mass of the human body, constituting in excess of 600 individual muscles. The most dramatic example of the crucial role of skeletal muscles in our survival comes from the "fight-or-flight" response, originally conceptualized by physiologist Dr. Walter Cannon. This term refers to a predictable strategy for perceived danger, namely engaging in combative action or fleeing from the stressful agent. Although these responses are not commonly used today, given the technological and social advances that have occurred, evolutionary theorists have proposed that the basic number, mass, architecture, and composition of our muscles represent the selective and adaptational consequences of preceding generations in which drawing and hauling were labors of survival and hostile encounters were common occurrences.

We each have inherited a large mass of tissue whose fundamental function is to contract and generate force and power. Muscles allow us to perform a diverse array of motor activities that range from the most delicate of tasks to those requiring large amounts of force and velocity. To allow us to function effectively and efficiently, an elaborate hierarchy of muscular organization has evolved. Muscles are organized in subsets, each with a common synergistic action and each able to contribute depending on the demands of the task. Muscle subgroups exist for posture, locomotion, vision, mastication, breathing, and a range of other functions.

Another level of organization allows selected muscles to perform specific mechanical functions efficiently and effectively. To a large degree, this capability can be attributed to the design and composition of the individual fibers that constitute the muscle. The individual muscle fibers can be distinguished by differences in the type and amount of proteins and substrates. The inherent ability to perform skilled motor activities not only is crucial to independent living, but also enriches our life by enabling us to interact with our physical environment.

Muscles also perform functions that either directly or indirectly contribute to our health and well-being. Shivering thermogenesis, for example, in which contracting muscles produce heat, can directly help us maintain body temperature in cold environments where heat loss is increased. Even at rest, the heat liberated by muscle significantly contributes to body temperature **homeostasis**. Muscles offer protection during falls and collisions, buffering the impact that could fracture bones and damage internal organs.

The control of body weight and adiposity can to a significant degree be managed by exercise-induced increases in caloric expenditure, which is primarily mediated by the contracting muscles. Increasing evidence suggests that regular exercise promotes bone health, reduces hypertension and lipid abnormalities, and can even improve cognitive function. Perhaps the best example of the direct role of muscles in disease prevention and management is type 2 diabetes. Type 2 diabetes, a disease in which blood glucose regulation and insulin resistance are abnormal, can to a large extent be controlled by regular physical activity. This is so because blood glucose disposal occurs primarily in skeletal muscle. The phenotypic character of trained muscle can also assist in managing chronic heart failure and chronic obstructive lung disease by attenuating the associated reduction in fiber area and fiber composition and thus the strain that accompanies these diseases. Trained muscle can also prevent excessive accumulation of the metabolic by-products of muscle contraction, such as lactic acid and **reactive oxygen species (ROS)**, which can negatively affect cardiovascular and ventilatory function.

We each inherit, as a characteristic of our species, a large mass of muscle tissue. Muscles are specialized to contract, enabling us to interact with each other and with our environment. To function optimally, muscles must contract regularly. The central role of muscle in health and well-being is perhaps best illustrated by recent research suggesting that the contracting muscle releases myokines, which can modulate function in other tissues and organs and are potentially important in the prevention and management of certain diseases.

Finally, an appealing and increasingly accepted hypothesis is that selected amino acids released by the working muscle can act as precursors for different neurotransmitters, the accumulation of which can dampen undesirable mood states such as depression.

Muscle Cell: Composition, Structure, and Function

The **mechanical properties** of a single muscle are, to a large degree, the product of its constituent muscle fibers. In humans, most muscles contain a mixture of different fiber types. Muscle fibers are regulated not independently, but rather in subgroups of fibers of the same type, with each subgroup activated by a single motor nerve. Each subgroup with its nerve constitutes a motor unit. The number of motor units in a muscle depends on its function. Muscles that are specialized for large power output, as in locomotion, have large numbers of motor units and a large number of fibers within each motor unit. In contrast, muscles that are specialized for very precise tasks, such as eye movement, display a small number of motor units with relatively few fibers in each motor unit. By modulating the impulse frequency, the motor unit pools recruited, and the timing of these events, the nervous system is able to generate purposeful, graduated motor responses.

Excitation and Contraction Processes

The muscle cell or muscle fiber is the fundamental force-generating entity, capable of translating instructions from a nerve into a desired mechanical event.

To do this, it has a variety of special structures that mediate specific processes (figure 8.1). The initiating event involves the conversion of the transmitter acetylcholine, released from the motor nerve into the neuromuscular junction, into an action potential. The action potential is, in turn, conducted onto the surface membrane of the fiber and ultimately, by invaginations of the surface membrane, into the interior of the fiber (figure 8.2). Repetitive action potentials, which are obligatory for tetanic contractions, require the protection of muscle membrane excitability, a function that is facilitated by restoration of transmembrane ionic gradients. A complex of special membrane channels and transporters accomplishes this by regulating the selective diffusion and transport of sodium, potassium, and chloride, all important in membrane potential.

Muscle fibers are covered by a membranous network (the sarcolemma) that penetrates into the interior of the fiber at regular intervals as transverse tubules (T-tubules). The T-tubules form a network that extends throughout the muscle fibers, encircling the individual myofibrils. Once inside the fiber via the T-tubules, the action potential must be able to induce the release of stored calcium ions (Ca^{2+}) from the sarcoplasmic reticulum (SR) if it is to increase the free calcium ion concentration ($[Ca^{2+}]_f$) enough to cause a tetanic contraction (approximately 100-fold) (Berchtold, Brinkmeier, and Muntener 2000). The SR is a system of tubes that extend longitudinally throughout the muscle fiber, connecting the lines of T-tubules and lying along the longitudinal surfaces of the myofibrils. The SR has three primary functions, namely to store Ca^{2+}, to release Ca^{2+}, and to sequester Ca^{2+} back into the SR. Calcium release from the SR is controlled by a protein called the Ca^{2+}–release channel (CRC) or ryanodine receptor (RyR). According to current evidence, the action potential in the T-tubule is sensed by a dihydropyridine receptor (DHPR). The resulting changes in the DHPR appear to lead to opening of the CRCs by both mechanical and chemical coupling. The opening of the CRCs leads to a rapid increase in $[Ca^{2+}]_f$ levels in the sarcoplasm (the gelatinous material between the myofibrils). Because the SR envelops the individual contracting myofibrils and allows close proximity between the T-tubules and CRCs, a near-synchronous opening of the CRCs occurs in a fiber in response to an action potential, allowing rapid increases in $[Ca^{2+}]_f$ and, consequently, rapid activation of the proteins constituting the contractile apparatus. A common term for this first category of signaling processes, which includes the sequence of events beginning with the action potential and culminating with an increase

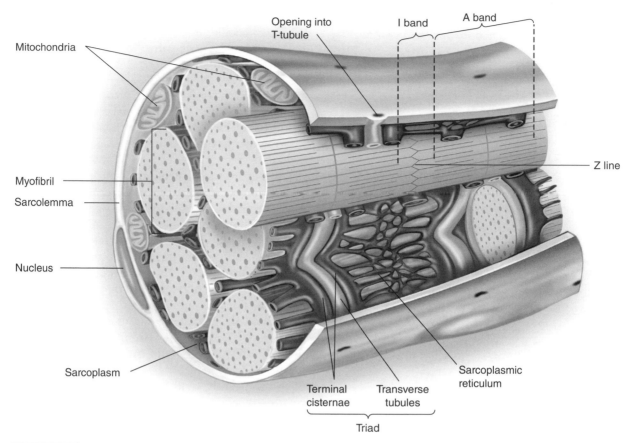

Mitochondria

Myofibril

Sarcolemma

Nucleus

Sarcoplasm

Opening into T-tubule

I band

A band

Z line

Terminal cisternae

Transverse tubules

Sarcoplasmic reticulum

Triad

FIGURE 8.1 Mammalian muscle fiber. The diagram depicts the constituents of a muscle fiber, which is composed of individual myofibrils. The sarcoplasmic reticulum (SR) has been described as a water jacket that surrounds each myofibril. The SR is composed of two regions, the terminal cisternae and the longitudinal reticulum. The terminal cisternae, which cross the myofibrils, lie on both sides of the T-tubules. A triad is composed of two terminal cisternae and one T-tubule. The longitudinal reticulum of the SR runs parallel to the myofibrils, inserting into the terminal cisternae. Most of the calcium ion (Ca^{2+})–release channels (CRCs) are located in the terminal cisternae, whereas Ca^{2+}–adenosine triphosphatase (ATPase), the enzyme responsible for pumping Ca^{2+} back into the SR, is mainly located in the longitudinal reticulum.

Reprinted, by permission, from J.H. Wilmore and D.L. Costill, 2008, *Physiology of sport and exercise*, 4th ed. (Champaign, IL: Human Kinetics), 28.

in $[Ca^{2+}]_f$ levels, is **excitation–contraction (E-C) coupling**.

The second category of signaling processes that mediate contraction involves the sensing of elevated $[Ca^{2+}]_f$ levels by the myofibrillar complex, which includes the contractile proteins actin and myosin and the regulatory proteins troponin and tropomyosin. The elevated $[Ca^{2+}]_f$ is sensed by a subunit of troponin, which induces movement of tropomyosin. This removes inhibition between actin and myosin by the thin filament, allowing these proteins to move from the dissociated or weakly bound state into a strong-binding, force-generating state. Movement is generated when the number of actin and myosin molecules in the force-generating state is sufficient to overcome an external load, allowing the individual myosin molecules to cycle to successive actin sites, which results in shortening of the muscle fiber. The

myosin molecule possesses a unique structure that allows it to interact with actin and to convert chemical energy into work. Each myosin molecule contains two heavy chains (HCs) and four light chains (LCs), all of which regulate the force–velocity characteristics of the fiber (Schiaffino and Reggiani 1996).

Relaxation from the force-generating state in the muscle fiber begins with termination of the motor nerve impulses and reductions in $[Ca^{2+}]_f$ levels. The reduction in $[Ca^{2+}]_f$ levels occurs as a consequence of both inhibition of Ca^{2+} release from the CRCs and uptake of Ca^{2+} back into the SR, where it is stored. As a result of the reduction in $[Ca^{2+}]_f$ concentration, Ca^{2+} is unloaded from the troponin subunit, allowing the thin filament to resume its inhibiting position, which permits actin and myosin to dissociate or assume a weak-binding, non-force-generating state. As with the rapid increase in $[Ca^{2+}]_f$ during activation, the reduction in $[Ca^{2+}]_f$ during relaxation also occurs rapidly, given the extensive SR network and the density of the Ca^{2+}–adenosine triphosphatase (ATPase), the enzyme responsible for pumping Ca^{2+} into the SR.

Muscles and muscle fibers also contain an elaborate network of connective tissue and cytoskeletal proteins that protect the architecture and organization of muscle, in addition to a host of other functions. Although these properties are not addressed in this chapter, it is recognized that they are vital to function and malleable in character.

Processes Involved in Energy Utilization

The processes involved in contraction and relaxation in the muscle cell are dependent on energy. The free energy is provided by the hydrolysis of adenosine triphosphate (ATP), which is generated from the combustion of foods such as fat and carbohydrate (CHO). The majority of the ATP consumed in working muscle is used for three specific functions: for transporting the cations sodium (Na^+) and potassium (K^+) across the sarcolemma and T-tubule, for sequestering Ca^{2+} into the SR, and for actomyosin cycling. To reestablish

excitability and allow for the regeneration of action potentials, the transport of Na^+ and K^+ is necessary following their flux into and out of the cell, respectively, during an action potential (Clausen 2003). An action potential begins with depolarization, which occurs when Na^+ channels open, allowing Na^+ to flow from outside the cell, where the concentration is greatest, to inside the cell. Repolarization occurs when K^+ channels open, allowing K^+ to flow from the inside of the cell down its concentration gradient to the outside of the cell. For action potentials to be generated at high frequencies, namely 40 to 50 impulses per second, which may be necessary to raise $[Ca^{2+}]_f$ to levels required to realize the full force potential of the cell, the transmembrane gradients for Na^+ and K^+ must be rapidly reestablished. This is accomplished via a specialized protein called Na^+-K^+-ATPase, which can both generate free energy from the hydrolysis of ATP and use the free energy to induce conformational changes in the protein and, in the process, transport Na^+ out of the cell and K^+ into the cell. Na^+-K^+-ATPase is a membrane protein composed of α and β subunits, each with several different molecular forms, called **isoforms**, that can regulate the exchange of Na^+ and K^+ transport by changing the catalytic rate.

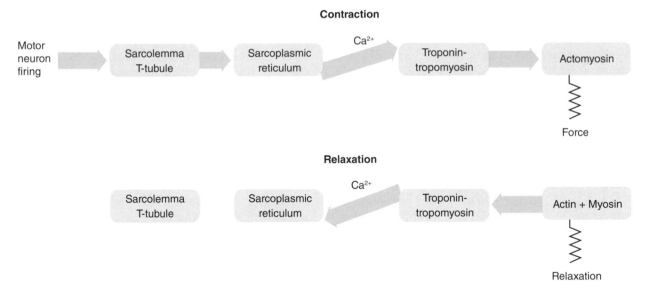

FIGURE 8.2 Excitation and contraction processes in muscle. The top panel illustrates the sequence of events involved in contraction beginning with the motoneuron impulses to the sarcolemma of the muscle fiber. The action potential is conducted to the interior of the fiber via the T-tubule, which is then transmitted to the calcium ion (Ca^{2+})–release channel of the sarcoplasmic reticulum (SR). With activation, the channels open, resulting in an increase in $[Ca^{2+}]_f$. The increase in $[Ca^{2+}]_f$ results in Ca^{2+} binding to a subunit of troponin, ultimately leading to movement of tropomyosin and removal of a preexisting inhibition between actin and myosin and to contraction. Relaxation occurs when the motoneuron impulses stop, and the Ca^{2+} is pumped back into the SR by Ca^{2+}–adenosine triphosphatase (ATPase), thereby lowering $[Ca^{2+}]_f$ and reestablishing the inhibition between actin and myosin.

Reprinted, by permission, from H.J. Green, 2004, "Mechanism and management of fatigue in health and disease," *Canadian Journal of Applied Physiology* 29(3): 264-273.

The transport of Ca^{2+} into the SR is also mediated by a specialized protein called the Ca^{2+}-ATPase. The Ca^{2+}-ATPase is similar to Na^+-K^+-ATPase in that the energy liberated by the hydrolysis of ATP causes a conformational change in the protein, allowing it to pump $[Ca^{2+}]_f$ back into the SR, where it can be stored in preparation for another contraction cycle. The regulation of $[Ca^{2+}]_f$ cycling is primarily dependent on both Ca^{2+} release and Ca^{2+} uptake. In addition to Ca^{2+}-ATPase and Na^+-K^+-ATPase, there is a third ATPase called myosin or actomyosin ATPase. The actomyosin ATPase uses the majority of energy in the working muscle cell. This ATPase forms part of the myosin HC component of the molecule. When activated, it provides the energy needed for actomyosin to cycle through a series of force-generating steps before dissociating to begin the next cycle. Other ATPases in the cell serve other functions; however, the ATP requirements of the other ATPases, although important, are not as pronounced.

Fuels and Metabolic Pathways Involved in Energy Production

To satisfy the work functions of the ATPases, a large increase in ATP regeneration must occur. Adenosine triphosphate is generated by an elaborate network of enzymes that form different metabolic pathways and segments (Hochachka 1994; Meyer and Foley 1996). These pathways are specialized to satisfy the energy requirements of the cell under special conditions or circumstances (figure 8.3). For example, if the requirements for ATP are relatively modest and if oxygen and other substrates are not limiting, the mitochondria, which contain the assembly of enzymes that form the citric acid cycle (CAC) and the electron transport chain (ETC), can use either fat or carbohydrate (CHO) to generate ATP and, in the process, produce the by-products water and carbon dioxide. Energy generated by mitochondrial respiration is termed **oxidative phosphorylation**. Alternatively, if the force and power generated by the muscle are large and consequently the demand for energy is high, ATP can be rapidly generated by another metabolic pathway, namely **glycolysis**. In this pathway, ATP regeneration is restricted to the use of CHO. Because this pathway does not use oxygen, it can also become the primary short-term source of energy for the working muscle in oxygen-deprived environments such as at altitude. Although the recruitment of this pathway to meet the energy demands of the cell has several appealing features, the trade-off is the final by-product that is produced, namely lactic acid. The accumula-

tion of lactic acid, which is a strong acid, can cause several undesirable effects unless properly buffered and disposed of.

Substrates, either fats or CHO, are essential fuels used by the metabolic pathways to respond to the energy needs of the cell. These substrates can be either stored endogenously in the cell as triglycerides and glycogen or delivered to the cell by the circulatory system in the form of fatty acids and glucose. The blood fatty acids are liberated from the adipose tissue, whereas blood glucose is released from the liver.

The transport of substrates and metabolites in and out of the working muscle cell must be carefully regulated to meet the prevailing demands and to protect the intracellular integrity of working muscle. In the case of substrates, a lack of availability can blunt ATP production depending on the metabolic pathway recruited. This is especially the case with CHO, which is in limited store in the liver and muscle. A depletion of this substrate from the liver, for example, can result in hypoglycemia and central nervous system dysfunction. It is also common for glycogen reserves to be essentially depleted in the working muscle. Depletion of muscle glycogen has often been associated with an inability to sustain a desired force or power (fatigue). Humans appear to be particularly dependent on glycogen as a fuel source during contractile activity.

An elaborate membrane network exists to precisely regulate what enters and leaves the cell. A barrier to the free diffusion of various constituents exists in the form of the surrounding membrane, which by virtue of its phospholipid composition is essentially impermeable to most molecular species. The transmembrane movement of various constituents is facilitated by numerous proteins embedded in the phospholipid bilayer that form channels, transporters, and pumps. This allows for the highly specific regulation of numerous molecules. Such is the case with glucose and fatty acids, both of which have specially designed carrier proteins that facilitate their transport across the membrane (Furuhashi and Hotamisligil 2008; Holloszy 2003). These proteins are expressed as isoform families of fatty acid–binding protein transporters (FABP) and glucose transporters (GLUT). We have also seen that special channels exist for Na^+ and K^+ diffusion and a wide range of other cations and anions. Special proteins also exist for the transport of major metabolic by-products of glycolysis, namely lactate and hydrogen ions (H^+). Movement of lactate and H^+ across the plasma membrane is facilitated by monocarboxylate transporters (MCT) (Juel and Halestrap 1999).

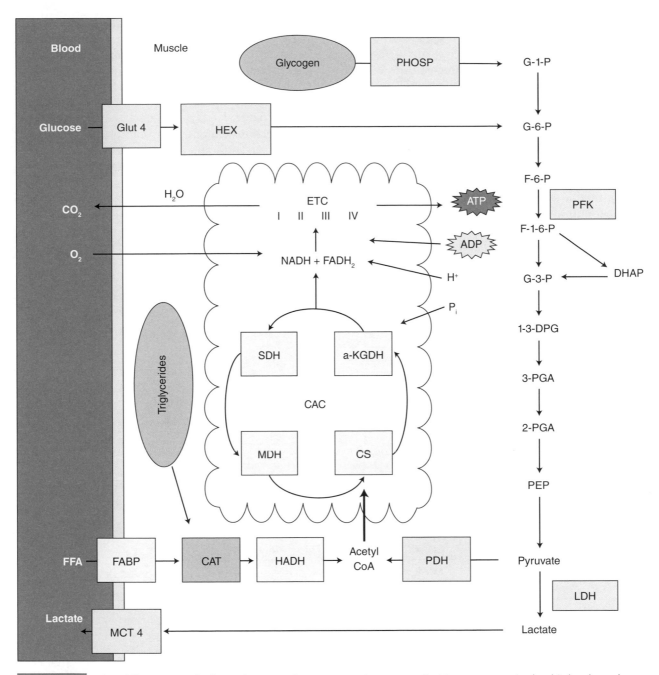

FIGURE 8.3 The different metabolic pathways with representative or rate-limiting enzymes (or both) that have been used to assess the potential of the pathway. Glycogenolysis, phosphorylase (PHOSP); glycolysis, phosphofructokinase (PFK); pyruvate oxidation, pyruvate dehydrogenase (PDH); glucose phosphorylation, hexokinase (HEX); oxidative potential, citrate synthase (CS), malate dehydrogenase (MDH); succinate dehydrogenase (SDH), election transport chain (ETC) fatty acid oxidation, β-hydroxacyl-CoA dehydrogenase (β-HADH); pyruvate reduction, lactate dehydrogenase (LDH). Also depicted are the transporters, namely the glucose transporter (GLUT4), the fatty acid transporter (FABP), and the lactate transporter (monocarboxylate transporter [MCT] 4). Carnitine acetyl transferase (CAT) is the enzyme involved in the transfer of fatty acids across the mitochondrial membrane.

Reprinted from H.G. Green and T.A. Duhamel. Unpublished data.

For muscle fibers to be able to contract and perform work, a close coordination must exist between those processes generating ATP and those processes using ATP. Moreover, this coordination must be precisely regulated because, at rest, ATP exists only at very low concentrations, sufficient for only a few contractions. Research has repeatedly shown that even with the most demanding tasks, ATP levels are extremely well protected, with reductions of greater than 25% rarely reported (figure 8.4). The protection of ATP during periods of increased demand depends on a close integration between the catalytic activity of the ATPases, the enzymes that use ATP, and the flux rate in the metabolic pathways that produce ATP (Meyer and Foley 1996). According to current theory, the catalytic rate at which each of the ATPases performs increased work involving cation transport and actomyosin cycling increases the rate at which ATP is hydrolyzed. Stored levels of phosphocreatine (PCr) provide short-term protection of ATP, because the near-equilibrium nature of the enzyme creatine phosphokinase makes the hydrolysis of PCr very sensitive to decreases in ATP. Production of ATP can also be supplemented by the adenylate kinase reaction, also a near-equilibrium enzyme. However, rapid activation of glycolysis and oxidative phosphorylation are essential if ATP is to be protected beyond the

first several contractions. Disturbances in the energy status of the cell are sensed via the accumulation the by-products of ATP hydrolysis, such as adenosine diphosphate (ADP), PCr, creatine, and inorganic phosphate (P_i), all of which are interrelated given the near-equilibrium nature of these reactions. Free ADP, in particular, is believed to be a major stimulus in the recruitment of oxidative phosphorylation and in glycogenolysis and glycolysis.

If the metabolic pathways are unable to respond with adequate production of ATP to meet demands, the activity of one or more of the ATPases is dampened and E-C is compromised, which in turn reduces ATP utilization. The dampening of the ATPases is believed to result from excessive accumulation, again, of one or more of the high-energy phosphate by-products, which in effect act as sensors of the energy status. In addition, other by-products such as H^+, ROS, and temperature (all of which can accu-

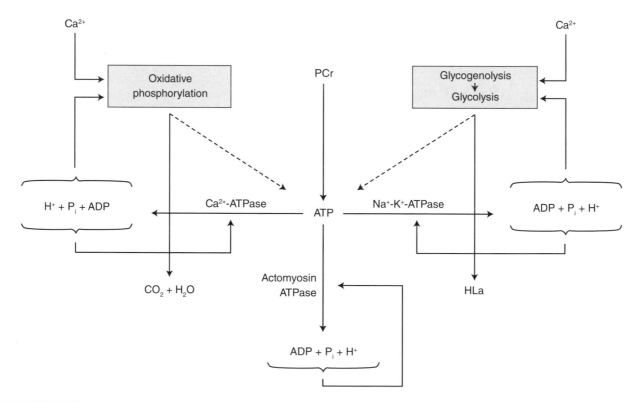

FIGURE 8.4 This schema depicts adenosine triphosphate (ATP) as the dependent variable whose concentration must remain protected. The hydrolysis of ATP by the different adenosine triphosphatases (ATPases) results in the production of metabolic by-products. The phosphorylation potential is sensed by the metabolic pathways (glycolysis, oxidative phosphorylation), which can stimulate increases in pathway flux rate and consequently increases in ATP production. Excessive accumulation of the by-products of ATP hydrolysis can inhibit the ATPases, resulting in reduced ATP utilization. Reductions in ATPase activity would be expected to result in fatigue. The terminal products for oxidative phosphorylation are carbon dioxide (CO_2) and water (H_2O), whereas lactic acid (HLa) is the terminal product of glycolysis.

Reprinted, by permission, from H.J. Green, 2000, Muscular factors in endurance. In *Encyclopedia of sports medicine: Endurance in sports*, edited by R. Shephard and P.O. Åstrand (Oxford, UK: Blackwell Science Ltd.), 179.

mulate with exercise) are potential inhibitory agents. It is important to emphasize that these by-products may also result in damage to proteins involved in E-C, which could depress ATP consumption as well. Biochemists use the term **phosphorylation potential** to refer to the concentration of high-energy compounds in muscle. The status of the phosphate energy system, given the importance of its by-products in regulating other metabolic pathways, is viewed as the principal integrator in coupling ATP demand to ATP supply.

An imbalance between ATP utilization and production ultimately limits the muscle mechanical response, resulting in fatigue (Allen, Lamb, and Westerblad 2008; Fitts 2008). Although fatigue, defined as an inability to produce a desired mechanical response, may be undesirable in the short term, it is protective in the long term, preventing the serious consequences associated with ATP depletion in the muscle cell. The need to protect ATP levels also offers some insight into the desired outcomes of training. A key question is, How can the phosphorylation potential of the cell, which is determined primarily by the levels of ATP and PCr, be protected while the muscle cell performs a greater amount of work following exercise training?

For muscle cells to contract, the neural signal must be conducted into the interior of the fiber, ultimately enabling the contractile proteins actin and myosin to associate in a force-generating state. The viability of the individual processes involved in excitation and contraction (E-C) depends on the activation of three enzymes, namely Na^+-K^+-ATPase, Ca^{2+}-ATPase, and actomyosin ATPase. The activation of these enzymes results in rapid increases in ATP utilization. The products of ATP utilization increase the flux in the metabolic pathways oxidative phosphorylation and glycolysis, which increase ATP production and defend ATP homeostasis. An inability of the metabolic pathways to produce sufficient ATP, as during heavy exercise, results in progressive accumulation of metabolic by-products that can inhibit the ATPases. Thus, excessive reductions in ATP are prevented at the expense of fatigue.

Muscle Fiber Types and Subtypes

The vast array of motor skills that we are capable of performing, ranging from the achievement of rapid velocities and large power outputs to sustained effort performed over many hours and even days, is possible because of differences in the structure and composition of the muscle cell. To achieve specialization in mechanical function, different fiber types and subtypes have evolved. A fiber type and its subtypes generally display a coordinated set of properties, each supporting a particular function. For example, fibers adept at generating large velocities and power outputs must first be able to rapidly convert the neural command into increases in $[Ca^{2+}]_f$ and then quickly translate the Ca^{2+} signal in rapid rates of cross-bridge attachment and detachment. Also crucial, given the rapid velocities, is the requirement for quick relaxation rates and consequently the ability to rapidly lower $[Ca^{2+}]_f$ levels. Muscle fibers specialized for high velocities and high power outputs also have unique energy needs. Metabolic pathways must be specialized for rapid activation and large flux rates to meet the demands for ATP generated by this type of work. The correlated expression of these basic properties in the muscle cell appears to exist across species. However, in the case of rodents, as an example, the potential of these individual properties is severalfold higher than in humans, resulting in much greater shortening velocities and ATP synthesis rates.

Fiber Type and Subtype Properties

Specialization at the level of distinct processes in the cell is primarily accomplished by the amount, type, and subcellular distribution of protein and protein isoforms and, particularly in the case of enzymes, the regulatory factors that govern the intrinsic activity (figure 8.5). A given protein can exist in several different isoforms or, in the case of enzymes, isozymes. Isoforms generally display only minor differences in composition. However, these differences in composition allow for greater specialization in function and a more appropriate response in different intracellular environments. To exploit the unique properties of the proteins and protein isoforms, differences in regulatory stimuli are needed. Regulatory stimuli can be localized in nature, generated specifically by events within the contracting cell or secondary to messages directed at the cell from external stimuli. Many of these external stimuli arise from changes in

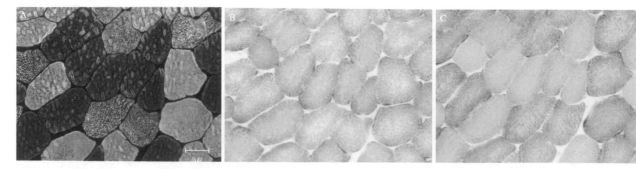

FIGURE 8.5 Cross section of fibers obtained from a tissue sample extracted from the vastus lateralis. Panel A: Muscle cross section incubated with primary antibodies against myosin HCI, HCIIa, and HCIIx, followed by incubation with isotype-specific fluorochrome-labeled secondary antibodies. Fibers are classified as type I (dark), type IIA (moderate), and type IIAX/X (moderate dark). Note that in humans the myosin HCIIx antibody shows some reactivity with HCIIa. Type IIAX represents a very small percentage of the fibers. Panel B: Serial cross section of panel A showing succinic dehydrogenase (SDH) activity staining. Panel C: Serial cross section of panels A and B showing glyceraldehyde phosphate dehydrogenase (GPDH) staining. SDH and GPDH are used as markers of oxidative and glycolytic potentials, respectively. Bar represents 50 μm. *Note:* These illustrations would normally be presented in different colors.

Used with permission from Joe Quadlrilatero, University of Waterloo, 2010.

the level of blood hormones and neurotransmitters. These hormones act, in large part, on specialized receptors embedded in the surface membrane of the cell, triggering a cascade of responses that ultimately allow increased or decreased function of the targeted protein. **Signal transduction** processes are vital in activating the metabolic pathways and E-C processes during exercise and in controlling the expression of specific phenotypes that result from training.

The classification of muscle fibers into distinct types and subtypes is not without controversy given the many potential properties that can be used for classification. Indeed, many schema are currently in use, producing a confusing array of terminology.

The most popular classification schema is based on the contractile protein myosin, specifically the HC component of the myosin that contains the ATPase (Reggiani, Bottinelli, and Steinen 2000). Each myosin molecule contains two HCs and four LCs. Myosin HCs contain several distinct isoforms (approximately nine in all) that are expressed in a species-specific and tissue-specific manner. In the case of human skeletal muscle, two major isoforms are recognized, namely HCI and HCII, which form the basis of the type I and type II classifications, respectively (table 8.1). In humans, HCII also exists as two isoforms, namely HCIIa and HCIIx. These fibers are called type IIA and type IIX, respectively.

TABLE 8.1 Light and Heavy Chain Composition in Histochemically Identified Fiber Types in Humans

Fiber type	Light chains	Heavy chains
IIA	$(LC1f)_2(LC2f)_2$	$(HCIIa)_2$
	$(LC1f)(LC3f)(LCS2f)_2$	$(HCIIa)_2$
	$(LC3f)_2(LC2f)_2$	$(HCIIa)_2$
IIX	$(LC1F)_2(LC2f)_2$	$(HCIIx)_2$
	$(LC1f)(LC3f)(LC2f)_2$	$(HCIIx)_2$
	$(LC3f)_2(LC2f)_2$	$(HCIIx)_2$
I	$(LC1s)_2(LC2s)_2$	$(HCI)_2$

Each myosin contains two heavy chains (HCs) and four light chains (LCs). In a given myosin, two of the LCs are essential (LC1, LC3), and two are regulatory or phosphorylatable (LC2). The HC isoforms include HCIIa, HCIIx, and HCI. Both fast (f) and slow (s) isoforms exist for the myosin LCs.

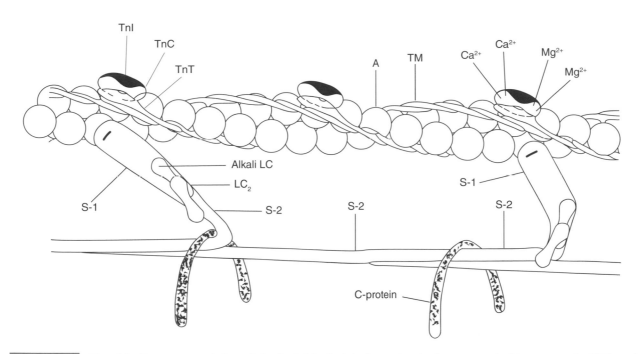

FIGURE 8.6 The thin filament consisting of the individual globular actin, the long-stranded tropomyosin (TM) and three troponin subunits (TnI, TnT, TnC), and the myosin molecule. The myosin contains the heavy chains (HCs), which consist of the S-1 and S-2 portions, and the different light chains (alkali or essential LC and LC2 or phosphorylatable LC).

Reprinted from R.L. Moss, G.M. Diffee and M.L. Greaser, 1995, "Contractile properties of skeletal muscle fibers in relation to myofibrillar protein isoforms," *Rev Physiol Biochem Pharmacol* 126: 1-63, with kind permission of Springer Science and Business Media.

It is also possible for various myosin HC isoforms to exist in a single fiber, which provides for additional subtypes such as type IIC, type IC, and type IIAX. Rodents contain a third HC isoform, namely HCIIb. Previously, HCIIx was erroneously labeled HCIIb in humans. A myosin molecule also contains two classes of LCs, namely the essential LCs (LC1, LC3) and the regulatory LCs (LC2), each of which have slow and fast isoforms.

The use of the myosin HC for fiber type and subtype classification provides insight into the functional nature of the fiber. The myosin HC isoform, in conjunction with the LC isoform, which acts as a secondary modulator, is the primary determinant of the velocity and power that can be generated (figure 8.6). It is for this reason that the type I and type II fibers have also been labeled slow-twitch (ST) and fast-twitch (FT), respectively (figure 8.7).

Using the myosin HC isoform as a basis for classification yields a set of properties generally consistent with differences in velocity and power characteristics of the different fiber types. For example, type IIA and type IIX fibers display a well-developed SR, with high densities of both the CRCs and Ca^{2+}-ATPase proteins. This allows for both rapid increases and decreases in $[Ca^{2+}]_f$ that are necessary for quick contraction and relaxation cycles (figure 8.8). In contrast, type I muscles, which display the HC isoform, are incapable

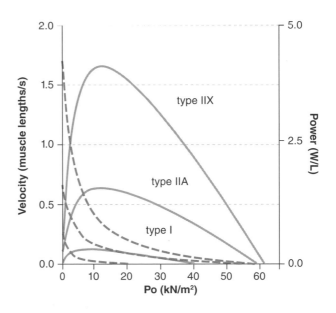

FIGURE 8.7 Force–velocity and power characteristics of selected fiber types. Velocity characteristics are represented by the dashed lines, whereas power (the product of force × velocity) is represented by the solid lines. Type I, type IIA, and type IIX represent the different fiber types.

Reprinted from the *Journal of Electromyography Kinesiology*, Vol. 9, R. Bottinelli, M.A. Pellegrino, M. Canepari, R. Rossi and C. Reggiani, "Specific contributions of various muscle fibre types to human muscle performance: An in vitro study," pp. 87-95, Copyright 1999, with permission from Elsevier.

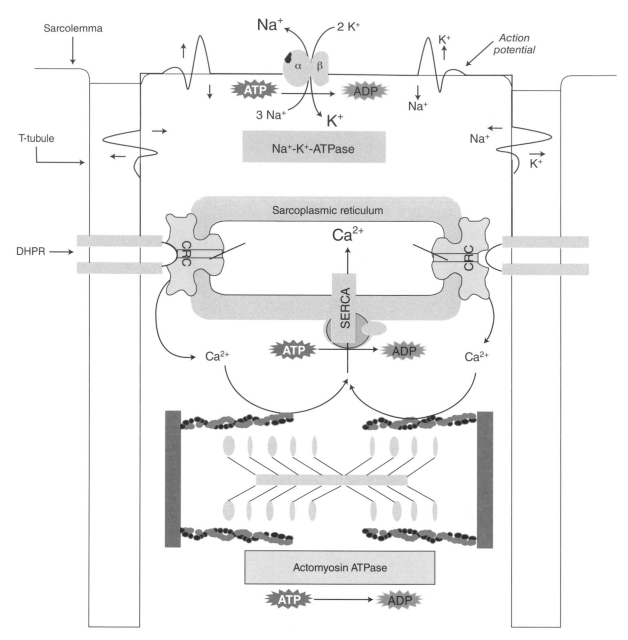

FIGURE 8.8 Essential elements of the Na+-K+-ATPase, Ca²⁺–release channels (CRCs), and Ca²⁺-ATPase (sarcoplasmic–endoplasmic reticulum calcium ATPase, or SERCA) of the sarcoplasmic reticulum (SR). The dihydropyridine receptor (DHPR), which is located in the T-tubules, is involved in communication between the T-tubules and CRC, as shown.

Reprinted from H.G. Green and T.A. Duhamel. Unpublished data.

of rapid cross-bridge cycling and are consequently unable to generate large velocities and large power outputs. As one would expect, the SR in these fibers is not well developed. Differences in SR between fiber types also extend to a complex of regulatory proteins and isoform types (table 8.2). In skeletal muscle, the Ca²⁺-ATPase exists as two sarcoplasmic–endoplasmic reticulum calcium ATPase (SERCA) isoforms. SERCA 1 is expressed in type II fibers, whereas SERCA 2 is expressed in type I fibers. These isoforms also convey properties consistent with the desired regulation of

the $[Ca^{2+}]_f$. The potential of the various metabolic pathways is also generally consistent with the HC composition of the fiber (table 8.3). Fast-contracting fibers have high activities of enzymes that allow rapid rates of high-energy phosphate transfer, glycogenolysis, and glycolysis. In contrast, slow-contracting type I fibers, which have relatively slow rates of ATP utilization as dictated by Ca²⁺-ATPase and actomyosin ATPase, do not have such well-developed pathways. The **oxidative potential** of type I fibers is high (as might be expected, given their relatively low energy

TABLE 8.2   Characteristics of Sarcoplasmic Reticulum in Human Fiber Types and Subtypes

	Fiber types		
	I	IIA	IIX
Ca^{2+}-ATPase			
Protein	MOD	MOD-HI	HI
Isoform	SERCA 2a	SERCA 1	SERCA 1
Maximal activity	MOD	MOD-HI	HI
Ca^{2+} uptake	MOD	MOD-HI	HI
Ryanodine receptor (RyR)			
Isoform	RyR I	RyR I	RyR I
Ca^{2+}-release rate	MOD	MOD-HI	HI

Specific characteristics are based on those actually measured in human single fibers. Where no experimental data were available on humans, the characteristics were projected based on rodent muscle. Ca^{2+} = calcium ion; ATPase = adenosine triphosphatase; MOD = moderate; HI = high; SERCA = sarcoplasmic–endoplasmic reticulum calcium ATPase. Ryanodine receptor (RyR) is also another term used for the Ca^{2+}-release channel.

TABLE 8.3   Potential of Metabolic Pathways and Segments in Human Fiber Types and Subtypes

	I	IIA	IIX
High-energy phosphate (CPK)	MOD	MOD-HI	HI
Glycogenolytic (PHOSPH)	MOD	MOD-HI	HI
Glycolytic (PFK)	MOD	MOD-HI	HI
Oxidative (CS, SDH)	HI	MOD	MOD-LO
β-Oxidation (HADH)	HI	MOD	MOD-LO

Relative pathway potential is based on maximal activities of rate-limiting or representative enzymes. MOD = moderate; HI = high; LO = low; CPK = creatine phosphokinase; PHOSPH = phosphorylase; PFK = phosphofructokinase; CS = citrate synthase; SDH = succinic dehydrogenase; HADH = hydroxyacyl–CoA dehydrogenase.

demands), providing for a high dependence on oxidative phosphorylation. Such is also the case with the type IIA fibers. However, these fibers also display a high glycolytic potential.

The specialization of the metabolic pathways to meet specific energy demands also extends to different isozymes. Both creatine phosphokinase (CPK) and lactate dehydrogenase, as examples, have multiple isoforms, allowing them to respond to unique challenges. Muscle fibers with a high oxidative potential are also liberally supplied with a vascular network of many capillaries and high levels of cellular perfusion, necessary to meet the demands for oxygen and other substrates to support mitochondrial respiration, as well as to remove the by-products generated from the contractile activity (table 8.4). Associated with the unique metabolic pathway structure in the different fiber types are the transporter proteins (table 8.5). GLUT4, MCT1, and FABP are highest in fibers with high oxidative potential, both type I and type IIA, whereas the lowest concentration is observed in type IIX. In contrast, MCT4 is higher in type IIX and type IIA compared with type I.

Somewhat at odds with the pattern of properties listed for defined fiber types is the Na$^+$-K$^+$-ATPase, the cation pump important in protecting membrane excitability (table 8.6). Na$^+$-K$^+$-ATPase is a heterodimer consisting of two subunits (α and β), each with several isoforms. The density and isoform composition of the α and β subunits of this protein appear to be scaled to the oxidative potential of the cell and not to the speed of contraction of the cell. This is curious because type II fibers, both type IIA and type IIX, need a high frequency of impulses to generate a maximal

TABLE 8.4　Fiber Type Distribution, Area, and Capillarization in Human Male Adult Muscle

	Fiber type			
	I	IIA	IIX	IIAX
Distribution (%)	50	33	12	5
Area (μm^2)	4,705	5,305	4,082	4,175
Caps (n)	5.37	5.13	3.65	4.19
Caps/area ($\mu m^2 \cdot 10^{-3}$)	1.17	0.99	0.90	1.0

Fiber type characteristics are based on tissue samples extracted from the vastus lateralis of untrained male adults. Area = average area of each fiber type; Caps = average number of capillaries around each fiber type; Caps/area = average area of the fiber supplied by a capillary.

From Green et al. Unpublished data.

TABLE 8.5　Membrane Transporters and Human Muscle Fiber Types and Subtypes

	Fiber type		
	I	IIA	IIX
Glucose			
GLUT1	LO	LO	LO
GLUT4	HI	HI	LO
Lactate			
MCT1	HI	MOD	LO
MCT4	LO	HI	HI
Fatty acids			
FABP	HI	MOD	LO

Relative concentrations for humans projected based on rodent muscle fiber types. HI = high; MOD = moderate; LO = low; GLUT = insulin-sensitive glucose transporter; MCT = monocarboxylate transporter; FABP = fatty acid-binding transporter.

TABLE 8.6　Characteristics of the Na$^+$-K$^+$-ATPase in Human Fiber Types and Subtypes

	Fiber types		
	I	IIA	IIX
Protein	HI	HI	MOD
Isoform	$\alpha_1\beta_1 / \alpha_2\beta_1$	$\alpha_1\beta_1 / \alpha_1\beta_2$ $\alpha_2\beta_2 / \alpha_2\beta_2$	$\alpha_2\beta_2$
Maximal activity	HI	HI	MOD

Specific characteristics are based on those actually measured in human single fibers. Where no experimental data were available on humans, the characteristics were projected based on rodent muscle. Isoform distribution represents the dominant α and β subunit distribution. Na$^+$ = sodium ion; K$^+$ = potassium ion; ATPase = adenosine triphosphatase; HI = high; MOD = moderate.

tetanic contraction. In a rat, the isoforms α_1, α_2, and β_1 are typically high in type I fibers, whereas the α_2 and β_2 predominate in type IIX fibers. Type IIA fibers contain a mixture of all isoforms. It is possible that the isoform combination is an important feature in meeting cell-specific demands.

Fiber Size

An additional property that determines the ability of the muscle cell to perform work is the area of the individual fibers. To a large degree, the fiber area reflects the number of myofibrils and consequently the number of actin and myosin molecules available for force generation. Although the fiber types and subtypes show large differences in area in rodents, such is not the case in humans. Type IIX fibers, which are more specialized for large power outputs, are comparable in area to type I and type IIA fibers in humans. In general, type IIX fibers, at least in the vastus lateralis, represent only 10% to 15% of the total fiber distribution in the untrained muscle. Regardless of fiber type differences, muscle mass is the most malleable of all properties, subject to pronounced changes with use and disuse, aging, and disease.

Muscle fibers exist in various types and subtypes. A common property used to identify fiber types is the myosin ATPase, which is located in the HC component of the molecule. Different types of fibers contain different HC isoforms, a feature now being exploited for fiber type classification using HC-specific antibodies. The HC isoforms are the primary determinant of the velocity that the fiber can generate. Accordingly, type I fibers are slow contracting, whereas type II fibers are fast contracting. The contractile speed of the fiber also correlates with a variety of other properties. For example, type II fibers contain a well-developed sarcoplasmic reticulum (SR), which allows for rapid regulation of $[Ca^{2+}]_f$ and rapid force development and relaxation. In this type of fiber, the potential for high-energy phosphate transfer and glycolysis is high, allowing for large and rapid production of ATP. In contrast, type I or slow-contracting fibers have a poorly developed SR and low high-energy phosphate transfer and

glycolytic potentials. Oxidative potential is always high in type I fibers and may be high in type II fiber subtypes such as type IIA compared with type IIX. Although myosin HC isoforms provide the most acceptable basis for classification, a considerable overlap appears to exist between fiber types and subtypes on many properties, which probably underlie the continuum of functional diversity that exists.

Regular exercise changes the expression of the various proteins and protein isoforms and their regulatory mechanisms. We next examine the cellular adaptations that occur with different types of training stimuli and the functional consequences of the adaptations, with an emphasis on human muscle. In this regard, it is important to recognize that for most voluntary activities, an orderly recruitment of motor units occurs, progressing from the type I to the type II fiber pools as the requirement for force and power increases.

Muscle Adaptation and Functional Consequences

A distinguishing feature of skeletal muscle is its ability to adapt to altered patterns of contractile activity (Petts 2002), a property often referred to as phenotypic plasticity. The extensive remodeling of the muscle cell that can occur extends to all levels of organization, including the morphological, ultrastructural, biochemical, and molecular levels, which can affect a wide range of processes. The specific changes that occur in cellular proteins, lipids, and carbohydrates (CHO) are to a large extent determined by the strain or challenge induced by the type of contractile activity, as modified by the environment in which the contractile activity is performed and the characteristics of the participant. Impressive advances have been made in recent years not only in detailing the specific cellular adaptations that occur with different training regimens, but also in unraveling the molecular signals involved in altering transduction and translation and consequently protein expression (Hawley 2009; Fluck 2006).

Studies investigating the muscular effects of regular exercise have mostly been performed over a relatively short time, normally not involving more than 10 to 12 weeks. The majority of these **acclimation** studies used three fundamentally different training models, each designed to **stress** different processes in the cell. The three models used are high-resistance training (HRT), consisting of brief periods of contraction involving high resistance or high power output; high-intensity intermittent training (HIIT); and prolonged endurance training (PET). The majority of published studies have generally emphasized the changes that occur in the metabolic pathways and segments involved in substrate delivery and ATP synthesis. It is only recently that studies examining the cellular adaptation to chronic activity have focused on the proteins and processes involved in E-C processes. Further understanding of the adaptations that occur at this level is essential if we are to fully comprehend the mechanisms regulating ATP homeostasis and work performance in the trained state.

Prolonged Endurance Training

Cellular adaptations to repetitive submaximal exercise include the promotion of energy metabolism and utilization of fatty acids and glucose, as well as changes in the proteins involved in E-C processes.

Energy Metabolism and Substrate Transport and Utilization

Prolonged submaximal exercise training promotes a variety of cellular adaptations in support of oxidative phosphorylation and lipid utilization (Holloszy and Booth 1976; Hood et al. 2006). These adaptations are not surprising given the dependency on aerobic-based ATP regeneration that occurs with this type of exercise (table 8.7). Among the most conspicuous changes are increases in the maximal activities of the enzymes of the CAC, the complexes of the respiratory chain, and fat oxidation (β-oxidation), which appear to occur secondary to increases in mitochondrial number and size and consequently enzyme protein content. These adaptations increase oxidative potential or the maximal potential for oxidative phosphorylation. As a result, a greater rate of ATP synthesis is possible, accompanied by increases in oxygen utilization.

Additional adaptations that might be predicted with extensive aerobic training involve downregulation of the maximal activities of the enzymes involved in high-energy phosphate transfer and in the glycogenolytic–glycolytic potential. Downregulation of the potentials of these metabolic pathways, in conjunction with the upregulation that occurs in the oxidative potential, would create a metabolic organization strongly poised for oxidative phosphorylation and lipid utilization similar to that observed for type I fibers. Yet few training studies, either with humans or animals, have been able to elicit a significant

downregulation of pathway potential for high-energy phosphate transfer and glycogenolysis-glycolysis. Such adaptations appear possible, however. Chronic, low-frequency electrical stimulation to low-oxidative, type II–based muscles, which parallels submaximal exercise, when sustained for 12 to 24 h per day over several weeks, can induce a metabolic reorganization similar to that observed in type I or ST fibers (Pette 2002). This finding suggests that voluntary exercise training programs have the potential to induce a response similar to that observed for chronic, low-frequency electrical stimulation. However, the volume of voluntary training needed is probably beyond what humans can reasonably accomplish.

The chronic adaptations to repetitive submaximal exercise also extend to the substrate used by the mitochondria for oxidative phosphorylation, namely fatty acids and glucose (table 8.8). Increases in facilitated diffusion of these substrates across the sarcolemma occur via increases in specific isoforms of their respective transport proteins, namely FABP and GLUT. Aerobic-based training elevates both FABP and GLUT4, which allows fatty acids and glucose to be more readily transported from the blood to the inside of the muscle cell (Holloszy 2003; Turcotte 2000). In the case of fatty acids, increases in the maximal activities of the enzymes involved in transport, activation, and **β-oxidation** allow the fatty acids to be more readily available as a substrate for mitochondrial respiration.

For glucose to be metabolized in the cell, it must be phosphorylated. This is accomplished by hexokinase, whose maximal activity increases soon after the onset of training. The formation of glucose-6-phosphate allows this substrate to enter the glycolytic pathway, resulting in the formation of pyruvate. Pyruvate dehydrogenase occupies a central position in allowing pyruvate to be more readily available to the CAC. An additional hallmark of endurance training is an increase of the muscle substrates, both triglyceride and glycogen, which are derived from the fatty acids and glucose precursors, respectively. Muscle glycogen supercompensation appears to be an early adaptive response to training promoted in part by increases in GLUT4, hexokinase, and glycogen synthase. This adaptation is to be expected, given glycogen's vital role as a substrate during exercise (even after training), in addition to its apparent role in E-C coupling. Additional muscular enzymatic adaptations allow ketone bodies such as β-hydroxybutyrate and acetoacetate, formed by the incomplete oxidation of fat by the liver, to be converted to forms that can also be used as mitochondrial substrates. Finally, adaptations in specific enzymes known as transaminases allow selected amino acids such as alanine and glutamine

TABLE 8.7 Adaptations in Metabolic Pathway and Segment Potentials With Different Types of Training

	PET	HRT	HIIT
High-energy phosphate transfer	NC	NC-SI	LI
Glycogenolysis	NC	NC-SI	LI
Glycolysis	NC	NC-SI	LI
Oxidative	LI	NC	MI-LI
β-Oxidation	LI	NC	NC
Glucose phosphorylation	LI	NC	MI-LI

PET = prolonged endurance training; HRT = high-resistance training; HIIT = high-intensity intermittent training; NC = no change; SI = small increase; MI = moderate increase; LI = large increase.

TABLE 8.8 Adaptations in Membrane Transporters to Different Types of Training

	PET	HRT	HIIT
Glucose			
GLUT1	SI	NC	MI-LI
GLUT4	LI	NC-SI	LI
Lactate			
MCT1	SI	NC-SI	LI
MCT4	LI	NC-SI	LI
Fatty acids			
FABP	LI	NC	NC-SI

PET = prolonged endurance training; HRT = high-resistance training; HIIT = high-intensity intermittent training; GLUT = glucose transporter; MCT = monocarboxylate transporter; FABP = fatty acid-binding transporter; NC = no change; SI = small increase; MI = moderate increase; LI = large increase.

to be converted into intermediates ready for entry into the CAC.

Regular submaximal exercise is also a powerful stimulus for angiogenesis, resulting in pronounced increases in vascular properties such as fiber capillarization (Egginton 2009). Since this type of training produces either no or moderate fiber hypertrophy, the capillary-to-fiber area ratio is enhanced, leading to an increased perfusion potential.

Excitation and Contraction

Multiple cellular adaptations are also observed in the proteins involved in E-C processes in response to regular, voluntary submaximal exercise. At the level of the contractile protein, a myosin shift is observed in the isoform type from HCIIx to HCIIa, which results in an increase in the type IIA fibers at the expense of type IIX (Baldwin and Haddad 2002) (table 8.9). This is a relatively minor transformation,

however, because muscles in untrained humans typically contain less than 15% of type IIX fibers. There is little consensus that training, regardless of type and particularly in humans, can significantly transform the myosin HC from the fast isoform types to slow, that is from HCII to HCI, and consequently the percent distribution of type I or ST fibers. Endurance training, at least in rats, is also accompanied by changes in the myosin LCs and, in particular, a reduction in LC1f, LC2f, and LC3f, which occurs in association with an increase in LC1s and LC2s. In rats, these changes are essentially confined to muscles that contain an abundance of type II or FT fibers. In humans, it has been difficult to isolate the fiber type population affected because the locomotor muscles, such as the vastus lateralis, which have been the most frequently examined, contain a mixture of fiber types and because measurements are usually performed on mixed tissue samples extracted by biopsy and

TABLE 8.9 Adaptations in Myosin Composition to Specific Types of Training

	PET	HRT	HIIT
Fiber type			
Type I	NC	NC	NC
Type IIA	SI	SI	SI
Type IIX	LD	LD	LD
Myosin HC			
HCI	NC	NC	NC
HCIIa	SI	SI	SI
HCIIx	LD	LD	LD

PET = prolonged endurance training; HRT = high-resistance training; HIIT = high-intensity intermittent training; HC = heavy chain; NC = no change; SI = small increase; LD = large decrease.

Data from Baldwin and Haddad 2002.

not at the single-fiber level. Low-intensity training can also induce a small hypertrophy in both the type I and type II populations as measured by the cross-sectional area of the fibers (table 8.10). The increase in cross-sectional area appears to be attributable to the increase in the myofibrillar content, the proteins that dominate the intracellular composition of the fiber.

Sustained activity places increased strain on the sarcolemma and T-tubule systems given the need to generate repetitive action potentials involved in conducting the neural signal to the interior of the fiber. Protection of membrane excitability is largely dependent on Na$^+$-K$^+$-ATPase, the enzyme responsible for restoring transmembrane gradients for Na$^+$ and K$^+$. Regular exercise is a potent stimulus for the upregulation of both the α and β subunits of Na$^+$-K$^+$-ATPase (Clausen 2003) (table 8.11). The increase in

these subunits is also accompanied by increases in the maximal catalytic activity (Vmax) of the enzyme, an adaptation that should increase the maximal transport rate of the cations. Current evidence, albeit limited, indicates that the α$_1$, α$_2$, and β$_1$ isoforms are among the initial properties to be upregulated with training. However, the changes that occur in specific isoforms with extended submaximal training remain unclear. Moreover, it is also uncertain which α and β isoform combinations are responsible for the increases in Vmax that are observed. It is possible, depending on the isoform population adapted with this type of training, that the sensitivity of the enzyme to K$^+$ or Na$^+$ could also be affected by changes in the properties affecting the intrinsic activity. Evidence indicates that training improves regulation of Na$^+$ and K$^+$ transport across the cell membrane and protects

TABLE 8.10 Adaptations in Fiber Area and Capillarization to Different Types of Training

	PET	HRT	HIIT
Area			
Type I	SI	MI	MI
Type IIA	SI	MI	MI
Type IIX	SI	MI	MI
Capillaries, area			
Type I	LI	NC	SI-MI
Type IIA	LI	NC	SI-MI

PET = prolonged endurance training; HRT = high-resistance training; HIIT = high-intensity intermittent training; NC = no change; SI = small increase; MI = moderate increase.

TABLE 8.11   Adaptations in Na⁺-K⁺-ATPase Characteristics to Specific Types of Training

	Fiber types		
	PET	**HRT**	**HIIT**
Maximal activity	LI	NC-SI	LI
Content	LI	NC-SI	LI
Isoforms			
α_1	LI	NC-SI	LI
α_2	LI	NC-SI	LI
β_1	LI	NC-SI	LI
β_3	SI	NC	SI

Na⁺ = sodium ion; K⁺ = potassium ion; ATPase = adenosine triphosphatase; PET = prolonged endurance training; HRT = high-resistance training; HIIT = high-intensity intermittent training; NC = no change; SI = small increase; LI = large increase.

Data from Clausen 2003.

membrane excitability during sustained low-intensity contractions. Although not yet commonly studied, adaptations might also occur in the many channels and transporters involved in the regulation of Na⁺, K⁺, and C1⁻, the main channels involved in depolarization and repolarization of the membrane.

The precise regulation of intracellular $[Ca^{2+}]_f$ is a prerequisite for purposeful mechanical responses in the cell, as well as a variety of other functions (Berchtold, Brinkmeier, and Muntener 2000; Brini and Carofoli 2009). If mechanical performance is to remain stable in the face of repetitive activation, the SR must be able to preserve Ca^{2+} release and Ca^{2+} uptake functions to maintain $[Ca^{2+}]_f$ levels and consequently the behavior of the myofibrillar proteins. Aerobic-based training decreases the rates of both Ca^{2+} release and Ca^{2+} uptake (Green 2000) (table 8.12). The decrease in both of these functions appears to be attributable to decreases in the density of CRC, the density of the Ca^{2+}-ATPase enzymes, and specifically the SERCA 1 isoform. In the case of the Ca^{2+}-ATPase, the lower protein content appears

TABLE 8.12   Adaptations in Sarcoplasmic Reticulum to Specific Types of Training

	Training type		
	PET	**HRT**	**HIIT**
Ca²⁺-ATPase activity	MD	NC-SI	MI
Ca²⁺ uptake	MD	NC-SI	MI
Isoforms			
SERCA 1	MD	NC	MI
SERCA 2	MI	NC	MD
Ca²⁺ release	MD	NC-SI	MI
RyR content	MD	NC-SI	MI
Isoform			
RyR 1	MD	NC-SI	MI

PET = prolonged endurance training; HRT = high-resistance training; HIIT = high-intensity intermittent training; Ca²⁺ = calcium ion; ATPase = adenosine triphosphatase; SERCA = sarcoplasmic–endoplasmic reticulum calcium; RyR = ryanodine receptor; NC = no change; SI = small increase; MI = moderate increase; MD = moderate decrease.

Data from Green 2000.

responsible for the decrease in the enzyme Vmax that appears to occur in the absence of changes in Ca^{2+} sensitivity. However, as with the Na^+-K^+-ATPase, the Ca^{2+}-ATPase and the CRC are subject to multiple regulatory factors, which may also change with training. Interestingly, the changes that result in Ca^{2+} uptake and Ca^{2+} release with training are consistent with a decrease in Ca^{2+} cycling with each contraction, while possibly protecting the $[Ca^{2+}]_f$ transient level. The reduction in the Ca^{2+} cycling potential may be beneficial given the low intensity of aerobic-based training programs. There is also evidence that the Ca^{2+} uptake rate is better defended during exercise following training, given the structural damage that occurs to the enzyme in the untrained state.

High-Resistance Training

Other training models designed to stress different elements of cellular function can also markedly alter the structure, composition, and function of muscle. Particularly intriguing are the adaptations that result from brief maximal or near-maximal contractions performed for relatively few repetitions per session. This type of resistance training, often classified as strength or power training, leads to a pronounced increase in muscle mass, reflected by increases in the cross-sectional area of both type I and type II fibers. The increase in protein synthesis mediated by high-resistance training (HRT) is believed to be primarily restricted to the regulatory and contractile proteins, increasing the number and size of myofibrils. As with prolonged submaximal exercise training in humans, HRT results in fiber subtype transformation that is restricted to the FT subtypes, with an increase in the percentage of type IIA fibers occurring in conjunction with a decrease in the percentage of type IIX fibers. These changes reflect a transformation of HCIIx toward HCIIa isoforms. Frequently, increases in type IC and type IIC fibers can also be identified, which indicates the presence of more than one myosin HC type in a fiber.

High-resistance training has minimal effects on the organization of the energy metabolic pathways and segments in the cell. Most studies report no changes in maximal activities of representative enzymes involved in high-energy phosphate transfer, glycogenolysis, glycolysis, or oxidative phosphorylation. Although there is some suggestion of decreases in capillaries per muscle fiber area with HRT given the increase in fiber cross-sectional area that occurs with resistance training, current evidence indicates that, at least with moderate resistance training, increases in capillarization can offset the approximate 15% to 20% increase in cross-sectional area of the fibers

that would be expected. Resistance training protocols vary widely and consequently provoke different neural and muscular challenges; thus, not all protocols produce the same adaptations. Few reports have addressed the effect of HRT on the membrane transporters and the E-C coupling processes. The current belief is that HRT has little impact on these properties.

High-Intensity Intermittent Training

Another training strategy that emphasizes the near extremes of physiologic function and consequently the unique capacity of muscle to remodel is high-intensity intermittent training (HIIT). This type of exercise includes elements of HRT, but the frequency of repetitions is increased in conjunction with the number of work bouts. In contrast to HRT, in which the brief duration of each contraction and the small number of repetitions primarily depend on stored ATP and PCr, HIIT demands a major involvement of high-energy phosphate transfer and glycogenolytic–glycolytic systems to rapidly supply large quantities of ATP to meet the energy needs of the different cellular ATPases. Given the relatively high force levels generated and the sustained nature of each work bout, the capacity of the E-C processes for signal transmission and for force development and relaxation is also challenged. Unique to HIIT is the need to maintain viability of the E-C processes given the limited reserve remaining and the accumulation of a range of metabolic by-products, which could inhibit one or more of the processes involved.

High-intensity intermittent training can result in a considerable reconfiguration of the energy metabolic machinery. Unlike aerobic-based training, HIIT can increase the maximal activities of rate-limiting enzymes involved in high-energy phosphate transfer, glycogenolysis, and glycolysis. Accompanying the increase in these anaerobic pathways is an increase both in the maximal activity of hexokinase, the enzyme used to phosphorylate glucose, and in GLUT4, the protein used to facilitate the transport of glucose across the sarcolemma. Depending on the specifics of the training protocol, this type of training may also increase oxidative potential and the ability to increase the rate of ATP production by mitochondrial respiration. Adaptations also occur at the level of glycogen synthase, the enzyme used to increase endogenous glycogen deposition in the cell. High-intensity intermittent training can markedly elevate glycogen deposition.

As expected, this kind of training leads to significant increases in the cross-sectional area of all fiber types. As with previously discussed training strategies, fiber type transformation appears to be primarily restricted to the transformation from type IIX to type IIA. This form of training generally also increases the number of capillaries surrounding each fiber, leading to increases in the capillary-to-fiber area ratio and increases in diffusion potential.

Increases in Na^+-K^+-ATPase protein expression are also observed with HIIT, which probably increases the Vmax of the enzyme. The increases extend to both the α and β subunits, with the specific isoforms affected as yet unclear. The increase in Vmax also appears to support the requirement for increased Na^+ and K^+ transport to protect membrane excitability, given the high frequency of action potentials required to perform this type of exercise. Studies have shown that the loss of K^+ from the working muscle to the blood is reduced following HIIT.

The sarcoplasmic reticulum (SR) also displays adaptations consistent with the strain imposed on Ca^{2+} cycling behavior. An upregulation in both the CRC and the Ca^{2+}-ATPase protein densities has been reported in response to sprint-type training. The increase in the CRC has been correlated with a more rapid and larger release of Ca^{2+} during activation. Increases in Ca^{2+}-ATPase protein, both SERCA 1 and SERCA 2, have also been reported. However, unexpectedly, increases in maximal Ca^{2+}-ATPase activity and Ca^{2+} uptake were not observed. The failure to find increases in Ca^{2+}-sequestering properties raises the possibility that the enzyme was structurally damaged or that regulatory signals controlling enzyme activation were deficient.

Additional adaptations relate to the management of the metabolic by-products that are generated by HIIT. One such by-product of major significance is lactic acid, which, as a strong acid, dissociates into H^+ and lactate. Pronounced increases in the lactate transporters MCT1 and MCT4 have been observed with HIIT, which would be expected to facilitate the transport of lactate and H^+ both out of the cell (MCT4) and into it (MCT1) (Juel 2006). This may be an adaptation in which the isoform affected is specific to a given fiber type. Several studies have reported increases in the buffering capacity of muscle after training, which could assist in the management of H^+ and consequently cellular acidity.

Functional Consequences of Training Adaptations

As a general rule, given the plasticity of skeletal muscle, the specific properties altered by training in large part reflect the strain imposed on the processes and proteins challenged. Accordingly, the type of task, intensity, rest-to-work ratios, and duration of each training session in conjunction with the training frequency all appear important in the outcome. Individual response patterns, both qualitative and quantitative, also reflect the capacity for adaptation as determined by age, gender, and genetic predisposition.

During prolonged submaximal exercise following prolonged endurance training (PET), compared with similar exercise before training, the phosphorylation potential of the cell is better protected as observed by a more preserved concentration of the adenine nucleotides such as ATP and PCr, two high-energy phosphate compounds (figure 8.9). In effect, the energy state or phosphorylation potential of the contracting cell is not as compromised, resulting in tighter metabolic control such that the same level of oxidative phosphorylation can be realized without as much of a disturbance in energy homeostasis. In addition to the increase in oxidative potential that occurs with this type of training, intrinsic mitochondrial changes take place, leading to a more efficient coupling between the sites of energy utilization (ATPases) and the mitochondria. As expected, there is less accumulation of the metabolic by-products P_i, free ADP, free adenosine monophosphate (AMP), and lactic acid. Time-dependent increases in the oxidation of fatty acids and decreases in CHO oxidation are also well-characterized effects of this type of training (Brooks 2009). The decrease in CHO oxidation occurs as a result of decreases in the utilization of both blood glucose and endogenous glycogen stored in the muscle cell. It is also generally believed that decreases in glycolysis and lactate production accompany this type of training; these, in conjunction with increased removal of hydrogen and lactate from the cell, contribute to a lower concentration of lactic acid as well.

The reduced utilization of CHO, both in blood glucose and in endogenous glycogen, has important functional consequences. Decreases in blood glucose utilization during exercise help protect levels of blood glucose, a substrate that is in limited supply and is a vital fuel for central nervous system function. The improved preservation of muscle glycogen levels during prolonged exercise has been linked frequently to improved endurance by mechanisms that are as yet unclear. Developing evidence suggests that CHO may be an essential substrate for the function of both Na^+-K^+-ATPase and Ca^{2+}-ATPase, the two enzymes involved in cation regulation and E-C coupling. A failure in either of these cation pumps could impair impulse transduction in the cell and result in an inability to develop and sustain force production. Reduction in the use of CHO and, in particular,

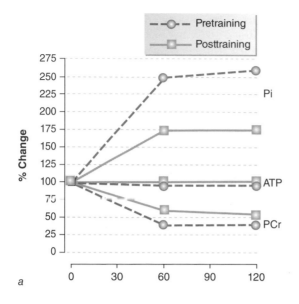

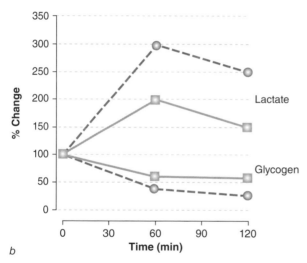

FIGURE 8.9 Percent changes in the high-energy phosphates (adenosine triphosphate [ATP], phosphocreatine [PCr]), and the related metabolites (a) inorganic phosphate (P$_i$), lactate, and glycogen (b) during submaximal exercise before and after training. Percent changes are based on measurements performed prior to exercise (0 min) and at 60 min or 120 min of exercise. Although not included in the graph, reduction with training also occurs in free adenosine diphosphate (ADP) and free adenosine monophosphate (AMP).

blood glucose during exercise also dampens the strategies invoked to keep glucose from approaching hypoglycemic levels. Acute increases in liver glycogenolysis and gluconeogenesis in conjunction with increases in lipolysis and fatty acid utilization by the working muscle are under elaborate hormonal regulation involving, in part, insulin, glucagon, and the catecholamines epinephrine and norepinephrine. The higher CHO levels observed during exercise after

training improve homeostasis and decrease the need for hormonal support. This places less strain on the endocrine systems.

The adaptation that occurs in the E-C coupling processes with PET may also confer an important functional benefit. Unlike the enzymes of metabolic pathways involved in ATP synthesis, whose maximal activity appears to be generally preserved during prolonged exercise, the maximal catalytic activities of both Na$^+$-K$^+$-ATPase and Ca^{2+}-ATPase are depressed during such exercise. The depression would be expected to reduce the rate of cation transport, particularly at high exercise intensities. In the case of Na$^+$-K$^+$-ATPase, training results in its increased abundance because of increases in the α and β subunit protein content. The associated increase in the maximal catalytic activity could allow more precise and sensitive regulation of the enzyme according to the demands of the exercise. This adaptation not only would improve protection of Na$^+$ and K$^+$ transmembrane gradients and membrane excitability during exercise but also would minimize the loss of K$^+$ from muscle.

In the case of the SR, the strain imposed by prolonged submaximal exercise training results in a downregulation of Ca^{2+}-ATPase activity and Ca^{2+} uptake. Because this type of training also decreases CRC content and Ca^{2+}-release kinetics, the Ca^{2+} cycling rate is reduced, a seemingly appropriate adaptation given the relatively low force levels generated by this type of training. In the absence of other regulatory changes, the decrease in cycling rate could decrease the energy costs of SR Ca^{2+} cycling and, in particular, Ca^{2+} uptake. There is some evidence that this type of training protects the CRC and the Ca^{2+}-ATPase proteins against the structural damage induced by repetitive exercise. The structural damage is believed to result from oxidation and nitrosylation processes, generated by reactive oxygen species (ROS) and **reactive nitrogen species (RNS)**. It is generally believed that although training cannot reduce the ROS produced from oxidative phosphorylation and purine nucleotide metabolism, which are the primary sources of production during exercise, training can more effectively scavenge ROS via adaptation of the **antioxidant** enzymes.

A potentially important factor in the adaptations that occur in skeletal muscle during aerobic-based training programs is the amount of muscle mass activated. In principle, the same adaptations should occur with small versus large muscle group protocols provided that the specific mechanical history of the given muscle is comparable. However, large muscle group activity places a greater strain on the respiratory and cardiovascular systems, potentially limiting blood flow and oxygen delivery to the working muscles. Regular aerobic exercise typically increases $\dot{V}O_2$max, also referred to as **peak aerobic power**, by 10% to 20%. At low to moderate exercise intensities, no differences are generally observed in $\dot{V}O_2$ during a steady state early in the exercise. This would imply that the net mechanical efficiency has not changed. At higher exercise intensities, however, $\dot{V}O_2$ can increase because of increases in $\dot{V}O_2$max, resulting in a higher level of oxidative phosphorylation and thereby reducing the contribution of glycolysis to ATP homeostasis.

With increases in $\dot{V}O_2$max, increases in $\dot{V}O_2$ kinetics also occur during the non-steady-state adjustment to exercise. Indeed, recent research has demonstrated that increases in $\dot{V}O_2$ kinetics during the non-steady state may also occur early in training before increases in $\dot{V}O_2$max and muscle oxidative potential. The consequence of these adaptations is that the muscle metabolic behavior observed during submaximal exercise resembles that of the trained state. By mechanisms that are as yet unknown, training also results in less upward drift in $\dot{V}O_2$ as the exercise is prolonged.

High-resistance training (HRT) programs offer the obvious benefit of increasing the force and power that can be generated by muscle. The increases in force and power result primarily from increases in the cross-sectional area of the fiber, because the transformations in the myosin isoforms with training are relatively modest. The increases in force and power permit the performance of a wider complexity of tasks, many of which may be essential to daily living and quality of life. The increases in maximal force and power that occur after HRT may also have benefits for the performance of prolonged submaximal exercise. Evidence exists that a training-like effect may occur in the metabolic behavior of the working muscle as indicated by the improved energy homeostasis and reduced by-product accumulation. This could occur as a result of the increase in cross-sectional area of the muscle, allowing the same submaximal absolute force to be generated for longer time periods after training compared with before training, by reducing the firing frequency to the muscle or reducing the number of motor units that need to be activated. It should be noted that some performance benefits of HRT appear to precede the muscular adaptations, apparently because of the optimization of neural recruitment. Currently it is uncertain what adaptations occur at the level of E-C coupling. The hypertrophy that is induced with HRT could effectively reduce the concentration of the complex of proteins involved, compromising function, unless coordinated change in expression occurs.

The potential of training to sculpt the muscle cell in a desired fashion is best realized with HIIT. This is the case because of the high intensity of HIIT and its ability to manipulate selective challenges via work-to-rest durations and ratios. Indeed, HIIT has been identified as a potent stimulus, capable of rapidly upregulating metabolic pathways such as the CAC and ETC. This property probably stems from its ability to disturb the energy status of the cell, which is sensed by specific proteins such as AMP-activated protein kinase (AMPK), which in turn alters the level of specific proteins by altering gene expression (McBride and Hardie 2009). The increase in the cross-sectional area of the muscle fibers, in conjunction with the increased capability of sarcolemma and T-tubule Na^+-K^+ exchange and SR Ca^{2+} cycling, provides for increased rates of signal transmission to the myofibrillar proteins and increases in the number and rate of actomyosin cycling and force generation. The energy needed for this type of adaptation is provided by the increase in the high-energy phosphate transfer reactions and glycogenolysis-glycolysis. Increases in cellular glycogen and in lactate transport and buffering mechanisms provide the fuel and the tolerance to by-product accumulation necessary to sustain this type of activity for longer periods following training. High-intensity intermittent training protocols can also be designed to benefit cellular aerobic metabolism, given the increase in mitochondrial potential and capillarization that can occur.

Skeletal muscles are capable of extensive adaptations in response to contractile activity if appropriately challenged. The adaptations to different proteins and processes that occur within the cell are highly dependent on the type of exercise and the demands imposed on E-C and metabolic pathways involved in energy production. Three different types of training programs illustrate the different types of muscle adaptations that can result, namely prolonged endurance training (PET), high-resistance training (HRT), and high-intensity intermittent training (HIIT). The best-characterized type of training is PET. This type of training, which primarily depends on energy supplied by aerobic-based metabolism, increases oxidative and β-oxidation potential; muscle fiber capillarization; and the proteins involved in glucose, fatty acid, and lactate transport.

This type of training is also accompanied by an upregulation of Na^+-K^+-ATPase and a downregulation of the sarcoplasmic reticulum Ca^{2+} cycling potential. Collectively, these adaptations increase aerobic capacity and result in less disturbance in phosphorylation potential while causing a greater dependency on fat oxidation and providing better management of metabolic by-products.

Skeletal Muscle Adaptation and Health Benefits

Yet to be addressed is how the muscular adaptations resulting from various training programs can contribute to health and well-being throughout the life span. The adaptations that occur can allow people to reclaim a more active lifestyle, permitting them to independently perform a greater diversity of the tasks of daily living. The adaptations can also have beneficial effects both at work and at leisure by increasing productivity and by allowing people a greater opportunity to interact with the environment. In effect, people can not only enjoy the benefits of greater task diversity but also experience less fatigue in the process. The reduction in strain that occurs in the different cellular processes as a result of training adaptations modulates the activation of neural and hormonal systems necessary to meet the various challenges of activity. For example, evidence demonstrates that feedback signals from the working muscles are reduced, which in turn results in less activation of hormonal, respiratory, and cardiovascular systems. The reduction in feedback is thought to occur as a result of the reduction in the accumulation of select metabolic by-products.

Some adaptations also have direct application to disease prevention. The depletion of muscle glycogen stores with exercise provides a sink for glucose disposal following meals. The increase in GLUT expression and glucose phosphorylation potential in adapted muscle facilitates the transfer and utilization of glucose, lessening insulin resistance and reducing the risk of type 2 diabetes.

Muscle contractile activity can also have a variety of indirect effects. Repetitive muscle contraction, particularly with large muscle groups, recruits a variety of support systems to satisfy the requirements of the working muscles. The recruitment of these systems in turn improves their behavior, reducing strain. Muscle activity also provides the vehicle for better bone health and improved immune and anti-inflammatory function, all of which may delay or prevent a number of diseases. Regular exercise has been shown to produce a more healthy blood lipid profile and to attenuate some forms of hypertension, both established risk factors in vascular disease. Recently it has been proposed that the primary effect of regular exercise in disease prevention is not via alteration in the established risk factors but in prevention of vascular endothelial dysfunction (Seals et al. 2009), resulting in a more responsive regulation of blood flow.

Aging Muscle: The Role of Training

Aging is accompanied by pronounced changes in skeletal muscle properties and physical capabilities. Aging results in reductions in muscle force, velocity, and power accompanied by reductions in muscle mass (**sarcopenia**) and changes in the intrinsic properties of the muscle fibers. For most people, the advancing years are also associated with large reductions in physical activity. These interrelationships suggest that the performance decrements observed with age can be mechanistically linked to alterations in muscle and that inactivity and disuse may be part of the etiology. Alternatively, the reduction in regular activity with age might occur as a consequence of the muscular changes. Several impressive reviews have been written in this area (Carmelli, Coleman, and Reznick 2002; Doherty 2003; Kent-Braun 2009; Margreth, Damiani, and Bortalosa 1999). These reviews indicate that many questions remain unanswered.

Changes in Muscle With Aging

It is known that aging reduces the cross-sectional area of muscle and the individual muscle cells, with the type II or FT fibers experiencing the greatest loss. By the eighth decade of life, sarcopenia may approach 40% of peak muscle mass. There is also evidence that in the advancing years, the proportion of type I fibers increases and the proportion of type II fibers decreases. The shift in fiber type proportions may reflect not only fiber type transformation, but also cell death as a result of a loss of neurons and type II motor units. Although these changes would undoubtedly compromise the mechanical function of muscle, it is not known whether this can explain the full age-associated decline observed, since intrinsic cellular changes are also known to occur with aging; researchers typically refer to these as changes in muscle quality.

The E-C coupling processes represent one area in which intrinsic changes in muscle are observed in animals. In muscle from a variety of species, age-dependent declines are observed in Na^+-K^+-ATPase pump content and in maximum capacity for cation transport across the sarcolemma. Similar declines with age, at least in pump content, have not been documented in human muscle. However, decreases in maximal enzyme activity and in maximal cation transport have been shown to occur with age in humans. These findings suggest that intrinsic structural alterations in the Na^+-K^+-ATPase enzyme itself or in one or more of the complex regulatory factors are modified with age. The expected effect of these changes would be to impair membrane excitability, which could result in loss of Na^+ and K^+ homeostasis and early fatigue.

Sarcoplasmic reticulum Ca^{2+} cycling also appears to be affected by age, because both Ca^{2+} uptake and Ca^{2+} release are decreased in older individuals. As with Na^+-K^+-ATPase, these decreases appear to be intrinsic because they are not accompanied by loss of either Ca^{2+}-ATPase or CRC protein content.

Evidence also indicates that the coupling between the T-tubule and CRC is altered in older muscles, given the reductions in DHPR and the reductions in the ratio of DHPR to CRC that have been observed. The loss of DHPR could reduce the direct coupling between DHPR and CRC and the Ca^{2+}-induced regulation of adjacent CRC. Another suggestion is that one or more of the regulatory proteins in T-tubule–CRC coupling are altered with age.

The age-associated changes that occur intrinsically also extend to the myosin, since it appears that both the force generated per unit of cross-sectional area and the maximal velocity of shortening are compromised (Canepari et al. 2010). These effects occur in fibers of like type and appear to be due to both a lower myosin concentration and alterations in myosin function.

There is consensus that the intrinsic changes in the proteins involved in E-C occur as a result of oxidation–nitrosylation reactions, which in turn result from excessive accumulation of ROS. Many of the proteins involved in muscle E-C processes are very prone to free radical–mediated damage. Free radical–mediated damage is also believed to be responsible for the structural damage that occurs to the cation pumps and the CRC with acute exercise.

Age-related changes in muscles also occur in the metabolic pathways involved in ATP production (Carmelli, Coleman, and Reznick 2002; Conley et al. 2007). The process most affected appears to be the aerobic process, given the pronounced reduction in maximal activity of a number of mitochondrial enzymes in both the CAC and the ETC. The maximal activities of many of the glycolytic enzymes also appear to be depressed by aging. In the case of the mitochondria, the age effect appears to involve more than a reduction in enzymatic activity since the P/O ratio, a measure of the ATP generated per molecule of oxygen consumed, is considerably lower in older mammals (Conley et al. 2007). The uncoupling that results is believed to occur secondarily to ROS-generated changes in the mitochondrial membrane, allowing protons (H^+) to leak out. The effect is that both the maximal ATP that can be generated by oxidative phosphorylation and the efficiency are reduced. Reductions in metabolic efficiency also mean that to attain a given level of ATP synthesis during submaximal exercise, mitochondrial respiration must be increased, presumably mediated by a greater elevation of the effectors, such as free ADP. Interestingly, reductions in capillary density exceed the reduction in the muscle fiber cross-sectional area that is typically observed in older individuals, resulting in a lower capillary-to-fiber area ratio. These changes suggest that coordinated reductions in blood flow may accompany the reduction in mitochondrial function.

As a consequence of the age-associated changes in muscle, both quantitative and qualitative, in combination with a multitude of other factors (e.g., muscle stiffness), the energetic cost of contractile activity is elevated with aging. Curiously, muscle fatigue, observed in tasks performed at a given relative intensity, appears to be more pronounced in young compared to older individuals (Kent-Braun 2009).

Training Adaptations in Aging Muscle

Because many of the muscle changes observed with age are similar to those observed with inactivity or disease, it has been speculated that activity level and not age is primarily responsible for the change. To some extent this may be true. Older individuals respond to aerobic-based training programs with increases in oxidative potential and in capillary-to-fiber area ratios, adaptations that are consistent with improved aerobic function. Older individuals also display increases in $\dot{V}O_2$max after training, suggesting linked adaptations in both central and peripheral processes.

Not unexpectedly, high-resistance training (HRT) in older individuals also increases fiber cross-sectional area and the force and power that can be generated by the trained muscle. Evidence is conflicting with respect to the effect of regular exercise, regardless of protocol, on the proteins and processes involved in E-C. Few studies have been performed, and most that have been performed involved nonhuman species. If, as suggested earlier, some aging-associated changes are attributable to intrinsic changes in the cell mediated by oxidation-nitrosylation, training adaptations may revolve around management of ROS and RNS with regard to the pathways involved in both production and scavenging. In this regard, emerging evidence demonstrates that the mitochondrial and myosin uncoupling that occur with age can both be attenuated with regular exercise. Proper exercise prescription may be essential to achieve desired benefits, because inactivity apparently is associated with increased ROS concentration, as are excessively demanding protocols of regular exercise.

There is little question that regular exercise can benefit the aging individual as assessed by improvements in both mechanical and aerobic capacities. Such improvements not only would permit performance of a greater number of daily tasks but also would effectively reduce the strain associated with many of the activities of daily living. The essential experimental question remains: What mechanisms underlie the improved physical capabilities? We can be encouraged by recent evidence that the beneficial effects of regular exercise and proper nutrition on health and well-being can extend well into the advancing years.

Summary

Skeletal muscle occupies a central role in both the prevention and the management of a variety of disease states. Muscles must be primarily viewed as a machine, specialized for translating chemical energy into physical movement. As a machine, muscle has a number of mass and fiber type properties that allow us to perform a vast array of motor tasks. However, this unique capability comes at a price. Muscles must be regularly challenged to contract or they lose their ability to function optimally. Inactive or sedentary muscles also affect a variety of other tissues and systems, compromising their functional potential and predisposing them to disease. Muscles are capable of extensive adaptations at all levels of organization. The nature of adaptations depends in large part on the type of contractile activity performed and the challenges imposed on the E-C processes, which use energy and metabolic pathways that supply the energy. This chapter describes the composition, structure, and functional properties of the various components of the muscle cell and the manner in

which they adapt to different training stimuli. The discussion provides insight into how these adaptations affect the performance of different motor tasks and, perhaps more importantly, how they defend against disease. An important feature of muscles is that they never lose their ability to adapt, even in advancing age.

Key Concepts

acclimation—Adaptation that occurs in response to a single factor, for example the changes occurring in response to a short-term, formalized training program.

acclimatization—Adaptations to the natural environment.

adaptation—Process of modification or adjustment to a changing environmental element to minimize the strain needed to maintain homeostasis.

antioxidants—Compounds capable of reacting with reactive oxygen species and neutralizing their potentially damaging effects. Examples include superoxide dismutase, catalase, and glutathione peroxidase.

β-oxidation—Process by which fatty acids, in the form of acyl-CoA molecules, are broken down by mitochondria to generate acetyl-CoA, which is used by the mitochondria as a substrate for oxidative phosphorylation.

excitation–contraction (E-C) coupling—Sequence of events starting with the spread of the action potential in the sarcolemma and culminating with the increase in free cytosolic Ca^{2+} levels.

glycolysis—For definition, see page 84.

homeostasis—Constancy of the internal environment, achieved by the maintenance of many properties, such as blood glucose, pH, and arterial oxygen tension, at fixed levels or set points.

isoforms—Separate proteins expressed in molecular forms that generally display only minor variations in composition. Multiple isoforms provide functional versatility in mechanical function and in different environments. Isozymes are a class of isoforms that catalyze the same biochemical reaction.

mechanical properties—Variety of properties of muscle that have been classed according to whether the activity is isometric or dynamic. The isometric potential is measured by maximal tetanic force, whereas the dynamic properties are based on the force–velocity properties. This dynamic allows for characterization of the power and work capabilities of the muscle.

oxidative phosphorylation—Phosphorylation of adenosine diphosphate (ADP) to adenosine triphosphate (ATP) using the energy provided by transport of electrons to molecular oxygen.

oxidative potential—Maximal potential rate at which the mitochondria can generate ATP using oxygen.

peak aerobic power—Measure of the maximal rate at which ATP can be generated by oxidative phosphorylation; also known as maximal oxygen uptake ($\dot{V}O_2$max). $\dot{V}O_2$max is usually measured using tasks involving large muscle groups in which intensity is progressively increased until fatigue.

phosphorylation potential—Content of high-energy bonds in the muscle; depends primarily on the level of the high-energy phosphagens, ATP and phosphocreatine.

reactive oxygen species (ROS)—By-products of oxidative metabolism as well as other reactions that can induce cellular dysfunction. Examples of ROS include superoxide radical anion (O_2^-), hydrogen peroxide (H_2O_2), and hydroxyl radical (OH^-). ROS production can be increased by both exercise and aging.

reactive nitrogen species (RNS)—Species derived from nitrous oxide, which reacts with other compounds such as peroxynitrate to form peroxynitrite ($ONOO^-$).

sarcopenia—Age-associated decline in muscle mass and function.

signal transduction—Process that involves the pathways beginning with the binding of a chemical to a receptor and terminating with a cellular response, usually by altering the phosphorylation status of a target protein.

strain—Effort that must be expended to resist stress forces.

stress—State of threatened balance to equilibrium or homeostasis.

Study Questions

1. Describe the different levels of organization that allow muscles to function effectively and efficiently.

2. Name the different types of myosin heavy chain isozymes in the different muscle fibers and outline how they regulate the force–velocity and power characteristic of the fiber.

3. What is meant by homeostasis, and how does exercise have the potential to disturb homeostasis?

4. Indicate the direct and indirect potential benefits of aerobic-trained muscle to health and well-being.

5. It has been stated that given the lifestyle of our ancestors, our natural state is one of physical activity, not inactivity. If this is the case, how would the composition of muscle have been different in our ancestors compared with individuals today?

6. Outline the processes and proteins in muscle fiber that are involved in translating a neural command into a mechanical event.

7. What is meant by phosphorylation potential? How can it be protected during heavy intermittent exercise?

8. Describe the adaptations that occur in the excitation and contraction proteins and processes in response to aerobic-based training, and discuss the implications of these adaptations to energy supply in submaximal exercise.

9. Prolonged endurance training, high-resistance exercise, and high-intensity intermittent exercise impose different degrees of strain on the excitation and contraction processes and the metabolic systems. Contrast the different forms of exercise in terms of the strain imposed on these systems.

10. Aging causes deteriorations in both muscle mass and muscle quality. Discuss the potential of aerobic-based training to offset these age-associated effects.

11. What is meant by intrinsic changes in muscle, and what is the potential impact on mechanical function?

References

Allen, D.G., G.D. Lamb, and H. Westerblad. 2008. Skeletal muscle fatigue: Cellular mechanisms. *Physiological Reviews* 88:287-322.

Baldwin, K.M., and F. Haddad. 2002. Skeletal muscle plasticity: Cellular and molecular responses to altered physical activity paradigms. *American Journal of Physical Medicine and Rehabilitation* 81(Suppl):S40-S51.

Berchtold, M.W., H. Brinkmeier, and M. Muntener. 2000. Calcium ion in skeletal muscle: Its crucial role for muscle function, plasticity and disease. *Physiological Reviews* 80:1216-1265.

Brini, M., and E. Carofoli. 2009. Calcium pumps in health and disease. *Physiological Reviews* 89:1341-1378.

Brooks, G.A. 2009. Cell-cell and intracellular lactate shuttles. *Journal of Physiology* 587:5591-5600.

Canepari, M., M.A. Pellegrino, G. D'Antona, and R. Bottinelli. 2010. Single muscle properties in aging and disease. *Scandinavian Journal of Medicine and Science in Sports* 20:10-19.

Carmelli, E., R. Coleman, and A.Z. Reznick. 2002. The biochemistry of aging muscle. *Experimental Gerontology* 37:477-489.

Clausen, T. 2003. Na^+-K^+-pump regulation and skeletal muscle contractility. *Physiological Reviews* 83:1269-1324.

Conley, K.E., S.A. Jubrias, C.E. Amara, and D.J. Marcinek. 2007. Mitochondrial dysfunction: Impact on exercise performance and cellular aging. *Exercise and Sport Sciences Reviews* 35:43-49.

Doherty, J.J. 2003. Invited review: Aging and sarcopenia. *Journal of Applied Physiology* 95:1717-1727.

Egginton, S. 2009. Invited review: Activity-induced angiogenesis. *Pflugers Archives: European Journal of Physiology* 457:963-977.

Fitts, R.H. 2008. The cross-bridge cycle and skeletal muscle fatigue. *Journal of Applied Physiology* 104:551-558.

Fluck, M. 2006. Functional, structural and molecular plasticity of mammalian muscle in response to exercise stimuli. *Journal of Experimental Biology* 209:2239-2248.

Furuhashi, M., and G.S. Hotamisligil. 2008. Fatty acid binding proteins: Role in metabolic diseases and potential drug targets. *Nature Reviews* 7:489-503.

Green, H.J. 2000. Muscular factors in endurance. In *Encyclopedia of sports medicine. Endurance in sports*, ed. R. Shephard and P-O. Åstrand (pp. 156-163). Oxford: Blackwell Science.

Hawley, J.A. 2009. Molecular responses to strength and endurance training: Are they compatible? *Applied Physiology Nutrition and Metabolism* 34:355-361.

Hochachka, P.W. 1994. *Muscles as molecular and metabolic machines.* Boca Raton, FL: CRC Press.

Holloszy, J.O. 2003. A forty-year memoir on the regulation of glucose transport into muscle. *American Journal of Physiology* 284:E453-E467.

Holloszy, J.O., and F.W. Booth. 1976. Biochemical adaptations to endurance training in muscle. *Annual Review of Physiology* 38:273-291.

Hood, D.A., I. Irrcher, V. Ljubicic, and A-M. Joseph. 2006. Coordination of metabolic plasticity in rat skeletal muscle. *Journal of Experimental Biology* 209:2265-2275.

Juel, C. 2006. Training-induced changes in membrane transport proteins of human skeletal muscle. *European Journal of Applied Physiology* 96:627-635.

Juel, J., and A.P. Halestrap. 1999. Lactate transport in skeletal muscle—role and regulation of monocarboxylate transporter. *Journal of Physiology* 517:633-642.

Kent-Braun, J.A. 2009. Skeletal muscle fatigue in old age: Whose advantage? *Exercise and Sport Sciences Reviews* 37:3-9.

Margreth, A., E. Damiani, and E. Bortalosa. 1999. Sarcoplasmic reticulum in aged muscle. *Acta Physiological Scandinavica* 167:331-338.

McBride, A.M., and D.G. Hardie. 2009. AMP-activated protein kinase – a sensor of glycogen as well as AMP and ATP. *Acta Physiologica* 196:99-113.

Meyer, R.A., and J.M. Foley. 1996. Cellular processes integrating the metabolic response to exercise. In *Handbook of physiology. Exercise: Section 12: Regulation and integration of multiple systems* (pp. 841-869). Bethesda, MD: American Physiological Society.

Pette, D. 2002. The adaptive potential of skeletal muscle fibers. *Canadian Journal of Applied Physiology* 27:423-448.

Reggiani, C., R. Bottinelli, and G.J.M. Steinen. 2000. Sarcomere myosin isoforms: Fine tuning a molecular motor. *News in Physiological Sciences* 15:26-33.

Schiaffino, S., and C. Reggiani. 1996. Molecular diversity of myofibrillar proteins: Gene regulation and functional significance. *Physiological Reviews* 76:371-423.

Seals, D.R., A.S. Walker, G.L. Pierce, and L A. Lesniewski. 2009. Habitual exercise and vascular aging. *Journal of Physiology* 587:5541-5549.

Turcotte, L.P. 2000. Muscle fatty acid uptake during exercise: possible mechanisms. *Exercise and Sport Science Reviews* 28:4-9.

9

Response of Liver, Kidney, and Other Organs and Tissues to Regular Physical Activity

Roy J. Shephard, MB, BS, MD (London), PhD, DPE, DLL

CHAPTER OUTLINE

A great deal has been written about the health implications of regular physical activity for the heart, lungs, and muscles, but the impact on such important body systems as the liver, kidneys, gut, and immune system has received much less attention. This chapter thus summarizes available knowledge concerning the effects of physical activity on the liver and gallbladder, the kidneys and the renal tract, the gastrointestinal tract, and the immune system. For each of these organs and systems, attention is directed first to the acute effects of a single bout of physical activity and then to the consequences of repeated physical activity; for the latter, we consider the benefits of several months of deliberate aerobic training, as well as the cumulative adverse effects of repeated bouts of exhausting exercise. A final section notes strengths and weaknesses of available evidence on these issues. There is little evidence of gender specificity in response, and most of the conclusions apply equally to the health of both men and women.

It should be emphasized at the outset that normal, moderate exercise has minor and generally positive effects on function in most of these regions of the body. Nevertheless, as described below, adverse consequences can arise from single bouts of very heavy, prolonged exercise and from periods of excessive training.

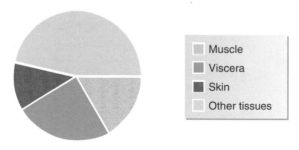

Rest, cardiac output 6 L/min

Muscle
Viscera
Skin
Other tissues

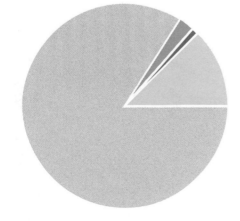

Maximum exercise, cardiac output 21.4 L/min

FIGURE 9.1 Approximate distribution of cardiac output under resting conditions and during maximal exercise.

Acute Effects of Physical Activity

In the short term, an intensive bout of physical activity causes large decreases in blood flow to the skin and viscera in an attempt to maintain blood pressure (Wade and Bishop 1962; Rowell 1986; Johnson 2000), particularly if the individual is simultaneously exposed to hot environmental conditions (figure 9.1). Thus, the total splanchnic blood flow of a young adult may decrease from a normal resting figure of perhaps 1.4 to 1.5 L/min to as little as 0.4 L/min (Rowell 1986). This may be enough to cause local **hypoxia** and changes in visceral function. The absolute redistribution of blood flow is somewhat less in elderly individuals, but this seems mainly attributable to their smaller resting visceral blood flow. A bout of intensive exercise can also increase the total energy requirements of the body 10- to 20-fold, with metabolic consequences for viscera such as the liver, kidneys, and gastrointestinal tract. Further, many physical pursuits subject the abdominal organs to considerable mechanical forces, disturbing their normal function.

Liver and Gallbladder

Vigorous exercise reduces blood flow not only in the hepatic arteries, but also in the hepatic vein, and these circulatory changes have been advanced to explain a decrease in the ability of the liver to clear test substances such as **sorbitol** from the bloodstream during exercise. By implication, a concomitant reduction in the clearance of pharmaceuticals, toxins, and metabolites occurs. There is also a major widening of the hepatic arteriovenous oxygen difference; the arterial oxygen content decreases little, but the oxygen content of blood in the hepatic veins may drop as low as 5 ml/L; this helps to maintain oxygen delivery to the liver in the face of decreased blood flow and thus permits the total cellular metabolism of the liver to remain unchanged or even to increase during vigorous physical activity.

The hepatic tissues play an important role in removing lactate from the circulation during and following a bout of anaerobic activity, although if visceral blood flow remains low, this may ultimately limit hepatic lactate metabolism. Vigorous physical activity also has a selective effect on the gluconeogenic activity of the hepatocytes, with perivenous cells upregulated to a greater extent than **periportal cells**. If blood glucose homeostasis is challenged by prolonged aerobic exercise and a depletion of muscle glycogen reserves, hepatic glucose production is enhanced as much as 14-fold; this is achieved by a depletion of hepatic glycogen stores and the formation of glucose from both lactate and amino acids (the process of **gluconeogenesis**) (Rowell 1986). A further important response to vigorous physical activity is an increased expression of **insulin-like growth factor (IGF) binding protein** messenger RNA; this persists for at least 12 h postexercise, with obvious implications for systemic levels of **IGF** and thus the process of skeletal muscle hypertrophy.

Information on dangerously exhausting bouts of physical activity in humans is limited to occasional clinical reports. However, it is possible to examine the effects of such activity in experimental animals. When rats perform a bout of very prolonged exercise, liver **glutathione** levels may be reduced to as little as 20% of the values found in a control group of the same species. This reflects in part the large increase in oxidative metabolism during exercise, but it also has implications for the total sulfhydryl content of the liver and thus its ability to buffer reactive species of oxygen and other toxic ions.

If an exhausting bout of physical activity is combined with excessive heat stress, as in a marathon run, acute hepatic failure may develop some 24 h after the event. The affected individual becomes clinically **jaundiced**, and laboratory tests of hepatic function show the picture typically associated with liver damage (elevated blood levels of the hepatic enzymes aspartate aminotransferase, alanine aminotransferase, gamma-glutamyl transferase, and alkaline phosphatase). Nevertheless, it has yet to be clearly established whether liver damage is the unique source of these enzymes. Studies are needed using more specific hepatic markers, such as glutathione S-transferase and cholinesterase. At the stage when blood enzyme levels are high, liver biopsies or postmortem specimens often show macroscopic evidence of tissue injury such as edema around the hepatic sinusoids and centrilobular **necrosis**.

Kidneys and Urinary Tract

There is also a drastic reduction of blood flow to the kidneys during a bout of vigorous physical activity. This has adverse effects on many aspects of renal function. Increased concentrations of circulating **antidiuretic hormone** indicate a marked antidiuretic response, an attempt to sustain plasma volume and thus systemic blood pressure. An increase in **aldosterone** secretion helps the reabsorption of sodium ions from filtered tubular fluid, thus compensating for a progressive loss of sodium ions in sweat. Despite large increases in plasma lactate concentration, the renal mechanisms involved in transcellular transfer of lactate quickly become saturated during a bout of heavy physical activity, and the urinary excretion of lactate remains a minor component of anaerobic metabolism (Poortmans and Vanderstraeten 1994).

Glomerular filtration is reduced by 30% or more during vigorous physical activity. Laboratory tests suggest that as many as half of marathon runners show a clinically significant depression of renal function immediately following an event, although in the majority of competitors, the disturbance is short-lived and normal function is restored within 24 h (Bellinghieri, Savica, and Santoro 2008; Lippi et al. 2008). Clinical tests of renal function have traditionally been based on the capacity of the glomeruli to clear **creatinine** from the bloodstream; but it is important to note that in athletes, the scores may be distorted from accepted norms by either an atypical muscle mass or the release of creatinine from a running-induced injury of the skeletal muscles. It has thus been suggested that sport physicians should base assessments of renal function on the serum levels of cytostatin C rather than creatinine. Use of this new approach suggests that the proportion of athletes adversely affected by endurance running is substantially smaller than previously believed.

If the athlete drinks an excessive volume of water or if excessive fluid is administered at a first aid station, the combination of impaired renal function and an increased output of antidiuretic hormones can occasionally lead to a clinically significant **hyponatremia** (a plasma sodium concentration less than 130 mM/L). If the physical activity is sufficient to

Impact of Vigorous Acute Exercise on Hepatic Metabolism

Total cellular metabolism unchanged or increased

Lactate clearance diminished

Clearance of hepatic function markers diminished

Glycogen stores depleted

Hepatic fat stores depleted

Gluconeogenesis upregulated (perivenous hepatocytes > periportal cells)

Expression of insulin-like growth factor binding protein-1 messenger RNA increased

Impact of Vigorous Acute Exercise on Renal Function

Antidiuresis attributable to increased secretion of antidiuretic hormone

Increased reabsorption of sodium ions attributable to increased secretion of aldosterone

Saturation of lactate transfer mechanisms that limit renal lactate excretion

Reduced glomerular filtration attributable to reduced glomerular blood flow

Proteinuria attributable to impaired glomerular electrostatic barrier, and decreased tubular reabsorption of protein

depress renal function, there is usually a substantial accompanying **proteinuria**, with protein leakage of 50 or more times resting values; this persists for about 4 h postexercise (Poortmans and Vanderstraeten 1994). The protein loss has been attributed to a constriction of the efferent arterioles of the renal glomeruli mediated by the alpha-sympathetic nerve supply. An increased leakage of protein through the glomerular wall occurs because the vigorous activity has reduced the glomerular electrostatic barrier, facilitating transfer of macromolecules into the renal tubules; reabsorption of protein in the renal tubules is also reduced. The process normally does not cause any great harm to the body, but the loss can be attenuated some 40% by administration of clonidine, an alpha-2-adrenergic antagonist that blocks the vasoconstrictor output from the brain stem.

The **prostaglandin** secretion associated with vigorous physical activity helps to counter renal vasoconstriction during exercise in the heat. If physical activity is pushed to exhaustion, **hemoglobinuria** and **myoglobinuria** can develop in association with a potentially fatal total cessation of renal function (heat **oliguria**, **anuria**, or acute tubular necrosis) (Clarkson 2007). The magnitude of the increase of serum **creatine kinase** following an exhausting bout of physical activity provides some indication of the extent of muscle damage and thus of individuals who are most at risk. Individuals with the rare condition **hereditary renal hypouricemia** are in particular danger. There may also be a linkage between risk and the individual's creatine kinase genotype. **Endotoxemia** and a generalized septic reaction may contribute to the failure of renal function in this situation, as discussed in the next section (Marshall 1998). Some reports have linked acute renal failure to the excessive use of **nonsteroidal anti-inflammatory drugs** and (particularly in those with preexisting renal conditions) to an overuse of creatine supplements. Particular caution is necessary with the use of anti-inflammatory prostaglandin inhibitors such

as ibuprofen, because such agents can exacerbate the exercise-induced depression of renal function.

Stomach and Intestinal Tract

Experimental evidence concerning the effects of exercise on esophageal mechanics is conflicting. One investigator found that treadmill running induced a modest increase in pressure at the lower esophageal sphincter, but reports from other laboratories have described decreases in esophageal contractions and sphincter pressures as the intensity of exercise was increased. Clinical reports note complaints of heartburn, but most forms even of high-intensity exercise seem to have little acute effect on the likelihood of esophageal reflux (Jozko et al. 2006; van Nieuwenhoven, Brouns, and Brummer 1999). One specific exception is the sport of surfboarding; problems have been noted particularly in athletes using short boards. Another potential exception is an increase in intra-abdominal pressure during the Valsalva maneuver (for instance, in weightlifters). Reflux also seems more likely to occur if athletes ingest carbohydrate-containing beverages rather than water. If a competitor is complaining of gastric reflux, the problem can sometimes be corrected by administration of histamine H_2 blockers.

The acute feelings of gastric nausea that many runners experience are probably related to an excessive ingestion of "replacement" fluids, a slowing of gastric emptying over the course of a long-distance race, or both. Immediate gastric emptying (particularly of solids) and small-intestinal motility may be unchanged or even enhanced by moderate physical activity, but both are depressed by a bout of intensive exercise (Moses 1994). Runners should limit their fluid intake to 600 ml/h, a volume that can empty from the stomach and be absorbed by the intestines during vigorous physical activity. Reports of abdominal cramping and diarrhea during endurance runs support the view that prolonged exercise subsequently speeds the overall rate of passage of food through the intestines (Moses 1994). An exercise prescription may thus be helpful to a person who is suffering from chronic constipation. The tendency toward an exercise-induced increase in gastrointestinal motility is sometimes exacerbated following abdominal irradiation (e.g., as a side effect of the treatment of prostate cancer). However, it is less clear that the diarrhea associated with intensive exercise is the same phenomenon as the faster transit time postulated in the regular moderate exerciser.

A prolonged and exhausting bout of physical activity can reduce blood flow to the intestines and thus cause ischemic damage. One important consequence is a decrease of the gastrointestinal barrier (demonstrated specifically in dogs that participated in a 1,770 km [2,849 mi] sled race); gastrointestinal bleeding is accompanied by a leakage of endotoxins from the gut into the bloodstream (Berg et al. 1999). One case report described exercise-induced bleeding from esophageal varices following administration of large doses of anabolic steroids, although it is unclear why anabolic agents should have such an effect except through an association with the repeated performance of high-intensity resistance exercise. Minor gastrointestinal bleeding seems a relatively common phenomenon among long-distance runners, particularly if they make frequent use of nonsteroidal anti-inflammatory drugs such as aspirin. The reported incidence of stools with a positive **occult blood** test is quite variable (from as low as 1% to as high as 85%, depending in part on the timing of stool sampling relative to competition and on the intensity of effort that has been undertaken) (Moses 1994; Peters et al. 2001). Both occult and visible bleeding seem to peak 24 to 48 h after a bout of exercise, and (probably because of differences in the intensity of effort) both are more prevalent following competition than after a practice run. The stomach is usually the prime site of blood loss, although the colon is sometimes affected.

Bleeding appears to be more common with running than with other types of sport. Although it was once blamed on a mechanical shaking of the viscera, the dominant reason seems to be ischemia of the gut wall, whether induced by the redistribution of visceral blood flow (Wade and Bishop 1962) or a subsequent septic reaction (Marshall 1998). Another contributing factor may be the secretion of substances such as vasoactive intestinal peptide, secretin, and peptide histidine-methionine (Brouns and Beckers 1993). In some athletes, another contributing factor may be ingestion of excessive quantities of nonsteroidal anti-inflammatory drugs, with a resulting increase in clotting time. The amount of blood loss is usually small. It may sometimes contribute to "athlete's anemia," but in most endurance athletes, a low hemoglobin concentration reflects an expansion of plasma volume rather than blood loss. There have been suggestions that bleeding can be prevented by prophylactic administration of antihistamine medications or proton pump inhibitors; however, this is not necessary in most competitors, since the blood loss generally resolves if the athlete takes two or three days of rest (table 9.1).

If circulating endotoxins are observed following a bout of exhausting exercise, they are from the cell walls of **Gram-negative bacteria** normally present in the gut. If hepatic function is also compromised, a portal vein endotoxemia can quickly progress to a systemic endotoxemia. The endotoxins stimulate release of the endogenous **pyrogens** tumor necrosis factor and interleukin-1 (IL-1); this sets in motion a chain of immunological disturbances that can progress to a form of endotoxic shock (Marshall 1998), widely disseminated intravascular coagulation, and the various clinical manifestations of exercise-induced heatstroke.

Circulation and Immune Function

The immediate effect of an acute bout of vigorous physical activity is a decrease in blood volume. This reflects a combination of fluid lost by sweating and the exudation of fluid into the active muscles; typi-

TABLE 9.1 Characteristics of Gastrointestinal Bleeding in Endurance Athletes

Features	Site	Likely cause
Occult blood in stools peaking 24 to 48 h postactivity	Esophageal varices	Mechanical trauma, NSAIDs, or both
Blood loss small, resolves in two to three days	Gastric mucosa	NSAIDs
Possible prevention by antihistamines	Intestinal ischemia	Bleeding and leakage of endotoxins into bloodstream

NSAIDs = nonsteroidal anti-inflammatory drugs.

cally, it develops progressively over the first 30 min of exercise. The magnitude of change depends on the environmental temperature but is often on the order of 10%. This contributes to a deterioration of circulatory function and has the important practical implication that blood cell counts and plasma concentrations of hormones and **cytokines** reported during and following exercise must be adjusted for this phenomenon.

It is difficult to evaluate overall responses of the human immune system to exercise, because most reports are based on changes in the numbers or the functional activity of blood constituents or both. Unfortunately, the bloodstream contains only 1% to 2% of the body's leukocytes, and apparent changes in functional status can arise quite quickly through hormone-induced alterations in the expression of adhesion molecules and thus exchanges of leukocytes between the bloodstream and sequestration sites (Shephard 1997). Moderate bouts of physical activity enhance some aspects of resting immune function. On the other hand, a very strenuous bout of physical activity leads to a transient (typically 2-72 h) decrease in both the count and the activity of circulating **natural killer (NK) cells**. There is also a suppression of lymphocyte proliferation, monocyte antigen presentation, and secretion of salivary and nasal secretory immunoglobulin (Ig) A and IgM for 3 to 24 h following physical activity (Gleeson 2000, 2009). Often an associated neutrophilia but a suppression of neutrophil function are observed. It has been argued that this pattern of changes provides a brief "open window" (Nieman 2000), when the susceptibility to acute viral or bacterial infections may be increased (Shephard 1997) (figure 9.2). Several studies have shown acute increases in the incidence of upper respiratory symptoms, although some of these symptoms may reflect an inflammation of the airway induced by cold, dry air, or by atmospheric pollutants rather than by a specific viral infection. It also remains a challenge to tease out the effect of possible immunosuppression from other factors increasing the individual's vulnerability to microorganisms, such as exposure to novel pathogens through travel, disturbances of sleep, mental stress, nutritional disturbances, and a depletion of key nutrients. It thus remains questionable whether a causal relationship has yet been established between vigorous exercise and increased susceptibility to infection.

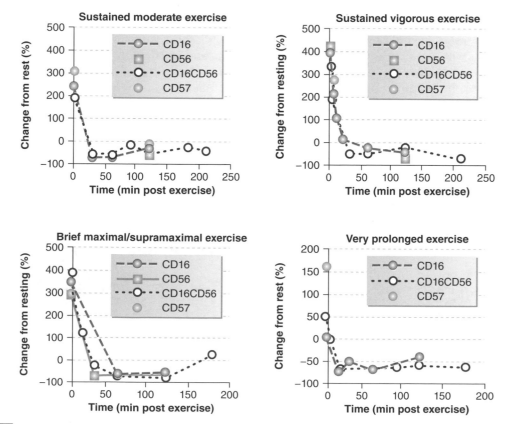

FIGURE 9.2 Changes in circulating concentration of various categories of natural killer cells (as identified by their CD numbers) following exercise bouts of varying duration and intensity.

Physical Activity and Immune Function

Following are effects of moderate to strenuous physical activity on immune function as seen 2 to 72 h after a bout of such exercise.

Decreased NK cell count

NK cell activity *per cell* unchanged

Decreased lymphocyte proliferation in response to mitogens

Decreased monocyte antigen presentation

Decrease of secretory IgA

Decrease of secretory IgM

Increased neutrophil count

Decreased neutrophil function

Chronic Effects of Physical Activity

Regular moderate physical activity such as prescribed exercise has a number of positive effects on the body systems considered in this chapter. However, adverse effects are sometimes seen following very intensive periods of preparation for international competition, and there is growing evidence of cumulative pathological consequences if the intensity of individual bouts of endurance exercise is sufficient to cause severe visceral hypoxia. It is thus important to monitor the enthusiasm of top-level coaches and individual competitors with type A personalities and to avoid both excessive training and repeated participation in extreme events.

Liver and Gallbladder

Regular moderate physical activity has no influence on the pharmacodynamics of drugs such as propanolol, but there is evidence of beneficial changes in several aspects of hepatic function. There have also been suggestions that the enhanced metabolic activity demanded by chronic endurance pursuits can stimulate a hypertrophy of hepatic tissues and thus facilitate recovery from hepatitis. However, repeated bouts of exhausting exercise can lead to permanent liver damage.

Experimental evidence has been garnered from such indices of hepatic function as aminopyrine metabolism, galactose elimination, and indocyanine green clearance. A small-scale, cross-sectional comparison between eight endurance runners and relatively sedentary medical students showed no significant intergroup differences in these measures. On the other hand, a three-month period of moderate training, sufficient to increase maximal oxygen intake by an average of 6%, has shown beneficial effects, including a 12% to 13% increase in the metabolism of antipyrine and aminopyrine; individual gains in maximal oxygen intake have a correlation of 0.6 to 0.7 with gains in these indices of hepatic metabolism. The cytochrome c oxidase activity of the liver is also enhanced by regular moderate endurance activity.

Information on IGF is conflicting; some studies have shown rising serum levels of IGF following regular aerobic training, suggesting that production is increased. Some cross-sectional studies also have suggested that resistance training, perhaps by stimulating anabolic activity, protects against the usually observed age-related decline in IGF. On the other hand, recent studies of older men have demonstrated that a combination of regular exercise and a low-fat diet increases serum concentrations of IGF binding protein, thus decreasing free concentrations of serum IGF-1. These changes in the IGF axis could influence the risk of developing a prostate cancer, although the evidence to date is conflicting. However, regular physical activity has no influence on the risk of hepatic or biliary tract tumors except through the prevention of obesity.

At the other end of the exercise spectrum, physical inactivity, with an increase in body mass index and waist circumference, has been associated with increases in two hepatic enzymes (alanine aminotransferase and gamma-glutamyl transferase), raising the possibility that a change in fat metabolism within the liver may contribute to the risk of diabetes mellitus. The fat content of the liver is correlated with an individual's fasting glucose level. Regular physical activity decreases the fat content, and this may contribute to the role of regular exercise in preventing type 2 diabetes mellitus in older individuals. Aging is commonly associated with a decreased glucagon signaling capacity and responsiveness. Stimulation of hepatic metabolism by regular physical activity can apparently offset this change, enhancing the capacity for glycolysis, gluconeogenesis, lipolysis, and insulin secretion. Mechanisms include a normalizing of the ratio of inhibitory to stimulatory G-protein and thus an increased activity of the "second messenger" adenylate cyclase.

Moderate aerobic training programs have been applied as a component of therapy in patients with chronic hepatic disease and fatty liver, apparently without adverse effect. In rats, at least, regular exercise decreased halothane-induced hepatotoxicity and attenuated the ethanol-induced decline in hepatic function.

Over-vigorous training programs or repeated bouts of exhausting exercise may have less positive effects. In one study in rats, the researchers noted that the increased production of nitric oxide associated with vigorous exercise augmented the activity of iron regulatory protein 1, thus downregulating one of the key enzymes of the Krebs cycle, cytosolic aconitase. In humans, it has been suggested that repeated bouts of hepatic ischemic hypoxia may predispose to areas of central lobular necrosis analogous to those seen in patients in whom hepatic blood flow is restricted by cardiac disease (Rowell 1986), Evidence of cellular damage has been inferred from the release of cellular constituents. Thus, a rigorous military training course induced both leukocytosis and the release of several hepatic enzymes. Liver disorders have frequently been suspected in extremely active individuals because of abnormally high serum concentrations of bilirubin and such enzymes as aspartate aminotransferase, alanine aminotransferase, and alkaline phosphatase. Laboratory findings of this type could indeed reflect chronic hepatitis; but in the case of the enzyme markers, it is difficult to be certain that a release from damaged muscle is not the primary cause. There are currently no reports directly linking a serum accumulation of hepatic enzymes or bilirubin with other evidence of adverse effects of intensive training on hepatic function. The abuse of steroids is a confounding factor in some classes of athletes; there have been occasional case reports that the abuse of large quantities of androgens not only increases hepatic enzyme levels but can also induce such conditions as peliosis hepatis (a benign form of purpura that affects the liver, with the development of blood-filled cysts), nodular hyperplasia (a rapid and disorganized but benign nodular growth of liver tissue), and (rarely) hepatic carcinoma.

Regular habitual physical activity reduces the risk of symptomatic gallbladder disease. One large epidemiological study showed an association between physical activity and a low incidence of gallbladder disease after adjusting for other risk factors, although the authors of a 16-year follow-up of Harvard alumni suggested that much of any such association might be mediated indirectly, through the influence of regular physical activity on obesity levels. Ultrasound studies have shown decreased evidence of gallstones in active individuals (de Oliveira and Burini 2009). There seems to be as much as a 40% decrease in the risk of gallbladder problems in active middle-aged women. Presumably, much of the advantage is gained from an increased metabolism of lipids and a decreased accumulation of cholesterol. Studies in mice bred for their susceptibility to gallbladder disease have shown that exercised animals also increase the expression of genes (*Ldlr* and *Scarb1*) involved in the hepatic clearance of cholesterol and upregulate a protein (Cyp27a1) associated with the production of bile acids.

Kidneys and Urinary Tract

There seems to be no evidence that repeated endurance exercise and associated dehydration have lasting ill effects on the kidneys. Indeed, regular physical activity increases glomerular filtration in healthy adults, and although a period of unaccustomed heavy training may induce some depression of renal function, a recent study of professional distance cyclists showed that these individuals had lower serum creatinine levels and better creatinine clearance rates than sedentary individuals. Repeated direct trauma to the kidneys in some sports (particularly boxing) may contribute to proteinuria and hematuria.

At the other end of the exercise spectrum, prolonged bed rest or exposure to microgravity increases the risk of developing **renal calculi**, and there is some evidence that deliberate exercise can protect astronauts against this hazard (Monga et al. 2006). On the other hand, exercise during a period of bed rest sometimes increases the risk of symptomatic calculi (Okada et al. 2008).

Physical activity has sometimes been restricted in patients with chronic renal disease, in part because of fears that a tubular reabsorption of glomerular filtered protein would worsen the nephropathy. However, such restrictions are generally unnecessary and may be counterproductive, and the exercise-induced proteinuria in such individuals is often no greater than that seen in athletes. In hypertensive rats with chronic renal failure, regular exercise appears to attenuate proteinuria and protect against progression of renal sclerosis. Likewise, a program of resistance exercise can be helpful in human chronic renal disease, helping to reverse the muscle wasting that may arise from a low-protein diet and possibly reducing the depression of mood state that is commonly associated with chronic renal disease. Currently it is regarded as a good practice to combine a program of moderate exercise with thrice-weekly dialysis programs for individuals with advanced renal disease. Obesity substantially increases the risk of renal tumors; and regular moderate physical activity, in addition to its role in controlling obesity, may have a weak independent effect in reducing this risk.

Reabsorption of sodium ions in the proximal tubules of the kidneys is increased during the early stages of training, when plasma volume is expanding. The expansion of plasma volume worsens edema; and the associated decrease in hemoglobin concentration exacerbates tissue hypoxia, thus increasing symptoms if a person with mountain sickness chooses to engage in vigorous training during the first few days at simulated or real high altitude.

Specific and intensive training of the pelvic floor muscles seems helpful to women suffering from stress incontinence (Hay-Smith et al. 2001, 2007, 2008).

Stomach and Intestinal Tract

The main effect of regular moderate physical activity on the stomach and intestinal tract seems to be a decrease in the risk of certain forms of cancer, mediated by a combination of altered motility and a reduced body fat content.

Debate continues concerning the mechanical effects of exercise on the gastrointestinal tract; both acute and chronic changes in gastrointestinal motility may occur. One uncertainty is whether running has an effect similar to that of seated exercise; several reports have noted little effect on overall gastrointestinal transit. Thus, the mouth-to-cecum transit time was unchanged by running 9.6 km (6 mi), and walking a distance of 4.5 km (2.8 mi) in 1 h did not change the total transit time in previously sedentary laboratory workers. Likewise, total colonic transit time was similar for laboratory technicians and soccer players who were training as much as 15 h per week. However, other studies have suggested a substantial effect of regular training. A six-week endurance training program reduced the total carmine transit time to a range of 35 to 24 h, whereas the transit for a control group remained at a sluggish 45 h. Likewise, one week of cycling or jogging at 50% of maximal oxygen intake almost doubled the

speed of intestinal transit in healthy young adults; and in a study of healthy and recreationally active elderly subjects, the deliberate two-week restriction of physical activity almost doubled the colonic transit time. Other types of exercise may also be beneficial. Thus, 13 weeks of resistance training sufficient to induce a 40% increase in the peak force developed by key muscle groups more than doubled the speed of total intestinal transit; this change was attributed almost exclusively to faster movement through the large intestine. Exercise-related changes in the type and quantity of food ingested may be an important issue. This idea is supported by the fact that no changes in bowel transit time were observed when training experiments were carried out in a metabolic laboratory, where subjects adhered to a controlled diet and sophisticated radioactive markers of fecal movement were used.

There is no evidence of any consistent relationship between regular physical activity and the incidence of gastric ulcers. However, cross-sectional analyses suggest a reduced risk of duodenal ulcers in individuals who are physically active, possibly through changes in immune function or a modulation of stress. Obesity increases the risk of neoplastic lesions in the esophagus and the gastric cardia; regular physical activity may alleviate this risk through control of obesity, but there are no reports suggesting that physical activity has an independent effect on the risk of neoplasia. Further study is also needed to examine the impact of regular physical activity on the risks of diverticulosis and inflammatory bowel disease, although one recent report showed an increase in neutrophil activation and an exacerbation of zinc deficiency when patients in remission from Crohn's disease undertook a moderate exercise program. Since physical activity stimulates the production of heat shock proteins, this also could be beneficial in the treatment of inflammatory bowel disorders (de Oliveira and Burini 2009).

A speeding of passage of food or mechanical bouncing (or both) or changes in segmentation of the large intestine, perhaps mediated by prostaglandins, could reduce exposure of the colon wall to toxins. This has been suggested as the basis of the reduced risk of cancers of the descending colon seen in those who engage in regular physical activity through either their occupation or leisure pursuits (Shephard and Futcher 1997). Other possible factors that could influence the risk of colonic cancer include exercise-induced reductions in gastrointestinal blood flow and neuroimmuno-endocrinological alterations.

The intensity of the training regimen was no more than moderate in many of the epidemiological studies showing that physical activity protected against cancer. Nevertheless, a change in local or general movement of the gut contents seems the most likely explanation for the reduction in risk of carcinoma by as much as 50% in a physically active person (figure 9.3). The need remains for further careful observations of the acute and chronic responses of both motility and cancer risk to appropriately graded intensities of effort.

Despite the occult hemorrhage frequently observed during acute bouts of endurance exercise (particularly running), a large-scale, three-year prospective study of the elderly showed that regular walking, gardening, or vigorous physical activity was associated with a substantially reduced risk of severe gastrointestinal hemorrhage, even after statistical adjustment for associated risk factors.

Circulation and Immune Function

Perhaps in part because of the depression of plasma volumes that occurs during an acute bout of exercise, a period of endurance training leads to an expansion of plasma volume. This effect is greater with upright than with supine exercise, probably because circulatory demands are greater in an upright position. Plasma volume expansion is accompanied by increases in serum albumin and sodium ion concentrations and a temporary decrease in the excretion of urine. One consequence of the plasma volume expansion is a tendency toward a reduction in hemoglobin concentration (so-called athlete's anemia); however,

there is an associated decrease in the viscosity of the blood, and tissue oxygen transport is usually well maintained. A true anemia is uncommon in athletes, although it can develop if food intake is inadequate or poorly balanced or if exercise-induced bleeding from the alimentary tract is frequent.

Athletes can induce temporary increases in their total hemoglobin by living at altitude or simulated hypoxic conditions and continuing to train as at sea level conditions. Any competitive advantage of oxygen transport gained in this way is lost within two to three weeks of return to sea level (Levine 1995). There has been an unfortunate tendency among endurance competitors such as distance cyclists to induce a similar increase in hemoglobin through erythropoietin doping, and several athletes have had medals revoked following detection of such abuses.

The impact of prolonged endurance training on the immune response has received only limited attention. Moderate aerobic training appears to augment the number and activity of NK cells, possibly reversing age-related decreases in both NK and **T-cell** function. Increased salivary concentrations of **immunoglobulins**, whether expressed as absolute concentrations or relative to grams of salivary protein, may also be seen. The main clinical evidence of improved immune function is a decrease in the symptoms associated with upper respiratory infections. A 12-month study of 547 healthy adults estimated a 20% reduction in risk of upper respiratory infection among individuals who engaged in moderate (>4 MET-hours per day) or vigorous (>12 MET-hours per

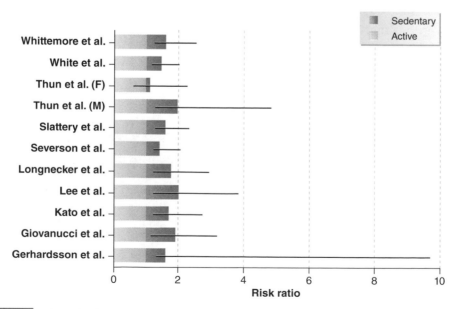

FIGURE 9.3 The risk of colon carcinoma seen in groups with a sedentary lifestyle relative to more active comparison groups. Findings from 11 surveys, as summarized by Shephard and Futcher (1997), are presented as risk ratios, with 95% confidence limits for the sedentary samples shown as black lines.

day) physical activity. However, in a second experiment, participants were deliberately inoculated with rhinovirus, and the severity and duration of illness did not differ between those who were undertaking moderate exercise (40 min at 70% of heart rate reserve [HRR] on alternate days) and sedentary controls.

Extreme conditioning programs such as the peak phases of preparation for international competition can temporarily reduce NK cell activity, neutrophil function, and serum, salivary, and membrane concentrations of IgA and IgM, particularly if the athlete's supply of nutrients and vitamins is inadequate. Such changes are in turn associated with an augmented susceptibility to acute respiratory infections (Mackinnon 2000). A deterioration of immune function is one of the markers of overtraining, although the immunological findings are less consistent than the associated deterioration in performance times and mood state (Shephard 1997). There have been suggestions that large doses of vitamin C may help to avert respiratory disease and sustain neutrophil function during periods of intensive training, possibly by countering an increased exercise-induced release of reactive oxygen species or by countering release of the immunosuppressant cytokine IL-6 (Fischer et al. 2004).

Obesity seems to be associated with an increase in leukemia and non-Hodgkin's lymphoma; as with other visceral tumors, regular physical activity may thus play a role in preventing such conditions.

One of the manifestations of aging is a deterioration of immune responses, with an increased risk of developing autoimmune diseases. Thus, one important benefit of regular physical activity may be a delaying of the autoimmune aspects of aging.

Strengths and Limitations of the Current Evidence

Scientific evidence concerning the effects of exercise on the viscera is less strong than for other systems such as the heart and the lungs. There are several reasons for this. The organs concerned are less readily accessible, and thus it is more difficult to evaluate their function. This is particularly true for human studies of immune function, since information on changes in leukocyte numbers, characteristics, and activity is usually garnered from peripheral blood samples, although only about 1% of the total cell population is resident in the bloodstream. Moreover, other organs have a single main function that can be summarized by a few simple parameters, for example, ejection fraction, maximal cardiac output, and maximal oxygen transport for the heart; and vital capacity,

forced expiratory flow rate, and pulmonary diffusing capacity for the lungs. However, the liver and kidneys have multiple functions, and commonly used tests assess only a single aspect of their performance. Finally, because information is easier to obtain, many more studies have examined the influence of physical activity on cardiorespiratory function than on visceral function. In the case of the viscera, it is thus more difficult to seek corroboration of findings in other populations and in other environments.

Human studies have been based largely on clinical case reports, small-scale experimental studies of the metabolism and excretion of test substances, and epidemiological studies of the risks of cancer in relation to habitual physical activity. The latter have often involved large populations, with careful sampling; but questionnaire estimates of physical activity have had limited absolute accuracy, and it has been difficult to make full adjustment for covariates such as smoking history.

Animal studies have allowed a more detailed analysis of internal organs, including biopsy and postmortem examination. It has also been possible to conduct experimental evaluations of interactions between physical activity and carcinogens or metastatic tumor cells. However, uncertainties remain concerning relationships between such experimental tumors and the spontaneous tumors that develop in humans over many years. The need to enforce exercise patterns, together with differences in body size and life span, have also hampered the extrapolation of animal findings to human subjects.

As discussed in several chapters of this book, surprisingly little information is available concerning either the minimum amount of physical activity needed to enhance health or the threshold beyond which health deteriorates rather than improves. The optimal dosage probably varies with the organ system or tissue under consideration, and dose–response information is particularly limited for the organs and tissues discussed in this chapter. There is an urgent need for better information on dose–response relationships than can be obtained from most types of questionnaires. This will require either long-term, objective assessments of physical activity patterns by modern motion sensors or the inference of cumulative physical activity in terms of the individual's attained level of physical fitness. Many of the visceral responses to physical activity seem to depend on the intensity of the effort undertaken; it is thus important to define the level of physical activity at which beneficial effects on metabolism, biological age, resistance to infection and cancer are replaced by the threats associated with exhausting physical activity, excessive training, or both.

Upper respiratory infections remain a major source of morbidity in otherwise healthy individuals. More research is thus needed to verify or disprove tantalizing suggestions that moderate physical activity can enhance immune barriers to such infections but that excessive training can exacerbate the risk. Such research should be designed to overcome the limitations of present data, obtaining objective rather than subjective evidence of viral infection and excluding artifacts from modulations of viral exposure, stress, and nutritional status plus physical changes in the respiratory membranes associated with the oral inhalation of increased volumes of cold, dry, or polluted air. There is also a need to confirm conflicting claims that an exercise-induced increase in the risk of upper respiratory infection can be countered by massive doses of antioxidants; again, this will require objective evidence of respiratory disease rather than a simple reporting of respiratory symptoms.

As family size decreases and longevity is extended by the countering of chronic disease, population aging is becoming an ever more pressing problem for developed societies. Little is yet known about interactions among exercise, aging, and the immune system; but the need exists to confirm small-scale studies suggesting that regular moderate exercise counters the age-related deterioration in overall immune function and disturbances of autoimmune responses.

Given suggestions that overly strenuous exercise can cause local tissue necrosis, a final need is for larger and longer-term studies comparing hepatic and renal function between those who have engaged in repeated ultra-endurance competition and those who have limited their physical activity to current population recommendations.

Summary

This chapter examines the responses of body systems often neglected by exercise scientists—the liver and gallbladder, the kidneys and urinary tract, the stomach and intestines, and the immune system. Both acute and chronic effects of physical activity are considered. As in many other parts of the body, single bouts of moderate physical activity induce correspondingly limited and beneficial responses. Periods of moderate-intensity training also result in adaptations that have positive consequences for both immediate physical performance and longer-term health, including modifications of the IGF axis, a decrease in the likelihood of developing some types of cancer, and a slowing of the decrements in immune response normally associated with aging. However, a single intense and exhausting bout of exercise, particularly if performed by vulnerable individuals under extreme environmental conditions, can have negative effects on health, occasionally progressing to such pathologies as hepatic and renal failure and systemic endotoxemia. Repeated bouts of exhausting exercise or an excessive intensity of training also carries potential hazards, including cumulative ischemic damage to the viscera and an increased susceptibility to infection associated with immunosuppression. When one is planning either single bouts of exercise or training programs, moderation is thus an important tenet for the viscera as for other body systems.

Key Concepts

aldosterone—A hormone produced in the adrenal cortex that stimulates the reabsorption of sodium ions in the kidneys, thus countering sodium loss in the sweat and conserving or increasing blood volume.

antidiuretic hormone—A pituitary hormone that reduces urine flow by stimulating water reabsorption in the kidneys. This hormone becomes active during prolonged physical activity, when large amounts of fluid are being lost in the sweat.

anuria—A rapidly fatal complete cessation of urine production, as may occur in severe cases of exhausting exercise and heat stress.

creatine kinase—An enzyme that leaks from muscle following damage, particularly after a period of prolonged eccentric exercise.

creatinine—An end product of creatine metabolism, normally cleared from the body in the urine. The rate of clearance of a test dose is used to assess changes in the efficiency of kidney function.

cytokines—Hormone-like compounds that act mainly on adjacent cells, modifying the proliferation and differentiation of cells, regulating immune responses, and initiating inflammation. Vigorous exercise greatly increases the production of cytokines by various types of white blood cells.

endotoxemia—Penetration of the gut wall by toxins from bacteria present in the intestines. The penetration, which is commonly attributable to a large decrease in the blood flow to the intestines during severe exercise, can cause a dangerous generalized septic reaction.

glomerular filtration—Rate at which the renal glomeruli can clear substances from the bloodstream, commonly measured by the clearance of a test dose of creatinine. Glomerular filtration is often compromised by prolonged endurance exercise.

gluconeogenesis—Production of glucose from noncarbohydrate precursors such as amino acids or lactate. This process is important during prolonged endurance activity as the carbohydrate reserves of the skeletal muscles are depleted and the blood glucose level falls.

glutathione—A sulfur-containing tripeptide that is a coenzyme but also acts as an antioxidant, protecting key enzymes of the body against degradation by reactive species of oxygen. Liver levels of glutathione drop during a prolonged bout of vigorous physical activity, attributable to a combination of a high rate of metabolism and an increased production of reactive species.

Gram-negative bacteria—A type of bacteria found in the intestines that does not retain the basic dye in its wall during Gram staining. Such bacteria produce the toxins that can give rise to endotoxemia if they penetrate into the bloodstream during exhausting physical activity.

hemoglobinuria—Appearance of hemoglobin in urine, sometimes associated with exposure of the kidneys to combinations of heat and exercise. The pigments and red cells may form casts that can block urinary flow, leading to kidney failure.

hereditary renal hypouricemia—A familial abnormality of uric acid excretion seen in Japanese and non-Ashkenazi Jews that can predispose to renal failure during bouts of exhausting physical activity.

hyponatremia—A clinically significant drop in plasma sodium ion concentration (values below 130 mM/L).

hypoxia—A lower than normal partial pressure of oxygen in a given tissue, usually leading to a depression of function in the affected organ.

immunoglobulins—Proteins secreted by lymphocytes that have specific antibody activity. A depression of IgA secretion by mucous membranes is thought to contribute to the vulnerability of overtrained individuals to upper respiratory tract infections.

insulin-like growth factor (IGF)—Homologs of insulin that are less important than insulin in the regulation of blood glucose but are more potent in the stimulation of growth and muscle hypertrophy.

insulin-like growth factor (IGF) binding protein—Protein that binds IGF in the serum and other body fluids, thus lowering effective plasma concentrations and regulating the action of IGF.

jaundice—A yellowing of the skin and the whites of the eyes, reflecting excessive concentrations of bilirubin in the blood. It may result from obstruction of the bile duct, excessive hemolysis of red cells, or impaired function of the liver.

myoglobinuria—Excretion of the red pigment of skeletal muscle (myoglobin) in the urine, associated with the muscle injury resulting from severe exercise.

natural killer (NK) cell—A type of white cell that provides the first line of defense against viruses and tumor cells, acting without any priming from other components of the immune system. It has been postulated that suppression of NK cell activity for 2 to 72 h following a heavy bout of exercise such as a marathon run leaves an "open window" when upper respiratory tract infections are more likely to develop.

necrosis—Cell breakdown and death, seen in the liver, for example, if the local blood flow becomes inadequate to sustain tissue oxygenation.

nonsteroidal anti-inflammatory drugs—Drugs such as aspirin and ibuprofen. These medications are commonly taken without prescription to relieve discomfort from minor injuries, but they slow blood clotting and thus can provoke gastrointestinal bleeding, particularly during exercise.

occult blood—Presence of pigments in the stools indicative of gastrointestinal bleeding. The pigments (breakdown products of hemoglobin) often give the stools a black color, but they may be detectable only by appropriate chemical tests.

oliguria—A greatly reduced flow of urine, as may occur with dangerous combinations of exhausting exercise and heat stress.

periportal cells—Liver cells grouped around the portal veins.

prostaglandins—A group of metabolites of arachidonic acid and related compounds secreted during exercise and injury. They are biologically very active. Their proinflammatory action can be countered by administration of nonsteroidal anti-inflammatory drugs such as ibuprofen.

proteinuria—Escape of plasma proteins into the urine, a phenomenon commonly seen during and immediately following bouts of prolonged and vigorous physical activity.

pyrogens—Substances that give rise to fever. Vigorous physical activity leads to the production of several pyrogens such as tumor necrosis factor and IL-1 within the body.

renal calculi—Stones formed in the urinary tract. The calculi are commonly a consequence of bone demineralization, for instance during bed rest or exposure to microgravity in space flight.

sorbitol—Sugar produced from glucose. The rate of clearance of a test dose of sorbitol from the bloodstream is used to assess changes in liver function.

T-cells—Class of lymphocyte responsible for cell-mediated immunity. The T-lymphocytes include cytolytic cells that break down cells identified for elimination and helper cells that facilitate this process. The number of T-cells diminishes with aging, but regular physical activity may help to slow this trend.

Study Questions

1. Can a bout of very heavy physical activity have adverse short-term consequences for organs such as the liver, kidneys, and intestines? If so, what are the dominant mechanisms, what environmental circumstances are likely to exacerbate these effects, and what precautions can be suggested to an athlete?

2. Would you agree that in the case of the liver, regular physical activity leads to an enhancement of metabolic function? If so, what are some of the more important reactions that are enhanced?

3. Why is it important to recommend regular physical activity to patients with chronic renal disease?

4. When regulating the preparation of athletes for top-level competition, would you look for immunological evidence of overtraining? If so, what alteration of immune function is the best indication that an athlete has been overtrained?

5. How convincing is the evidence that heavy exercise creates an open window in which athletes become vulnerable to upper respiratory infections?

6. Many exercise scientists focus mainly on the cardiorespiratory effects of physical activity. Do you think more attention should be directed to the viscera, and if so, why?

7. How far can habitual physical activity modify the risk of visceral cancers? What mechanisms have been suggested as responsible for the benefits observed?

8. How would you assess the strength of evidence concerning interactions between physical activity and the viscera, relative to knowledge of effects on the heart and lungs?

References

Bellinghieri, G., V. Savica, and D. Santoro. 2008. Renal alterations during exercise. *Journal of Renal Nutrition* 18:158-164.

Berg, A., H.M. Müller, S. Rathmann, and P. Deibert. 1999. The gastrointestinal system—an essential target organ of the athlete's health and physical performance. *Exercise Immunology Review* 5:78-95.

Brouns, F., and E. Beckers. 1993. Is the gut an athletic organ? Digestion, absorption and exercise. *Sports Medicine* 15:242-257.

Clarkson, P.M. 2007. Exertional rhabdomyolysis and acute renal failure in marathon runners. *Sports Medicine* 37:361-363.

de Oliveira, E.P., and R.C. Burini. 2009. The impact of physical exercise on the gastrointestinal tract. *Current Opinion in Clinical Nutrition and Metabolic Care* 12:533-538.

Fischer, C.P., N.J. Hiscock, M. Penkowa, S. Basu, B. Vessby, A. Kallner, L.B. Sjöberg, and B.K. Pedersen. 2004. Vitamin C and E supplementation inhibits the release of interleukin-6 from contracting human skeletal muscle. *Journal of Physiology* 558:633-645.

Gleeson, M. 2000. Mucosal immune responses and risk of respiratory illnesses in elite athletes. *Exercise Immunology Review* 6:5-42.

Gleeson, M. 2009. Exercise immunology. In *The Olympic textbook of science in sport*, ed. R.J. Maughan (pp. 149-162). Oxford: Blackwell Scientific.

Hay-Smith, E.J., K. Bø, L.C. Berghmans, H.J. Hendriks, R.A. de Bie, and E.S. van Waalwijk van Doorn. 2001, 2007, 2008. Pelvic floor muscle training for urinary incontinence in women. *Cochrane Database Systematic Reviews* 2001;(1):CD001407; 2007;(1):CD001407; 2008;(3):CD001407.

Johnson, J.M. 2000. Endurance exercise and the regulation of visceral and cutaneous blood flow. In *Endurance in sport* (2nd ed.), ed. R.J. Shephard and P-O. Åstrand. Oxford: Blackwell Scientific.

Jozko, P., D. Waskow-Czopnik, M. Medras, and L. Paradowski. 2006. Gastoesophageal reflux disease and physical activity. *Sports Medicine* 36:385-391.

Levine, B. 1995. Training and exercise at high altitudes. In *Sport, leisure and ergonomics,* ed. G. Atkinson and T. Reilly (pp. 74-92). London: E & FN Spon.

Lippi, G., G. Banfi, G.L. Salvagno, M. Franchini, and G.C. Guidi. 2008. Glomerular filtration rate in endurance athletes. *Clinical Journal of Sports Medicine* 18:286-288.

Mackinnon, L.T. 2000. Chronic exercise training effects on immune function. *Medicine and Science in Sports and Exercise* 32(Suppl 7):S369-S376.

Marshall, J.C. 1998. The gut as a potential trigger of exercise-induced inflammatory responses. *Canadian Journal of Physiology and Pharmacology* 76:479-484.

Monga, M., B. Macias, E. Groppo, M. Kostelec, and A. Hargens. 2006. Renal stone risk in a simulated microgravity environment: Impact of treadmill exercise with lower body negative pressure. *Journal of Urology* 176:127-131.

Moses, F.M. 1994. Physical activity and the digestive processes. In *Physical activity, fitness and health,* ed. C. Bouchard, R.J. Shephard, and T. Stephens. Champaign, IL: Human Kinetics.

Nieman, D.C. 2000. Special feature for the Olympics: Effects of exercise on the immune system: Exercise effects on systemic immunity. *Immunology and Cell Biology* 78:496-501.

Okada, A., H. Ohshima, Y. Itoh, T. Yasui, K. Tozawa, and K. Kohri. 2008. Risk of renal stone formation induced by long-term bed rest could be decreased by premedication with bisphosphonate and increased by resistive exercise. *International Journal of Urology* 15:630-635.

Peters, H., W.R. De Vries, G. Vanberge-Henegow, and L. Akkermans. 2001. Potential benefits and hazards of physical activity and exercise on the gastrointestinal tract. *Gut* 48:435-439.

Poortmans, J.R., and J. Vanderstraeten. 1994. Kidney function during exercise in healthy and diseased humans. *Sports Medicine* 18:419-437.

Rowell, L.B. 1986. *Human circulation: Regulation during physical stress.* New York: Oxford University Press.

Shephard, R.J. 1997. *Physical activity, training and the immune response.* Carmel, IN: Cooper.

Shephard, R.J., and R. Futcher. 1997. Physical activity and cancer: How may protection be maximized? *Critical Reviews in Oncogenesis* 8:219-272.

van Nieuwenhoven, M.A., F. Brouns, and R.J. Brummer. 1999. The effect of physical activity on parameters of gastrointestinal function. *Neurogasteroenterology and Motility* 11:431-439.

Wade, O.L., and J.M. Bishop. 1962. *Cardiac output and regional blood flow.* Oxford: Blackwell Scientific.

PART III

Physical Activity, Fitness, and Health

You are now moving into the central topic of this book—the relationships among physical activity, fitness, and health. In part I, you learned about the evolution of the concepts concerning the relationships among sedentary time, physical activity, physical fitness, and health. You have also reviewed some of the known differences and similarities between women and men and among ethnic groups, as well as the changes that take place with age. In part II, you learned about the acute and chronic effects of physical activity on the heart, lungs, blood vessels, skeletal muscles, endocrine and other hormones, kidneys, and other organs and tissues.

This part of the book provides information that underscores sedentary living habits, low physical activity levels, and poor fitness as major public health problems. The growing recognition of the threat to health, function, and well-being posed by inactivity largely fueled the development of exercise science in the latter half of the 20th century. Nine chapters are devoted to this key area. Chapter 10 deals with the contribution of a sedentary lifestyle and low levels of fitness to the risk of dying prematurely. Chapter 11 examines the effects of physical activity and fitness on cardiac, vascular, and pulmonary morbidities. Chapter 12 focuses on obesity, and chapter 13 deals with diabetes mellitus. These two chapters are particularly important given the current dramatic increases in the prevalence of obesity and type 2 diabetes mellitus in industrialized countries. Chapter 14 reviews the evidence for a role of physical activity and fitness in the risk of developing a number of cancers. Chapter 15 addresses bone health and joint disorders, and chapter 16 reports on muscular fitness and disorders. Chapter 17 covers fitness and activity in children, whereas chapter 18 discusses the risks of adverse musculoskeletal and cardiac events associated with regular physical activity. You will need to have a good grasp of this corpus of knowledge to take full advantage of the new material presented in subsequent parts of the book.

10

Physical Activity, Fitness, and Mortality Rates

Michael J. LaMonte, PhD, MPH*; and Steven N. Blair, PED

*Corresponding Author

CHAPTER OUTLINE

As reviewed in chapter 1, human evolution has been dependent on a physically active lifestyle. Thus, existence in a modern world where physical activity has largely been engineered out of daily living has led to aberrations from our evolutionary constitution. Logically, then, a sedentary way of life should be unhealthy for our species. Speaking to this point on the eve of the 1996 Olympiad in Atlanta, Georgia, while announcing the release of the seminal U.S. Surgeon General's report on physical activity and health, Dr. David Satcher remarked that few will ever achieve the level of athletic performance displayed by the athletes competing in the Olympic games, but that *everyone* can benefit from an active way of living (U.S. Department of Health and Human Services [DHHS] 1996). Indeed, the October 2008 release by the U.S. government of the first formal federal guidelines on physical activity for Americans established physical activity promotion at the population level as a major public health target (Physical Activity Guidelines Advisory Committee [PAGAC] 2008).

Mortality is perhaps the single best measure of the force that a given risk factor or disease exerts on population health. Sedentary habits and low cardiorespiratory fitness are among the strongest predictors of premature mortality and thus are a major threat to public health in the first part of the 21st century, especially in industrialized populations in which **prevalence** of inactivity is high. An important distinction is that **physical activity** refers to a behavior, specifically body movement that results in increased **energy expenditure of activity** above resting levels, and **physical fitness** is a set of physiological attributes that is determined largely by recent physical activity habits. Although the term *physical fitness* represents a broad spectrum of physiological attributes such as aerobic power, muscular strength and endurance, body composition, musculoskeletal flexibility, speed, and balance, this chapter focuses only on **cardiorespiratory fitness** (aerobic power) and **muscular fitness** (muscular strength and endurance).

■ Physical activity is a behavior, specifically body movement that results from skeletal muscle contraction. Physical fitness is a set of physiological attributes that result from participation in physical activity and, to some degree, from genetic influences.

In the first edition of this textbook, we summarized epidemiological findings in support of the hypothesis that sedentary lifeways predispose to premature mortality and decreased longevity. We drew heavily from the rich body of work by two pioneering physical activity epidemiologists, Professors Jeremy Morris (1910-2009) and Ralph Paffenbarger Jr. (1922-2007). Seminal prospective cohort studies conducted by Morris in British civil servants (circa 1950) and by Paffenbarger in San Francisco longshoremen (circa 1970), along with studies by Professor Henry Taylor (1919-1983) in U.S. railroad employees (circa 1960), showed that men with higher levels of occupational physical activity (e.g., double-decker bus conductors, dockworkers, railroad switchmen) experienced **rates of mortality from coronary heart disease (CHD)** that were at least 50% lower than those seen in their less active coworkers (e.g., bus drivers, desk clerks).

Because classifying physical activity levels by occupational duties provides a crude assessment of overall daily physical activity **exposure**, and because widespread reductions in occupational energy expenditure were occurring throughout the 20th century, Morris and Paffenbarger expanded their research in physical activity **epidemiology** to include examination of leisure-time activity in relation to health **outcomes**. Study findings by Morris in male civil servants and by Paffenbarger in male Harvard alumni consistently showed that higher levels of leisure-time activity were associated with significantly lower mortality from all causes and cardiovascular disease (CVD). The suggestion in all these studies that vigorous-intensity activity may be the principal component of overall activity conferring mortality benefits focused ensuing research (and scientific debate) on the relative roles of activity intensity and energy expenditure as predictors of mortality and morbidity, and this debate continues today. Collectively, the seminal studies of Morris, Taylor, and Paffenbarger provided the first prospective, systematic examination of physical inactivity and mortality and established the value of epidemiological research in exercise science, as well as the value of physical activity habits in the context of public health. Detailed summaries of these and other studies on physical activity and health can be found elsewhere (DHHS 1996; PAGAC 2008).

■ Seminal population-based studies on the relationships of occupational and leisure-time physical activity habits with mortality outcomes were conducted by Professor Jeremy Morris among London busmen and postal workers, Professor Henry Taylor among U.S. railroad workers, and Professor Ralph Paffenbarger Jr. among San Francisco dockworkers and Harvard alumni. These were the first large-scale epidemiological studies to show that physical activity is an important and independent predictor of mortality from all causes and CVD. Together, Morris, Taylor, and Paffenbarger introduced the application of epidemiological research to the exercise sciences.

The primary objective of this chapter is to summarize recent evidence relating physical activity or cardiorespiratory fitness with mortality. We present information on activity or fitness and mortality in several groups according to their demographic and health status. To illustrate key points, we draw data from selected epidemiological studies that are frequently cited in consensus statements and systematic reviews, with an emphasis on studies published since 2000. We defer discussion of potential biological mechanisms that may help explain the lower death rates observed among active and fit individuals, compared with their sedentary and unfit counterparts, to other chapters in this book that cover these issues in detail.

Physical Activity and Mortality

Clear and strong evidence supports an inverse association between levels of physical activity or cardiorespiratory fitness and mortality from all causes, CVD, and all cancers combined (DHHS 1996; PAGAC 2008). These associations generally exhibit a curvilinear **dose–response** gradient; are temporally sequenced and biologically plausible; are seen in women and men, older and younger individuals, and healthy individuals and those with existing disease; and remain significant after potential **confounding factors** are accounted for. Moreover, in observational **prospective studies**, individuals whose physical activity or fitness level increases across serial assessments, even later in life, experience substantial mortality benefit as compared to peers who are persistently inactive and unfit. Meeting these criteria leaves little doubt that the association between inactivity and mortality is causal. A comprehensive systematic review of studies published since the 1996 Surgeon General's report indicated that the minimal dose of activity currently recommended for health benefits (≈8-10 **metabolic equivalent [MET]** hours [MET-hours] per week in activities of at least moderate intensity) is associated with a significant 20% to 30% lower risk of all-cause mortality (PAGAC 2008). An alternative way to express the minimal recommended dose of activity is 150 min per week of moderate-intensity activity, 75 min per week of vigorous-intensity activity, or a combination of the two intensities, with 1 min of vigorous intensity equal to 2 min of moderate intensity. For those individuals who already are regularly active at these levels, additional mortality benefit can be obtained through participation in greater amounts of physical activity.

© Simone van den Berg

Consistent exercise engagements, such as a group spinning class, help reduce the risks of dying from CVD and cancer.

Federal guidelines on physical activity and health indicate that mortality risk is 20% to 30% lower among adults who achieve recommended amounts of physical activity, which equate to a weekly energy expenditure of about 8 to 10 MET-hours per week in moderate- to vigorous-intensity activities. Alternatively, this recommended level of activity could be expressed as 150 min per week in moderate-intensity activity, 75 min per week in vigorous-intensity activity, or a combination of the two intensities, with 1 min of vigorous intensity equal to 2 min of moderate intensity. Greater health benefit is obtained from higher amounts of activity, such as 300 min per week of moderate-intensity or 150 min per week of vigorous-intensity activities. Participation in resistance exercises at least two days a week also contributes to health benefits, likely including lower mortality risk.

Recently investigators have attempted to refine the understanding of specific aspects of the activity–mortality relationship, including evaluation of distinct exposures such as sitting time or meeting current recommended activity levels; use of objective measures in addition to questionnaires to quantify activity exposures; and the roles of activity type, intensity, and pattern in relation to mortality. We next briefly highlight key aspects from some of these recent studies.

Sitting Time

Prolonged sitting is becoming recognized as a distinct component of sedentary habits that may reflect not merely an adverse exposure at the extreme low end of the physical activity continuum (PAGAC 2008). Only recently have epidemiological studies been conducted to investigate the relationship of sitting time with mortality outcomes. Katzmarczyk and colleagues (2009) reported on 17,013 adults, 18 to 90 years of age, who completed questionnaire assessments on physical activity and sedentary habits in the 1981 Canada Fitness Survey and were then followed up for mortality during an average of 12 years. After adjustment for differences in age, sex, and several other potential confounding factors, **relative risks** across incremental categories of daily sitting time (almost none of the time, one-fourth of the time,

half of the time, three-fourths of the time, and almost all of the time) were 1.00 (referent), 1.11, 1.36, and 1.54 (*p*-value for trend, $p < 0.0001$) for all-cause mortality and 1.00, 1.01, 1.22, 1.47, and 1.54 (trend, $p < 0.0001$) for cardiovascular mortality. Sitting time was not significantly associated with cancer mortality in this study. The investigators further investigated the sitting time–mortality relationship according to leisure-time physical activity level, using a cutoff value of <7.5 and ≥7.5 MET-hours per week to define inactive and active individuals, respectively. The findings, shown in figure 10.1, demonstrate a dose–response gradient for higher mortality rates across incremental categories of daily sitting time among both inactive and active individuals. These data support the hypothesis that sedentary habits may confer mortality risk through pathways related to but separate from physical activity levels. Clinicians and public health practitioners might therefore be prudent not only in promoting recommended levels of moderate- to vigorous-intensity activity but also in discouraging excessive sitting time.

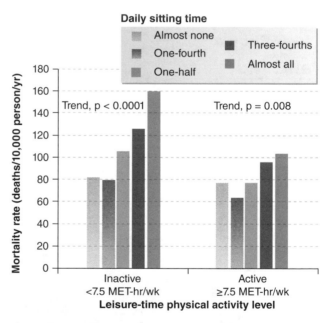

FIGURE 10.1 Age-standardized rates of all-cause mortality according to categories of daily sitting time and leisure-time physical activity among 17,013 adults in the 1981 Canada Fitness Survey who were followed up for an average of 12 years (1,832 deaths). Sitting categories (from left to right) are defined as follows: almost none of the time (referent), one-fourth of the time, half of the time, three-fourths of the time, and almost all of the time. Mortality rates are shown as deaths per 10,000 person-years.

Adapted, by permission, from P.T. Katzmarzyk, T.S. Church, C.L. Craig and C. Bouchard, 2009, "Sitting time and mortality from all causes, CVD, and cancer," *Medicine and Science in Sports and Exercise* 41:998-1005.

Intensity of Physical Activity

As public health recommendations evolved to emphasize a broad spectrum of benefits through regular participation in moderate-intensity activity (Blair, LaMonte, and Nichaman 2004), a keen interest arose in studies having the capacity to examine intensity-specific health effects while accounting for potential confounding by total energy expenditure. This follows from the possibility that lower risk of mortality or morbidity associated with vigorous activity may simply reflect greater total energy expenditure through vigorous activity rather than higher intensity per se. Paffenbarger and coworkers in the Harvard Alumni Health Study again spearheaded discovery in this important area of work. In their study, 13,485 men reported their activity habits in 1977 and were followed for mortality through 1992 (Lee and Paffenbarger 2000). First, reported activity levels were grouped according to absolute intensities of light (<4 METs), moderate (4-5.9 METs), and vigorous (≥6 METs). Then, within each stratum of absolute intensity, mortality risk was computed across categories of weekly energy expenditure ranging from <150 kcal/week in the lowest group to ≥1,500 kcal/week in the highest group. After adjustment for potential confounding factors, including each of the *other activity components*, light-intensity activity was not associated with mortality benefit (trend across categories of energy expenditure, $p = 0.72$), whereas moderate-intensity (trend, $p = 0.07$) and vigorous-intensity (trend, $p < 0.001$) activities were associated with lower risks of all-cause mortality, though the association was somewhat stronger for vigorous activity.

Evidence for health benefits from moderate-intensity activity also has been seen in women, as recently reported in the Nurses' Health Study (Manson et al. 1999). To isolate the contributions of moderate- and vigorous-intensity activity, investigators compared intensity-specific risks of combined fatal and nonfatal CHD at a specified amount of energy expenditure. Results showed that CHD benefit was no greater for vigorous- compared with moderate-intensity activity once energy expenditure was held constant. Each 5 MET-hours per week increment in moderate-intensity activity (e.g., 90 min of brisk walking) and in vigorous activity (e.g., 45 min of jogging) was associated with a 14% and 6% lower risk of CHD, respectively. Collectively, the findings from the Harvard Alumni Health Study and the Nurses' Health Study indicate that total energy expenditure, rather than intensity, likely is the key factor underlying activity-related mortality benefits. For adults who are sedentary or who are averse to strenuous activity, promoting increased energy expenditure at moderate intensities confers substantial benefit; for those who prefer and are able to perform vigorous-intensity activity, such activity may be a more time-efficient lifeway for prolonged longevity.

In addition to accounting for total energy expenditure when studying intensity-specific associations with mortality outcomes, it is necessary to consider the role of relative intensity. Even though the absolute energy cost (intensity) of a given activity generally is the same among apparently healthy individuals, differences in cardiorespiratory and muscular fitness due to inactivity or aging result in a higher relative intensity (e.g., higher percentage of maximal aerobic power) required to complete the same activity for less fit or older individuals compared with their younger and more fit peers. For example, brisk walking (e.g., 5.6 km/h or 3.5 mph) on level ground is a moderate-intensity activity (≈4 METs) on an absolute scale. However, on a relative scale, this same activity would be light intensity for a young, fit adult whose maximal capacity is 12 METs and would be quite strenuous for an older adult whose maximal capacity is 6 METs.

Paffenbarger and coworkers studied the relationship between relative intensity and mortality in 7,337 Harvard alumni (mean age 66 years) who in 1988 completed questionnaire assessments on physical activity habits, including the Borg scale for perceived exertion during activity as a proxy of relative intensity, and were followed through 1995 for combined fatal and nonfatal CHD (Lee et al. 2003). Borg scale categories of perceived effort were "nothing to weak (referent), moderate, somewhat strong, and strong to maximal." This investigation yielded two key findings. First, multivariable-adjusted relative risks for fatal or nonfatal CHD across the categories of relative intensity were 1.00 (referent), 0.86, 0.69, and 0.72 (trend, $p = 0.02$), indicating a dose–response gradient for lower CHD risk across incrementally higher levels of relative intensity. Second, even men who were physically active during leisure time at less than recommended levels of total energy expenditure or absolute intensity experienced significantly lower **incidence** of fatal or nonfatal CHD. The adjusted relative risk for CHD for the lowest and highest categories of relative effort was 0.62 ($p < 0.05$) in men expending <1,000 kcal/week and 0.41 ($p < 0.05$) in men whose absolute intensity was less than moderate (<3 METs). These findings suggest that, particularly in older adults, relative intensity may be an important component in the causal pathway between physical activity and mortality risk and may account for some of the variation in observed associations between populations.

Meeting Activity Recommendations

Current guidelines for physical activity and public health are based on a rigorous scientific review that gave rise to conclusions on the type and amount of activity required to affect a variety of health outcomes (PAGAC 2008). These conclusions were not based on any single study alone, but instead were inferred from plausible patterns of findings on activity dose that were consistently seen in the reviewed literature. Recently, investigators have tested the hypothesis that participation specifically in the activity dose currently recommended for public health would be associated with reduced mortality risk. In one of the largest studies ever conducted on physical activity and mortality, a total of 252,925 U.S. adults enrolled in the National Institutes of Health (NIH)–American Association of Retired Persons Diet and Health Study completed a baseline physical activity questionnaire in 1995-1996 and were followed up for mortality outcomes through 2002 (Leitzmann et al. 2007). Participants were categorized as inactive (reported no activity), insufficiently active to meet recommended activity levels, meeting only recommended levels of moderate-intensity activity ($\geq$150 min/week), meeting only recommended levels of vigorous-intensity activity ($\geq$60 min/week), or meeting both moderate- *and* vigorous-intensity recommendations.

The primary findings are shown in figure 10.2. Risks of mortality from all causes and CVD were about 50% lower, and from cancer were about 30% lower, in those who met recommended activity levels compared with their inactive peers. Mortality benefit did not materially differ for those who met recommendations through moderate-intensity compared with vigorous-intensity activities. Another interesting finding emerged from this study: Participants who reported being active but at levels insufficient to meet current recommendations also demonstrated significantly lower mortality risks compared with their sedentary peers. This suggests that avoiding sedentary lifeways may confer health benefits for some individuals, even at activity levels below the recommended minimal dose, and supports the guideline that "some physical activity is better than none" (PAGAC 2008). Further examination and clarification of potential health benefits at activity doses lower than currently recommended should be a primary focus in future studies on physical activity and health.

Another area of interest pertains to the pattern of activity through which an individual achieves recommended activity levels. Given ever-increasing work and family demands, individuals may choose to compress their activity time into less frequent ses-

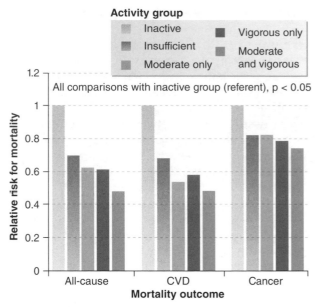

FIGURE 10.2 Multivariable-adjusted relative risk of mortality outcomes according to recommended levels of physical activity in 252,925 adults enrolled in the National Institutes of Health–American Association of Retired Persons Diet and Health Study, who were followed up during the interval 1995-1996 to 2002 (7,900 deaths). Activity categories (from left to right) were defined as follows: inactive (referent), insufficiently active to meet recommended levels, meeting moderate-intensity recommendation only ($\geq$150 min/week), meeting vigorous-intensity recommendation only ($\geq$60 min/week), and meeting both the moderate- and vigorous-intensity recommendations. For each mortality outcome (all-cause, CVD, and cancer), relative risks are significant at the 0.05 level for all activity groups compared with the inactive group. Risk estimates were adjusted for age, sex, race, education, body mass index, smoking, family history of cancer, hormone therapy use (women only), aspirin use, multivitamin use, and dietary intake. The moderate-only and vigorous-only categories were mutually adjusted.

Data from Leitzmann et al. 2007.

sions, for example on the weekends. Whether or not mortality benefits are similar in adults who achieve activity recommendations across two weekend days compared to their peers who are regularly active on a greater number of days is largely unclear. An investigation in the Harvard Alumni Health Study has begun to clarify this issue (Lee et al. 2004). The investigators reported on 8,421 men who completed activity questionnaires in 1988 and 1993 and were followed up for mortality through 1997. Weekend warriors were defined as men expending $\geq$1,000 kcal/week during one or two sessions, and mortality rates in these men were compared to those observed in inactive men (<500 kcal/week). Among men without major chronic disease risk factors, the multivariable-adjusted mortality risk was 59% lower ($p < 0.05$) in weekend warriors compared with inactive men. No mortality benefit was seen in weekend warriors who had existing risk factors. This finding suggests that achieving the recommended activity dose, and not necessarily the frequency of activity through which this occurs, is the principal determinant of mortality risk reduction in apparently healthy adults. Because the beneficial effect of physical activity on metabolic risk factors tends to be short-lived, more frequent physical activity may be necessary to delay mortality in those with manifest metabolic dysfunction or familial disposition for chronic disease.

Type of Physical Activity

Prominent in recent recommendations on physical activity and public health is a broadening of the exercise paradigm, from one that focuses almost exclusively on traditional exercise prescription to enhance physical fitness to one that promotes accumulating recommended activity levels through a variety of activities including, but not limited to, formal exercise (Blair, LaMonte, and Nichaman 2004). Two recent studies reported mortality risk in relation to both leisure-time or exercise activities and nonleisure or nonexercise activities, the latter defined as household chores, stair climbing, and walking or cycling for transportation (Matthews et al. 2007; Arrietta and Russell 2008). In both studies, higher amounts of nonleisure or nonexercise activities were significantly associated with lower risk of all-cause mortality, even after adjustment for differences in age, other relevant confounders, and *leisure-time* or *exercise activity*. For example, among 67,143 Chinese women ages 40 to 70 years followed up for an average of six years (1,091 deaths), multivariable-adjusted mortality risk was 34% lower ($p < 0.05$) in the highest (≥ 18.1 MET-hours/day) compared with the lowest (≤ 9.9 MET-hours/day) quartile of daily nonexercise activity (Matthews et al. 2007). Mortality risk also was lower across incremental categories of walking (trend, $p = 0.07$) and cycling (trend, $p < 0.05$) for transportation analyzed as separate exposures. These findings support public health emphasis on the benefits of overall activity habits, which are particularly relevant to individuals who lead active daily lives but who do not engage in formal exercise programs.

Objective Measures of Physical Activity

Self-reported physical activity assessments are prone to substantial misclassification because of response bias and other problems, which can threaten the accuracy of associations between activity exposures and mortality outcomes. Recent studies have examined mortality risk in relation to objective activity assessments using doubly labeled water and accelerometers. Among 302 older adults, the six-year rates of all-cause mortality (55 deaths) across incremental tertiles (<521, 521-770, >770 kcal/day) of total daily activity-related energy expenditure measured with doubly labeled water were 40.9, 28.6, and 19.4 per 1,000 **person-years** (trend, $p = 0.03$) (Manini et al. 2006). After adjustment for age, sex, and numerous other covariables, each 1-standard deviation (SD) (287 kcal/day) unit increase in energy expenditure was associated with a 30% lower mortality risk. In another study on 225 older adults who had existing peripheral artery disease and were followed for an average of five years (75 deaths), the multivariable-adjusted risk of all-cause mortality was more than threefold higher ($p = 0.02$) in the lowest (<540 activity counts) than in the highest ($\geq 1,113$ activity counts) quartiles of accelerometer counts averaged across seven days of observation (Garg et al. 2006). In this study, the adjusted relative risk of mortality was 1.84 ($p = 0.04$) among those in the lowest compared with the highest quartiles of self-reported walking activity.

Patterns of association between mortality and objective measures of physical activity appear generally consistent with the patterns of association reported in studies using questionnaire assessments of activity, though few studies have included both objective and questionnaire assessments of activity to allow for direct comparison. The magnitude and precision of mortality risk estimates may be higher for objective assessments because of reduced measurement error. However, the studies just discussed included relatively small sample sizes and a small number of deaths for analysis. A variety of feasibility issues have thus far limited use of objective activity measures in large epidemiological studies. Additional data in large, diverse samples that include both questionnaire and objective activity assessments, as well as adequate follow-up for sufficient numbers of end-point events, would likely help to further refine current understanding of the dose response between physical activity and mortality or other study outcomes.

The Role of Genetics in Associations Between Activity and Mortality

Both physical activity habits and longevity are known to have heritable components. Quantifying the extent to which the association between physical activity and mortality is explained by genetics is relevant when limited public health resources are allocated for the development of targeted strategies to intervene on modifiable risk factors. Targeting exposures that have been related to outcomes largely via genetic selection likely would not be a cost-effective widespread approach to improving population health, but instead may be a more feasible intervention target in population subgroups with specific genetic compositions. Recent findings in the Swedish Twin Registry study (Carlsson et al. 2007) addressed this issue as related to physical activity and mortality. Physical activity questionnaires were completed in 1972 by 13,109 twin pairs (40% monozygotic twin pairs), who then were followed for mortality through 2004 (1,800 deaths). As expected, in the overall study population, all-cause mortality risk was 25% and 36% lower among women and men, respectively, in the high- compared with low-activity groups. When analyses were conducted on physical activity–discordant monozygotic twin pairs, the twin with higher physical activity had a 20% lower risk ($p <$ 0.05) of mortality compared with the less active twin. Similar mortality risk reduction in high- compared with low-activity groups was seen when monozygotic twin pairs were analyzed separately according to sex. Associations between activity and mortality were weaker and not significant among dizygotic twins. The significant protective association between physical activity and mortality observed in monozygotic twins tempers arguments that the association is accounted for principally by genetic selection and supports the hypothesis that the association is one of causality.

It is important to recognize that complex, multifactorial webs of causation underlie most etiological pathways linking lifestyle and other exposures with morbidity and mortality; genetic components likely are involved in several aspects therein, although these effects may be difficult to assess at present. As discussed in greater detail elsewhere in this textbook, ongoing progress in the technological capacity for refined examination of complex gene–gene and gene–environment interactions, as well as the emerging field of epigenetics, will no doubt lead to an even better understanding of the role of genetics in the numerous established health benefits associated with physical activity.

Fitness and Mortality

As mentioned earlier, physical fitness includes a wide variety of physiological responses, but in this chapter, the discussion of fitness is limited to cardiorespiratory fitness and muscular fitness. *Cardiorespiratory fitness* as used here refers to maximal physical working capacity or aerobic power, which is an integrated assessment of the function of the heart, lungs, vascular system, and skeletal muscles. *Muscular fitness* as used here refers to strength and endurance of the skeletal muscles. Cardiorespiratory and muscular fitness are objective, reproducible measures that reflect the combined functional influences of physical activity habits, genetics, and disease status. Because fitness exposures are less prone to misclassification, they may more accurately reflect the consequences of a sedentary or irregularly active lifestyle than do self-reported activity exposures. Compared with physical activity, however, fewer published reports are available on associations of cardiorespiratory or muscular fitness with mortality. The cost and administrative burden of exercise testing have limited the inclusion of fitness exposures in epidemiological studies. Identifying a minimum threshold level of fitness for mortality benefit is therefore difficult. Some studies have shown that cardiorespiratory fitness levels of <5 METs tend to be associated with higher risk of mortality from all causes and CVD, whereas ≥10 METs is associated with excellent survival. Because the distribution of cardiorespiratory fitness differs between women and men, and age-related declines are not linear across the adult age range, the minimal level of cardiorespiratory fitness required for health benefits likely is sex and age specific (Blair et al. 1989). At present, available data are insufficient to identify a minimum level of health-related muscular strength, power, or endurance.

Cardiorespiratory Fitness

Data on cardiorespiratory fitness and mortality were brought to the forefront of epidemiological research and public health through findings reported in the Aerobics Center Longitudinal Study (ACLS), for which one of the authors of this chapter (SNB) served as principal investigator from the early 1980s until 2006. Low-, moderate-, and high-fitness groups were defined as the lowest one-fifth, middle two-fifths, and upper two-fifths of the distribution of age- and sex-standardized maximal times on a treadmill exercise test. Steep inverse gradients for age-adjusted rates of all-cause mortality across these incremental fitness groups are seen in women (39.5, 16.4, and 7.4 deaths per 10,000 person-years) and in men (64.0, 26.3, and 20.3 deaths per 10,000 person-years) (Blair et al. 1989). We see a substantial mortality benefit of at least 50% when comparing rates between the moderate- and low-fitness groups; we see an additional but smaller benefit when comparing rates between the high and moderate groups. After extensive adjustment for potential confounders, mortality risk remained 37% lower among men and 48% lower among women in the combined moderate- and high-fitness groups compared with their low-fit peers. The asymptote of the dose–response curve between fitness and mortality was about 9 METs in women and about 10 METs in men; these might be reasonable targets for public health, though many individuals may obtain benefits at lower fitness levels. Similar inverse patterns of association with fitness generally are seen in ACLS participants when cardiovascular, stroke, and total cancer mortality are the study end points, though the strongest associations tend to be observed with all-cause and cardiovascular outcomes.

Patterns of association between objectively measured cardiorespiratory fitness and mortality from cardiovascular and all causes in the ACLS and other studies generally are consistent with extant data on associations between physical activity exposures and mortality outcomes (PAGAC 2008). The magnitude of association with mortality outcomes tends to be higher for fitness than for activity exposures (PAGAC 2008); this is likely attributable to the greater objectivity and lower likelihood of misclassification of exposure with laboratory-based quantification of fitness compared to self-reported recall of past physical activity habits.

Recent work in the ACLS and other studies expands earlier reported findings on the mortality benefits of higher fitness to two population subgroups for whom functional capacity has both clinical and public health relevance: the elderly and asymptomatic individuals with low pretest probability of CVD. The rapid expansion of the older U.S. population underscores a need to preserve functional independence and quality of life at advanced ages previously not encountered in usual preventive health care. Recent findings from the ACLS demonstrate a powerful mortality benefit in older adults (Sui et al. 2007). A total of 4,060 adults ages 60 to more than 80 years completed symptom-limited maximal treadmill exercise tests during 1971 and 2001 and were followed for mortality during an average of 13.6 years. Low- (referent), moderate-, and high-fitness groups were defined as already described. Inverse gradients for mortality risk from several causes were seen across fitness categories. After adjustment for age, sex, and several other clinical and behavioral covariables, relative risks across the incremental fitness groups were 1.00, 0.71, and 0.59 (trend, $p < 0.001$) for all-cause mortality; 1.00, 0.77, and 0.57 (trend, $p = 0.001$) for CVD mortality; and 1.00, 0.92, and 0.84 (trend, $p = 0.45$) for total cancer mortality. Additional study findings are shown in figure 10.3. When grouped into age strata of 60 to 69, 70 to 79, and ≥80 years,

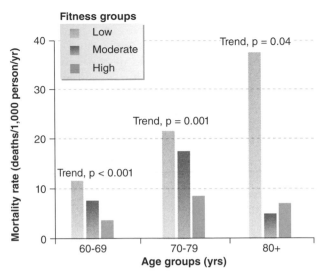

FIGURE 10.3 Age- and sex-adjusted rates (deaths per 1,000 person-years) of CVD mortality according to categories of age and cardiorespiratory fitness in 4,060 adults, ages 60 years and older, enrolled in the Aerobics Center Longitudinal Study, who were followed up for an average of 13.6 years (989 deaths). Fitness groups (from left to right) were defined as low (referent), moderate, and high according to sex-specific cutoff at the lowest 20%, middle 40%, and upper 40% of maximal METs during treadmill exercise testing. These cutoff values were 7.2 and 9.5 METs for men and 5.8 and 7.6 METs for women.

Adapted, by permission, from X. Sui et al., 2007, "Estimated functional capacity predicts mortality in older adults," *Journal of the American Geriatrics Society* 55: 1940-1947.

age- and sex-adjusted rates of all-cause mortality were significantly inversely associated with fitness levels in each age group. Within each age group, adults with high fitness had lower mortality rates than did their low-fit peers in the youngest age group. In fact, the difference in mortality rates between 60- to 69-year-old adults with low fitness (12 deaths per 1,000 person-years) and adults 80 years and older with high fitness (7 deaths per 1,000 person-years) suggests a 20-year benefit of being fit at advanced ages. These results in the ACLS are consistent with other published study findings showing substantial mortality benefit among older adults who lead an active and fit way of life. Indeed, health care providers should be vigilant in assessing and promoting physical activity and fitness among their elderly patients.

Historically, exercise stress testing has principally been used as a clinical tool to evoke electrocardiographic evidence diagnostic of obstructive coronary artery disease in individuals with suspected disease because of exertional chest pain during activities of daily living. Low diagnostic value of exercise testing in women and in asymptomatic individuals with low pretest probability of CVD based on normal risk factor profiles has limited the widespread use of exercise testing. However, growing evidence suggests that nonelectrocardiographic exercise test information, such as data on cardiorespiratory fitness, may have sufficient prognostic value to enhance individual CVD risk assessment in women and certain clinical subgroups. Findings in the Lipid Research Clinics Prevalence Study illustrate this point (Mora et al. 2005). At baseline, 6,126 adults who were on average 45 years of age and were without known CVD completed a maximal treadmill exercise test. All participants had low to moderate pretest probabilities of CVD (<20% probability of experiencing a first CVD event over 10 years) based on Framingham risk scores computed from their CVD risk factor profile. Fitness levels were based on categories of maximal METs and postexercise heart rate recovery (beats per minute [bpm]). Higher heart rate recovery reflects a better cardiorespiratory response to exercise stress testing. MET and heart rate recovery categories were based on median cutoffs separately for men (10.7 METs; 55 bpm heart rate recovery) and women (7.5 METs; 56 bpm heart rate recovery).

During a 20-year follow-up, there were 145 CVD deaths in men and 101 CVD deaths in women. After adjustment for covariables, including Framingham risk score, relative risks of CVD mortality in those below median MET levels compared with those above were 3.09 and 2.59 in women and men, respectively ($p < 0.001$ each). Likewise, comparisons for heart rate recovery were 3.25 in women and 1.99 in men ($p < 0.001$ each). Next, participants were grouped by CVD risk (very low, low, and intermediate) based on Framingham risk scores of <6%, 6% to 9%, and 10% to 19%, respectively. Within each stratum of Framingham risk score, age-adjusted rates of CVD mortality were computed according to fitness groups defined by combined METs and heart rate recovery (figure 10.4). Fitness additionally stratified CVD risk across all Framingham risk categories in women and in men. Cardiovascular disease death rates were highest in those with low METs and heart rate recovery and were lowest in those with high fitness parameters, regardless of Framingham risk group. The greatest benefit of fitness was seen in the intermediate CVD risk group (10-19% Framingham risk score); in treating this group, clinicians often struggle with the costs and benefits of ordering additional clinical assessments, such as costly nuclear imaging tests, to guide the initiation and intensity of primary prevention therapy. The Lipid Research Clinics Prevalence Study findings suggest that individual CVD risk assessment should not stop with personal and family history interviews and calculated Framingham risk scores but should also include measures of fitness as a means of improving CVD risk stratification in asymptomatic women and men.

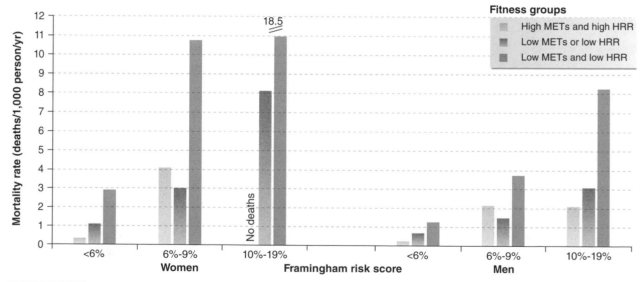

FIGURE 10.4 Age-adjusted rates of CVD mortality according to Framingham risk score and cardiorespiratory fitness in 6,126 adults (mean age 45 years; 46% women) enrolled in the Lipid Research Clinics Prevalence Study. At baseline exercise testing, participants were without known CVD and had multivariable predicted 10-year Framingham risk scores <20%, which indicated low to moderate short-term probability of experiencing a primary cardiovascular event. During a 20-year follow-up, there were 145 cardiovascular deaths in men and 101 in women. Fitness was defined according to combined categories of maximal METs and postexercise heart rate recovery using sex-specific median cutoffs (men: 10.7 METs and 55 bpm heart rate recovery; women: 7.5 METs and 56 bpm heart rate recovery).

Data from Mora et al. 2005.

■ Cardiorespiratory fitness is a measure of the efficiency of the heart, lungs, vasculature, and skeletal muscles to deliver and use oxygen during heavy, dynamic physical activity. The primary determinant of cardiorespiratory fitness is participation in moderate- to vigorous-intensity physical activities; however, genetics is also influential to some degree. Because cardiorespiratory fitness measures are more objective and reproducible than self-reported recall of past physical activity, these measures provide an accurate assessment of sedentary and irregularly active lifestyles for use as exposures in epidemiological studies of mortality outcomes.

■ Data from the Aerobics Center Longitudinal Study indicate that moderately fit men and women have approximately half the risk (e.g., 50% lower) of all-cause mortality compared with their low-fit counterparts. Mortality risk is 10% to 15% lower when highly fit study participants are compared with their moderately fit peers.

Muscular Fitness

Most studies on fitness and mortality have used cardiorespiratory fitness as the exposure; however, a few studies used some aspect of muscular fitness as the predictor (PAGAC 2008). Most of these latter studies used grip strength as the exposure and were conducted in elderly populations. The findings generally show an inverse association between grip strength and mortality. One issue with these studies is the concern that grip strength in elderly populations may be an overall marker for frailty and health status and that strength per se is not a causal factor. Two recent studies shed additional light on this topic of increasing interest among geriatricians and public health practitioners.

In the ACLS, 8.762 men ages 20 to 80 years completed baseline assessments of muscular strength and cardiorespiratory fitness and were followed up for mortality during an average of 20 years (Ruiz et al. 2008). Participants were grouped into thirds of muscular strength based on combined scores for age- and weight-standardized maximal leg press

and bench press. There were 503 deaths in 165,251 person-years of observation. After adjustment for age, smoking, reported physical activity habits, body size, and other covariables, relative risks across incremental thirds of muscular strength were 1.00 (referent), 0.72, and 0.77 (trend, $p = 0.01$) for all-cause mortality, were 1.00, 0.74, and 0.71 (trend, $p = 0.09$) for CVD mortality, and were 1.00, 0.72, and 0.68 (trend, $p < 0.03$) for total cancer mortality. Further adjustment for cardiorespiratory fitness slightly attenuated these associations between muscular strength and mortality. When men were grouped into age strata of <60 and ≥60 years, significant inverse gradients for age-adjusted rates of all-cause and cancer mortality were seen across incremental thirds of muscular strength in younger and older age groups.

Findings for mortality benefit from greater muscular fitness in the ACLS are corroborated by results in the Dynamics of Health, Aging and Body Composition (Health ABC) Study (Newman et al. 2006). In Health ABC, 2,292 elderly adults (52% women), ages 70 to 79 years, completed handgrip strength testing and isokinetic testing for quadriceps leg strength; had their body composition measured using dual-energy X-ray absorptiometry (DEXA); and had their thigh muscle area measured using computed tomography at baseline. They were followed for mortality during an average of four years (286 deaths). For each 1-SD unit (38 N/m) decrement in quadriceps muscle strength, the unadjusted relative risk of all-cause mortality was 1.51 (95% **confidence interval [CI]**: 1.28-1.79) in men and was 1.65 (95% CI: 1.19-2.30) in women. These associations were materially unchanged after adjustment for age, smoking, chronic conditions, inflammatory biomarkers, body composition, thigh muscle area, and *aerobic physical activity habits*. A similar pattern of results was observed when grip strength was used as the exposure. Interestingly, muscle size determined by computed tomography or DEXA was not predictive of mortality. These findings indicate that muscle quality (e.g., muscle function) may be more important than muscle quantity as a determinant of successful aging and longevity in older adults.

The limited published data on the association between muscular fitness and mortality suggest that muscular fitness may confer health benefits through mechanisms that are independent of cardiorespiratory fitness levels and aerobic physical activity habits. Whether the protective effect is attributable to maximal muscular strength per se or to regular participation in resistance exercises is not known and should be a focus in future studies examining muscular fitness and health outcomes.

■ Limited data on the association between muscular fitness and mortality suggest that muscular fitness may confer health benefits through mechanisms that are independent of cardiorespiratory fitness levels and aerobic physical activity habits.

Activity or Fitness and Mortality in Adults With Existing Diseases

Improving physical activity and cardiorespiratory fitness is becoming a cornerstone element of secondary and tertiary prevention programs aimed at improving prognosis in various disease subsets. Historically, the richest supporting data were from studies on CHD patients, but recent research has expanded to show mortality benefit among active and fit adults with a variety of chronic diseases that are prevalent or becoming prevalent worldwide and thus exact considerable toll on population health.

Associations between physical activity or cardiorespiratory fitness and mortality in individuals with diabetes (Church et al. 2004), hypertension (Hu et al. 2007), metabolic syndrome (Katzmarczyk, Church, and Blair 2004), CVD (Kavanaugh et al. 2003), breast cancer (Holmes et al. 2005), and overweight or obesity (Lee and Paffenbarger 2000) are summarized in table 10.1. Each of these prospective studies included a large sample of adults with manifest chronic diseases diagnosed using clinical criteria, with follow-up intervals exceeding five years for mortality from all causes or CVD. When possible, cutoff values for activity (e.g., kcal/week) and fitness (e.g., METs) exposure groups are provided. The relative risk estimates and 95% confidence intervals presented are from the fullest multivariable-adjusted model reported in the original papers.

TABLE 10.1 Prospective Studies of the Association Between Physical Activity or Cardiorespiratory Fitness and Mortality in Individuals With Clinically Manifest Disease

	Population	Follow-up (deaths)	Exposure	Main findings for PA or CRF RR (95% CI)*
Diabetes				
Church et al. 2004	2,196 U.S. men, ages 23-79 years, diabetes by physician diagnosis or FPG >126 mg/dl Outcome: All-cause mortality	14.6 years (275)	CRF from a maximal exercise test	Quartiles of METs >11.71: RR = 1.00 (referent) 10.09-11.71: RR = 1.60 (0.93-2.76) 8.83-10.08: RR = 2.77 (1.65-4.66) ≤8.82\: RR = 4.49 (2.64-7.64) Trend, $p < 0.0001$
Hypertension				
Hu et al. 2007	26,643 Finnish adults (46% women), ages 25-64 years, hypertension by physician diagnosis, medication use, or measured SBP ≥140 mmHg or DBP ≥90 mmHg Outcome: CVD mortality	19.9 years (3,743)	PA from questionnaire (leisure PA)	Women Low: RR = 1.00 (referent) Moderate: RR = 0.78 (0.70-0.87) High: RR = 0.76 (0.60-0.97) Trend, $p < 0.001$ Men Low: RR = 1.00 (referent) Moderate: RR = 0.84 (0.77-0.92) High: RR = 0.73 (0.62-0.86) Trend, $p < 0.001$

	Population	Follow-up (deaths)	Exposure	Main findings for PA or CRF RR (95% CI)*
Metabolic syndrome				
Katzmarzyk et al. 2004	3,757 U.S. men, ages 20-83 years, metabolic syndrome by ATP-III criteria Outcome: CVD mortality	9.5 years (59)	CRF from a maximal exercise test	Tertiles of METs T1: RR = 1.00 (referent) T2: RR = 0.48 (0.21-1.11) T3: RR = 0.29 (0.13-0.66) Trend, p = 0.002
CVD				
Kavanagh et al. 2003	2,380 medically stable U.S. women, mean age 59.7 years, ≈14 weeks post-myocardial infarction or CABG Outcome: All-cause mortality	6.1 years (209)	CRF from a maximal exercise test	Peak METs <3.7: RR = 1.00 (referent) ≥3.7: RR = 0.71 (0.53-0.95) p = 0.02
Breast cancer				
Holmes et al. 2005	2,987 U.S. women, ages 30-55 years, with stage I, II, or III breast cancer Outcome: All-cause mortality	8 years (463)	PA from questionnaire (leisure PA, MET-hours/week)	MET-hours/week <3.0: RR = 1.00 (referent) 3.0-8.9: RR = 0.71 (0.56-0.89) 9.0-14.9: RR = 0.59 (0.41-0.84) 15.0-23.9: RR = 0.56 (0.41-0.77) ≥24.0: RR = 0.65 (0.48-0.88) Trend, p = 0.003
Overweight/Obesity				
Lee and Paffenbarger 2000	13,485 U.S. men, mean age 57 years, weight status defined by BMI: normal <25 and overweight/obese ≥25 kg/m² Outcome: All-cause mortality	15 years (2,539)	PA from questionnaire (leisure PA, kcal/week)	BMI ≥25 <1,000 kcal/week: RR = 1.00 (referent) ≥1,000 kcal/week: RR = 0.80 (0.71-0.91) BMI <25 <1,000 kcal/week: RR = 0.90 (0.79-1.02) ≥1,000 kcal/week: RR = 0.67 (0.60-0.75)

*In each study, the relative risks (RR) and 95% confidence intervals (CI) were adjusted for age, sex (where applicable), and a variety of risk predictors (space precludes listing each of the covariables used in the respective studies; interested readers can refer to the published papers for greater detail). For each study, we show estimates for the most fully adjusted model reported. For consistency with our approach in the text of the chapter, when possible, we show the data with low PA or CRF as the referent category, in some instances taking the reciprocal of the published estimates. RR >1.0 indicates higher mortality risk; RR <1.0 indicates lower mortality risk; 95% CI that does not include 1.00 indicates statistically significant findings.

Abbreviations: PA, physical activity; CRF, cardiorespiratory fitness; ATP, Adult Treatment Panel; BMI, body mass index; CABG, coronary artery bypass graft surgery; CVD, cardiovascular disease; SBP, systolic blood pressure; DBP, diastolic blood pressure; FPG, fasting plasma glucose; METs, metabolic equivalents (1 MET = oxygen uptake of 3.5 ml · kg⁻¹ · min⁻¹).

Each disease listed in the table is associated with at least a doubling of mortality risk from all causes and CVD. However, it is clear that higher levels of activity and fitness confer substantial protection against mortality in adults at high risk for death because of existing chronic disease. In each study, after adjustment for smoking habits, family history of disease, and other important confounders, mortality risk was significantly lower in those with higher activity or fitness compared with their inactive and low-fit peers. In several of the studies, confounding and reverse causation (i.e., selectively higher death rates among the least active and fit individuals because of more severe existing disease) were further addressed by adjusting for medication or diet therapy and by subgroup analysis that eliminated early deaths. The prospective design of these studies provides a stronger test of causation between activity or fitness and mortality than is possible in other epidemiological designs commonly used to study specific disease conditions, such as **case–control studies**. In support of a causal association, several of these associations demonstrated an inverse dose–response gradient. Several of these conditions (e.g., diabetes, hypertension, metabolic syndrome) are particularly prevalent in obese adults. Overweight or obese men who are active at levels approximately equivalent to current federal recommendations (e.g., 1,000 kcal/week or more) had a 20% lower risk ($p < 0.05$) of all-cause mortality compared with their obese, inactive counterparts. Mortality risk among normal-weight, inactive men was not significantly different from that for their inactive, overweight or obese peers.

Quantifying the Population Mortality Burden of Inactivity

Physical inactivity and low cardiorespiratory and muscular fitness are inversely associated with mortality outcomes in a variety of population groups, as demonstrated by the relative risks for each exposure reviewed previously. However, relative risk provides only a measure of the strength of association between an exposure and outcome. The societal burden of a given exposure depends both on the exposure prevalence and on its strength of association with a health outcome. **Population attributable risk (PAR)** is a measure used in public health to estimate the fraction of a given outcome (e.g., CVD mortality) that is attributed to a defined risk factor exposure and, assuming a causal relationship, that could be prevented if exposure to the risk factor were eliminated. PAR increases as a direct function of both exposure prevalence and relative risk. PAR is a theoretical but useful measure to assess strategies for modifying risk exposures, the consequences of which in turn guide public health decision making regarding policy and financing for health promotion programs.

The PAR for all-cause mortality was examined in relation to inactivity, low cardiorespiratory fitness, and several other conventional risk factor exposures among 12,943 (491 deaths) women and 40,842 (3,333 deaths) men in the ACLS who were followed up for 15 years. The major findings of this work have been summarized elsewhere (Blair 2009); however, we provide additional detail here to further illustrate their relevance. Baseline exposure prevalences among decedents for low fitness (lowest 20% of age- and sex-specific maximal METs), sedentary lifestyle, obesity (body mass index $\geq$30), current smoking, and diagnosis of three conditions (hypertension, diabetes, and hypercholesterolemia) were 35%, 48%, 10%, 20%, 31%, 6%, and 28% in women and were 37%, 52%, 17%, 26%, 48%, 12%, and 36% in men. Age-adjusted relative risks for mortality for each exposure in the same order were 2.22, 1.17, 2.11, 1.84, 1.49, 2.16, and 1.14 in women and were 2.25, 1.24, 1.66, 1.57, 1.61, 1.74, and 1.24 in men. Using standard epidemiological computations, the PARs (as a percentage) of mortality attributed to each exposure were 19%, 7%, 5%, 9%, 10%, 3%, and 3% in women and were 20%, 10%, 7%, 9%, 18%, 5%, and 7% in men. Because both the prevalence and relative risk of low fitness are large in ACLS participants, the PAR for mortality attributed to low fitness is large and on an order of magnitude that is at least as high as, or higher than, that for traditional risk factors, including inactivity. Low fitness may be a better indicator of the adverse health consequences of sedentary living habits. If the association between fitness and mortality is causal, and if all unfit women and men in the ACLS were to have become fit, about one in five of the deaths might have been avoided. When stratified on age (18-40, 40-60, $\geq$60 years) and body mass index-defined weight categories (18.5-24.9, 25.0-29.9, $\geq$30 kg/m^2) in both women and men, PARs for mortality associated with low fitness were similar in younger ($\approx$15%) and older adults ($\approx$15-20%) but were markedly higher in obese ($\approx$40-60%) compared with normal-weight and overweight participants ($\approx$10-15%). Clearly, obese adults with low fitness are a particularly high-risk subgroup whose mortality burden would be substantially lower with improved fitness levels.

About 70 million U.S. adults (of a total population of about 300 million) report being sedentary. Because of the large number of people at risk and because of the high relative risks for a variety of adverse health outcomes, including premature mortality, in sedentary individuals, the population health burden attributed to sedentary habits and low fitness is substantial and on an order of magnitude comparable to or exceeding traditional population risk indicators such as smoking and high cholesterol.

Summary

Human beings evolved to live a physically active lifestyle, but modern men and women live in a world where physical activity has largely been engineered out of daily living. Compelling evidence from many prospective epidemiological studies shows that an unfit and sedentary way of life increases risk of morbidity and mortality from chronic diseases. The associations between inactivity or low fitness and mortality are strong, graded, and temporally consistent, and they remain after adjustment for numerous potentially confounding variables, suggesting that the association is one of causality. Low cardiorespiratory fitness accounts for a substantial fraction of deaths in the population, carrying a PAR on an order of magnitude that is as high as or higher than that of other risk predictors that receive considerable attention in public health. Sedentary habits and low fitness are among the strongest predictors of mortality and pose a major threat to population health and well-being in most countries of the world. Fitness and good health are not destinations, but rather a lifelong journey. A major vehicle for travel along this journey is regular physical activity throughout the life span.

We dedicate this chapter to two remarkable epidemiologists, Ralph S. Paffenbarger Jr. and Jeremy Morris, both of whom recently passed away. They have provided direction, inspiration, counsel, and leadership to us and the overall discipline of physical activity epidemiology. It is often said that standing on the shoulders of giants is a key to realization of the future. Indeed, Professors Paffenbarger and Morris provided such shoulders to many of their colleagues and students through the years. They were not only marvelous examples for scientific achievement but also for their kindness and warm human spirit. They are sorely missed.

Physical activity has the potential to affect nearly every system in the body. An active lifestyle is not a fad; rather, it is a return to our evolutionary way of living. Sedentary existence is unnatural and results in maladaptive changes in our anatomical, physiological, and biochemical constitution that increase the likelihood of disease and premature death.

Key Concepts

cardiorespiratory fitness—For definition, see page 50.

case–control study—A type of observational study in which participants are selected on the basis of having (cases) and not having (controls) a specific disease or condition at a defined point in time. Investigators collect information on previous exposures of interest, often by questionnaire, and examine whether exposure frequency is higher (positive association) or lower (negative association) in cases compared with controls. This allows the investigators to determine, for example, whether participants with a disease or condition have a lower prevalence of the exposure compared with control participants.

confidence interval (CI)—A method of determining statistical significance of an observed measure of association (e.g., relative risk). If the null value (relative risk = 1.00) is within the confidence interval, then the observed relative risk is considered not to be statistically significant—that is, the association likely is due to chance. If the null value is not within the confidence interval, then the association is considered statistically significant and unlikely explained by chance alone. For example, if the relative risk for coronary mortality is 2.00 for sedentary compared to active adults and the 95% confidence interval is 1.89 to 2.76, then the association between physical activity and coronary mortality would be considered statistically significant (not due to chance alone) because the null value is not within the confidence interval. Specifying that the confidence interval is at the 95% level indicates the degree of certainty at which to interpret the test of significance. The width of the confidence interval reflects the reliability (reproducibility) of the observed association. Narrower intervals reflect greater reliability, whereas wider intervals reflect less reliability.

confounding factor—An extraneous variable that is related to both the exposure and outcome under analysis and that accounts for part or all of the exposure–outcome association. Statistical methods are used to control or adjust for confounding factors so that the true association between the exposure and outcome may be examined.

dose response—Relationship wherein a change in the dose of exposure (e.g., MET-hours/week of physical activity) is associated with a graded change (increase or decrease) in outcome (e.g., mortality). Dose response often is characterized in terms of its pattern (e.g., linear or curvilinear) and its statistical significance (e.g., *p*-value for test of trend).

energy expenditure of activity—Net transfer of energy required to support skeletal muscle contraction during physical activity. This term is used to quantify the volume or *dose* of physical activity, computed as the product of frequency, duration, and intensity of a specified physical activity. Energy expenditure typically is expressed as kilocalories per week (kcal/week) or metabolic equivalent hours per week. (MET-hours/week).

epidemiology—For definition, see page 50.

exposure—Agents or factors that can affect health. In epidemiological analyses, these factors are examined for their association, both causal and noncausal, with study outcomes. Exposures may be behavioral factors, such as physical activity; environmental factors, such as air pollution; biological factors, such as blood cholesterol concentration; or social factors, such as household income. Exposure variables also are referred to as independent variables, predictor variables, or explanatory variables.

incidence—Occurrence of new cases that have developed in a specified population during a defined time interval. An example would be the number of new heart attacks occurring in U.S. men during the interval 1987-2001, for which there may have been 4 cases for every 1,000 men at risk in the population (4 cases per 1,000 in population at risk).

metabolic equivalent (MET)—For definition, see page 19.

muscular fitness—Component of physical fitness that encompasses the expression of maximal skeletal muscle strength and skeletal muscle endurance at submaximal workloads.

outcome—Effect on a health parameter that results from an exposure. Outcomes may be biological responses, such as changes in physical fitness; occurrence of nonfatal disease, such as developing diabetes; or mortality from all causes or from specific causes, such as cardiovascular mortality. Outcome variables also are referred to as dependent variables or response variables.

person-years—Summary of the amount of follow-up in a prospective epidemiological study. One person followed for one year would provide one person-year of follow-up, and 10 persons followed for 10 years would provide 100 person-years of follow-up. This method of determining the amount of exposure in a study is necessary when the length of follow-up is different for different individuals. It is not necessary to use person years if all study participants are enrolled at the same time and are followed for the same length of time

physical activity—For definition, see page 19.

physical fitness—For definition, see page 19.

population attributable risk (PAR)—A measure used in public health to quantify the population burden of a health outcome (e.g., mortality) attributed to a risk factor exposure (e.g., inactivity). Computation of PAR takes into account both the exposure prevalence and strength of association with the outcome and is interpreted as the percentage of deaths attributed to exposure that, if causally associated, might be prevented if the exposure was eliminated.

prevalence—Proportion of cases within a population at a particular point in time. For example, in 2003 the prevalence of stroke (individuals who have had a documented stroke) might be 2/100, which would mean that of every 100 persons in the population, two had a stroke. Therefore, the prevalence of stroke within the population at risk in 2003 would be reported as 2%.

prospective study—Study in which the exposure variables are collected at baseline and the population is followed over time with monitoring for outcomes. For example, deaths occurring in both sedentary and physically active groups would be counted, and incidence for the two groups would be calculated.

***p*-value**—A measure of statistical significance that represents the probability that an observed finding (e.g., higher mortality risk in sedentary compared with active adults) is due solely to chance. The smaller the *p*-value, the greater the statistical significance and the less likely that observed associations are chance findings. For example, $p < 0.05$ means that the probability of chance explaining the findings is less than 5 in 100; for $p < 0.001$, the probability of chance findings is less than 1 in 1,000.

rates—Way of expressing exposures and outcomes in epidemiology; rates are derived from a numerator (number of cases or individuals with the characteristic) and a denominator (number of individuals in the population from which the cases are derived). Rates in epidemiology may be expressed as a proportion, for example four cases of heart attack per 1,000 men at risk; or as a rate per unit time, for example, 15 deaths per 1,000 person-years of follow-up.

relative risk (RR)—A measure of the direction and strength of association between an exposure and outcome, computed as a ratio of the risk for an outcome (e.g., mortality) among the exposed group (e.g., sedentary) to the risk among the nonexposed (referent) group (e.g., active). Direction of association is based on whether the relative RR is above or below 1.0. When RR = 1.0, there is no difference in risk between exposed and nonexposed groups, which indicates no association. When RR > 1.0, risk in the exposed is higher than in the nonexposed, which indicates a positive or adverse association. When RR < 1.0, risk in the exposed is less than in the nonexposed, which indicates a negative (inverse) or protective association. Strength of association is determined by how much greater or smaller the observed relative risk is relative to 1.0 (the null value). RR = 1.10 is a small association indicating a 10% higher risk in the exposed than nonexposed; RR = 1.50 is a moderate association (50% higher risk among the exposed); and RR = 2.00 is a strong association (twofold higher risk among the exposed).

Study Questions

1. Describe the general components and study design of an epidemiological investigation of physical activity and mortality.

2. List three epidemiological studies on the association between physical activity (or cardiorespiratory fitness) and mortality.

3. Describe what is meant by the terms *confounding* and *bias*. How are these issues typically addressed in epidemiological studies of exposures and outcomes?

4. Why are associations between cardiorespiratory fitness and mortality apparently stronger than associations between self-reported physical activity and mortality?

5. True or false: Protection against mortality by physical activity and cardiorespiratory fitness exposures is seen only among young men, individuals who are normal weight, and individuals without diseases such as diabetes.

References

Arrieta, A., and L.B. Russell. 2008. Effects of leisure and non-leisure physical activity on mortality in U.S. adults over two decades. *Annals of Epidemiology* 18:889-895.

Blair, S.N. 2009. Physical inactivity: The biggest public health problem of the 21st century. *British Journal of Sports Medicine* 43:1-2.

Blair, S.N., H.W. Kohl, R.S. Paffenbarger Jr., D.G. Clark, K.H. Cooper, and L.W. Gibbons. 1989. Physical fitness and all-cause mortality. *Journal of the American Medical Association* 262:2395-2401.

Blair, S.N., M.J. LaMonte, and M.Z. Nichaman. 2004. The evolution of physical activity recommendations: How much is enough? *American Journal of Clinical Nutrition* 79(Suppl):913S-920S.

Carlsson, S., T. Andersson, P. Lichtenstein, K. Michaelsson, and A. Ahlbom. 2007. Physical activity and mortality: Is the association explained by genetic selection? *American Journal of Epidemiology* 166:255-259.

Church, T.S., Y.J. Cheng, C.P. Earnest, C.E. Barlow, L.W. Gibbons, E.L. Priest, and S.N. Blair. 2004. Exercise capacity and body composition as predictors of mortality among men with diabetes. *Diabetes Care* 27:83-88.

Garg, P.K., L. Tian, M.H. Criqui, K. Liu, L. Ferrucci, J.M. Guralnik, J. Tan, and M.M. McDermott. 2006. Physical activity during daily life and mortality in patients with peripheral arterial disease. *Circulation* 114:242-248.

Holmes, M.D., W.Y. Chen, D. Feskanich, C.H. Kroenke, and G.A. Colditz. 2005. Physical activity and survival after breast cancer diagnosis. *Journal of the American Medical Association* 293:2479-2486.

Hu, G., P. Jousilahti, R. Antikainen, and J. Tuomilehto. 2007. Occupational, commuting, and leisure-time physical activity in relation to cardiovascular mortality among Finnish subjects with hypertension. *American Journal of Hypertension* 20:1242-1250.

Katzmarczyk, P.T., T.S. Church, and S.N. Blair. 2004. Cardiorespiratory fitness attenuates the effects of the metabolic syndrome on all-cause and cardiovascular disease mortality in men. *Archives of Internal Medicine* 164:1092-1097.

Katzmarczyk, P.T., T.S. Church, C.L. Craig, and C. Bouchard. 2009. Sitting time and mortality from all-causes, cardiovascular disease and cancer. *Medicine and Science in Sports and Exercise* 41:998-1005.

Kavanagh, T., D.J. Mertens, L.F. Hamm, J. Beyene, J. Kennedy, P. Corey, and R.J. Shephard. 2003. Peak oxygen intake and cardiac mortality in women referred for cardiac rehabilitation. *Journal of the American College of Cardiology* 42:2139-2143.

Lee, I.M., and R.S. Paffenbarger Jr. 2000. Associations of light, moderate, and vigorous intensity physical activity with longevity. *American Journal of Epidemiology* 151:293-299.

Lee, I.M., H.D. Sesso, Y. Oguma, and R.S. Paffenbarger Jr. 2003. Relative intensity of physical activity and risk of coronary heart disease. *Circulation* 107:1110-1116.

Lee, I.M., H.D. Sesso, Y. Oguma, and R.S. Paffenbarger Jr. 2004. The "weekend warrior" and risk of mortality. *American Journal of Epidemiology* 160:636-641.

Leitzmann, M.F., Y. Park, A. Blair, R. Ballard-Barbash, T. Mouw, A.R. Hollenbeck, and A. Schatzkin. 2007. Physical activity recommendations and decreased risk of mortality. *Archives of Internal Medicine* 167:2453-2460.

Manini, T.M., J.E. Everhart, K.V. Patel, D.A. Schoeller, L.H. Colbert, M. Visser, F. Tylavsky, D.C. Bauer, B.H. Goodpaster, and T.B. Harris. 2006. Daily activity energy expenditure and mortality among older adults. *Journal of the American Medical Association* 296:171-179.

Manson, J.E., F.B. Hu, J.W. Rich-Edwards, G.A. Colditz, M.J. Stampfer, W.C. Willett, F.E. Spiezer, and C.H. Hennekens. 1999. A prospective study of walking as compared with vigorous exercise in the prevention of coronary heart disease in women. *New England Journal of Medicine* 341:650-658.

Matthews, C.E., A.L. Jurj, X. Shu, H. Li, G. Yang, Q. Li, Y. Gao, and W. Zheng. 2007. Influence of exercise, walking, cycling, and overall nonexercise physical activity on mortality in Chinese women. *American Journal of Epidemiology* 165:1343-1350.

Mora, S., R.F. Redberg, A.R. Sharrett, and R.S. Blumenthal. 2005. Enhanced risk assessment in asymptomatic individuals with exercise testing and Framingham risk scores. *Circulation* 112:1566-1572.

Newman, A.B., V. Kupelian, M. Visser, E.M. Simonsick, B.H. Goodpaster, S.B. Kritchevsky, H.F.A. Tylavsky, S.M. Rubin, and T.B. Harris. 2006. Strength, but not muscle mass, is associated with mortality in the Health, Aging and Body Composition Study cohort. *Journal of Gerontology: Medical Sciences* 61A:72-77.

Physical Activity Guidelines Advisory Committee. 2008. *Physical Activity Guidelines Advisory Committee report, 2008.* Washington, DC: U.S. Department of Health and Human Services.

Ruiz, J.R., X. Sui, F. Lobelo, J.R. Morrow Jr., A.W. Jackson, M. Sjostrom, and S.N. Blair. 2008. Association between muscular strength and mortality in men: prospective cohort study. *British Medical Journal* 337:a439.

Sui, X., J.N. Laditka, J.W. Hardin, and S.N. Blair. 2007. Estimated functional capacity predicts mortality in older adults. *Journal of the American Geriatrics Society* 55:1940-1947.

U.S. Department of Health and Human Services. 1996. *Physical activity and health: A report of the Surgeon General.* Atlanta: U.S. Department of Health and Human Services, Centers for Disease Control and Prevention.

11

Physical Activity, Fitness, and Cardiac, Vascular, and Pulmonary Morbidities

Ian Janssen, PhD

CHAPTER OUTLINE

Low Physical Activity and Low Cardiorespiratory Fitness as Risk Factors for Cardiovascular Morbidities

Specific Cardiovascular Morbidities

Special Considerations

Low Physical Activity and Low Cardiorespiratory Fitness as Risk Factors for Pulmonary Morbidities

Biological Mechanisms

Hypertension

Coronary Heart Disease, Stroke, and Peripheral Vascular Disease

Asthma

Role of Physical Activity in Patients with Cardiac, Vascular, and Pulmonary Morbidities

Cardiovascular Disease

Chronic Obstructive Pulmonary Disease and Asthma

Summary

Review Materials

In this chapter we explore the role of physical activity and fitness in preventing and treating **cardiovascular disease** and lung disease. Cardiovascular disease—including **coronary heart disease**, **stroke**, **hypertension**, rheumatic fever, congenital heart defects, congestive heart failure, and **peripheral vascular disease**—is the leading cause of death in men and women in industrialized countries. This chapter focuses on four specific forms of cardiovascular disease that are particularly important in the context of physical activity and fitness: coronary heart disease, stroke, hypertension, and peripheral vascular disease. As with cardiovascular disease, lung disease is a leading cause of hospitalization, disability, and death. This chapter focuses on two specific forms of lung disease that are important in the context of physical activity and fitness: **chronic obstructive pulmonary disease** and **asthma**.

Low Physical Activity and Low Cardiorespiratory Fitness as Risk Factors for Cardiovascular Morbidities

The first part of the chapter examines the role of physical activity and cardiorespiratory fitness in the prevention of cardiac and vascular diseases. The relationship between physical activity and fitness outcomes is discussed, as is the dose–response relationship of physical activity and fitness to cardiovascular disease. The influence of age, gender, and race on these associations is also covered.

Specific Cardiovascular Morbidities

We begin by examining the role of physical activity and fitness for each of the major forms of cardiovascular disease. Research conducted over the past 30 years has provided a substantial knowledge base for most of the cardiovascular morbidities.

Coronary Heart Disease and Stroke

A vast scientific literature has examined the role that physical activity (both leisure time and occupational) and cardiorespiratory fitness play in the risk of coronary heart disease. These studies indicate that low physical activity and low cardiorespiratory fitness are causally linked to an increased risk of coronary heart disease (Physical Activity Guidelines Advisory Committee [PAGAC] 2008). That is, physically inactive individuals and individuals with a low cardiorespiratory fitness are at greater risk of developing

coronary heart disease than physically active and fit individuals. Furthermore, both occupational and leisure-time physical activity levels affect coronary heart disease risk. Results from a review of literature conducted on 30 studies published since 1996 suggest that moderately active men have a 19% reduction in the risk of developing coronary heart disease compared to men with a low activity level, and that highly active men have a 32% reduction in that risk compared to men with a low activity level (PAGAC 2008). Similarly, moderately active women have a 22% reduction in the risk of developing coronary heart disease compared to women with a low activity level, and highly active women have a 39% reduction in that risk compared to women with a low activity level (PAGAC 2008).

There is a dose–response relationship between physical activity level and coronary heart disease risk, as illustrated in figure 11.1. This figure shows an inverse, curvilinear association between total weekly physical activity energy expenditure—whether it be strenuous sports or nonvigorous activities—and the **relative risk** of having a heart attack (Paffenbarger, Wing, and Hyde 1978). Two conclusions can be drawn from this dose–response curve. First, there is an increasing protective effect of physical activity throughout the energy expenditure range such that the most active individuals have, on average, the lowest risk of developing coronary heart disease. Second, the reduction in risk is greatest at the lower end of the physical activity spectrum, implying that the largest impact on coronary heart disease risk

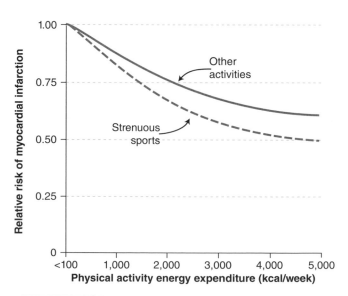

FIGURE 11.1 Dose–response relationship between physical activity level and coronary heart disease risk.

Reprinted from R.S. Paffenbarger Jr., A.L. Wing, and R.T. Hyde, 1978, "Physical activity as an index of heart attack risk in college alumni," *Am J Epidemiol* 108(3): 161-75, by permission of Oxford University Press.

in the population would be achieved by having sedentary individuals perform modest amounts of physical activity.

Low physical activity and low cardiorespiratory fitness are causally linked to an increased risk of transient ischemic attacks and stroke (PAGAC 2008). The protective effects of physical activity and fitness are apparent for both **ischemic strokes** and **hemorrhagic strokes**. Thus, physically inactive individuals and individuals with a low cardiorespiratory fitness have a greater risk of having a stroke than do physically active and fit individuals. Results from a literature review of 19 studies published since 1996 suggest that moderately and highly active men have about a 30% reduction in the risk of having a stroke compared to men with a low activity level, and that moderately and highly active women have about a 25% reduction in the risk of having a stroke compared to women with a low activity level (PAGAC 2008). There appears to be a curvilinear association between physical activity level and stroke risk, which is similar in shape to the dose–response relationship observed between physical activity and coronary heart disease risk (see figure 11.1).

Hypertension

In-depth reviews of the published scientific literature have concluded that a low physical activity and low cardiorespiratory fitness are causally linked to an increased risk of hypertension (Katzmarzyk and Janssen 2004; Pescatello et al. 2004). Thus, physically inactive and unfit individuals are at greater risk of developing hypertension than physically active and fit individuals. A **meta-analysis** concluded that physically active individuals have a 24% lower risk of developing hypertension than physically inactive individuals (Katzmarzyk and Janssen 2004).

Peripheral Vascular Disease

The influence of physical activity and fitness on peripheral vascular disease risk has not been thoroughly examined. However, because about 75% of the deaths in individuals with peripheral vascular disease are caused by coronary heart disease or stroke, and because low physical activity and low fitness are risk factors for coronary heart disease and stroke as discussed previously, it is reasonable to assume that low physical activity and low fitness are also risk factors for peripheral vascular disease.

In one of the few studies that have examined the relationship between physical activity and peripheral vascular disease risk, an inactive lifestyle was reported to be an independent risk factor for *asymptomatic* peripheral vascular disease—that is, inactive men and women had a 60% greater risk of developing asymptomatic peripheral vascular disease than physically active men and women (Hooi et al. 2001). However, in that study a sedentary lifestyle was not an independent risk factor for *symptomatic* peripheral vascular disease—that is, low physical activity was not a risk factor for symptomatic peripheral vascular disease after the researchers considered the effects of other risk factors such as hypertension and diabetes mellitus. Although low physical activity did not have an effect on symptomatic peripheral vascular disease that was independent of these other risk factors, low physical activity is a well-established risk factor for hypertension and diabetes. Thus, participation in physical activity may indirectly protect against the development of symptomatic peripheral vascular disease by protecting against the development of the primary risk factors for this disease, such as hypertension and diabetes.

Special Considerations

This next section addresses several questions. Is the relationship between physical activity and fitness and cardiovascular morbidities consistent across age, gender, and race? Do both aerobic and resistance types of physical activity have a protective effect on cardiovascular risk? What is the dose–response relationship between physical activity, fitness, and cardiovascular disease?

Consistency of Findings for Physical Activity and Cardiorespiratory Fitness

When we compare the effects of cardiorespiratory fitness and physical activity as risk factors for cardiovascular disease, we see a noticeably stronger and more consistent effect for cardiorespiratory fitness than for physical activity (Blair, Cheng, and Holder 2001). An example of this comparison is shown in figure 11.2. The figure represents the relative risk for myocardial infarction over a five-year follow-up in a sample of 1,453 Finnish men aged 42 to 60 years (Lakka et al. 1994). This study showed a strong dose–response relationship between fitness and heart attack risk; the effects of physical activity were not as strong or as clear as those for fitness.

The finding that cardiorespiratory fitness is a better predictor of cardiovascular disease morbidity and mortality than is physical activity warrants consideration. Cardiorespiratory fitness and physical activity are closely related in that an individual's fitness level is in large measure determined by her physical activity participation over recent weeks or months. Although no consensus exists, most experts believe that the beneficial health effects of cardiorespiratory fitness are largely mediated by physical activity

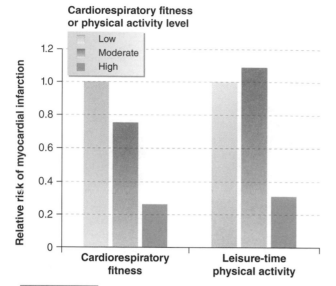

FIGURE 11.2 Relationships among cardiorespiratory fitness, physical activity level, and risk of myocardial infarction in men.

Data from Lakka et al. 1994.

level. If cardiorespiratory fitness is determined by physical activity, and if the protective effects of fitness are mediated by physical activity, why then is physical activity not a stronger predictor of cardiovascular disease risk than cardiorespiratory fitness? The answer to this question is likely explained by the fact that fitness is measured using precise and objective measures (e.g., $\dot{V}O_2$max exercise test), whereas physical activity is usually assessed using imprecise and subjective measures (e.g., self-reported questionnaire). The imprecise physical activity measures are therefore subject to higher rates of misclassification than the precise fitness measures, and the greater misclassification rates for physical activity likely explain the poorer associations with cardiovascular disease outcomes (Blair, Cheng, and Holder 2001).

> Cardiorespiratory fitness level is a stronger predictor of cardiovascular disease than is physical activity level. This is explained by the greater degree of measurement error for physical activity.

Consistency of Findings for Aerobic-Based and Resistance-Based Physical Activities

A plethora of scientific studies have examined the influence of aerobic-based physical activity and cardiorespiratory fitness on cardiovascular disease risk. The relationships between physical activity and

fitness and cardiovascular disease risk that have been covered to this point in the chapter for the most part come from aerobic-based measures of physical activity and fitness. Only a handful of scientific studies have examined the influence of strength-based physical activities or musculoskeletal fitness on cardiovascular disease risk. The available information suggests that, as with aerobic-based physical activity, resistance activities protect against cardiovascular disease. For instance, in the Health Professional's Follow-Up Study of more than 40,000 American men, those who participated in 30 min or more of weight training per week had a 23% reduction in coronary heart disease risk compared with those who performed no weight training (Tanasescu et al. 2002). Although the effect of resistance training on the development of hypertension has not been studied, exercise training studies ranging in duration from 6 to 30 weeks have shown that resistance exercise reduces resting systolic and diastolic blood pressures by an average of 3 mmHg (Kelley and Kelley 2000), further suggesting a protective effect of strength training.

Influence of Age, Gender, Race, and Changes in Physical Activity and Fitness

The available data indicate that the relationship between physical activity and fitness and cardiovascular morbidities is relatively consistent in men and women and in young, middle-aged, and older adults (PAGAC 2008). However, a major unresolved issue is the effects of race on the observed relationship. At present, studies are somewhat sparse in non-Caucasian races, and few studies have made direct racial comparisons. When the influence of physical activity and fitness on cardiovascular morbidities has been evaluated in non-Caucasian populations outside the United States (e.g., Japan, China, India), consistent evidence of a protective effect of physical activity has been found (PAGAC 2008). However, one of the few studies to make direct racial comparisons within the United States did not report such findings. This was the Atherosclerosis Risk in Communities Study, a follow-up study of over 7,000 African American and white men and women aged 46 to 65 years. In this cohort, the influence of physical activity on the development of hypertension (Pereira et al. 1999) and clinical cardiovascular disease events (Folsom et al. 1997) was much stronger and more consistent in whites than in African Americans. The authors of these papers suggested that the lack of an association in African Americans may have been due to the limited number who reported engaging in vigorous physical activity. Additional research on racial differences is needed before definitive conclusions can be drawn.

The vast majority of published studies used a single baseline measure of physical activity or fitness as it related to the risk for the cardiovascular disease end point, which at times was measured more than 25 years after the baseline exam (PAGAC 2008). To put this into context, envision measuring physical activity level in a large cohort of research participants in 1985 and then relating that single physical activity measure to the risk of cardiovascular disease up to the year 2010. Examination of a single baseline time point does not allow consideration of the changes in physical activity and other behaviors and risk factors that occurred during the follow-up period. Thus, in the example just mentioned, some of the subjects who were highly active in 1985 would have become inactive over the 25-year follow-up, whereas some of the subjects who were inactive in 1985 would have become physically active over the 25-year follow-up. An important question is whether these changes in physical activity influence cardiovascular disease risk.

Limited research data are available on changes in physical activity and fitness as they relate to the development of coronary heart disease, stroke, hypertension, and peripheral vascular disease. Figure 11.3 presents the data from one of the few studies that have examined this issue. In this study of 9,777 men aged 20 to 82 years, cardiorespiratory fitness was assessed at baseline and after an average five-year follow-up period. Study participants were then classified into one of four groups: (1) those who were

unfit at baseline and unfit after follow-up, (2) those who were unfit at baseline but fit after follow-up, (3) those who were fit at baseline but unfit after follow-up, and (4) those who were fit both at baseline and after follow-up. After the second fitness test was performed, participants were followed for an additional five years, on average, to determine the influence of baseline fitness and changes in fitness on the risk for cardiovascular disease mortality. As shown in figure 11.3, men who maintained a high level of fitness were the least likely to die from cardiovascular disease, whereas men who were persistently unfit were the most likely to die from cardiovascular disease (Blair et al. 1995). More importantly, an increase in fitness in unfit men was associated with a decrease in cardiovascular mortality risk, whereas a decrease in fitness in fit men was associated with an increase in cardiovascular mortality risk. An important public health message can be drawn from this study: Fit individuals should remain physically active, and unfit individuals should become physically active.

Low Physical Activity and Low Cardiorespiratory Fitness as Risk Factors for Pulmonary Morbidities

This section examines the role of physical activity and cardiorespiratory fitness in the primary prevention of asthma and chronic obstructive pulmonary disease. It must be noted that scientific evidence for a role of physical activity and fitness in lung function and disease is sparse. In a 25-year follow-up of 429 men, the least active had a 30% greater decline in lung function than the most active (Pelkonen et al. 2003). One of the few studies that examined the role of physical activity or fitness as it pertains to the development of asthma examined 757 asymptomatic children (aged 8-11 years at baseline) over a 10-year follow-up period. During this follow-up period, 7% of the children developed asthma. There was a clear, inverse dose–response relationship between cardiorespiratory fitness level at baseline and the risk of developing asthma during adolescence, such that 16% of the least fit subjects developed asthma, 11% of the moderately fit subjects developed asthma, and only 4% of the most fit subjects developed asthma.

As for the risk of developing chronic obstructive pulmonary disease, the limited information available suggests that physical activity participation does not offer a protective effect. In a 22-year follow-up of 3,686 longshoremen (men who load and unload

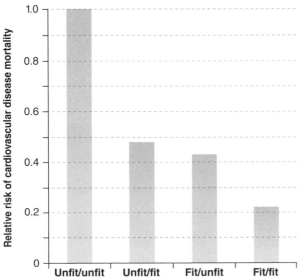

FIGURE 11.3 Influence of baseline fitness and changes in fitness on cardiovascular disease mortality.

Data from Blair et al. 1995.

ships at a seaport), cigarette smoking was a risk factor for death from chronic obstructive pulmonary disease and cardiovascular disease, as would be expected. On the other hand, physical inactivity was a risk factor for cardiovascular disease–related mortality but not for chronic obstructive pulmonary disease–related mortality (Paffenbarger et al. 1978).

It is unknown whether physical activity and fitness have comparable effects on pulmonary disease risk; whether the effects are different for aerobic and resistance types of physical activity; and whether age, gender, or race (or some combination of these) influences the relationships. Additional research is required in the area of physical activity, fitness, and pulmonary disease risk.

Biological Mechanisms

This part of the chapter briefly reviews the potential biological mechanisms that help us interpret the cause-and-effect relationship of physical activity and fitness with various forms of cardiovascular and lung disease.

Hypertension

Blood pressure is determined by the cardiac output (volume of blood pumped by the heart) and the total peripheral resistance of the blood vessels (determined by blood viscosity, length of blood vessels, and radius of blood vessels). Reductions in cardiac output do not occur after exercise training, indicating that a reduction in total peripheral resistance is the primary mechanism by which physical activity reduces resting blood pressure. The changes in total peripheral resistance in turn are primarily mediated by changes in blood vessel diameter. A number of neural and local changes occur in response to chronic physical activity participation that reduce the vasoconstrictive state of the peripheral vasculature and in so doing decrease total peripheral resistance and blood pressure. These changes include less sympathetic neural influence on the peripheral blood vessels and local vasodilator influences on the blood vessels from molecules such as nitric oxide.

Coronary Heart Disease, Stroke, and Peripheral Vascular Disease

There are several plausible biological mechanisms by which physical activity reduces coronary heart disease, stroke, and peripheral vascular disease risk; the mechanisms are similar for these three forms of cardiovascular disease. Physical activity reduces

blood pressure, improves the blood lipid profile (e.g., decreases triglycerides, increases high-density lipoprotein or "good" cholesterol), and decreases systemic inflammation (e.g., decreases blood levels of C-reactive protein [CRP]) and in so doing decreases damage and **atherosclerosis** of the cardiac, cerebral, and peripheral blood vessels. Physical activity also improves endothelial function (e.g., improves the vasodilation and vasoconstriction properties of the blood vessels) and has an antithrombotic effect (e.g., reduces blood clotting), which further reduces the risk of adverse cardiac and cerebrovascular events.

Asthma

There is no accepted biological mechanism that explains the cause-and-effect relationship of physical activity and fitness with **asthma**. It has been proposed that high fitness could increase the threshold for respiratory symptoms and increase the level at which respiratory discomfort develops. Furthermore, a lower ventilation rate and volume during more intense aerobic exercise in fit individuals, which would occur consequent to a training-induced increase in the ventilatory threshold (e.g., the exercise intensity at

Maintaining physical fitness improves blood pressure and asthma.

which ventilation starts to increase at a quicker rate with further increases in intensity), would result in a smaller ventilatory stimulus to induce an asthma attack. In short, increased physical activity during childhood may positively influence the lungs and decrease the risk of developing asthma (Rasmussen et al. 2000).

Role of Physical Activity in Patients with Cardiac, Vascular, and Pulmonary Morbidities

In this section, we consider the role of physical activity in reducing morbidity and mortality in patients with cardiac, vascular, and pulmonary morbidities. Patients with preexisting cardiovascular disease (coronary heart disease, stroke, peripheral vascular disease) or lung disease (chronic obstructive pulmonary disease) need to participate in a medical evaluation, including an exercise stress test, before beginning a physical activity program. Furthermore, decisions regarding the degree of supervision and monitoring during the physical activity program and the nature of the physical activity program itself need to be made by a team of health care professionals. The physical activity program for cardiovascular disease and lung disease patients also needs to be geared toward the unique problems of the individual, because persons with cardiovascular and lung disease vary greatly in their clinical (e.g., severity of disease, existence of other diseases) and functional (e.g., impairments that limit physical activity) status. Detailed discussion of physical activity prescription for individuals with cardiovascular and lung disease, though the topic is important, is beyond the scope of the chapter and is not presented here.

In general, individuals with coronary heart disease, stroke, peripheral vascular disease, and chronic obstructive pulmonary disease have a number of additional illnesses and diseases. For instance, many people who have suffered a stroke have also had a heart attack. Functional status is often severely impaired in individuals with the more severe forms of cardiovascular and lung disease, and many of these individuals have difficulty performing simple activities of daily living such as walking up a flight of stairs. The illness, disease, and reduction in function often lead to depression and social isolation. In short, most individuals with cardiac and pulmonary morbidities have a reduced quality of life. Thus, in addition to focusing on the symptoms of the disease, primary goals of cardiac and pulmonary rehabilita-tion programs are improving cardiorespiratory and musculoskeletal fitness, decreasing functional disability, improving psychological well-being, and improving the overall quality of life.

■ Individuals with cardiovascular and lung disease often have many illnesses, diseases, functional problems, and psychosocial issues. Thus, physical activity programs for these patients need to focus on improving overall health and well-being and not merely the symptoms of the specific disease.

Cardiovascular Disease

Physical activity programs are effective treatments for improving exercise tolerance, functional status, cardiovascular risk factor profile, psychological status, and quality of life in individuals with peripheral vascular disease (Regensteiner and Hiatt 2002), coronary heart disease (Leon et al. 2005), stroke (Gordon et al. 2004), and hypertension (Pescatello et al. 2004). In addition, physical inactivity and low cardiorespiratory fitness are risk factors for mortality in individuals with cardiovascular disease. Figure 11.4 presents an illustration of this effect for all-cause mortality. In this study of 772 men with established coronary heart disease, various types of physical activity were

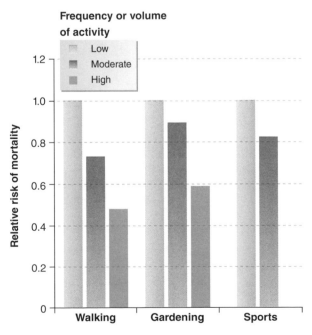

FIGURE 11.4 Influence of physical activity on all-cause mortality in men with coronary heart disease.

Data from Wannamethee, Shaper, and Walker 2000.

related to all-cause (as shown in figure 11.4) and cardiovascular disease mortality risk over a five-year period (Wannamethee, Shaper, and Walker 2000). The effect of low- to moderate-intensity physical activities, such as walking and gardening, was particularly apparent in this study.

The observation that low- to moderate-intensity activities had the greatest benefit is consistent with current exercise rehabilitation programs, which recommend low- to moderate-intensity physical activity for individuals with cardiovascular disease. More specific physical activity recommendations based on current clinical guidelines for coronary heart disease (Leon et al. 2005; Williams et al. 2007), hypertension (Pescatello et al. 2004; Williams et al. 2007), and stroke (Gordon et al. 2004) are provided in table 11.1.

Chronic Obstructive Pulmonary Disease and Asthma

Pulmonary rehabilitation programs including physical activity reduce disease symptoms while increasing exercise tolerance, functional status, psychological status, and quality of life in individuals with chronic obstructive pulmonary disease (American Thoracic Society 1999). These beneficial changes occur despite the fact that participation in physical activity has little or no effect on the degree of airway obstruction in chronic obstructive pulmonary disease. Given the overlying theme of this chapter, it should not be surprising that cardiorespiratory fitness level is negatively related to mortality risk among individuals with chronic obstructive pulmonary disease. In fact, low cardiorespiratory fitness is a better predictor of

TABLE 11.1 Physical Activity Guidelines for Individuals With Selected Cardiovascular and Pulmonary Diseases

Disease	Exercise type	Frequency (days/week)	Duration and intensity
Coronary heart disease	Aerobic	3-7	30 to 60 min (continuous or intermittent) at a moderate intensity (40-85% $\dot{V}O_2$max)
	Strength	2-3	One set of 10 to 15 repetitions of 8 to 10 exercises
Hypertension	Aerobic	5-7	30 or more minutes (continuous or intermittent) at a moderate intensity (40-60% $\dot{V}O_2$max)
	Strength	2-3	One set of 10 to 15 repetitions of 8 to 10 exercises
Stroke	Aerobic	3-7	20 to 60 min (continuous or intermittent) at a moderate intensity (40-70% $\dot{V}O_2$max)
	Strength	2-3	One to three sets of 10 to 15 repetitions of 8 to 10 exercises
	Flexibility	2-3	Each stretch held for 10 to 30 s
	Neuromuscular	2-3	Coordination and balance exercises
Chronic obstructive pulmonary disease*	Aerobic	2-5	20 to 30 or more minutes (continuous or intermittent) at a moderate intensity (~60% $\dot{V}O_2$max)
	Strength	2-3	Loads ranging from 50% to 80% of one-repetition maximum

*For chronic obstructive pulmonary disease, the exercise type, frequency, and duration or intensity reflect current practices rather than specific exercise clinical guidelines.

mortality in patients with chronic obstructive lung disease than more traditional markers of disease severity such as measures of lung function and smoking (Oga et al. 2003).

Current clinical guidelines recommend that physical activity training be included in the management of patients with moderate to severe chronic obstructive pulmonary disease (American Thoracic Society 1999). Pulmonary rehabilitation programs emphasize aerobic exercise with a supplement of resistance training. More specific details are provided in table 11.1.

As with more severe forms of lung disease, asthma is characterized by a wide range of severity both within and between patients. In severe cases of asthma, when airflow is highly obstructed with no reversibility, the physical activity capabilities and exercise rehabilitation with the disease are similar to those for chronic obstructive pulmonary disease (Satta 2000). However, in patients with mild to moderate asthma, the physical activity capabilities are relatively comparable to those of healthy, asthma-free persons. In fact, when free of symptoms, asthmatic and nonasthmatic people have the same physiological response to physical activity. With appropriate training and medication, individuals with mild to moderate asthma can successfully train and compete in high-intensity endurance events. However, regardless of fitness level and the degree of asthma, there remains the possibility of an exercise-induced asthma attack, a transient airway obstruction that sometimes occurs immediately after exercise. Although treatable with medication, repeated exercise-induced asthma attacks can alter psychosocial behaviors of asthmatics, leading to a restriction in exercise and a negative attitude toward physical activity. This may result in an unnecessarily inactive lifestyle, even in asthmatics who rarely experience exercise-induced asthma (Satta 2000).

Summary

Cardiovascular and lung diseases are highly prevalent and are leading causes of disability and death. The research findings covered in this chapter provide strong support for the notion that low physical activity and low fitness are leading risk factors for the development of most forms of cardiovascular disease and lung disease. According to current public health guidelines, adults should accumulate 150 min per week of moderate-intensity physical activity or 75 min per week of vigorous aerobic physical activity to help prevent these diseases (PAGAC 2008).

It is also clear that participation in physical activity has many beneficial effects on disease symptoms, functional mobility, overall health and quality of life, and mortality risk in individuals with cardiovascular disease and lung disease. Physical activity recommendations for individuals with these diseases vary depending on the type of disease, severity of disease, and needs and characteristics of the patient.

Key Concepts

asthma—A chronic disease affecting the airways of the lungs in which the inside walls of the airways are inflamed (swollen), making the airways very sensitive to normal irritants or allergens. When the airways react, they become narrow, causing symptoms like wheezing, coughing, chest tightness, and trouble breathing. Asthma cannot be cured, but for most patients it can be controlled.

atherosclerosis—Process in which deposits of fatty substances, cholesterol, cellular waste products, calcium, and other substances build up in the inner lining of an artery to form plaque.

cardiovascular disease—Dysfunctional conditions of the heart, arteries, and veins that supply oxygen to vital life-sustaining areas of the body like the brain, the heart itself, and other vital organs.

chronic obstructive pulmonary disease—A lung disease in which the lung is damaged and the airways are partly obstructed, making it difficult to breathe.

coronary heart disease—A narrowing of the coronary arteries that feed the heart (atherosclerosis), resulting in an insufficient blood supply to the heart muscle and causing angina (chest pain) or a myocardial infarction (heart attack).

hemorrhagic stroke—Stroke that occurs when a blood vessel bursts inside the brain. Bleeding irritates the brain tissue, causing swelling and forming a mass (hematoma), which displaces normal brain tissue.

hypertension—High blood pressure. Typically defined as a resting systolic blood pressure ≥ 140 mmHg or a resting diastolic blood pressure ≥ 90 mmHg. Hypertension is preceded by prehypertension, which is typically defined as a resting systolic blood pressure ranging from 120 to 139 mmHg or a resting diastolic blood pressure ranging from 80 to 89 mmHg. Blood pressure is dependent on output from the heart, blood vessel flexibility and resistance to blood flow, volume of blood, and blood distribution to organs, which are in turn influenced by hormones and the nervous system.

ischemic stroke—Stroke that occurs when too little blood reaches an area of the brain, usually because a clot has blocked a blood vessel.

meta-analysis—A quantitative, systematic review of the scientific literature in which the results from many studies dealing with the same topic are combined. This statistical method of combining the results of a number of studies allows more accurate estimations of effects than can be derived from individual studies.

peripheral vascular disease—Disease of the blood vessels outside the heart and brain. It is often seen as a narrowing of vessels that carry blood to the legs and in rare cases to the arms, stomach, or kidneys.

relative risk—For definition, see page 183.

stroke—Damage to part of the brain caused by interruption to its blood supply or leakage of blood outside the vessel walls. Sensation, movement, or function controlled by the damaged area is impaired. Strokes are fatal in about one-third of cases.

Study Questions

1. What is the difference between ischemic stroke and hemorrhagic stroke? Is physical inactivity a risk factor for ischemic stroke, hemorrhagic stroke, or both?

2. Explain the biological mechanisms by which high physical activity and fitness protect against the development of hypertension.

3. Provide examples of variables that are typically improved with physical activity participation in patients with severe forms of cardiovascular disease or lung disease, other than the symptoms of the disease itself.

4. Does physical activity or cardiorespiratory fitness have a stronger effect on cardiovascular disease risk? Why?

5. Explain the dose–response relationship between physical activity level and the risk of coronary heart disease.

6. Are there any racial or gender differences in the protective effects of physical activity on coronary heart disease, hypertension, and stroke risk? If so, what are the differences, and what are some possible explanations for these differences?

7. What are the current public health recommendations for physical activity that are aimed at reducing the risk of chronic diseases such as cardiovascular and lung disease?

8. What frequency, duration, and intensity of aerobic physical activity are recommended for individuals who have suffered a stroke?

References

American Thoracic Society. 1999. Pulmonary rehabilitation-1999. *American Journal of Respiratory and Critical Care Medicine* 159(5 Pt 1):1666-1682.

Blair, S.N., Y. Cheng, and J.S. Holder. 2001. Is physical activity or physical fitness more important in defining health benefits? *Medicine and Science in Sports and Exercise* 33(6 Suppl):S379-399; discussion S419-420.

Blair, S.N., H.W. Kohl 3rd, C.E. Barlow, R.S. Paffenbarger Jr., L.W. Gibbons, and C.A. Macera. 1995. Changes in physical fitness and all-cause mortality. A prospective study of healthy and unhealthy men. *Journal of the American Medical Association* 273(14):1093-1098.

Folsom, A.R., D.K. Arnett, R.G. Hutchinson, F. Liao, L.X. Clegg, and L.S. Cooper. 1997. Physical activity and incidence of coronary heart disease in middle-aged women and men. *Medicine and Science in Sports and Exercise* 29(7):901-909.

Gordon, N.F., M. Gulanick, F. Costa, G. Fletcher, B.A. Franklin, E.J. Roth, and T. Shephard. 2004. Physical activity and exercise recommendations for stroke survivors: An American Heart Association scientific statement from the Council on Clinical Cardiology, Subcommittee on Exercise, Cardiac Rehabilitation, and Prevention; the Council on Cardiovascular Nursing; the Council on Nutrition, Physical Activity, and Metabolism; and the Stroke Council. *Circulation* 109(16):2031-2041.

Hooi, J.D., A.D. Kester, H.E. Stoffers, M.M. Overdijk, J.W. van Ree, and J.A. Knottnerus. 2001. Incidence of and risk factors for asymptomatic peripheral arterial occlusive disease: A longitudinal study. *American Journal of Epidemiology* 153(7):666-672.

Katzmarzyk, P.T., and I. Janssen. 2004. The economic costs associated with physical inactivity and obesity in Canada: An update. *Canadian Journal of Applied Physiology* 29(1):90-115.

Kelley, G.A., and K.S. Kelley. 2000. Progressive resistance exercise and resting blood pressure: A meta-analysis of randomized controlled trials. *Hypertension* 35(3):838-843.

Lakka, T.A., J.M. Venalainen, R. Rauramaa, R. Salonen, J. Tuomilehto, and J.T. Salonen. 1994. Relation of leisure-time physical activity and cardiorespiratory fitness to the risk of acute myocardial infarction. *New England Journal of Medicine* 330(22):1549-1554.

Leon, A.S., B.A. Franklin, F. Costa, G.J. Balady, K.A. Berra, K.J. Stewart, P.D. Thompson, M.A. Williams, and M.S. Lauer. 2005. Cardiac rehabilitation and secondary prevention of coronary heart disease: An American Heart Association scientific statement from the Council on

Clinical Cardiology (subcommittee on Exercise, Cardiac Rehabilitation, and Prevention) and the Council on Nutrition, Physical Activity, and Metabolism (subcommittee on Physical Activity), in collaboration with the American Association of Cardiovascular and Pulmonary Rehabilitation. *Circulation* 111:369-376.

Oga, T., K. Nishimura, M. Tsukino, S. Sato, and T. Hajiro. 2003. Analysis of the factors related to mortality in chronic obstructive pulmonary disease: Role of exercise capacity and health status. *American Journal of Respiratory and Critical Care Medicine* 167(4):544-549.

Paffenbarger, R.S. Jr., R.J. Brand, R.I. Sholtz, and D.L. Jung. 1978. Energy expenditure, cigarette smoking, and blood pressure level as related to death from specific diseases. *American Journal of Epidemiology* 108(1):12-18.

Paffenbarger, R.S. Jr., A.L. Wing, and R.T. Hyde. 1978. Physical activity as an index of heart attack risk in college alumni. *American Journal of Epidemiology* 108(3):161-175.

Pelkonen, M., I.L. Notkola, T. Lakka, H.O. Tukiainen, P. Kivinen, and A. Nissinen. 2003. Delaying decline in pulmonary function with physical activity: A 25-year follow-up. *American Journal of Respiratory and Critical Care Medicine* 168(6):494-499.

Pereira, M.A., A.R. Folsom, P.G. McGovern, M. Carpenter, D.K. Arnett, D. Liao, M. Szklo, and R.G. Hutchinson. 1999. Physical activity and incident hypertension in black and white adults: The Atherosclerosis Risk in Communities Study. *Preventive Medicine* 28(3):304-312.

Pescatello, L.S., B.A. Franklin, R. Fagard, W.B. Farquhar, G.A. Kelley, and C.A. Ray. 2004. American College of Sports Medicine position stand. Exercise and hypertension. *Medicine and Science in Sports and Exercise* 36(3):533-553.

Physical Activity Guidelines Advisory Committee. 2008. *Physical Activity Guidelines Advisory Committee report, 2008.* Part G. Section 2: Cardiorespiratory health (pp. G2-1–Gs-57). Washington, DC: U.S. Department of Health and Human Services.

Rasmussen, F., J. Lambrechtsen, H.C. Siersted, H.S. Hansen, and N.C. Hansen. 2000. Low physical fitness in childhood is associated with the development of asthma in young adulthood: The Odense schoolchild study. *European Respiratory Journal* 16(5):866-870.

Regensteiner, J.G., and W.R. Hiatt. 2002. Current medical therapies for patients with peripheral arterial disease: A critical review. *American Journal of Medicine* 112(1):49-57.

Satta, A. 2000. Exercise training in asthma. *Journal of Sports Medicine and Physical Fitness* 40(4):277-283.

Tanasescu, M., M.F. Leitzmann, E.B. Rimm, W.C. Willett, M.J. Stampfer, and F.B. Hu. 2002. Exercise type and intensity in relation to coronary heart disease in men. *Journal of the American Medical Association* 288(16):1994-2000.

Wannamethee, S.G., A.G. Shaper, and M. Walker. 2000. Physical activity and mortality in older men with diagnosed coronary heart disease. *Circulation* 102(12):1358-1363.

Williams, M.A., W.L. Haskell, P.A. Ades, E.A. Amsterdam, V. Bittner, B.A. Franklin, M. Gulanick, S.T. Laing, and K.J. Stewart. 2007. Resistance exercise in individuals with and without cardiovascular disease: 2007 update: A scientific statement from the American Heart Association Council on Clinical Cardiology and Council on Nutrition, Physical Activity, and Metabolism. *Circulation* 116(7):572-584.

12

Physical Activity, Fitness, and Obesity

Robert Ross, PhD, FACSM; and Ian Janssen, PhD

CHAPTER OUTLINE

©Bananastock

Obesity is a leading risk factor for premature mortality and numerous chronic health conditions that reduce the overall quality of life. The prevalence of obesity has increased to epidemic proportions in both developed and developing countries during the past two to three decades, and this condition affects virtually all ages, races, and socioeconomic groups and both sexes. Obesity reflects a continued positive energy balance, which is accompanied by unhealthy weight gain and is linked to physical inactivity.

In this chapter we explore the role that physical activity plays in preventing and treating obesity. The first part of the chapter provides a definition and assessment system for overweight and obesity, examines prevalences and trends in overweight and obesity from a global perspective, and discusses the impact of obesity on health risk. In the second part, the role of specific fat depots in determining obesity-related health risk is examined. This is followed by an examination of the role of physical activity in the etiology of obesity from a population perspective. The third part of the chapter reviews the relationship between excess weight and physical activity and fitness, and the fourth section identifies the role of physical activity in the prevention and treatment of obesity.

Definition and Problem of Overweight and Obesity

Obesity is a condition of excessive fat accumulation that may impair health. The underlying disease reflects a continued positive energy balance, which is accompanied by undesirable weight gain. However, the degree of excess fat, its distribution within the body, and associated health consequences vary considerably among obese individuals.

Overweight and obesity are commonly assessed in the research and clinical setting using the **body mass index (BMI)**, a simple index of weight for height, calculated as weight in kilograms divided by the square of height in meters (kg/m^2). The globally accepted BMI classification system for adults is shown in table 12.1 (World Health Organization 1998). This classification system is based on the relationships among BMI, mortality, and chronic disease. Note that the BMI values are age independent and are the same for both men and women. At present, these BMI values can be used for all racial and ethnic groups, with the exception of Asian populations, for whom a BMI of 23 kg/m^2 should be used to denote overweight and a BMI of 27 kg/m^2 should be used to define obesity (World Health Organization Expert Consultation 2004). The lower overweight BMI cut point for Asians reflects the fact that body fat and health risk are higher for a given BMI in these subgroups (Deurenberg-Yap and Deurenberg 2003).

In addition to BMI, **waist circumference** can be used as a simple anthropometric index of overweight and obesity. There is currently no consensus on what waist circumference thresholds should be used to denote increased health risk. The most commonly used cut points in men are ≥94 cm, which denotes a moderately increased risk of obesity-related complications, and ≥102 cm, which denotes a substantially increased risk of obesity-related complications. The

TABLE 12.1 Classification of Overweight in Adults According to Body Mass Index (BMI)

Classification	BMI (kg/m²)	Morbidity and mortality risk	
		Low waist Men ≤ 102 cm Women ≤ 88 cm	High waist Men > 102 cm Women > 88 cm
Underweight	<18.5	Low[a]	NA
Normal range	18.5-24.9	Low	Increased
Overweight	≥25		
Preobese	25-29.9	Increased	High
Obese class I	30-34.9	High	Very high
Obese class II	35-39.9	NA	Very high
Obese class III	≥40	NA	Extremely high

NA = not applicable. All underweight individuals have a low waist circumference, and all class II and class III obese individuals have a high waist circumference. [a]Low risk for obesity-related complications, but increased risk for some other health complications.

Data from National Institutes of Health, National Heart, Lung, and Blood Institute 1998; World Health Organization, 1998.

corresponding waist circumference thresholds in women are 80 and 88 cm, respectively (World Health Organization 1998).

Ideally, waist circumference should be used in combination with BMI as an indicator of health risk, because waist circumference explains an additional component of obesity-related morbidity and mortality. An example of this effect is illustrated in figure 12.1. This figure shows the incidence rate of coronary heart disease according to BMI and waist circumference in a longitudinal follow-up study of 44,702 female American nurses. Within the low-, moderate-, and high-BMI categories, larger waist circumference values were associated with increased risk of coronary heart disease.

Organizations such as the U.S. National Institutes of Health (NIH) (NIH National Heart Lung and Blood Institute 1998) have proposed that waist circumference values of ≥102 cm in men and ≥88 cm in women can be used within the BMI categories listed in table 12.1 to differentiate between those with and without abdominal obesity. For example, an obese class I male with a waist < 102 cm would be considered to have a lower abdominal fat mass and a "high" obesity-related risk, whereas an obese class I male with a waist > 102 cm would be considered to have a higher abdominal fat mass and a "very high" obesity-related health risk.

Waist circumference can be used alone, or in combination with BMI, as an indicator of obesity-related health risk.

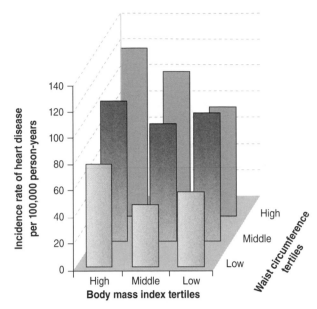

FIGURE 12.1 The incidence rate of coronary heart disease according to body mass index and waist circumference in a longitudinal study of 44,702 female American nurses.

Adapted from *JAMA*, 1998, 280(21): 1843-1848. Copyright © 1998 American Medical Association. All rights reserved.

In children, BMI changes substantially with age, rising steeply during infancy, falling during the preschool years, and then rising again continuously into adulthood. This effect is illustrated in figure 12.2, which shows the mean BMI by age and gender in six nationally representative data sets. Because of these changes, overweight and obesity in children

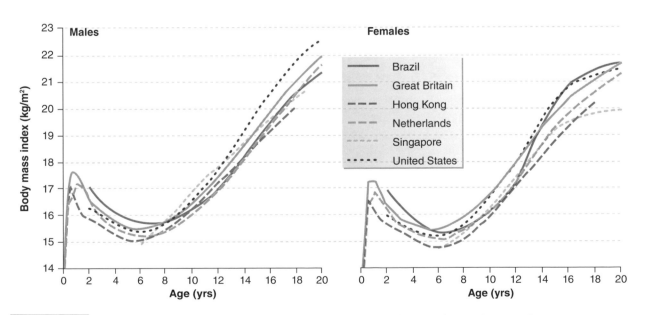

FIGURE 12.2 Six nationally representative data sets with children aged 0 months to 18 years of age.

Reprinted from T.J. Cole, M.C. Bellizzi, K.M. Flegal, and W.H. Dietz, 2000, "Establishing a standard definition for child overweight and obesity worldwide: International survey," *British Medical Journal* 320: 1240-1243, with permission from the BMJ Publishing Group.

and adolescents are determined using age-specific BMI thresholds. The level of agreement about the classification of overweight and obesity by BMI in the pediatric population is not the same as for adults. A number of countries have produced nationally representative BMI-for-age growth references, which allow an individual's BMI to be expressed as an age- and sex-specific percentile. Historically, the 85th and 95th percentiles have been used to determine overweight and obesity status, respectively (World Health Organization 1998). International BMI growth references for defining overweight and obesity in children and adolescents have also been developed through regression of the adult BMI cut points of 25 kg/m² and 30 kg/m² at age 18 back through the growth curve (Cole et al. 2000). In 2006, the World Health Organization (WHO) released a new set of BMI cut points for children that relies on growth *standards* rather than growth references (Borghi et al. 2006). The WHO growth standards, which were developed using healthy children living in favorable conditions, describe how children should grow in order to minimize morbidity risk.

Impact of Overweight and Obesity on Health

Overweight and obesity are leading risk factors for premature mortality and numerous chronic health conditions that reduce the overall quality of life. These chronic health conditions include type 2 diabetes, coronary heart disease, hypertension, stroke, certain forms of cancer, gallbladder disease, and osteoarthritis (NIH National Heart Lung and Blood Institute 1998; World Health Organization 1998). Table 12.2 summarizes the extent of the morbidity and mortality risk for the average obese individual compared with the average individual who has a normal body weight.

Global estimates indicate that approximately 58% of diabetes cases, 21% of ischemic heart disease cases, and 8% to 42% of certain cancers are directly attributable to excess body weight (World Health Organization 1998). Conservative estimates indicate that obesity alone, not including overweight, accounts for 2% to 7% of total direct health care costs in developed countries (World Health Organization 1998).

TABLE 12.2   Health Risks Associated With Obesity

Health condition	Extent of increased risk in obese individuals
Premature mortality	↑
Coronary artery disease	↑↑
Stroke	↑
Hypertension	↑↑
Congestive heart failure	↑
Pulmonary embolism	↑↑↑
Type 2 diabetes	↑↑↑↑
Colorectal cancer	↑
Postmenopausal breast cancer	↑
Kidney cancer	↑↑
Pancreatic cancer	↑↑
Gallbladder disease	↑↑
Osteoarthritis	↑↑↑
Chronic back pain	↑↑↑
Asthma	↑

↑ Risk for condition increased by approximately 25% to 50%; ↑↑ risk for condition increased by approximately 200%; ↑↑↑ risk for condition increased by approximately 350%; ↑↑↑↑ risk for condition increased by over 400%.

Data from Anis et al. 2010.

Indirect costs, which are far greater than the direct costs, include lost work days, disability pensions, impaired quality of life, and premature mortality. Indeed, in industrialized countries, obesity is one of the leading causes of death.

Global Prevalence and Secular Trends in Obesity

Overweight and obesity are at epidemic proportions globally. It was estimated that there were 937 million overweight (preobese) adults and an additional 396 million obese adults across the globe in 2005 (Kelly et al. 2008). Obesity affects both sexes and virtually all ages, races, and socioeconomic groups. Even in developing countries, obesity coexists along with malnutrition. Current obesity levels in adults range from less than 5% in China, other Asian countries, and many African nations to greater than 20% in high-income countries such as the United States, Canada, and most European nations (Kelly et al. 2008). Within high-income countries, the combined prevalence of overweight and obesity in adults is approximately 50% (Kelly et al. 2008), implying that about half of the adult population in these countries is at increased health risk from excess body weight. Perhaps even more troubling, estimates suggest that the prevalence of school-age children who were overweight in 2010 exceeded 45% and that about 15% were obese (Wang and Lobstein 2006). The following evidence indicates that the situation in the 21st century is likely to be significantly worse.

The distribution of BMI is shifting upward in most child and adult populations, and the prevalence of obesity has increased at an alarming rate in both developed and developing countries during the past two to three decades. Here we use recent findings from Canada to illustrate this effect. Between 1981 and 2008, the average BMI of middle-aged Canadian men increased from 26.1 kg/m² to 28.3 kg/m², and the prevalence of obesity more than doubled from 12% to 27% (Shields et al. 2010). The sudden increase in the prevalence of obesity appears to have started during the 1980s. Although limited representative population data are available, it appears that abdominal obesity as reflected by waist circumference is also high and increasing. For instance, in middle-aged

> Overweight and obesity are highly prevalent conditions in both developed and undeveloped nations. There has been a dramatic increase in the prevalence of obesity in recent years.

Canadian women, the average waist circumference increased by 10 cm, and the prevalence of a high waist circumference (>87 cm) increased from 18% to 47%, between 1981 and 2008.

Fat Depots

In this section we consider the importance of specific fat depots. A vast number of scientific studies conducted during the course of the past half century have provided clear evidence that the distribution of fat is a more important determinant of obesity-related health risk than is the degree of overall adiposity.

Total Fat

The relationship between BMI and obesity-related health risk is explained in large measure by the ability of BMI to predict total fat. Although BMI is a good marker of total fat on a population basis, the relationship between BMI and adiposity is marked by large interindividual variation, and somewhere on the order of 25% to 50% of the between-individual variation in total fat is not accounted for by BMI (Janssen, Heymsfield, and Ross 2002). Gender, race, age, genetic factors, and fitness level all influence the relationship between BMI and total fat. More direct measures of body fat (e.g., skinfolds) provide more accurate measures of total fat and obesity-related health risk than does BMI (Janssen, Heymsfield, and Ross 2002). Waist circumference is also a good marker of total fat and, interestingly, is as strong a correlate of total fat as is BMI (Janssen, Heymsfield, and Ross 2002).

Abdominal Subcutaneous and Visceral Fat

In addition to the degree of excess fat, the distribution of fat within the body is an important determinant of the health consequences associated with obesity. In particular, the two abdominal fat depots—**abdominal subcutaneous fat** and intra-abdominal or **visceral fat**—are associated with the pathogenesis of numerous cardiovascular disease and diabetes risk factors (NIH National Heart Lung and Blood Institute 1998; World Health Organization 1998). The accumulation of excess visceral fat is of particular relevance for obesity-related health risk.

Anatomically defined, visceral fat consists of **adipocytes** contained within the visceral peritoneum. The visceral peritoneum is the membrane that covers the abdominal organs of the gastrointestinal tract, and it extends from about the 11th thoracic vertebra to about the fifth lumbar vertebra (figure 12.3). This definition excludes perirenal adipocytes that compose the extraperitoneal fat surrounding the kidneys. However, the peritoneum cannot be seen on magnetic resonance imaging (MRI) or computed tomography (CT) images; therefore in obesity studies, visceral fat is typically defined as the sum of the fat contained within (intraperitoneal, ~80% of total visceral fat) and behind (extraperitoneal, ~20% of total visceral fat) the peritoneum.

The mechanisms that explain the association between visceral fat and obesity related health risk are incompletely understood. Free fatty acids, glycerol, and hormones released from adipose tissue within the visceral peritoneum are drained into the hepatic portal vein that travels directly to the liver. A common postulate is that sustained exposure of the liver to an increased flux of free fatty acids via the portal circulation is the antecedent to many of the disturbances in glucose and lipid metabolism that

are associated with abdominal obesity (Björntorp 1990). This is often referred to as the "portal theory." However, emerging evidence provides strong support for the notion that adipose tissue-derived **cytokines** may be the mechanism that links visceral fat with metabolic health risk. Adipose tissue releases over 100 hormones and autocrine and paracrine factors. The majority of these factors are cytokines, and many of these are involved in the pathogenesis of atherosclerosis, hypertension, and insulin resistance. Most of these factors are secreted to a different extent in different fat depots, and for the most part, visceral fat is a more active producer of these factors than is subcutaneous fat (Matsuzawa 2002). Furthermore, the portal drainage of visceral fat through the liver and other visceral organs may amplify the effect of the cytokines released by this fat depot.

Subcutaneous fat is the layer of adipocytes that lies directly between the dermis layer of the skin and extends over the whole body. No commonly accepted

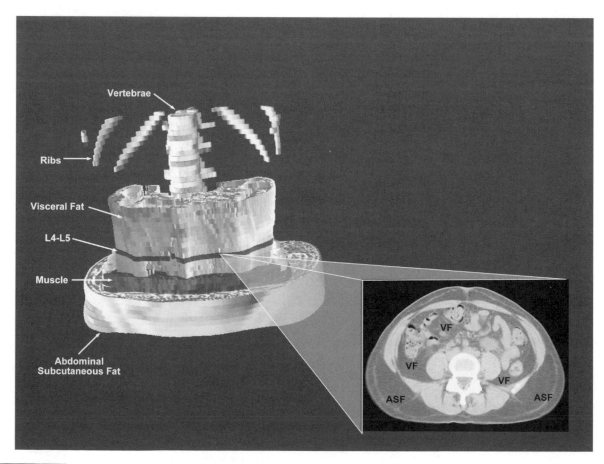

FIGURE 12.3 Tissue distribution in the abdominal cavity. The two-dimensional image on the lower right represents a cross-sectional computed tomography image obtained at the level of the intervertebral disc between the fourth and fifth lumbar vertebrae. The abdominal subcutaneous fat (ASF) is the dark gray tissue located directly beneath the skin. The visceral fat (VF) is the dark gray tissue located underneath the skeletal muscle and surrounding the visceral organs. The three-dimensional image is a computerized recreation of the various tissue layers in the abdomen.

boundaries define the proximal and distal borders of abdominal subcutaneous fat. It is well established, however, that there are regional differences in **lipolysis** between subcutaneous adipocytes in the abdomen and those in the leg. Subcutaneous adipocytes in the abdominal region have an increased lipolysis rate and deposit free fatty acids into the systemic circulation at a greater rate than subcutaneous adipocytes in the appendicular regions (Arner et al. 1990). On a metabolic level, the contribution that abdominal subcutaneous fat makes to obesity-related health risk may result from the increased spillover of free fatty acids into the systemic circulation, which are subsequently delivered to other tissues such as the liver, pancreas, and skeletal muscle.

A number of factors influence fat distribution. For a given level of total fat, men have more abdominal and visceral fat than women; older adults have more abdominal and visceral fat than younger adults; Caucasians have more abdominal and visceral fat than African Americans; and physically inactive and unfit individuals have more abdominal and visceral fat than physically active and fit individuals.

Radiological imaging methods such as CT and MRI are the most accurate means available for in vivo quantification of body composition at the tissue level. Although access and cost are obstacles to routine use, these imaging approaches are now used extensively in body composition research and are the methods of choice for measuring abdominal fat. Computed tomography and MRI systems generate cross-sectional images of human anatomy, as illustrated in figure 12.3. Image analysis programs are used to determine the area (cm²) of abdominal subcutaneous fat and visceral fat in these cross-sectional images, with a greater area representing a greater tissue size. Typically, a single-image strategy is used; that is, one image is obtained at a predetermined level of the abdomen, usually at the level corresponding to the intervertebral disc between the fourth and fifth lumbar vertebrae. The area measures from the single cross-sectional image are used as an index of tissue size. However, multiple cross-sectional images can also be obtained across the abdominal region, either contiguously or with gaps between two consecutive images, and the area values derived from these multiple images can be used to calculate the volume or mass of the entire abdominal subcutaneous and visceral fat depots. Recent advances in MRI make it possible to acquire contiguous images of the entire abdomen during a routine breath hold (Hu et al. 2010).

Unfortunately, CT and MRI are not appropriate techniques for assessing abdominal fat in the clinical setting because of their high cost and limited accessibility. On the other hand, waist circumference is a convenient and simple measurement that is an approximate index of abdominal subcutaneous and visceral fat. Furthermore, changes in waist circumference reflect changes in abdominal obesity and risk factors for cardiovascular disease (Janssen, Heymsfield, and Ross 2002; Ross et al. 2000). Thus, waist circumference in combination with BMI is a useful clinical tool that can identify individuals at increased health risk attributable to abdominal obesity (NIH National Heart Lung and Blood Institute 1998; World Health Organization 1998).

■

The accumulation of excess visceral fat is of particular concern in overweight and obese individuals. In the research setting, visceral fat can be measured using sophisticated imaging techniques, whereas in the clinical setting, waist circumference is used as an index of visceral obesity. Although visceral fat tends to increase with increasing BMI, other factors such as age, race, physical activity, gender, and genetics influence visceral fat accumulation.

Nonadipose Tissue Fat

In recent years, research has focused on the relevance of nonadipose tissue fat, otherwise known as **ectopic fat**, in determining obesity-related health risk. The mechanisms responsible for the increase of lipid in ectopic tissues are unclear. It is postulated that ectopic lipid accumulation may occur secondary to the development of dysfunctional (insulin resistant), peripheral, subcutaneous adipocytes that lose the ability to buffer excess energy intake (Després and Lemieux 2006). Of primary concern are the ectopic fat depots in skeletal muscle and the liver.

In skeletal muscle, lipids can be stored within the muscle fibers (these are referred to as intramyocellular lipids) and outside the muscle fibers (these are referred to as extramyocellular lipids). The intramyocellular lipids are primarily stored in droplets close to the mitochondria, where they serve as a readily available energy source. The extramyocellular lipids accumulate to form adipocytes between the muscle fibers and bundles, where they serve as a long-term energy storage site. When numerous extramyocellular adipocytes accumulate, they are seen as marbled fat within the skeletal muscle.

Obesity is associated with an increase in both intramyocellular lipids and extramyocellular lipids, although the increase in intramyocellular lipids is of greater concern for obesity-related health risk. Most notably, intramyocellular lipids have been linked with insulin resistance, a leading risk factor for type 2 diabetes and cardiovascular disease (Kelley, Goodpaster, and Storlein 2002). In fact, the relationship between intramyocellular lipids and insulin resistance is independent of total and abdominal fat content.

In some situations, however, the accumulation of intramyocellular lipids is not related to insulin resistance. Most notably, a hallmark adaptation to chronic exercise is an increase in lipid oxidation, and, accordingly, intramyocellular lipids levels are increased in aerobically trained athletes. Despite high intramyocellular lipid levels, athletes are characterized by elevations in insulin sensitivity (Goodpaster et al. 2001). The periodic depletion and repletion of intramyocellular lipids that occur with regular exercise may disrupt the mechanistic link between intramyocellular lipids and insulin resistance that occurs in obesity. That is, high levels of intramyocellular lipids do not appear to have adverse metabolic consequences in skeletal muscle that has the capacity for efficient lipid utilization, as in aerobic athletes (Goodpaster et al. 2001). Further, more recent evidence suggests a repartitioning of intramyocellular lipids in response to exercise, such that the increase in intramyocellular triglycerides is countered by a decrease in cytosolic ceramides and diacylglycerides (Dubé et al. 2008). The latter are thought to be involved in the attenuation of insulin-mediated glucose uptake. Thus, the role of intramyocellular lipid accumulation in explaining obesity-related health risk is confounded by decreasing levels of physical activity. This suggests that the ability to oxidize intramyocellular lipids and the distribution of lipid intermediates may be more important than the storage of lipids per se. If so, physically active obese individuals would be at less risk for a given level of intramyocellular lipid accumulation than their sedentary counterparts.

A second ectopic fat depot of interest is liver fat. In obesity studies, liver fat is often measured by CT. The density or attenuation of muscle and liver in CT images is an indication of the lipid content of these tissues—the greater the lipid content, the lower the tissue density and the darker the color of the liver on the CT images. The limitations inherent to CT-measured tissue quality confound interpretation, and thus it is generally accepted that proton magnetic resonance spectroscopy is the optimal noninvasive method to assess fatty liver. Determination of liver fat may provide a mechanistic link between abdomi-

nal obesity, in particular visceral fat, and increased obesity-related health risk. It has been hypothesized that the metabolic importance of visceral fat may be mediated by the delivery of free fatty acids into the hepatic portal vein, exerting potent and direct effects on the liver. Consistent with this position, liver fat content is correlated with total and visceral fat content, and reductions in total and visceral fat induced by caloric restriction are associated with corresponding reductions in liver fat (Malnick, Beergabel, and Knobler 2003). Evidence indicates that liver fat is related to metabolic risk factors such as plasma triglyceride levels and insulin resistance and that the effects of liver fat are at least in part independent of visceral fat content, suggesting that the amount of fat within the liver carries an independent health risk. Preliminary evidence suggests that habitual physical activity is associated with lower levels of liver fat accumulation (Perseghin et al. 2007).

> Ectopic fat depots in skeletal muscle and the liver increase with obesity, and evidence suggests that these fat depots partially explain the effect of obesity on health risk. More research is needed to clarify the role of obesity and physical activity on ectopic fat content and related health risks.

Relationships Among Excess Weight, Physical Activity, and Fitness

We now consider the role that physical activity plays in the development of obesity. In so doing, we consider the following questions. Do population-wide changes in physical activity levels explain the recent obesity epidemic? Are physically inactive individuals more likely to develop obesity than physically active individuals? Does participating in physical activity increase hunger and energy intake?

Role of Physical Activity in the Etiology of Obesity

Obesity results from a chronic energy imbalance whereby intake exceeds expenditure. Thus, the hypothesis that physical inactivity has contributed to the obesity epidemic is very reasonable. However, the relative contributions of physical inactivity versus dietary consumption to the obesity epidemic are unclear and are difficult to assess due to temporal

Physical activity or inactivity is just one contributing factor to the obesity epidemic.

changes in the ways these behaviors have been measured in health surveys.

Ecological information obtained at the population level suggests that, between 1970 and 2000, energy intake remained relatively stable in American children and youth and increased slightly in American adults (Briefel and Johnson 2004). During the same time frame, leisure-time physical activity decreased in children and youth and increased in adults (Knuth and Hallal 2009). These ecological data are not congruent with the large increase in the prevalence of adult and childhood obesity that occurred during the past few decades. There are a number of possible explanations for this inconsistency. Among these are the methodological and technical problems of reporting physical activity and food intake that may have changed over time. Another factor to consider is the increased public awareness of the health benefits of physical activity and of what "counts" as physical activity. In recent years, knowledge has increased that activities such as walking "count" as physical activity, whereas in the past people may have viewed only vigorous activities such as running or swimming as physical activity. The most likely explanation, however, is that population trends in physical activity participation have focused solely on leisure-time physical activity and thus are not sensitive to any changes that have occurred in non-leisure-time physical activity levels. This is essential, because leisure-time physical activity accounts for a small proportion of the day (<1 h) for most individuals in most countries around the world and consequently a small proportion of total daily energy output. Thus, when we consider obesity, it is also necessary to take into account the energy expenditure during the remainder of the day.

Unlike leisure-time physical activity, non-leisure-time physical activity has decreased substantially in the past 20 to 30 years (Knuth and Hallal 2009). Physical activity has been engineered out of daily life by increasing mechanization at work and in the home. In the workplace, for example, there are far fewer blue-collar manual labor jobs and far more white-collar desk jobs in the new millennium than there were in the 1970s. Walking and biking as a means of transportation have declined, whereas the use of automobiles has steadily increased. At home, inactive forms of entertainment have been on the rise. Digital video disc players, cable and satellite television, and home computers and the Internet are technologies that have become widely available to the public only in the past two decades. The accessibility to countless labor-saving devices has also increased in recent years, such as remote controls, scrub-free cleaning products, lighter and more efficient tools, and dishwashing machines, just to name a few. Consistent with these changes in technology, adults in industrialized nations spend approximately 20% less time on housework now than they did 30 years ago (Knuth and Hallal 2009). These changes have fostered sedentary habits and have greatly reduced energy expenditure in the portion of the day not spent on leisure-time physical activity.

What is the implication of a decrease in non-leisure-time physical activity? Because leisure-time physical activity levels have changed minimally, the implication is that total physical activity–induced energy expenditure at the population level has likely decreased during the past 30 years. Even a 10 kcal/day reduction in total daily physical activity would be substantial when added up over time. An individual who is 42 J (10 kcal) per day above his weight maintenance energy requirement would gain about 0.5 kg (1 lb) of body fat per year. In fact, the average young to middle-aged adult in North America gains about 0.5 kg per year in body weight (Williamson et al. 1993).

■ **Population-based data suggest that dietary intake and leisure-time physical activity levels have changed minimally during the recent obesity epidemic. Recent technological advances have engineered non-leisure-time physical activity out of the daily routine of most individuals, and these changes have likely contributed to the obesity epidemic.**

Given the potential importance of the decrease in non-leisure-time physical activity, it is reasonable to ask whether the general population is going to abstain from using these labor-saving devices to curtail the obesity epidemic. The answer to this question is almost certainly no, because these devices make life easier and more enjoyable and free up additional leisure time. The implication is that the public must be willing to compensate by decreasing caloric intake or increasing leisure-time physical activity levels. This fundamental public health message is not novel. Indeed, Jean Mayer, the nutritionist who founded the School of Nutrition at Tufts University, made the following observation in 1955:

> In many cases adaptations to modern conditions without development of obesity implies that the person will have either to step up his activity or endure mild or acute hunger all his life. . . . If the first alternative, stepping up activity, is difficult, it is well remembered that the second alternative, life-time hunger, is so much more difficult that to rely on it for weight control in cases of sedentary overweight can only continue the fiascos of the past. (Mayer 1955)

Relationships Among Excess Weight, Physical Activity, and Fitness

Observational or cross-sectional data on the relationships between physical activity, cardiorespiratory fitness level, and body weight and obesity have, in general, shown an inverse association between these measures. That is, physically active and fit individuals are considerably less likely to be obese than physically inactive and unfit individuals. Population-based longitudinal studies have also shown an inverse relationship between physical activity level and increase in weight and body fat or the prevalence of overweight or obesity. That is, physically active and fit individuals are less likely to develop obesity than physically inactive and unfit individuals.

Interactions Among Physical Activity, Energy Intake, and Body Weight

It is commonly held that physical activity increases hunger and food intake, thereby compromising its utility as a strategy for controlling body weight and obesity. The scientific evidence, however, does not support this contention. To the contrary, adult men and women can tolerate daily energy deficits of up to 4,187 J (1,000 kcal) for two weeks when engaging in physical activity programs without any influence on hunger or **ad libitum** food intake (Blundell et al. 2003). Thus, the short-term effect of participating in a physical activity program is weight loss, although the weight loss would be minimal over a two-week period.

In the long term, about two weeks after the beginning of a physical activity program, food intake begins to increase. On average, the increase in ad libitum energy intake compensates for about 30% of the physical activity–induced energy expenditure. Thus, the average individual will likely continue to lose weight in the weeks and months after commencing a physical activity program. However, there is a large variation in the level of compensation that occurs between subjects. Some individuals do not compensate at all, whereas others compensate almost 100% (Blundell et al. 2003). In other words, some individuals show no increase in ad libitum energy intake even weeks after commencing a physical activity program, whereas the ad libitum energy intake in some people will increase by an amount comparable to what they expended during their physical activity session. These compensators, who unfortunately cannot be readily identified, must make a concerted effort not to increase food intake if weight loss is a goal.

■ **A physical activity program does not result in an increase in food intake in the short term; over the long term, the increase in caloric intake compensates on average for only 30% of the physical activity–induced energy expenditure.**

With regard to physical activity and diet interactions, it is important to consider the alternative. That is, what happens to energy intake in physically active individuals who become physically inactive? This situation does not induce a compensatory reduction in ad libitum energy intake and leads to a markedly positive energy balance, most of which is stored as fat (Blundell et al. 2003). Thus, if physically active individuals suddenly become inactive, they will in all likelihood gain body weight and fat unless they make a concerted effort to reduce their food and caloric intake.

Role of Physical Activity in Prevention and Treatment of Excess Weight

From the preceding discussion, it is clear that a decrease in physical activity contributes to the increased prevalence of obesity worldwide. Accordingly, it is intuitive to suggest that an increase in physical activity levels would be associated with a decrease in obesity. Indeed, evidence from population-based studies with long-term follow-up confirms that age-related weight gain is attenuated in physically active adults compared with sedentary adults (Saris et al. 2003). However, while the experts agree that an increase in physical activity is associated with a lower prevalence of obesity, precisely how much physical activity is required to prevent age-related weight gain is the subject of considerable debate.

Physical activity guidelines for adults were initially formulated for the prevention of morbidity and mortality. Indeed, there is now a large body of evidence suggesting that the accumulation of 30 min or more of moderate-intensity physical activity on most days of the week provides substantial benefits across a broad range of health outcomes. Although 30 min of daily physical activity may prevent unhealthy weight gain in some individuals, it is now generally reported that this volume of physical activity may be insufficient for the prevention of the age-related weight gain in many if not most adults. Table 12.3 summarizes the reports from various expert groups that have considered how much physical activity is required to prevent weight gain. In general, the expert

TABLE 12.3 Current Recommendations for Prevention of Weight Gain

Expert group (year)	Recommendation for prevention of weight gain
World Health Organization (1998)[a]	Men and women should achieve a PAL of 1.75.
U.S. Surgeon General (2001)[b]	Adults should get at least 30 min of moderate physical activity on most days of the week. Children should aim for 60 min.
International Obesity Task Force (2002)[c]	Men and women should achieve a PAL of 1.8 to prevent unhealthy weight gain. Vigorous activity is more clearly linked to weight stability.
Institute of Medicine (2002)[d]	All adults should accumulate 60 to 90 min of daily physical activity. This corresponds to a PAL greater than 1.6.
Stock Conference (2003)[e]	Men and women should engage in moderate-intensity physical activity for about 45 to 60 min per day or achieve a PAL of 1.7.

PAL = physical activity level.

[a]World Health Organization. 1998. *Obesity: Preventing and managing the global epidemic* (WHO/NUT/NCD/98.1.1998). Geneva: WHO.

[b]U.S. Surgeon General. 2001. Call to action to prevent and decrease overweight and obesity. Washington, DC: U.S. Department of Health and Human Services.

[c]Erlichman, J., A.L. Kerbey, and W.P. James. 2002. Physical activity and its impact on health outcomes. Paper 2: Prevention of unhealthy weight gain and obesity by physical activity: An analysis of the evidence. *Obesity Reviews* 3(4):273-287.

[d]Institute of Medicine. 2002. Dietary reference intake for energy, carbohydrate, fiber, fat, fatty acids, cholesterol, protein and amino acids. Washington, DC: National Academy Press.

[e]Saris, W.H., S.N. Blair, M.A. van Baak, S.B. Eaton, P.S. Davies, L. Di Pietro, M. Fogelholm, A. Rissanen, D. Schoeller, B. Swinburn, A. Tremblay, K.R. Westerterp, and H. Wyatt. 2003. How much physical activity is enough to prevent unhealthy weight gain? Outcome of the IASO 1st Stock Conference and consensus statement. *Obesity Reviews* 4(2):101-114.

groups derived their recommendations from analysis of longitudinal, population-based studies that related estimates of self-reported physical activity levels over time to corresponding changes in body weight. The principal exception was the report from the Institute of Medicine, which based its recommendations on a cross-sectional analysis of data that used the doubly labeled water method to estimate total energy expenditure. Combined with the measurement of basal metabolic rate, the total energy expenditure values derived by doubly labeled water can be used to calculate an individual's **physical activity level (PAL)**. The PAL is also used in population-based studies as a way of standardizing the various approaches used to determine physical activity energy expenditure. In short, the higher the PAL, the higher the level of physical activity performed on a daily basis.

The PAL is defined as the ratio of total energy expenditure to 24-h basal energy expenditure. Thus, PAL depends to a certain degree on body size and age, because these variables contribute to basal energy expenditure. Individuals can be placed into one of four activity categories based on their PAL, as shown in table 12.4.

The four PAL categories in the table correspond roughly to quartiles in the population. Thus, the Sedentary category is the lowest 25% of the population, whereas the Very Active category is the highest 25%. The Sedentary category was defined according to basal energy expenditure, the thermic effect of food, and the energy expended in physical activities that are required for independent living. Incorporating about 40 min per day of walking at a speed of 4.8 to 6.4 km/h (3 to 4 mph), in addition to the activities that are part of daily living, raises the PAL to the Low Active level in the average 70 kg (154 lb) person. To reach a PAL of 1.7 to 1.8—which is currently recom-

mended for the prevention of age-related weight gain—the average 70 kg person must incorporate about 2 h per day of walking at 3 to 4 mph, in addition to the activities that are part of daily living.

The distances and times required to move to the more active categories vary considerably by body weight. Thus, more walking is required for lean individuals and less for obese individuals. People can reduce the times substantially by walking faster or performing other, more vigorous activities. For example, if the average person walked 30 min per day at 6.4 km/h (4 mph), cycled moderately for another 25 min, and played tennis for 40 min, the PAL would increase to about 1.75 (Active).

The consensus opinion at present is that the prevention of weight gain in both developed and undeveloped countries is associated with a PAL of about 1.7 to 1.8 (table 12.3). To achieve a PAL of 1.8 would require a physical activity habit equivalent to walking 8 to 11 km (5 to 7 mi) per day at 4.8 km/h (3 mph) in addition to the habitual activity required by a sedentary lifestyle. For most inactive people, this would require adding more than 60 min of physical activity to their daily routine.

On the other hand, the guidelines for prevention of weight gain have been derived in large measure from population-based cohorts of men and women, and thus the implementation of the guidelines may vary substantially among individuals. In other words, some people will maintain body weight by accumulating only 30 min of daily physical activity, whereas others may find it necessary to accumulate 60 to 120 min or more to maintain energy balance and to prevent weight gain. The point is that the quantity of physical activity required to maintain body weight (e.g., energy balance) varies depending on the individual. Also note that the guidelines for prevention

TABLE 12.4   Physical Activity Level (PAL)

		Walking equivalence, at a pace of 3 to 4 mph, for individuals of various weights		
PAL category	PAL value	44 kg (97 lb)	70 kg (154 lb)	120 kg (264 lb)
Sedentary	1.0-1.39	0	0	0
Low active	1.4-1.59	~2.9 mi	~2.2 mi	~1.5 mi
		43-58 min	33-44 min	22-30 min
Active	1.6-1.89	~9.9 mi	~7.3 mi	~5.3 mi
		148-198 min	109-146 min	79-106 min
Very active	1.9-2.5	~22.5 mi	~16.7 mi	~12.3 mi
		337-450 min	250-334 min	184-246 min

of weight gain through physical activity are derived from studies that used primarily Caucasian adults. Accordingly, the potential influence of race on these guidelines is unknown.

■ **On average, a PAL of about 1.75, which is equivalent to about 60 to 90 min of daily leisure-time physical activity, is recommended to prevent age-related weight gain.**

Treatment of Obesity

The independent role of physical activity as a treatment strategy for obesity has received considerable attention. Early reviews of the literature suggested that the reduction in body weight (1-2 kg [2.2-4.4 lb]) associated with physical activity alone (e.g., no caloric restriction) was marginal, and thus physical activity in the absence of caloric restriction was not a particularly useful strategy for the treatment of obesity (NIH National Heart Lung and Blood Institute 1998). Subsequently, a careful inspection of the exercise studies revealed that, for the most part, few of the early studies prescribed an exercise program that would be expected to lead to meaningful weight loss (Ross and Janssen 2001). On the other hand, in studies in which the prescribed exercise program did result in a meaningful negative energy balance, weight loss was substantial (Ross and Janssen 2001). In other words, weight loss is positively related to the volume of physical activity performed. This point is illustrated in figure 12.4, showing a dose–response

relationship between caloric expenditure and the time spent exercising with the corresponding reductions in body weight and total fat. However, daily exercise is not always associated with reductions in body weight or body fat. Some investigators report a resistance to weight loss in response to daily exercise performed for about 30 to 40 min for several months (Donnelly et al. 2003). Nevertheless, the majority of studies suggest that regular physical activity without restriction of caloric intake is associated with weight loss and a reduction in total fat in overweight men and women.

From figure 12.4, it is also clear that exercise performed for as little as 200 min per week is associated with weight loss. In fact, weight loss on the order of 0.5 kg (1 lb) per week is achieved in response to exercise performed for between 300 and 400 min per week or about 50 min per day. This observation is consistent with the position of the American College of Sports Medicine, which recommends that overweight and obese persons seeking weight loss should exercise between 200 and 300 min per week, the equivalent of about 8,374 J (2,000 kcal) per week (American College of Sports Medicine 2001).

Treatment of Abdominal Obesity

Whether an increase in physical activity is associated with a significant reduction in abdominal obesity is an important question. Minor reductions (~2 cm [~0.8 in.]) in waist circumference (a surrogate for abdominal fat) are observed in response to exercise-induced weight loss on the order of 2 to 3 kg (4.4-6.6 lb) (NIH National Heart Lung and Blood Institute

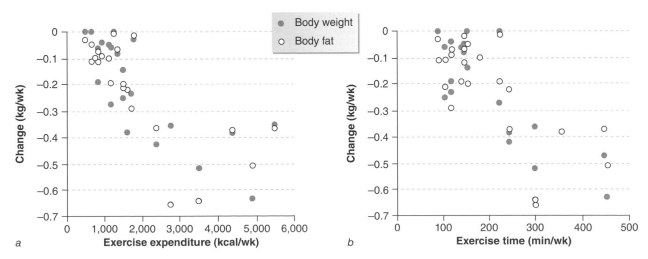

FIGURE 12.4 The dose–response relationship between (a) the weekly exercise caloric expenditure or (b) weekly minutes of exercise and the corresponding changes in body weight (black data points) and total fat (gray data points). Each data point represents the mean of one study.

Data from Ross and Janssen 2001.

1998). In other words, similar to body weight, waist circumference is reduced a small amount in response to a small amount of physical activity. Although it is unclear whether a dose–response relationship exists between the amount of physical activity and the reduction in waist circumference, it is evident that larger reductions in waist circumference are observed in response to a significant amount of daily exercise. Indeed, exercise performed for 300 to 400 min per week, or about 60 min per day, is associated with reductions of about 0.5 cm per week. In fact, exercise performed for about 60 min per day for three to four months is associated with reductions in waist circumference that approach 5 to 6 cm (2-2.4 in.) in both men and women, as illustrated in figure 12.5.

Whether exercise-induced weight loss is associated with corresponding reductions in abdominal subcutaneous and visceral fat has also been considered (Ross and Janssen 2001). It is generally observed that exercise is associated with a substantial reduction in abdominal subcutaneous and visceral fat independent of gender and age. An example of this effect is

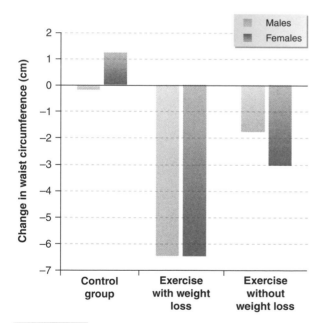

FIGURE 12.5 Changes in waist circumference in obese men and women after a 12- to 14-week program. The mean values from three treatment groups are shown. The control group did not exercise or change caloric intake. The "exercise with weight loss" group and "exercise without weight loss" group participated in daily exercise that consisted of about 60 min of vigorous walking. The "exercise with weight loss group" did not change caloric intake and lost ~7% of their body weight. The "exercise without weight loss" group increased caloric intake to compensate for the exercise-induced energy expenditure and had no change in body weight.

Data from Ross et al. 2000.

shown in figure 12.6, where a 10% reduction in body weight is associated with a reduction in abdominal subcutaneous and visceral fat that approximates 25% and 35%, respectively. Figure 12.6 also shows that the greater the exercise level expressed in minutes per week, the greater the reduction in both abdominal subcutaneous and visceral fat.

Because visceral fat is such an important predictor of health risk, practitioners have questioned whether this fat depot is selectively reduced in response to exercise-induced weight loss. The answer depends on how the reduction is presented. That is, for a given weight loss, a greater reduction in abdominal subcutaneous fat is observed if the reduction is expressed in absolute values (e.g., cm^2 at the L4-L5 image); the reason is that most adults have more abdominal subcutaneous fat than visceral fat. On the other hand, if the reduction in abdominal subcutaneous and visceral fat is expressed in relative terms (e.g., relative to the initial size of the depot), then the reduction in visceral fat is greater than the reduction in subcutaneous fat.

Exercise-Induced Reduction in Adiposity Without a Change in Body Weight

Emerging evidence suggests that regular exercise can reduce total and abdominal obesity in the absence of any change in body weight. This is supported by at least two lines of evidence. First, for any given level of BMI between 18 and 35 kg/m^2, adults who are physically active (e.g., have a higher level of cardiorespiratory fitness) have a lower waist circumference and lower levels of abdominal subcutaneous and visceral fat compared with their sedentary counterparts (lower level of cardiorespiratory fitness) (Janssen et al. 2004). Second, results from well-controlled, randomized trials reveal that obese men and women who participate in exercise programs for three to four months can experience significant reductions in both waist circumference (figure 12.5) and abdominal subcutaneous and visceral fat despite no change in BMI (Ross et al. 2000, 2004). These observations are important because they suggest that those who seek obesity reduction by increasing physical activity should be educated about the possibility that reductions in waist circumference, total fat, and abdominal fat can occur with or without a corresponding weight loss. On the other hand, it is equally important to note that the reduction in both total and abdominal fat depots is much greater in response to exercise with weight loss than to exercise without weight loss (Ross and Bradshaw 2009). These observations highlight

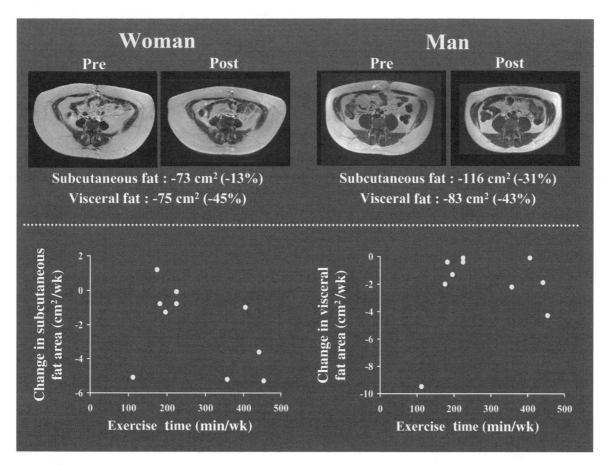

FIGURE 12.6 Exercise-induced reductions in abdominal subcutaneous and visceral fat. The images on the top are the pre- and posttraining images of a woman (left) and man (right) who lost substantial amounts of abdominal fat in response to an aerobic exercise program that consisted of about 60 min of daily moderate-intensity walking. The absolute reduction (cm²) in subcutaneous fat was greater than the absolute reduction in visceral fat, but the relative (%) reduction in subcutaneous fat was less than the relative reduction in visceral fat. The figures on the bottom illustrate the dose–response relationship between the weekly minutes of exercise and the corresponding reductions in abdominal subcutaneous fat (left) and visceral fat (right). Each data point represents the mean of one study.

Data from Ross and Janssen 2001.

the importance of monitoring obesity reduction using both BMI and waist circumference.

■

Exercise in the absence of weight loss is associated with significant reductions in total, abdominal, and visceral fat in obese men and women. These reductions are, however, smaller than those associated with exercise-induced weight loss.

Summary

The prevalence of obesity is already high and is increasing worldwide. This poses a major threat to public health; and innovative, multidisciplinary strategies are required to combat the problem. The information presented in this chapter provides strong support for the recommendation that physical activity should be an integral component in the strategies developed to both prevent and treat the obesity epidemic. Current guidelines suggest that adults should accumulate about 60 min of moderate-intensity physical activity daily to prevent unhealthy weight gain. Results from shorter obesity treatment studies in which dietary intake was carefully controlled suggest that 60 min of moderate-intensity exercise without a change in energy intake is associated with substantial reductions in total and abdominal obesity in obese men and women.

Although it is now clear that an increase in daily physical activity is required for most individuals, the challenge that remains is how to engage in and maintain a physically active lifestyle. Increasing physical activity to the levels recommended for obesity prevention and reduction will require a multidisciplinary approach that includes such components as educating allied health care providers about the benefits of physical activity, reestablishing daily physical education programs in our school systems, and working with urban planners to develop environments that encourage physical activity. Although the societal challenge to increase physical activity levels to an appropriate amount to combat the obesity epidemic is immense, the benefits are many, and thus the problem must be approached with vigor, step by step.

Key Concepts

abdominal subcutaneous fat—Layer of fat that lies directly underneath the skin in the abdominal region.

adipocyte—An adipose tissue or fat cell that stores lipids.

ad libitum—At one's pleasure; as one wishes.

body mass index (BMI)—A simple index of weight for height, calculated as weight in kilograms divided by the square of height in meters (kg/m²), that is commonly used to determine overweight and obesity status in research and clinical settings.

cytokines—For definition, see page 160.

ectopic fat—Fat that is stored outside of the adipose tissue depots.

lipolysis—Lipid breakdown reaction in adipocytes whereby the triglyceride molecule is hydrolyzed in the cell's cytosol into its components glycerol and three fatty acid molecules.

obesity—A condition of excessive fat accumulation to the extent that health may be impaired.

physical activity level (PAL)—Total daily caloric expenditure divided by total calories from resting metabolism. This term is being increasingly used as an overall indicator of energy expenditure.

visceral fat—Internal fat in the abdominal region that surrounds the organs of the gastrointestinal tract. Visceral fat consists of omental and mesenteric adipocytes and is contained within the visceral peritoneum.

waist circumference—A measurement of abdominal circumference commonly obtained at the top of the iliac crest. Waist circumference is used to characterize levels of abdominal obesity in research and clinical settings.

Study Questions

1. What is BMI, how is it calculated, and what cut points are used to define overweight and obesity in adult men and women?

2. List five major chronic diseases that are associated with obesity.

3. Provide two examples of ectopic fat deposition in obesity, and explain their relationship to obesity-related disease.

4. Describe how the average dietary intake, average leisure-time physical activity levels, and average total physical activity levels have changed in the past three decades and how these changes have contributed to the obesity epidemic.

5. What is PAL, how is it calculated, and what levels are currently recommended for the prevention of unhealthy weight gain?

6. What changes occur in abdominal subcutaneous and visceral fat in obese individuals in response to daily exercise performed for about 60 min at a moderate intensity?

7. Discuss the importance of waist circumference in monitoring success in obesity reduction programs.

8. What changes occur in waist circumference in response to exercise with or without weight loss?

References

American College of Sports Medicine. 2001. Appropriate intervention strategies for weight loss and prevention of weight regain for adults. *Medicine and Science in Sports and Exercise* 33(12):2145-2156.

Arner, P., E. Kriegholm, P. Engfeldt, and J. Bolinder. 1990. Adrenergic regulation of lipolysis in situ at rest and during exercise. *Journal of Clinical Investigation* 85(3):893-898.

Björntorp, P. 1990. "Portal" adipose tissue as a generator of risk factors for cardiovascular disease and diabetes. *Arteriosclerosis* 10(4):493-496.

Blundell, J.E., R.J. Stubbs, D.A. Hughes, S. Whybrow, and N.A. King. 2003. Cross talk between physical activity and appetite control: Does physical activity stimulate appetite? *Proceedings of the Nutrition Society* 62(3):651-661.

Borghi, E., M. de Onis, C. Garza, J. Van den Broeck, E.A. Frongillo, L. Grummer-Strawn, S. Van Buuren, H. Pan, L. Molinari, R. Martorell, A.W. Onyango, and J.C. Martines. 2006. Construction of the World Health Organization child growth standards: Selection of methods for attained growth curves. *Statistics in Medicine* 25(2):247-265.

Briefel, R.R., and C.L. Johnson. 2004. Secular trends in dietary intake in the United States. *Annual Review of Nutrition* 24:401-431.

Cole, T.J., M.C. Bellizzi, K.M. Flegal, and W.H. Dietz. 2000. Establishing a standard definition for child overweight and obesity worldwide: International survey. *British Medical Journal* 320(7244):1240-1243.

Després, J.P., and I. Lemieux. 2006. Abdominal obesity and metabolic syndrome. *Nature* 444:881-887.

Deurenberg-Yap, M., and P. Deurenberg. 2003. Is a re-evaluation of WHO body mass index cut-off values needed? The case of Asians in Singapore. *Nutrition Reviews* 61(5 Pt 2):S80-S87.

Donnelly, J.E., J.O. Hill, D.J. Jacobsen, J. Potteiger, D.K. Sullivan, S.L. Johnson, K. Heelan, M. Hise, P.V. Fennessey, B. Sonko, T. Sharp, J.M. Jakicic, S.N. Blair, Z.V. Tran, M. Mayo, C. Gibson, and R.A. Washburn. 2003. Effects of a 16-month randomized controlled exercise trial on body weight and composition in young, overweight men and women: The midwest exercise trial. *Archives of Internal Medicine* 163(11):1343-1350.

Dubé, J.J., F. Amati, M. Stefanovic-Racic, F.G. Toledo, S.E. Sauers, and B.H. Goodpaster. 2008. Exercise-induced alterations in intramyocellular lipids and insulin resistance: The athlete's paradox revisited. *American Journal of Physiology: Endocrinology and Metabolism* 204:E882-E888.

Goodpaster, B.H., J. He, S. Watkins, and D.E. Kelley. 2001. Skeletal muscle lipid content and insulin resistance: Evidence for a paradox in endurance-trained athletes. *Journal of Clinical Endocrinology and Metabolism* 86(12):5755-5761.

Hu, H.H., H.W. Kim, K.S. Nayak, and M.I. Goran. 2010. Comparison of fat-water MRI and single-voxel MRS in the assessment of hepatic and pancreatic fat fractions in humans. *Obesity* 18:841-847.

Janssen, I., S.B. Heymsfield, and R. Ross. 2002. Application of simple anthropometry in the assessment of health risk: Implications for the Canadian Physical Activity, Fitness and Lifestyle Appraisal. *Canadian Journal of Applied Physiology* 27(4):396-414.

Janssen, I., P.T. Katzmarzyk, R. Ross, A.S. Leon, J.S. Skinner, D.C. Rao, J.H. Wilmore, T. Rankinen, and C. Bouchard. 2004. Fitness alters the associations of BMI and waist circumference with total and abdominal fat. *Obesity Research* 12(3):525-537.

Kelley, D.E., B.H. Goodpaster, and L. Storlien. 2002. Muscle triglyceride and insulin resistance. *Annual Review of Nutrition* 22:325-346.

Kelly, T., W. Yang, C.S. Chen, K. Reyonlds, and J. He. 2008. Global burden of obesity in 2005 and projections to 2030. *International Journal of Obesity* 32:1431-1437.

Knuth, A.G., and P.C. Hallal. 2009. Temporal trends in physical activity: A systematic review. *Journal of Physical Activity and Health* 6(5):548-559.

Malnick, S.D., M. Beergabel, and H. Knobler. 2003. Non-alcoholic fatty liver: A common manifestation of a metabolic disorder. *QJM* 96(10):699-709.

Matsuzawa, Y. 2002. Importance of adipocytokines in obesity-related diseases. *International Journal of Obesity* 26(Suppl 1):S63.

Mayer, J. 1955. The physiological basis of obesity and leanness. I. *Nutrition Abstracts and Reviews. Series A Human and Experimental* 25(3):597-611.

National Institutes of Health National Heart Lung and Blood Institute. 1998. Clinical guidelines on the identification, evaluation, and treatment of overweight and obesity in adults: The evidence report. *Obesity Research* 6(S2):S51-210.

Perseghin, G., G. Lattuada, F. De Cobelli, F. Ragogna, G. Ntali, A. Esposito, E. Belloni, T. Canu, I. Terruzzi, P. Scifo, A. Del Maschio, and L. Luzi. 2007. Habitual physical activity is associated with the intra-hepatic fat content in humans. *Diabetes Care* 30:683-688.

Ross, R., and A. Bradshaw. 2009. The future of obesity reduction: Beyond weight loss. *Nature Reviews Endocrinology* 5(6):319-325.

Ross, R., D. Dagnone, P.J. Jones, H. Smith, A. Paddags, R. Hudson, and I. Janssen. 2000. Reduction in obesity and related comorbid conditions after diet-induced weight loss or exercise-induced weight loss in men. A randomized, controlled trial. *Annals of Internal Medicine* 133(2):92-103.

Ross, R., and I. Janssen. 2001. Physical activity, total and regional obesity: Dose-response considerations. *Medicine and Science in Sports and Exercise* 33(6 Suppl):S521-S527; discussion S528-529.

Ross, R., I. Janssen, J. Dawson, A.M. Kugl, J. Kuk, S. Wong, T.B. Nguyen-Duy, S.J. Lee, K. Kilpatrick, and R. Hudson. 2004. Exercise with or without weight loss is associated with reduction in abdominal and visceral obesity in women. A randomized controlled trial. *Obesity Research* 12(5):789-798.

Saris, W.H., S.N. Blair, M.A. van Baak, S.B. Eaton, P.S. Davies, L. Di Pietro, M. Fogelholm, A. Rissanen, D. Schoeller, B. Swinburn, A. Tremblay, K.R. Westerterp, and H. Wyatt. 2003. How much physical activity is enough to prevent unhealthy weight gain? Outcome of the IASO 1st Stock Conference and consensus statement. *Obesity Reviews* 4(2):101-114.

Shields, M., M.S. Tremblay, M. Laviolette, C.L. Craig, I. Janssen, and S. Connor Gorber. 2010. Fitness of Canadian adults: Findings from the 2007-2009 Canadian Health Measures Survey. *Health Reports* 21(1):21-35.

Wang, Y., and T. Lobstein. 2006. Worldwide trends in childhood overweight and obesity. *International Journal of Pediatric Obesity* 1(1):11-25.

Williamson, D.F., J. Madans, R.F. Anda, J.C. Kleinman, H.S. Kahn, and T. Byers. 1993. Recreational physical activity and ten-year weight change in a US national cohort. *International Journal of Obesity and Related Metabolic Disorders* 17(5):279-286.

World Health Organization. 1998. *Obesity: Preventing and managing the global epidemic.* Report of a WHO consultation on obesity. WHO/NUT/NCD/981. Geneva: WHO.

World Health Organization Expert Consultation. 2004. Appropriate body-mass index for Asian populations and its implications for policy intervention. *Lancet* 363:157-163.

13

Physical Activity, Fitness, and Diabetes Mellitus

Roeland J. Middelbeek, MD, MS; and Laurie J. Goodyear, PhD

CHAPTER OUTLINE

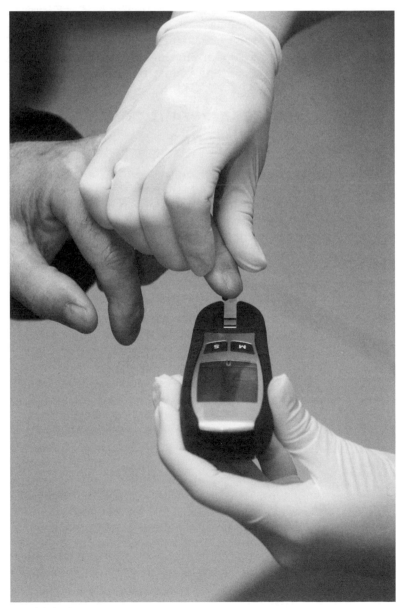

©Picsfive/fotolia

Diabetes mellitus is a chronic disease that encompasses a heterogeneous group of disorders; the predominant form is type 2 (or non-insulin-dependent) diabetes. Even though the symptoms of diabetes mellitus have been known for centuries, only in the past few decades has research begun to unravel the various causes of diabetes. Physical activity provides remarkable health benefits to those with type 2 diabetes. Moderate-intensity, regular physical activity has been demonstrated to prevent and delay the onset of diabetes in high-risk subjects. Changes in lifestyle that include introducing physical activity have a very positive impact on public health and can reduce the enormous economic burden of diabetes and its long-term complications. In this chapter, we discuss several aspects of this endocrine disease and its complications, focusing primarily on type 2 diabetes. We then discuss the basis of this disease at a molecular level and, finally, present the epidemiological body of evidence that supports the important role of physical activity in preventing and treating type 2 diabetes.

Diabetes: Definitions and Prevalence

Diabetes, the most common endocrine disorder, affects multiple organs and body functions, causing serious health complications such as renal failure, heart disease, nerve damage, stroke, and blindness. The body cannot control the level of circulating blood glucose because of either insufficient **insulin** production or inadequate response by organs to circulating levels of insulin, the major hormone controlling the body's **glucose homeostasis**. Some of the most characteristic symptoms associated with the onset of diabetes are frequent urination, excessive thirst, unexplained weight loss, and fatigue, although symptoms can be mild in type 2 diabetes. Diabetes is commonly diagnosed when the level of glucose in the blood is equal to or greater than 126 mg/dl or 7.0 mmol/L after overnight fasting or when the blood glucose level exceeds 200 mg/dl 2 h after a standardized **oral glucose tolerance test** (OGTT). Three major types of diabetes have been defined: type 1 diabetes, type 2 diabetes, and gestational diabetes.

In type 1 diabetes, which is an autoimmune disorder, the cells responsible for releasing insulin, the pancreatic β-cells, are mistakenly recognized by the immune system as "foreign" and selectively destroyed. As a result, pancreatic β-cells are virtually erased from the body, and circulating insulin levels in the blood dramatically decrease or even disappear. Regular insulin injections become necessary to sus-

tain life. This type of diabetes occurs more frequently in populations descended from northern European countries rather than southern European countries, the Middle East, or Asia. In the United States, approximately 3 people per 1,000 develop type 1 diabetes.

Type 2 diabetes, unlike type 1, is mainly a metabolic disorder in which insulin levels may be normal, elevated (hyperinsulinemia), or even decreased (hypoinsulinemia), although hyperinsulinemia characterizes most patients. At least in the early stages, most people with type 2 diabetes can produce insulin but cannot use it effectively. Two major components act in this pathology: pancreatic β-cell dysfunction, in which these cells gradually fail to adequately produce insulin in response to circulating levels of glucose in the blood, or **insulin resistance**, in which insulin-sensitive tissues such as skeletal muscle, fat, and liver become, to some extent, desensitized to insulin. Insulin resistance appears to precede, and at least in part to be responsible for, pancreatic β-cell dysfunction. The onset of type 2 diabetes typically occurs with advancing age, but as we discuss later in this chapter, it is increasing alarmingly in young individuals.

Type 2 diabetes is frequently found in obese people, which suggests a metabolic cause. The negative consequences of a sedentary lifestyle and obesity may be exacerbated by genetic components in individuals and in people of certain ethnic backgrounds. For example, type 2 diabetes is more common in people of Native American, Hispanic, Asian American, and African American descent than it is in other populations. It is the most common form of diabetes worldwide and in the United States, where approximately 90% of diabetes cases are classified as type 2. This disorder typically has a slow onset, usually developing over several years. Treatment usually combines diet with exercise, oral medication (e.g., metformin, sulfonylureas), and sometimes insulin injections. The major features of types 1 and 2 diabetes are highlighted in table 13.1.

Epidemiology, Etiology, and Complications of Type 2 Diabetes

The alarming increase in the incidence of type 2 diabetes worldwide makes it imperative that the health care community understand the dimensions of the increase, the factors that lead to individuals' developing the disease, and the health consequences that result if the condition is not diagnosed and controlled.

TABLE 13.1   Characteristics of the Predominant Types of Diabetes

	Type 1 diabetes (IDDM)	Type 2 diabetes (NIDDM)
Type of disorder	Autoimmune disorder	Metabolic disorder
Insulin level	Hypoinsulinemia	Hyperinsulinemia
Age of onset	Predominantly in youth	Predominantly after age 40
Genetic component	Weak	Strong
Proportion of diabetes patients	5-10%	90-95%
Insulin dependence	Permanent	Permanent only in subset of patients
Insulin resistance	Low	High
Onset	Acute and potentially severe	Mostly mild, insidious
Other	Often normal body weight	Frequently linked to obesity

IDDM = insulin-dependent diabetes mellitus; NIDDM = non-insulin-dependent diabetes mellitus.

The third major type, gestational diabetes, usually develops during the second or third trimester of pregnancy and resolves after the baby is delivered. Treatment of this condition focuses on controlling diet and measuring blood glucose levels, and it sometimes requires insulin injections. Women who have diabetes during pregnancy are at higher risk for developing type 2 diabetes later in their lives.

Epidemiology of Type 2 Diabetes

Type 2 diabetes ranks among the world's most common chronic diseases, with a remarkably high economic impact in developed countries. The term *prevalence* indicates the number of people with a particular condition, whereas *incidence* refers to the number of new cases per year. The prevalence of diabetes is increasing dramatically in the United States and worldwide, causing some health organizations and researchers to consider it an epidemic.

■ **The incidence of type 2 diabetes has been rising and has now reached epidemic levels. It accounts for 90% of the diabetic cases in the United States.**

Epidemiological surveys for diabetes are complicated to perform and prone to underestimating the real magnitude of the problem. Recent estimates by the International Diabetes Federation and the World Health Organization suggest that globally the number of persons with diabetes will have increased from 171 million in the year 2000 to 366 million by 2030. This rate of increase is predicted to occur in virtually every country throughout the world. However, the greatest increases for the next decade are expected to occur in developing countries, particularly in Asia. Recent years have already seen a sharp increase in the incidence of type 2 diabetes in these countries, including China and India.

The Centers for Disease Control and Prevention estimated that in 2007 a total of 24 million people in the United States were diagnosed with diabetes, which equaled 8% of the population. However, another 57 million people are estimated to have prediabetes, which carries an increased risk for diabetes and cardiovascular disease (CVD) (Centers for Disease Control and Prevention 2008). One of the most important recent population-based studies from the United States (Third National Health and Nutrition Examination Survey, NHANES III) showed a marked increase in the prevalence of diabetes. As some excellent reviews have discussed in depth (Boyle et al. 2001; Mokdad et al. 2000; Shaw, Sicree, and Zimmet 2010), epidemiological predictions drawn from NHANES III are consistent with the trend seen in virtually every developed Western country. It was predicted that the number of people diagnosed with type 2 diabetes in the United States would increase by 165% over 50 years, from 11 million in 2000 to 29 million by the year 2050. The highest increases are expected to happen among people aged 75 years and over (336%) and among African Americans (275%). According to the Centers for Disease Control, during the period of 2005-2008, there was a large increase in the prevalence of diabetes in almost all parts of the United States (see figure 13.1). The greatest morbidity and mortality rates from type 2 diabetes occur in the elderly and minority groups in the United States, a trend that epidemiological studies predict will not

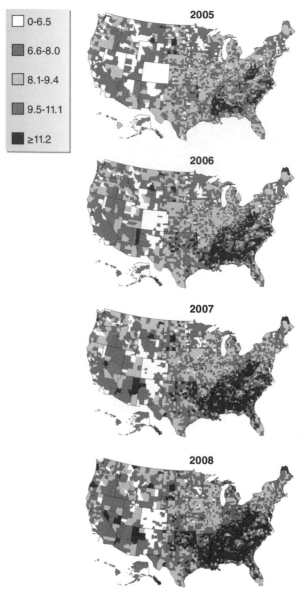

0-6.5	
6.6-8.0	
8.1-9.4	
9.5-11.1	
≥11.2	

2005

2006

2007

2008

FIGURE 13.1 Epidemiological data maps developed by the Centers for Disease Control and Prevention, showing trends for prevalance of type 2 diabetes throughout several years in the United States. For more information about U.S. diabetes data and trends, visit http://apps.nccd.cdc.gov/DDT_STRS2/NationalDiabetes PrevalenceEstimates.aspx.

Reprinted from the Centers for Disease Control and Prevention.

change in the near future. Another trend, and an enormous challenge for the future, is the emergence of an obesity-induced diabetes epidemic in children and adolescents.

Among the major reasons behind the expected dramatic increase in the incidence of type 2 diabetes are the anticipated world population growth, mostly in developing countries; the increase in longevity in most Western countries; and certain environmental factors. In particular, rapidly changing and unhealthy dietary patterns, along with increasingly sedentary lifestyles, lead to obesity, a major risk factor in the development of type 2 diabetes.

Fortunately, even small changes in lifestyle can be very significant for preventing and treating type 2 diabetes and for limiting the incidence of this metabolic disorder in the long term. In contrast, unhealthy diets rich in saturated fats, together with reduction in physical activity even at early ages, may result in a public health problem even greater than predicted to date by epidemiological studies.

Mechanisms Leading to Type 2 Diabetes

Type 2 diabetes is mainly caused by a combination of three different defects: (1) impaired insulin secretion by pancreatic β-cells, (2) insulin resistance in the peripheral tissues, and (3) increased glucose production by the liver.

It is widely accepted that insulin resistance plays a central role in the pathogenesis of type 2 diabetes. The term insulin resistance refers to the subnormal response to a given concentration of insulin by the major insulin-sensitive organs of the body, that is, skeletal muscle, liver, and adipose tissue. In other words, an adequate level of insulin is released at mealtimes, but its effect (glucose disposal) is impaired. After a period of compensated insulin resistance, **impaired glucose tolerance (IGT)** eventually develops despite elevated insulin concentrations. Finally, after maintenance of a hyperinsulinemic state over time, pancreatic β-cell failure results in decreased insulin secretion. This leads to the onset of overt clinical type 2 diabetes, when insulin resistance and impaired β-cell function occur simultaneously, resulting in fasting hyperglycemia.

Aside from the influence of some important genetic factors, insulin resistance is a common feature of individuals who are obese—mainly when the excess fat is concentrated in the abdominal region (i.e., central adiposity)—and particularly in individuals who are physically inactive. In fact, a primary mechanism causing insulin resistance is weight gain. It has been suggested that insulin resistance may represent a feedback mechanism to prevent further weight gain above a certain threshold.

Figure 13.2 highlights the so-called two-step model leading to type 2 diabetes. This model proposes that insulin resistance precedes and contributes to β-cell failure, which ultimately causes the symptoms of

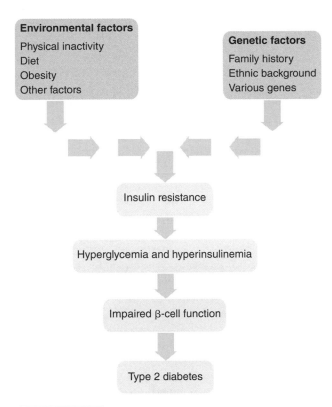

FIGURE 13.2 Multiple environmental and genetic factors lead to type 2 diabetes. These factors are not fully defined, and there are likely to be more elements involved in the development of this disease.

Data from the Centers for Disease Control and Prevention.

diabetes. The first change from normoglycemia to IGT is largely attributable to insulin resistance, whereas the second, from IGT to type 2 diabetes, arises from β-cell dysfunction and subsequent declining insulin secretion. In studies performed on a variety of ethnic groups who developed IGT from normal glucose tolerance, the insulin resistance phenomenon preceded the defect in insulin secretion.

Individuals who have been diagnosed with type 2 diabetes show a marked reduction in early insulin secretion (i.e., the phase of insulin secretion that occurs immediately after eating, also referred to as the "cephalic" phase), a consequence of malfunctioning of the pancreatic β-cells. These cells progressively lose their ability to "sense" and adequately respond to changes in the concentration of blood glucose. The detailed molecular and signaling mechanisms underlying this defect are not yet fully understood. However, it is known that progression in the decay of β-cell function takes a long time, so by the time type 2 diabetes is diagnosed, β-cell dysfunction may have been occurring for over a decade.

The rising hyperglycemic state characteristic of this metabolic disorder is exacerbated in the long term by the cytotoxic effects of high blood glucose concentrations. Animal studies suggest that chronic hyperglycemia itself is detrimental to insulin secretion and may also induce insulin resistance, and it has been shown that this glucotoxicity may eventually lead to a permanent loss of β-cell function. Thus, improving blood glucose control is essential to preventing further deterioration in β-cell function and consequent progression of diabetes.

◼ **Major causes of type 2 diabetes are obesity, sedentary lifestyle, and lack of physical activity, combined with genetic predisposition.**

Complications Associated With Type 2 Diabetes

People with type 2 diabetes are at risk for a variety of serious long-term complications (see "Type 2 Diabetes Complications"), attributable to the wide scope of insulin action on glucose, lipid, and protein metabolism. In some cases, there is also a risk for acute emergency complications such as diabetic hyperglycemic hyperosmolar coma, which is characterized by severe dehydration, decreased consciousness, and extreme hyperglycemia. Usually the kidneys compensate for high glucose levels in the blood by excreting excess glucose in the urine. However, under dehydration, the kidneys conserve fluid, and blood glucose levels increase greatly.

The long-term complications of type 2 diabetes are frequently broken down into microvascular and macrovascular complications.

- *Microvascular complications* are those that damage organs through their effects on small blood vessels. The most common microvascular complications are retinopathy, nephropathy, and neuropathy. Diabetic retinopathy is caused by damage to blood vessels of the retina and is the leading cause of blindness in the United States and in most developed countries. Diabetic nephropathy causes kidney damage and is the most common cause of chronic kidney failure in the United States. The earliest step of this pathology consists of thickening of the renal glomerulus, reducing the kidney's filtration capacity. Increasing numbers of glomeruli are destroyed with time. As a consequence, the kidney allows more albumin than normal in the urine, leading to a condition called microalbuminuria, which heralds the onset of diabetic nephropathy and may result in the need

Type 2 Diabetes Complications

Acute Complications
Hyperglycemic coma (unusual)

Chronic Complications
Diabetic retinopathy
Diabetic nephropathy
Diabetic neuropathy
Cardiovascular diseases

for dialysis or a kidney transplant. Finally, diabetic neuropathy is caused by peripheral nerve damage, as a result of high blood glucose levels, and can lead to decreased sensation and neuropathic pain over time.

- *Macrovascular complications* are those that affect large blood vessels and result in CVD, such as coronary heart disease and stroke, and peripheral vascular disease. Peripheral vascular disease causes arteriosclerosis of the extremities, characterized by narrowing of the arteries that supply the legs and feet. This decreases blood flow, which can injure nerves and other tissues in the extremities. Type 2 diabetes is a major risk factor for CVD, and statistics predict that about 80% of type 2 diabetic patients will die because of this complication. On average, type 2 diabetic patients will die 5 to 10 years earlier than their nondiabetic counterparts, mainly because of CVD. The treatment of CVD accounts for a large part of the huge health care costs attributed to type 2 diabetes. The fact that some of the complications associated with type 2 diabetes often begin to develop well before the disorder is diagnosed contributes to this economic burden.

Impact of Physical Activity on Insulin and Glucose Metabolism

In previous sections, we discussed the concept of diabetes, its associated conditions, and the worldwide problem of its prevalence. We now approach this disorder from a physiological and cellular perspective and focus on the molecular mechanisms in skeletal muscle underlying the beneficial aspects of physical activity. More in-depth information on this topic is available in previous reviews (Jessen and Goodyear 2005; Hawley and Lessard 2008).

Glucose Metabolism and Type 2 Diabetes

To maintain whole-body glucose homeostasis, coordination of three different metabolic events is required: adequate secretion of insulin by pancreatic β-cells, suppression of hepatic glucose production, and stimulation of glucose uptake by insulin-sensitive tissues, primarily muscle. During an acute bout of exercise, the increased need for metabolic fuel is met by increases in both carbohydrate and fat utilization in the skeletal muscle. Glucose is taken up from blood into the working skeletal muscles via glucose transporters. In people who do not have diabetes, unless the exercise is of extremely long duration, blood glucose concentrations do not decrease appreciably. This is the case because glucose output by the liver is precisely matched to glucose uptake in the muscle and because insulin secretion by the β-cells of the pancreas is reduced in response to lower blood glucose levels. In people with type 2 diabetes who have moderate hyperglycemia and who are insulin resistant, glucose concentrations can still be effectively decreased with moderate-intensity exercise, an important health benefit of exercise.

Skeletal muscle is the major organ in the body responsible for glucose disposal and consequently is of prime importance in metabolic disorders such as IGT and type 2 diabetes. Three potential rate-controlling steps and molecules for insulin-stimulated muscle glucose metabolism (i.e., synthesis of glycogen from glucose) have been identified: glucose transporter 4 (GLUT4), hexokinase, and glycogen synthase. Each of these steps is impaired in people with type 2 diabetes (see figure 13.3).

The major physiological stimulators of muscle glucose uptake are exercise and insulin. These stimuli enhance glucose transport into the muscle cells where it can be used for adenosine triphosphate (ATP) production or stored in the form of glycogen. Glucose transport in skeletal muscle occurs primarily by facilitated diffusion, using glucose transport carrier proteins. In mammalian tissues, glucose transporters constitute a family of structurally related proteins (isoforms) with tissue-specific expression patterns. There are 13 different glucose transporter isoforms; GLUT4 is the most abundant isoform present in skeletal muscle. Glucose transport is the rate-limiting step in muscle glucose utilization. In response to exercise or insulin, GLUT4 moves from an intracellular location to the plasma membranes and the transverse tubules (translocation). The amount of GLUT4 in the plasma membrane is tightly regulated by exercise and insulin, exerting a fine control on glucose uptake and metabolism within the muscle.

Muscle cell

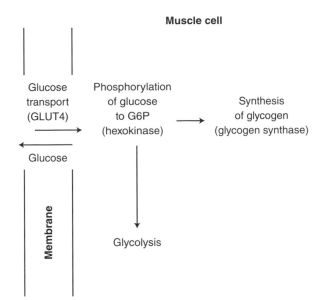

FIGURE 13.3 Glucose is transported in the skeletal muscle fibers via glucose transporter 4 (GLUT4). Glucose is phosphorylated to glucose-6-phosphate (G6P) via the enzyme hexokinase. Depending on the energy needs of the muscle, G6P can be stored as glycogen via the rate-limiting enzyme glycogen synthase or can be used as substrate for glycolysis.

Glycogen synthesis is the primary pathway for nonoxidative glucose disposal in nondiabetic persons. The rate of glycogen formation in people with diabetes is decreased, representing 40% of the rate typical of control subjects. It is believed that defects in muscle glycogen synthesis play a significant role in the insulin resistance that precedes the development of type 2 diabetes. In addition, impairment of GLUT4 translocation and hexokinase phosphorylation, causing defects in glucose transport, have been shown to be early factors in the pathogenesis of type 2 diabetes rather than a consequence.

Insulin Signaling in Skeletal Muscle

Understanding type 2 diabetes requires a closer view of the signaling mechanisms triggered by insulin in skeletal muscle (see figure 13.4). The cascade of intracellular signaling events stimulated by insulin involves multiple effector proteins that orchestrate diverse cellular responses. Insulin binds to the extracellular portion (α-subunit) of the transmembrane insulin receptor, and this event leads to activation of the transmembrane β-subunits and further autophosphorylation of the insulin receptor. The signal is next transduced through phosphorylation of a family of closely related proteins, which are referred to as insulin receptor substrates (IRS).

In addition to the IRS, the insulin receptor may phosphorylate and activate the Src homology 2 domain containing transforming protein (SHC), which ultimately results in the activation of mitogen-activated protein kinase. Insulin receptor substrate molecules contain multiple tyrosine phosphorylation sites that, after becoming phosphorylated by insulin stimulation, bind additional downstream signaling molecules. IRS1 appears to be the predominant isoform that mediates signal transduction in skeletal muscle, whereas IRS2 appears to be important in β-cell development. Both isoforms are important for insulin regulation of glucose metabolism in the liver. Once IRS1 is activated in muscle, the signal is transduced by activation of the phosphatidylinositol-3-kinase (PI3-K). PI3-K is a heterodimer protein consisting of a regulatory subunit (p85) associated with a catalytic subunit (p110). Interaction of phosphorylated IRS with the p85 subunit of PI3-K ultimately activates the enzyme. The p110 catalytic subunit of PI3-K uses phosphatidylinositol-4,5-bisphosphate as substrate, resulting in the phosphorylated lipid phosphatidylinositol-3,4,5-trisphosphate (PIP_3). PIP_3 is required to activate the membrane-associated enzyme 3-phosphoinositide-dependent protein kinase 1 (PDPK1). On activation, PDPK1 phosphorylates and activates the protein kinase Akt (also known as protein kinase B) and also the atypical protein kinase C isoforms, ζ and λ. Akt can phosphorylate TBC1 domain family, member 4 (TBC1D4) and its paralog TBC1D1, and these proteins likely play an important role in transducing the insulin signal to the translocation of GLUT4. Many additional signals downstream of TBC1D4/TBC1D1 are thought to be involved in movement of the GLUT4 vesicles to the plasma membranes and transverse tubules, although the identification and function of these are still under investigation.

Exercise Signaling in Skeletal Muscle

As mentioned earlier, exercise and insulin are the major mediators of glucose transport activity in muscle, and both stimuli cause GLUT4 translocation. In fact, insulin and exercise share many similar biological effects in skeletal muscle, as both stimuli can increase glucose transport, amino acid uptake, and glycogen synthesis. Because of these similarities, it was first believed that insulin and exercise used similar signaling cascades. However, it has been demonstrated that activity of one of the major insulin signaling molecules, PI3-K, is not increased immediately after exercise or muscle contraction. The lack of activation of PI3-K is consistent with findings

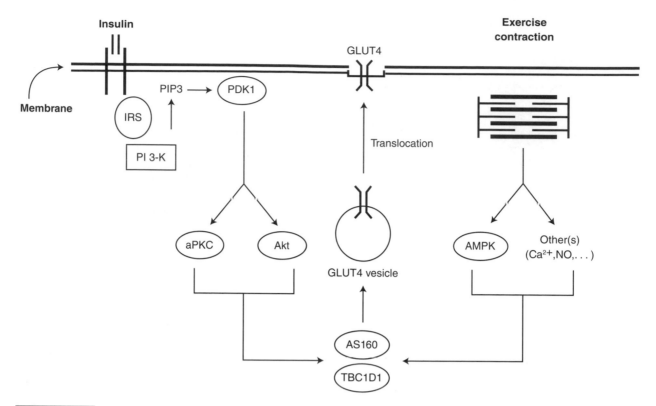

FIGURE 13.4 Insulin increases glucose transport in skeletal muscle through the translocation of the glucose transporter 4 (GLUT4) protein. Signaling to the GLUT4 vesicle involves insulin binding to its receptor, phosphorylation of insulin receptor substrate proteins (IRS), activation of phosphatidylinositol-3-kinase (PI3-K) generation of phosphatidylinositol-3,4,5-triphosphate (PIP$_3$), and activation of atypical protein kinase C (aPKC) and the protein kinase Akt via the 3-phosphoinositide-dependent protein kinase 1 (PDPK1) protein. Mechanisms downstream of aPKC and Akt are not fully understood. The signaling for exercise-stimulated glucose transport is poorly understood. Although there is some evidence that adenosine monophosphate (AMP)–activated protein kinase (AMPK) may be involved, activation of this molecule cannot fully explain how exercise increases GLUT4 translocation.

that insulin and contraction use differential signaling leading to glucose uptake and glycogen synthesis in skeletal muscle, and that contraction-stimulated glucose uptake occurs through a PI3-K-independent mechanism. However, whether insulin and exercise signaling converge farther downstream at a point distal to PI3-K is unknown. Recently, it has been suggested that TBC1D4 and its paralog TBC1D1 function as a convergence point for exercise and insulin signals, controlling the translocation of intracellular vesicles containing GLUT4.

■ **Insulin and physical activity both stimulate glucose uptake in the muscle but via distinct mechanisms.**

In recent years, there also has been a vast amount of research on the role of the 5'-adenosine monophosphate (AMP)–activated protein kinase (AMPK)

in skeletal muscle glucose transport. AMPK is a member of a metabolite-sensing protein kinase family that acts as a fuel gauge monitoring cellular energy levels. When AMPK "senses" decreased energy storage by a change in the AMP-to-ATP ratio within the cell, it switches off ATP-consuming metabolic pathways and switches on alternative pathways for ATP regeneration. AMPK activity increases in response to muscle contraction or exercise, an event that has been correlated with GLUT4 translocation and glucose transport in skeletal muscle. However, recent evidence suggests that although AMPK may be part of an insulin-independent, redundant signaling mechanism for exercise-mediated muscle glucose uptake, additional AMPK-independent signaling mechanisms may contribute to the regulation of glucose uptake in skeletal muscle in response to exercise. Further studies are necessary to unravel more detailed molecular pathways involved in exercise signaling in muscle.

Given that exercise increases glucose metabolism by insulin-independent signaling cascades, activation of this pathway provides an alternative strategy to increase glucose transport in insulin-resistant skeletal muscle of people with diabetes and prediabetes. This is beneficial in the management of type 2 diabetes, given that exercise-induced AMPK activity is not impaired in insulin-resistant skeletal muscle. Lessons learned from the AMPK signaling cascade and other insulin-independent mechanisms in skeletal muscle may become very valuable in developing drugs to treat patients with diabetes.

Insulin Resistance and Physical Activity

As defined in a previous section, insulin resistance is the inability of peripheral target tissues (especially muscle, adipose tissue, and liver) to respond properly to normal circulating concentrations of insulin. Insulin resistance can be found in patients more than a decade before diabetes appears and is the best predictor for later development of the disease. One treatment to improve insulin resistance is exercise. A single bout of exercise can increase skeletal muscle glucose transport and metabolism and can also have profound effects on glycogen metabolism. These metabolic changes occur in the skeletal muscle and can improve glucose homeostasis in persons with insulin resistance. In addition, these metabolic changes may be responsible for the ability of physical activity to prevent or delay the onset of type 2 diabetes.

Insulin sensitivity is related to the degree of physical activity. Exercise training improves glucose tolerance and insulin action in insulin-resistant people and patients with type 2 diabetes. This stems from adaptations in multiple tissues including the pancreas, liver, adipose tissue, and skeletal muscle. Exercise training leads to increased expression of GLUT4 in skeletal muscle, which has been correlated with improved insulin action on glucose metabolism. These findings are clinically relevant because insulin-stimulated tyrosine phosphorylation of IRS1 and activity of PI3-K are reduced in skeletal muscle from patients with type 2 diabetes.

Another beneficial effect of acute exercise is the ability to improve insulin secretion from the pancreatic β-cells. A recent study showed improved insulin secretion in patients with type 2 diabetes, as well as IGT subjects, after low-intensity exercise (Michishita et al. 2008). In adipose cells, training increases the capacity to store and mobilize free fatty acids. Furthermore, studies show increases in glucose uptake in different adipose tissue depots after exercise training.

The effects of chronic exercise training in the liver include decreased hepatic glucose production for a given workload of exercise, most likely attributable to increased ability of muscle to use fatty acids for a cellular fuel source. However, the maximal capacity of liver for hepatic glucose production is increased, a function of increased gluconeogenesis. This enhanced liver capacity allows for increased exercise duration with higher-intensity exercise. Less is known about the effects of exercise on hepatic insulin resistance and glucose production. A recent study, however, did show improved hepatic insulin sensitivity in adults after exercise with and without weight loss (Coker et al. 2009). The exact mechanisms underlying these improvements remain to be elucidated. The responses of both the liver and adipose tissue to acute bouts of exercise provide fuel to the contracting muscle. The liver provides glucose (derived from glycogenolysis and gluconeogenesis), and adipose tissue hydrolyzes triglycerides and releases non-esterified ("free" or unsaturated) fatty acids into the bloodstream. The increase in lipolysis and availability of fatty acids in the bloodstream are coordinated by neuronal, hormonal, and circulatory events in order to optimally deliver nutrients to muscle.

In summary, exercise training is an important therapeutic strategy to partially restore insulin sensitivity in people with type 2 diabetes. Indeed, recent epidemiological studies have determined that regular physical exercise can reduce one's risk of developing type 2 diabetes.

Epidemiological Evidence Indicating Benefits of Physical Activity in Preventing Type 2 Diabetes

It is well accepted that a physically active lifestyle plays an important role in preventing a variety of chronic diseases, including type 2 diabetes. In particular, exercise induces metabolic changes that significantly affect both high-risk individuals with IGT and patients with diabetes. As a consequence of the physiological benefits of regular exercise, active individuals show better insulin and glucose profiles. The epidemiological evidence in favor of exercise as a major lifestyle component in preventing or delaying the onset of type 2 diabetes is quite convincing, irrespective of ethnicity, gender, or age group (table 13.2).

In a **cohort study** (Helmrich et al. 1991) performed on approximately 6,000 male participants ranging from 45 to 55 years, physical activity during

TABLE 13.2   Epidemiological Data on the Effects of Exercise on Prevention of Type 2 Diabetes

Study population	Main findings	Reference
Cohort studies		
Males age 45-55	Physical activity level was inversely related to risk of developing type 2 diabetes especially in men at high risk for type 2 diabetes.	Helmrich et al. 1991
Females age 34-59	Eight-year follow-up study with vigorous exercise at least once per week showed a 16% lower risk of type 2 diabetes.	Manson et al. 1991
Males age 40-84	Five-year follow-up study with vigorous exercise at least once per week showed a 29% lower risk of type 2 diabetes.	Manson et al. 1992
Feasibility studies		
Men with IGT age 47-49	Five-year follow-up study showed that OGTT normalized in >50% of the subjects who exercised regularly.	Eriksson and Lindgarde 1991
Men and women with IGT age 25-74	Six-year follow-up study showed 8.3 cases of type 2 diabetes per 100 person-years in the exercise group versus 15.7 cases in the control group.	Pan, Li, and Hu 1995

IGT = impaired glucose tolerance; OGTT = oral glucose tolerance test.

leisure time was inversely related to the development of type 2 diabetes. In particular, for each 500 kcal/week increment in energy expenditure, the risk for developing type 2 diabetes decreased by 6%. Interestingly, this association was unaltered by other factors, such as obesity, hypertension, and a family history of diabetes. Another interesting conclusion derived from this study was that the beneficial effects of physical activity were strongest in participants at highest risk for type 2 diabetes, that is, those individuals with high **body mass index (BMI)**, hypertension, or a family history of diabetes.

Another epidemiological study (Manson et al. 1991) focused on women and examined the association between regular vigorous exercise and the incidence of type 2 diabetes. Participants were approximately 87,000 nondiabetic women aged 34 to 59 years. The follow-up duration of this study was eight years. Women who engaged in vigorous exercise at least once per week showed a 16% lower relative risk of developing type 2 diabetes. As with the study by Helmrich and colleagues (1991), adjustments for age, BMI, family history of diabetes, and other variables did not alter the beneficial effect of exercise.

In another prospective cohort study, Manson and colleagues (1992) evaluated approximately 21,000 male participants aged 40 to 84 years who were initially free of diagnosed type 2 diabetes mellitus. After five years of follow-up, the outcome showed an inverse correlation between the incidence of type 2 diabetes and the frequency of vigorous exercise. Men who engaged in vigorous physical activity at least

once per week had a 29% lower risk of developing diabetes compared with those who performed no vigorous exercise. In agreement with previous studies (Helmrich et al. 1991), the inverse relationship between exercise and risk of this metabolic disorder was particularly pronounced among overweight men. However, there was still a significant reduction after data adjustment for BMI as well as for age.

In nonrandomized feasibility studies performed in Sweden and China, physical activity was instituted as part of an intervention to prevent the development of diabetes among persons with IGT. The Swedish feasibility study (Eriksson and Lindgarde 1991) included 41 subjects aged 47 to 49 at the early stage of type 2 diabetes and 181 additional participants with IGT. The aim was to test the feasibility of long-term intervention, with emphasis on changes in lifestyle. The intervention program consisted of dietary treatment or an increase in physical activity. After a five- to six-year follow-up period, OGTTs showed normalization in more than half of the participants with IGT, and more than 50% of the participants with diabetes went into remission. As expected, the improvement in glucose tolerance was correlated with body weight reduction and increased physical activity.

Pan, Li, and Hu (1995) focused on individuals with IGT. Participants aged 25 to 74 were matched according to their BMI and assigned to four different intervention groups: control, diet, exercise, and diet plus exercise. After a six-year follow-up period, the incidence of diabetes was 8.3 cases per 100 person-years versus 15.7 cases in the control group. In agreement with the previously cited studies, the exercise intervention and the incidence of type 2 diabetes were inversely related.

All of these epidemiological approaches unambiguously indicate that physical activity is a powerful means to prevent or delay type 2 diabetes among high-risk people. As mentioned before, physically active people have better profiles of blood insulin and glucose concentrations than their sedentary counterparts, partly attributable to an exercise-induced increase in insulin sensitivity in peripheral tissues. In addition, some of the beneficial effects of physical activity on insulin sensitivity may be an indirect consequence of weight loss or changes in body composition (decreased adiposity).

Whereas insulin functions as the ultimate diabetes therapy, exercise has been proven to prevent and treat type 2 diabetes in the general population.

Summary of Randomized Controlled Trials on the Prevention of Type 2 Diabetes

Randomized controlled trials (RCTs) are used to evaluate the effectiveness of particular interventions on health indicators. Randomized controlled trials are extremely useful tools to evaluate preventive and public health measures, pharmacological treatments, physical and psychological therapies, and more. These trials need to be large enough and of sufficient duration to allow for a proper evaluation of the intervention programs tested. The main advantage of a random allocation is that intervention groups are comparable in terms of all factors that might influence the outcome. Therefore, any differences in outcome can be attributed to a particular intervention program.

In 1997, the Oslo Diet and Exercise Study in Norway (Torjesen et al. 1997) addressed the effect of diet and exercise intervention on insulin resistance for one year. Participants were randomly allocated to the following groups: control, diet only, exercise only, and diet plus exercise. The diet intervention consisted of reduced total fat intake; the exercise protocol entailed endurance training three times a week. The exercise intervention program did not notably change insulin resistance, unlike the diet-only and diet plus exercise programs. This outcome could have been attributable to the nature of the exercise program,

the short duration of the study, or compliance rates of participants with the various treatment arms.

The Da Qing IGT and Diabetes Study in China (Pan et al. 1997) drew more convincing conclusions. This study focused on individuals with IGT, a major risk factor for developing type 2 diabetes. A large number of participants (577) were distributed among the same intervention groups applied in the Oslo study protocol: control, diet only, exercise only, and diet plus exercise. The follow-up evaluation was conducted at two-year intervals over six years, and the risk for developing type 2 diabetes was assessed. After six years, the incidence of diabetes was 67.7% in the control group, 43.8% in the diet group, 41.1% in the exercise group, and 46.0% in the diet plus exercise group. This study demonstrated that lifestyle interventions significantly reduced the risk for developing type 2 diabetes in a high-risk population group.

The Finnish Diabetes Prevention Study (Tuomilehto et al. 2001) focused on middle-aged, overweight participants with IGT. This study analyzed two different groups, a control and an intervention group. Intervention was aimed at reducing body weight and total fat intake while increasing intake of fiber and physical activity. The follow-up duration of the interventional study was 3.2 years. The cumulative incidence of diabetes was 11% in the intervention group and 23% in the control group. In other words, the risk of developing type 2 diabetes was reduced by 58% in the intervention group. This study and the Da Qing study both demonstrate that changes in lifestyle are critical to prevent type 2 diabetes in high-risk persons.

In a more recent study, the Diabetes Prevention Program (Knowler et al. 2002) compared the beneficial effects of diet, physical activity, and weight loss with metformin treatment. Metformin is an antihyperglycemic drug widely used in the management of type 2 diabetes. To date, metformin has been the most efficient single pharmacological treatment for the disorder, acting by increasing the sensitivity of peripheral tissues to insulin. Participants in this study had IGT and thus were a high-risk population group for developing type 2 diabetes. Individuals were randomly assigned to three different groups: placebo (control), metformin administration, and diet plus exercise. The diet plus exercise intervention consisted of reducing body weight by at least 7% and incorporating a moderate-intensity exercise protocol (minimum of 150 min per week). The follow-up evaluation was conducted over an average period of 2.8 years. The physical activity plus diet program was the most effective intervention to prevent the progression of IGT to type 2 diabetes. The combined exercise, diet, and weight loss intervention reduced the incidence of diabetes by 58%, whereas metformin reduced the risk by 31%.

Recently published results for the 10-year follow-up of the Diabetes Prevention Program showed that subjects who were previously treated with metformin or lifestyle intervention maintained a reduced risk of developing diabetes. Compared with the placebo group, the former lifestyle group showed a reduction on average of 34%, while the former metformin-treated group showed a reduction of 18%. The initial weight loss of the lifestyle intervention group was not maintained over the course of the follow-up (Diabetes Prevention Program Research Group et al. 2009). A similar observation was made in the follow-up of the Finnish Diabetes Prevention Study (Lindström et al. 2006). In this study, individuals with a high risk of diabetes showed improvements in glucose tolerance and a reduced incidence of diabetes, even after the lifestyle intervention had ended.

From most of the major RCTs aimed at preventing type 2 diabetes, the conclusion can be drawn that lifestyle changes play a major role. In particular, moderate physical activity, alone or in combination with diet, seems to be the most effective intervention to reduce the incidence of diabetes in persons at high risk. Whether this benefit is attained solely through the associated weight loss has not been fully clarified.

Importance of Regular Physical Activity for People With Type 2 Diabetes

The therapeutic benefits of regular exercise in the treatment of type 2 diabetes have long been recognized. As early as 1919, it was reported that exercise lowered blood glucose concentrations and improved glucose tolerance in diabetic patients. In the 1935 edition of *The Treatment of Diabetes Mellitus* by Joslin and colleagues, exercise was recommended in the "everyday treatment of diabetes." From the epidemiological studies discussed in the previous sections, it is clear that regular moderate-intensity exercise can be an important part of a regimen to prevent and treat type 2 diabetes. Regular physical activity potentiates the effects of diet and oral antihyperglycemic therapy (e.g., metformin and sulfonylureas) to lower glucose levels and improve insulin sensitivity in obese people with type 2 diabetes.

Many of the health benefits that regular physical activity provides in the prevention of chronic metabolic diseases such as type 2 diabetes may be attributable to the overlapping actions of individual exercise sessions and long-term adaptations to exer-

©Phototom

Prescribed use of insulin monitors helps people with type 2 diabetes see how regular physical activity can lower glucose levels.

cise training. As mentioned earlier, acute exercise produces major effects on whole-body glucose disposal and skeletal muscle glucose uptake and metabolism. However, the elevated insulin-stimulated glucose disposal rates, responsible for the improved insulin sensitivity, tend to disappear after about five to seven days of inactivity. Hence, the effects of exercise training in increasing insulin action are transient and require a regular and constant practice of physical activity.

In addition to improving glucose tolerance and insulin resistance, exercise training has other beneficial effects in people with diabetes, such as improved cardiovascular fitness, lowered blood pressure, improved blood lipid profiles, weight loss and in particular reduced abdominal and intra-abdominal fat (a major risk factor for insulin resistance), improved mental health, and promotion of a sense of well-being. Accordingly, regular physical activity improves morbidity and mortality in people with type 2 diabetes. Multiple factors may modulate the response to exercise training in subjects with diabetes, such as the degree of insulin resistance and insulin deficiency, the frequency and intensity of exercise, adherence to diet, and weight loss. Insulin sensitivity and the rate of glucose disposal are related to cardiorespiratory fitness even in older persons. The additional potential beneficial effects of exercise training to lower cardiovascular risk in people with type 2 diabetes may reduce the risk of macrovascular or atherosclerotic complications typical of diabetes. Physical activity also improves endothelial function,

Strengths and Limitations of the Evidence

Strengths:
Overwhelming data demonstrate the rising worldwide incidence of type 2 diabetes and its associations with obesity and lack of physical activity. Large clinical trials have proven that exercise has beneficial effects on the incidence of diabetes, on control of the disease, and on reduction of its complications.

Limitations:
Although the mechanisms of insulin action on muscle glucose uptake seem to be well understood, much less is known about the mechanisms underlying exercise-stimulated glucose uptake in muscle. Elucidating these pathways will indicate molecular targets for novel pharmacological therapies to prevent diabetes and to control its complications. Finally, turning evident clinical and scientific evidence into behavioral changes in the general population remains difficult.

but the long-term effects on microvascular disease are unclear.

The recommendations provided by the Centers for Disease Control and Prevention and the American College of Sports Medicine (Haskell et al. 2007) encourage sedentary people to increase their level of physical activity in a moderate and feasible manner. The physical activity program should be flexible and should fit the life demands of individuals regardless of income, race, or other socioeconomic factors. For instance, walking for about 30 min on most days—averaging approximately 150 min of moderate physical activity per week—seems to be sufficient in many individuals. Ideally, these changes in lifestyle should be maintained over the years to prevent the risk of developing type 2 diabetes or to reduce its associated complications.

Summary

The prevalence of type 2 diabetes has increased dramatically in recent years and has now reached epidemic levels. The increase in this disease correlates with increased rates in obesity and sedentary behavior and decreases in physical activity. There is now strong epidemiological evidence that regular physical exercise can prevent or delay the onset of type 2 diabetes. The mechanism for this beneficial effect is thought to be attributable to the effects of repeated bouts of exercise to improve overall glucose homeostasis. As examples, exercise increases glucose uptake and subsequently reduces insulin secretion from the pancreas. Future research should focus on understanding how these important metabolic changes occur at the molecular level, which may provide novel tools for developing antidiabetic drugs.

Key Concepts

body mass index (BMI)—For definition, see page 212.

cohort study—A type of epidemiologic study design in which participants are grouped on the basis of their self-selected exposure of interest. They are then followed over time for the development of the disease of interest. For example, three groups of participants—inactive, moderately active, and highly active—may be followed for the development of colon cancer. The rates of colon cancer in the three groups are then compared to assess whether higher levels of physical activity are associated with lower rates of colon cancer.

diabetes mellitus—A chronic disorder characterized by a deficiency of insulin secretion or insulin action, which impairs the body's ability to regulate the levels of blood glucose. It is diagnosed when the level of glucose in the blood is greater than 7.0 mmol/L (fasting) or greater than 11.1 mmol/L (random).

glucose homeostasis—Every process involved in maintaining an internal equilibrium of glucose within the organism.

impaired glucose tolerance (IGT)—Condition in which blood glucose during the oral glucose tolerance test is higher than normal but not high enough for a diagnosis of diabetes. This prediabetic state is associated with insulin resistance. People with IGT are at a greater risk of developing type 2 diabetes.

insulin—A polypeptide hormone secreted by the β-cells of the pancreatic islets. Insulin is one of the most important hormones in maintaining glucose homeostasis and also regulates the metabolism of fats and proteins.

insulin resistance—Inability of peripheral target tissues, that is, muscle, fat, and liver, to respond properly to normal insulin concentrations present in blood.

oral glucose tolerance test—Test to measure the body's ability to use or metabolize glucose, administered to diagnose diabetes. The individual drinks a 75 g glucose solution, and blood glucose concentrations are measured for up to 2 h after ingestion.

randomized controlled trial—Trial in which patients are randomly assigned to two groups: one treatment group and one control group. Assigning patients at random reduces the risk of bias and increases the probability that differences between the groups can be attributed to the treatment.

Study Questions

1. What is the importance of insulin in regulating normal physiology in the body?

2. Explain the major differences between type 1 and type 2 diabetes.

3. Name the major reasons for the worldwide increase in the incidence of type 2 diabetes.

4. Why is the economic burden associated with type 2 diabetes so high? Name at least four health complications associated with this metabolic disorder.

5. Name some of the biological effects that insulin and exercise share in skeletal muscle.

6. Discuss why unraveling the cell signaling events occurring with exercise may be of great relevance in the treatment of diabetes.

7. Identify the metabolic changes occurring in response to physical activity that may prevent the onset of type 2 diabetes.

8. Drawing on epidemiological evidence, discuss whether ethnic group, gender, or age is a factor in determining the beneficial effects of physical activity on the prevention of type 2 diabetes.

9. Explain the benefits of using randomized controlled trials to draw epidemiological conclusions.

10. Summarize how regular physical activity can benefit people at high risk of developing type 2 diabetes.

References

Boyle, J.P., A.A. Honeycutt, K.M. Narayan, T.J. Hoerger, L.S. Geiss, H. Chen, and T.J. Thompson. 2001. Projection of diabetes burden through 2050: Impact of changing demography and disease prevalence in the U.S. *Diabetes Care* 24:1936-1940.

Centers for Disease Control and Prevention. 2008. National diabetes fact sheet: General information and national estimates on diabetes in the United States, 2007. Atlanta: U.S. Department of Health and Human Services, Centers for Disease Control and Prevention.

Coker, R.H., R.H. Williams, S.E. Yeo, P.M. Kortebein, D.L. Bodenner, P.A. Kern, and W.J. Evans. 2009. The impact of exercise training compared to caloric restriction on hepatic and peripheral insulin resistance in obesity. *Journal of Clinical Endocrinology and Metabolism* 94:4258-4266.

Diabetes Prevention Program Research Group, W.C. Knowler, S.W. Fowler, R.F. Hamman, C.A. Christophi, H.J. Hoffman, A.T. Brenneman, J.O. Brown-Friday, R. Goldberg, E. Venditti, and D.M. Nathan. 2009. 10-year follow-up of diabetes incidence and weight loss in the Diabetes Prevention Program Outcomes Study. *Lancet* 374:1677-1686.

Eriksson, K.F., and F. Lindgarde. 1991. Prevention of type 2 (non-insulin-dependent) diabetes mellitus by diet and physical exercise. *Diabetologia* 34:891-898.

Haskell, W.L., I.M. Lee, R.R. Pate, K.E. Powell, S.N. Blair, B.A. Franklin, C.A. Macera, G.W. Heath, P.D. Thompson, A. Bauman, American College of Sports Medicine, American Heart Association. 2007. Physical activity and public health: Updated recommendation for adults from the American College of Sports Medicine and the American Heart Association. *Circulation* 116:1081-1093.

Hawley, J.A., and S.J. Lessard. 2008. Exercise training-induced improvements in insulin action. *Acta Physiologica (Oxford)* 192:127-135.

Helmrich, S.P., D.R. Ragland, R.W. Leung, and R.S. Paffenbarger. 1991. Physical activity and reduced occurrence of non-insulin-dependant diabetes mellitus. *New England Journal of Medicine* 325:147-152.

Jessen, N., and L.J. Goodyear. 2005. Contraction signaling to glucose transport in skeletal muscle. *Journal of Applied Physiology* 99:330-337.

Joslin, E.P., H.F. Root, P. White, and A. Marble. 1935. *The treatment of diabetes mellitus.* 5th ed. Philadelphia: Lea & Febiger.

Knowler, W.C., E. Barrett-Connor, S.E. Fowler, R.F. Hamman, J.M. Lachin, E.A. Walker, D.M. Nathan; Diabetes Prevention Program Research Group. 2002. Reduction in the incidence of Type 2 diabetes with lifestyle intervention or metformin. *New England Journal of Medicine* 346:393-403.

Lindström, J., P. Ilanne-Parikka, M. Peltonen, S. Aunola, J.G. Eriksson, K. Hemiö, H. Hämäläinen, P. Härkönen, S. Keinänen-Kiukaanniemi, M. Laakso, A. Louheranta, M. Mannelin, M. Paturi, J. Sundvall, T.T. Valle, M. Uusitupa, J. Tuomilehto; Finnish Diabetes Prevention Study Group. 2006. Sustained reduction in the incidence of type 2 diabetes by lifestyle intervention: Follow-up of the Finnish Diabetes Prevention Study. *Lancet* 368:1673-1679.

Manson, J.E., D.M. Nathan, A.S. Krolewski, M.J. Stampfer, W.C. Willett, and C.H. Hennekens. 1992. A prospective study of exercise and incidence of diabetes among US male physicians. *Journal of the American Medical Association* 268:63-67.

Manson, J.E., E.B. Rimm, M.J. Stampfer, G.A. Colditz, W.C. Willett, A.S. Krolewski, B. Rosner, C.H. Hennekens, and F.E. Speizer. 1991. Physical activity and incidence of non-insulin dependent diabetes mellitus in women. *Lancet* 338:774-778.

Michishita, R., N. Shono, T. Kasahara, and T. Tsuruta. 2008. Effects of low intensity exercise therapy on early phase insulin secretion in overweight subjects with impaired glucose tolerance and type 2 diabetes mellitus. *Diabetes Research and Clinical Practice* 82:291-297.

Mokdad, A.H., E.S. Ford, B.A. Bowman, D.E. Nelson, M.M. Engelgau, F. Vinicor, and J.S. Marks. 2000. Diabetes trends in the U.S.: 1990-1998. *Diabetes Care* 23:1278-1283.

Pan, X., G. Li, and Y. Hu. 1995. Effect of dietary and/or exercise intervention on incidence of diabetes in 530 subjects with impaired glucose tolerance from 1986-1992. *Chinese Journal of Internal Medicine* 34:108-112.

Pan, X.R., G.W. Li, Y.H. Hu, J.X. Wang, W.Y. Yang, Z.X. An, Z.X. Hu, J. Lin, J.Z. Xiao, H.B. Cao, P.A. Liu, X.G. Jiang, Y.Y. Jiang, J.P. Wang, H. Zheng, H. Zhang, P.H. Bennett, and B.V. Howard. 1997. Effects of diet and exercise in preventing NIDDM in people with impaired glucose tolerance. The Da Qing IGT and Diabetes Study. *Diabetes Care* 20:537-544.

Shaw, J.E., R.A. Sicree, and P.Z. Zimmet. 2010. Global estimates of the prevalence of diabetes for 2010 and 2030. *Diabetes Research and Clinical Practice* 87:4-14.

Torjesen, P.A., K.I. Birkeland, S.A. Anderssen, I. Hjermann, I. Holme, and P. Urdal. 1997. Lifestyle changes may reverse development of the insulin resistance syndrome. The Oslo Diet and Exercise Study: A randomized trial. *Diabetes Care* 20:26-31.

Tuomilehto, J., J. Lindstrom, J.G. Eriksson, T.T. Valle, H. Hamalainen, P. Ilanne-Parikka, S. Keinanen-Kiukaanniemi, M. Laakso, A. Louheranta, M. Rastas, V. Salminen, and M. Uusitupa. 2001. Prevention of type 2 diabetes mellitus by changes in lifestyle among subjects with impaired glucose tolerance. *New England Journal of Medicine* 344:1343-1350.

14

Physical Activity, Fitness, and Cancer

I-Min Lee, MBBS, MPH, ScD

©Duncan Noakes

Today we have clear evidence that physical activity decreases the risk of developing many chronic diseases. As we look back on the history of epidemiologic studies investigating the health benefits of physical activity, we find that many of the early studies focused on cardiovascular disease (or risk factors for this disease) as the health outcome of interest. This was appropriate because cardiovascular disease is the leading cause of death globally. It was not until the mid-1980s that researchers also began to place emphasis on the question of whether physical activity plays any role in the prevention of cancer, the second leading cause of death around the world.

Over the subsequent two decades, many studies addressed the relationship between physical activity or physical fitness and cancer development. The large body of evidence that accumulated led two cancer agencies to make recommendations specifically targeting physical activity as a cancer preventive measure for the first time in 2002. The American Cancer Society (ACS) periodically publishes guidelines on healthy nutrition to prevent cancer. In 2002, the ACS guidelines included physical activity for the first time, recommending that regular physical activity be undertaken to decrease the risk of developing colon and breast cancers (Byers et al. 2002). That same year, in one of its handbooks on cancer prevention, the International Agency for Research on Cancer (IARC) of the World Health Organization also recognized the importance of physical activity in reducing the risk of colon and breast cancers and possibly other cancers (IARC 2002). More recently, the importance of physical activity for preventing certain cancers was confirmed by an international panel of experts convened by the World Cancer Research Fund and the American Institute for Cancer Research (World Cancer Research Fund/American Institute for Cancer Research 2007) and an expert panel convened by the U.S. government to comprehensively examine the scientific base for making physical activity recommendations (Physical Activity Guidelines Advisory Committee [PAGAC] 2008). In this chapter, we review the evidence on physical activity and cancer prevention.

Importance of Cancer

In the United States, cancer is a leading cause of mortality and morbidity. In fact, cancer ranks as the second leading cause of death among both U.S. males and females, trailing only deaths from heart disease (figure 14.1) (Jemal et al. 2010). In 2007, 2.4 million persons died in the United States; of these deaths, 25% were attributed to heart disease,

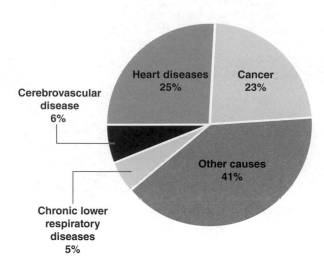

FIGURE 14.1 Leading causes of death in the United States, 2006.

Data from Jemal et al. 2009.

and 23%—more than half a million deaths—were attributed to cancer. Looked at in a different way, these data indicate that approximately one of every four deaths in 2007 was caused by cancer. The most common fatal cancers occurring among U.S. men in 2010 were estimated to be, in order of frequency, cancers of the lung, prostate, colorectum, pancreas, and liver (table 14.1). Among U.S. women, they were expected to be cancers of the lung, breast, colorectum, pancreas, and ovary. Heart disease and cancer are important chronic diseases not only in the United States but also globally; it was estimated that of the 58 million deaths worldwide in 2005, 30% were attributable to cardiovascular diseases and 13% to cancer (Strong et al. 2005). (As a comparison, 30% of worldwide deaths in 2005 were estimated to be attributable to communicable diseases, maternal and perinatal conditions, and nutritional deficiencies.)

In addition to deaths occurring from cancer, new diagnoses of cancer are made each year in a large number of males and females. The ACS estimated that 1.5 million new cases of cancer would be diagnosed in the United States in 2010 (not counting the common basal and squamous cell cancers of the skin) (Jemal et al. 2010). The most common sites of new cancers occurring in men in 2010 were expected to be, in order of frequency, the prostate, lung, colorectum, bladder, and skin (melanoma) (table 14.1). Among women, they were expected to be breast, lung, colorectum, uterus, and thyroid. You will note that the lists of the most common newly diagnosed cancers are similar, but not identical, to the lists of the most common fatal cancers. The reason is that some cancers have a better prognosis than others, partly because of the ability to diagnose them at an

TABLE 14.1 Most Common Cancers in U.S. Men and Women

Most common fatal cancers		Most common newly occurring cancers	
Men	**Women**	**Men**	**Women**
Lung (29%)	Lung (26%)	Prostate (28%)	Breast (28%)
Prostate (11%)	Breast (15%)	Lung (15%)	Lung (14%)
Colon/rectum (9%)	Colon/rectum (9%)	Colon/rectum (9%)	Colon/rectum (10%)
Pancreas (6%)	Pancreas (7%)	Bladder (7%)	Uterus (6%)
Liver (4%)	Ovary (5%)	Melanoma (5%)	Thyroid (5%)

Cancer—A Major Cause of Death

Globally, as well as in the United States, cancer is the second leading cause of death, after cardiovascular disease.

How Physical Activity and Physical Fitness Decrease the Risk of Developing Cancer

Although the exact mechanisms underpinning lower cancer rates among active compared with less active persons are unknown, several broad categories of plausible mechanisms have been proposed to explain why physically active men and women may be at lower risk of developing cancer (Sternfeld and Lee 2009).

First, sex hormones may play a role. These hormones have powerful mitogenic and proliferative effects and are important in the development of reproductive cancers. Investigators have proposed that physical activity reduces the risk of developing breast cancer through its effects on menstrual function and female sex hormone levels. Girls who participate in physical activity and sport tend to be older at menarche and are more likely to have cycles that are anovulatory. These effects can decrease breast cancer risk because later age at menarche is associated with lower breast cancer rates, whereas anovulatory cycles are associated with lower levels of estrogen. Physical activity also has been associated with changes in female sex steroid hormones in adult women. Among

earlier (and hence more treatable) stage and also partly because some cancers are more amenable to treatment than others. For example, both prostate and breast cancers are more easily diagnosed at an earlier stage than lung cancer. Also, these cancers are more successfully treated than lung cancer.

Because cancer poses a major public health burden, in the United States as well as globally, there has been a great deal of interest in searching not only for a cure for cancer but also for its causes. Cancer is multifactorial in its origin, with contributions coming from both genetic and environmental components. From a public health perspective, modifiable determinants of cancer are important because such factors are amenable to change. One determinant, or risk factor, that has received a great deal of attention recently and is the focus of this chapter, is physical inactivity.

Plausible Mechanisms for Reduced Cancer Risk With Physical Activity

- Modulation of reproductive hormone levels
- Decrease in body weight and adiposity
- Change in levels of insulin-like growth factors and their binding proteins
- Decrease in inflammation
- Decrease in intestinal transit time
- Enhanced immune function

adult pre- and postmenopausal women, higher levels of physical activity have been correlated with lower levels of estrogen and progesterone. Additionally, increased concentrations of sex hormone binding globulin have been observed in women who are physically active. These globulins bind to estrogens in the circulation, leading to lower concentrations of the free, active hormones. Such changes in estrogen levels with physical activity also can be expected to decrease the risk of developing endometrial cancer among physically active women, because higher levels of estrogen strongly predict higher rates of endometrial cancer.

In men, changes in androgen levels with physical activity have been postulated to decrease the risk of prostate cancer. Although androgen levels may be acutely elevated after a session of aerobic exercise, basal levels appear lower, within physiological range, among highly trained men compared with sedentary men. The amount of physical activity associated with lower androgen levels is high (e.g., elite marathon runners), and it is not clear whether more moderate levels of physical activity can lower testosterone levels.

A second major pathway through which physical activity may influence the risk of cancer is via its influence on weight and adiposity. Because physical activity is associated with lower body weight and fat, it may reduce the risk of developing several obesity-related cancers, such as postmenopausal breast cancer, endometrial cancer, and colorectal cancer. Among obese, postmenopausal women compared with lean women, levels of estrogen are higher because adipose tissue can convert estrogen precursors to estrogen. As already discussed, higher levels of estrogen increase the risk of female reproductive cancers.

Physically inactive persons who take up physical activity may be more likely to lose abdominal fat. Obesity, in particular abdominal obesity, is associated with insulin resistance, hyperinsulinemia, hypertriglyceridemia, and higher levels of insulin-like growth factors. Insulin and insulin-like growth factors have been implicated in the etiology of several cancers, such as breast, prostate, and colon cancers. Thus, this may represent a third pathway through which physical activity has the potential to influence cancer development.

Inflammation may play a role as well; several chronic inflammatory diseases increase the risk of cancer (Puntoni et al. 2008). For example, persons with inflammatory bowel disease are at increased risk of colorectal cancer, while those with chronic hepatitis experience higher rates of hepatocellular cancer. Elevated levels of inflammatory markers such as C-reactive protein (CRP) and decreased levels of anti-inflammatory markers such as adiponectin have

been linked with increased cancer risk (Schottenfeld and Beebe-Dimmer 2006). Physical activity is associated with a reduced inflammatory state, and this also may be the result of decreased adiposity (Church et al. 2010).

Another commonly cited mechanism for lower rates of specifically colon cancer among physically active persons relates to change in intestinal transit time (Sternfeld and Lee 2009). It has been proposed that physical activity speeds up transit time within the colon, decreasing exposure to carcinogens, cocarcinogens, or promoters in the fecal stream. However, although some studies have shown faster transit time among physically active persons, not all studies have supported this finding.

Finally, physical activity may reduce the risk of developing cancer by enhancing the innate immune system, which is responsible for regulating susceptibility to cancer development (Sternfeld and Lee 2009). In general, the available evidence suggests that moderate levels of physical activity can enhance the immune system. However, prolonged and intense exercise (e.g., running a marathon) may have the opposite effect, leading to a temporary period of immunosuppression that lasts perhaps days to as long as two weeks.

How We Study Whether Physical Activity and Physical Fitness Decrease the Risk of Developing Cancer

Although several plausible mechanisms have been proposed to explain decreased cancer rates among physically active or fit persons, epidemiologic studies are needed to provide direct evidence of a protective effect of physical activity on cancer risk. **Epidemiology** is the study of the distribution and determinants of disease in human populations. Several epidemiologic study designs are available; next we describe three that have the most relevance to this chapter.

The **randomized clinical trial (RCT)** is generally considered the gold standard of epidemiologic study designs. In this study design, investigators take a group of eligible participants and randomly assign them to "treatment" groups. In an investigation of physical activity, for example, researchers might assemble a group of eligible participants—such as individuals who are sedentary—for study. The researchers might then randomly assign these participants to exercise at three levels: none, moderate, and high. The participants are then followed over time to assess the outcome of interest.

Evidence for a Role of Physical Activity in Preventing Cancer

The evidence for a role of physical activity in preventing certain cancers comes primarily from epidemiologic studies. The findings from such studies are supported by plausible biologic mechanisms.

This study design is considered the gold standard because the investigator assigns the exposure: Participants do not select their own physical activity. Self-selection of physical activity may lead to **confounding** by other health habits because healthy behaviors tend to cluster. For example, active persons also are likely to smoke less and to follow healthy diet patterns. Because of the random assignment by the investigator, this clustering of healthy behaviors is unlikely to occur.

Although this study design is considered the most rigorous, it may not always be feasible, or even desirable. A major factor is cost: RCTs are by far the most expensive study design. Another factor to consider is compliance with the assigned treatment (or intervention). For results from a RCT to be valid, compliance has to be high. It is not difficult to imagine that in the example just mentioned, previously sedentary individuals who are assigned to exercise may drop out of their exercise program, particularly if the study lasts for many years. However, to examine directly whether physical activity is associated with lower cancer rates, a study of long duration is required because cancer takes years to develop.

Thus, primarily for reasons of cost and compliance, no randomized trials of physical activity and cancer rates have been conducted. Instead, shorter-term studies—which cost less and are more likely to engender high compliance from participants—have assessed predictors of cancer (e.g., body fat) (Irwin et al. 2003) rather than cancer occurrence itself. One limitation of RCTs, whether long- or short-term, is the characteristics of participants being studied. The criteria for inclusion in a clinical trial tend to be strict; participants often need to be in good health to enter the study, must agree to be randomized to the different treatment (or intervention) groups, and must agree to remain committed to the study protocol for the duration of the trial. This leads to selection of individuals who tend not to be representative of the population of interest (e.g., all women). Although participants for the other study designs, described next, also may not be representative, the lack of representativeness generally is more pronounced for RCTs because participants have to agree to be randomized to different treatment groups as opposed to merely being observed for their usual habits.

The direct epidemiologic evidence that we have on the association of physical activity and cancer risk comes, instead, from two other study designs: cohort and case–control studies. In a **cohort study** design, participants are grouped on the basis of their self-selected exposure. For example, in a cohort study of physical activity and colon cancer, investigators might study three groups of participants who are free of colon cancer: individuals who have chosen to be inactive, moderately active, and highly active. The three groups are followed over time for the development of the disease of interest, in this example, colon cancer. The rates of colon cancer in the three groups are then compared to assess whether higher levels of physical activity are associated with lower rates of colon cancer.

Not all types of physical activity are feasible in RCTs. Researchers conducting an RCT must choose which activity is suitable for randomizing participants to, in order to obtain good compliance.

Because participants in this study design are selecting their physical activity, confounding is a concern. This concern can be minimized through the use of statistical methods in data analyses to take into account differences in other health habits that may be associated with physical activity and that independently predict the occurrence of cancer.

In a **case–control study** design, participants are grouped on the basis of the outcome of interest, for example, colon cancer. Participants with cancer are referred to as cases. A comparison group of participants without colon cancer is needed; they are referred to as controls. Both groups are assessed for their physical activity in the past, and case participants are assessed for physical activity that occurred before the onset of cancer. The prevalence of different levels of physical activity in the two groups is then assessed. This allows the investigators to determine whether people with colon cancer have a lower prevalence of physical activity compared with control participants (i.e., are case participants less active?).

As with the cohort study design, because physical activity is selected by participants and not assigned at random by the investigator, we are concerned with confounding by other risk factors for cancer. Similarly, this concern can be minimized by statistical methods in data analyses that adjust for differences in other risk factors that may be associated with physical activity and that independently predict the occurrence of cancer.

In the interpretation of the findings from epidemiologic studies, several concepts are important. First, the term **relative risk** (often abbreviated as RR) is a measure of association used in these studies. The relative risk measures the magnitude of association between the exposure (in our example, physical activity) and the disease (in our example, cancer). Relative risk essentially compares the rates of cancer among persons with different levels of physical activity. A relative risk of 0.5 for cancer associated with physical activity, compared with inactivity, indicates that active persons have 0.5 times (or 50%) the risk of developing cancer compared with those who are inactive (who serve as the reference group).

A second important concept is that of the **confidence interval** (often abbreviated as CI). When we calculate a relative risk, we can also calculate a band of uncertainty, the confidence interval, around our estimate of the relative risk. Typically, 95% confidence intervals are used in epidemiologic studies. For example, if we estimate the relative risk for cancer associated with physical activity, compared with inactivity, to be 0.5 and the associated 95% confidence interval to be 0.3 to 0.8, this means we are 95% certain that the true estimate of the relative risk lies between 0.3 and 0.8. The narrower the width of the confidence interval, the more precise the estimate of the relative risk. Additionally, if the 95% confidence interval does not span a value of 1.0, this implies that the findings are statistically significant at $p < 0.05$.

Finally, epidemiologic studies use the term **dose response** to describe the phenomenon of a graded effect of exercise. In the case of physical activity and cancer, the presence of an inverse dose response indicates that the higher the level of physical activity, the lower the rates of cancer. The shape of the dose–response curve can take many forms—for example, linear or curvilinear. Most commonly, epidemiologic studies test for a curve that is linear.

Physical Activity, Physical Fitness, and Site-Specific Cancers

Next we discuss findings from epidemiologic studies of physical activity or physical fitness and cancer risk (Sternfeld and Lee 2009). The studies all had cohort or case–control study designs. The available data primarily come from studies of physical activity; there have been relatively few studies of physical fitness or markers of physical fitness (e.g., heart rate). Because the evidence indicates that the association of physical activity or physical fitness with cancer differs for different cancers, we discuss the individual cancers separately.

Colorectal Cancer

The large intestine comprises both the colon and the rectum; the rectum is continuous with the end of the colon. Although they are closely linked anatomically, the epidemiologic literature suggests that physical activity has different associations with colon and rectal cancer. Thus, we discuss colon cancer and

Epidemiologic Study Designs

There are three major epidemiologic study designs:

1. Randomized clinical trial
2. Cohort study
3. Case–control study

rectal cancer separately. Studies that have looked at colorectal cancer as a single outcome are not considered because combining the two separate cancers may obscure the findings for each individual cancer site.

Colon Cancer

More than 50 published studies have examined the association between physical activity and the risk of developing colon cancer. These studies have been conducted in North America, Europe, Asia, and Australia and New Zealand. Additionally, a few studies have assessed physical fitness in relation to colon cancer risk. The evidence from these cohort and case–control studies indicates that physically active (or fit) men and women have a lower risk of developing colon cancer compared with inactive (or unfit) individuals. Because few studies have included participants from racial or ethnic minority groups, the data are primarily from studies of white persons. Additionally, most of the data relate to physical activity during middle age and later years; there have been few studies of physical activity during childhood and adolescence, and we have almost no data on changes in physical activity over time.

How much of an effect does physical activity have? A meta-analysis, which averaged the results across all studies, reported a summary relative risk of 0.76 (95% confidence interval: 0.72-0.81) associated with physical activity, that is, a risk reduction of 24% (Wolin et al. 2009). Although the data are limited, the available information suggests that approximately 30 to 45 min per day of moderate-intensity physical activity is needed to reduce risk, as indicated by the findings from men in the Harvard Alumni Health Study (Lee, Paffenbarger, and Hsieh 1991) and women in the Nurses' Health Study (Wolin et al. 2007). As shown in figure 14.2a, significantly lower rates of colon cancer were observed among men expending 4,200 kJ (1,000 kcal) or more per week. This amount of energy expenditure can be accomplished with 30 to 45 min per day of moderate-intensity physical activity. In figure 14.2b, significantly lower rates of colon cancer were seen among women exercising > 4 h per week, or approximately 40 min per day.

Many of the studies of physical activity and colon cancer risk have provided some information on a dose–response relationship, as they investigated participants in at least three levels of physical activity. However, because the dose–response relationship was not often studied in detail, the available information suggests only that an inverse dose response is likely. There are not enough data to provide details on how much additional protection physical activity might confer at increasingly higher levels of activity.

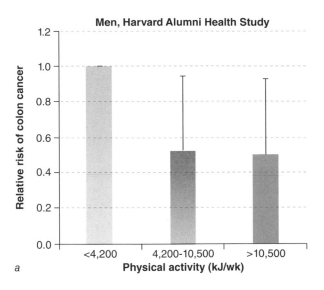

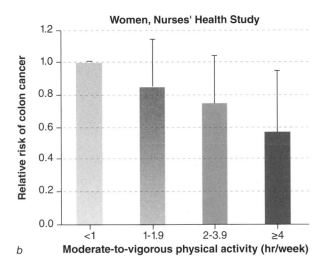

FIGURE 14.2 Relationship between physical activity and colon cancer risk among (a) men in the Harvard Alumni Health Study and (b) women in the Nurses' Health Study. Bars represent upper limit of 95% confidence interval.

Data from Lee, Paffenbarger, and Hsieh 1991.

Previously, we discussed concerns about confounding arising from cohort and case–control studies. Many of the investigators who studied physical activity and colon cancer risk did adjust for potential confounders in the data analyses, including controlling for differences in age, body mass index, smoking, and diet. Even after these adjustments, physical activity was found to be associated with lower colon cancer risk. Thus, it is unlikely that confounding by other risk factors for colon cancer can explain the inverse relationship seen between physical activity and the risk of developing colon cancer.

Rectal Cancer

More than 30 studies conducted in North America, Europe, Asia, and Australia have investigated the relationship between physical activity and the risk of developing rectal cancer. Although some researchers have reported lower rectal cancer rates among active individuals, the data on the whole do not support any relationship between physical activity and rectal cancer risk. Across all studies, the median relative risk when comparing most active to least active participants is 1.0. That is, on average, the data indicate that the rates of rectal cancer among most and least active persons are the same.

Breast Cancer

Like the studies of colorectal cancer, almost all of the epidemiologic studies of breast cancer have addressed physical activity rather than physical fitness. More than 60 published studies, conducted in North America, Europe, Asia, and Australia, have examined whether physical activity or fitness is associated with a lower risk of developing breast cancer in women. Overall, the data are reasonably consistent in supporting an inverse relationship between physical activity and breast cancer incidence rates.

Although individual studies have indicated that active women may have half, or even less than half, the incidence rates of breast cancer compared with inactive women, the data on the whole suggest a smaller magnitude of association than that observed for colon cancer. A meta-analysis, averaging results across studies, showed a summary relative risk of 0.81 (95% confidence interval: 0.73-0.89) associated with physical activity, that is, a 19% risk reduction (Monninkhof et al. 2007). Because risk factors for breast cancer in premenopausal women and postmenopausal women may be different, several studies have examined these groups separately. Physical activity appears to have a somewhat larger effect for post- than for premenopausal women.

Data on how much physical activity is needed to decrease the risk of breast cancer are relatively few. It appears that some 4 to 7 h per week of moderate- to vigorous-intensity physical activity is required, similar to the amount observed for colon cancer. For example, in a Norwegian study, significantly lower rates of breast cancer were observed among women undergoing regular physical activity, defined as exercising to keep fit for at least 4 h per week or participating in competitive sport (figure 14.3) (Thune et al. 1997). In this study, higher levels of physical activity on the job also were associated with lower risk. Note that the magnitude of reduced risk

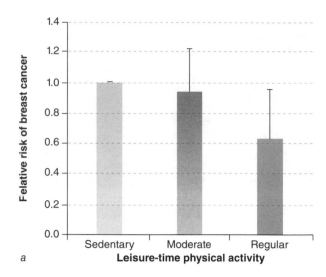

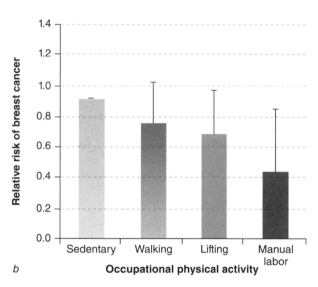

FIGURE 14.3 Relationship between (a) leisure-time physical activity or (b) occupational physical activity and breast cancer risk in a Norwegian study. Bars represent upper limit of 95% confidence interval.

Data from Thune et al. 1997.

in this study—on the order of 40% to 50%—is larger than that seen in other studies, such as the Women's Health Study (Lee et al. 2001) or the Women's Health Initiative (McTiernan et al. 2003), where the degree of risk reduction observed was closer to the overall average of 19%. The data from the Women's Health Initiative are noteworthy because a sizeable number of participants in this study were women belonging to minority racial and ethnic groups.

The timing of physical activity, in order for it to decrease breast cancer risk, is unclear. Most studies have assessed physical activity only at one time in relation to breast cancer rates. However, in several studies the aim has been to estimate physical activity throughout a woman's life, including physical activity carried out during adolescence. As discussed previously, this has a biological rationale, because physical activity during adolescence is associated with favorable changes in menstrual function and female sex steroid hormones. Although some studies have suggested that physical activity throughout a woman's life is more strongly protective, others have not supported this, indicating a similar inverse relation to breast cancer risk even if carried out during middle age and older ages.

As with the colon cancer studies, the dose–response relationship between physical activity and breast cancer risk has not been investigated in detail. The data suggest an inverse dose–response relationship between physical activity and the risk of breast cancer, with higher levels of activity associated with increasingly lower risks. However, we do not have enough information to indicate how much additional decrease in risk might be observed with increasingly higher levels of activity.

Although there have been no RCTs, the data that we have for breast cancer are unlikely to be confounded by risk factors associated with a physically active lifestyle. Many studies have continued to show a lower risk of breast cancer among active women even after adjustment for other risk factors such as age, body mass index, alcohol intake, use of oral contraceptives and hormone therapy, reproductive variables (ages at menarche and menopause, menopausal status, parity, age at first birth, breastfeeding), benign breast disease, and family history of breast cancer.

Prostate Cancer

The findings for prostate cancer have been less consistent than those for colon and breast cancer. More than 40 studies, conducted in North America, Europe, and Asia, have examined the relationship between physical activity (the majority of studies) or physical fitness and the risk of developing prostate cancer.

Although individual studies have indicated that physical activity or fitness is associated with a lower risk of prostate cancer, on the whole the data from epidemiologic studies do not provide strong support for an inverse relationship between physical activity and the risk of this cancer. The median relative risk across all studies comparing most active with least active men is 0.9. This median value is close to a relative risk of 1.0, which indicates no difference in prostate cancer incidence rates between active and inactive men.

It is unclear why the findings have been inconsistent. Apart from age and race, few risk factors for prostate cancer have been established. Thus, it is unlikely that the different findings reflect uncontrolled confounding by risk factors related to both physical activity and prostate cancer risk. It has been postulated that the age at which prostate cancer develops is important: Early-onset prostate cancer may be more likely influenced by genetic factors, with physical activity less important, and the opposite may be the case for prostate cancers occurring at older ages. However, studies of physical activity and prostate cancer occurring at different ages have not provided clear evidence supporting this hypothesis. Another potentially relevant factor is the level of physical activity that may be needed to reduce prostate cancer risk. As discussed previously, one postulated mechanism for lowered risk of this cancer is the modulation of male sex steroid hormones. Changes in androgen levels with physical activity have generally been observed in studies where the level of activity was very high. Epidemiologic studies of physical activity and prostate cancer risk have typically been conducted in the general population, where there is a low prevalence of very high levels of physical activity. Perhaps these studies have not been able to document an inverse relationship between physical activity and prostate cancer rates because the level of physical activity among participants was not high enough to effect changes in hormone levels and hence subsequent prostate cancer risk.

The issue of screening for prostate cancer may explain the discrepant findings. A screening test that can detect very early-stage prostate cancer, called prostate-specific antigen (PSA) screening, gained widespread use in the United States from the late 1980s through the mid-1990s. If physically active men also are more health conscious and undergo more frequent PSA screening, higher rates of prostate cancer may be observed because of increased detection of early-stage cancers. This may obscure the protective effect of physical activity because of the detection of early-stage cancers that would not have been diagnosed clinically. Some support for this argument comes from two studies conducted as part of the Harvard Alumni Health Study (Lee, Sesso, and Paffenbarger 2001; Lee, Paffenbarger, and Hsieh 1992). In the first study of physical activity and prostate cancer, the cases of interest were prostate cancer diagnosed in 1988 or earlier. Investigators reported an almost halving of prostate cancer incidence rates among men aged 70 years or older who expended 16,800 kJ (4,000 kcal) or more per week (equivalent to some 5 h per week of vigorous activity), compared

Lee

with sedentary men expending <4,200 kJ (1,000 kcal) or more per week. However, in an updated analysis of these men that examined prostate cancer diagnosed after 1988, the earlier observations were not replicated. In this later study, no differences in prostate cancer rates were observed among men with different physical activity levels. These different findings may have been attributable to increased screening for prostate cancer among the most active men in the later study.

Finally, physical activity may reduce the risk of advanced prostate cancers but not early-stage cancers. In the Health Professionals Follow-Up Study, investigators observed an approximate halving of risk of metastatic prostate cancer among very active men compared with sedentary men (Giovannucci et al. 1998). However, when total prostate cancers—including early-stage cancers—were examined, no differences in cancer rates were seen among men with different physical activity levels.

Lung Cancer

In the past decade, several studies have suggested an intriguing inverse relationship between physical activity or physical fitness and the risk of developing lung cancer (the data coming primarily from studies of physical activity). More than 20 studies, conducted in North America and Europe, have examined this hypothesis; and the authors of a meta-analysis reported a summary relative risk of 0.70 (95% confidence interval: 0.62-0.79) when comparing most active with least active participants (Tardon et al. 2005). That is, on average, the studies have indicated that people who are physically active have a 30% lower risk of developing lung cancer than those who are sedentary.

Although the data supporting a role of physical activity in lowering lung cancer rates are promising, a major concern, given that the findings are derived only from cohort and case–control studies, is confounding by cigarette smoking. As discussed previously, physically active persons tend to be health conscious and thus are less likely to smoke cigarettes, a major risk factor for lung cancer. Although almost all studies have adjusted for cigarette smoking in the analyses (typically, the number of cigarettes smoked and the duration of smoking), it is difficult to be certain that the effect of cigarette smoking was completely controlled. Residual confounding, as well as confounding by other smoking-related factors (e.g., use of low-tar cigarettes or filter tips, depth of inhalation during smoking, passive smoking), might still be present.

To be certain that the inverse relationship between physical activity and lung cancer rates does not reflect confounding by smoking, a study could be conducted only among never-smokers. However, lung cancer occurs very infrequently among never-smokers, which would make this a difficult study to conduct. Several studies have examined nonsmokers (never- and past smokers) separately, with similar physical activity lung cancer associations observed for nonsmokers and smokers. For example, in the Harvard Alumni Health Study, similar results were observed among all participants, which included both smokers and nonsmokers, as well as among nonsmokers alone (Lee, Sesso, and Paffenbarger 1999). The magnitude of risk reduction observed among the most active participants compared with the least active was 39% (figure 14.4a). When nonsmokers were examined separately, the corresponding risk reduction was 46%.

One study was large enough to allow examination of never-smokers as a separate group. In this case–control study conducted in Canada (Mao et al. 2003), an inverse relationship between physical activity and lung cancer risk was observed among all participants. The most active men had a 26% lower risk of developing lung cancer compared with least active men. For women, the corresponding risk reduction was 28%. The number of lung cancers in never-smokers only (men and women combined) was 126. Although this number may have been too small to yield statistically significant findings, it is noteworthy that the magnitude of risk reduction among never-smokers expending 34.4 metabolic equivalent (MET) hours (MET-hours) per week or more in recreational physical activity was 32% compared with that among the least active participants expending <6.3 MET-hours per week (figure 14.4b). This risk reduction was similar to that seen among past smokers (34% reduction) and current smokers (41% reduction) exercising at that level in the same study.

A few studies have addressed lung cancers of different histological types. Although cigarette smoking increases the risk of all types of lung cancer, the increase in risk is stronger for certain histological types such as squamous cell and small cell carcinomas than for others such as adenocarcinomas. In a Norwegian study, investigators reported significant inverse associations between physical activity and small cell carcinomas and adenocarcinomas of the lung, but not for squamous cell carcinomas, in men. This may provide some indirect evidence that the association observed in the study did not reflect confounding by smoking. However, in the

240

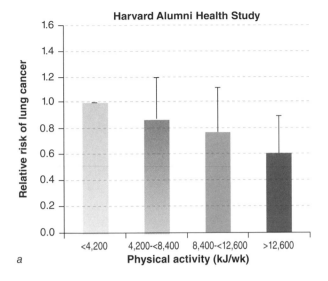

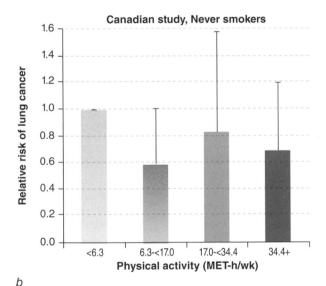

FIGURE 14.4 Relationship between physical activity and lung cancer risk in (a) the Harvard Alumni Health Study and (b) never-smoking participants in a Canadian study. Bars represent upper limit of 95% confidence interval.

Data from (a) Lee. Sesso, and Paffenbarger Jr. 1999, and (b) Mao et al. 2003.

Canadian study described previously, the strongest inverse relationships were noted between physical activity and squamous cell carcinomas in men but between physical activity and small cell carcinomas in women. Thus, on balance, although the data indicate an inverse relationship between physical activity or physical fitness and lung cancer risk, the findings should be considered preliminary.

Other Cancers

Several other cancer sites have been investigated for a relationship with physical activity or physical fitness. These include endometrial, ovarian, testicular, pancreatic, kidney, and bladder cancer as well as hematopoietic cancers. The data suggest an inverse relationship between physical activity and endometrial cancer risk, consistent with the hypothesis that physical activity can lower estrogen levels, which are related to increased risk of this cancer. However, many of the studies of endometrial cancer have not controlled for the use of postmenopausal estrogen therapy, an important risk factor for this cancer. With regard to the other cancers, the data are too sparse to allow any conclusions regarding whether physical activity is associated with decreased rates of occurrence.

Physical Activity and Cancer Survivors

The studies described in this chapter focused on physical activity or fitness in the primary prevention of cancer. In recent years, a few studies have looked at whether physical activity improves the prognosis in cancer patients (Sternfeld and Lee 2009). The sparse data available suggest that physical activity may be associated with lower mortality rates among patients with colorectal and breast cancers.

Summary

Many published studies have addressed the role of physical activity or physical fitness in preventing cancer (table 14.2). The data are clearest in support of physical activity or fitness as a means of preventing colon and breast cancers. It also appears reasonably clear that physical activity or fitness does not influence rectal cancer rates. For prostate cancer, the data have been equivocal, whereas for lung and endometrial cancers, there are suggestive, although not definitive, data to support inverse associations.

Thus this chapter discusses an important benefit of physical activity—decreasing the risk of developing certain types of cancer. The evidence discussed in the chapter provides yet another compelling argument for adopting and maintaining a physically active way of life.

TABLE 14.2   Summary of Epidemiologic Data on the Association of Physical Activity With Risk of Developing Cancer

Cancer site	Epidemiologic data
Colon	Inverse association; likely dose response
Rectum	No association
Breast	Inverse association; likely dose response
Prostate	Equivocal data
Lung	Possible inverse association
Endometrium	Possible inverse association
Other cancers	Limited data

In Review

1. Cancer is one of the leading causes of mortality in the United States as well as worldwide.
2. Biologically, it appears plausible that higher levels of physical activity or physical fitness result in lower rates of cancer.
3. The data from epidemiologic studies indicate that higher levels of physical activity or physical fitness are associated with lower rates of colon and breast cancers.

Key Concepts

cancer—A chronic disease that is multifactorial in its etiology, with contributions from both environmental factors (not only factors such as radiation but also smoking, physical inactivity, diet, viruses, etc.) and genetic factors. The disease arises when excessive, uncontrolled, and purposeless proliferation of cells occurs in the absence of physiological stimuli. These cells invade the surrounding tissues and spread by means of the lymphatics and blood vessels to give rise to secondary cancers, or metastases.

case–control study—For definition, see page 181.

cohort study—For definition, see page 227.

confidence interval—For definition, see page 181.

confounding—A phenomenon that may exist in the data from epidemiologic studies and cloud the interpretation of findings. As an example, say we compare the rates of colon cancer that develop among a group of athletes and among a group of sedentary persons. We find that the rates are lower among the athletes compared with the sedentary group. Can we conclude that physical activity lowers the risk of developing colon cancer? If we look more closely at the groups, we might find, perhaps, that the athletes are much younger than the sedentary group. Cancer rates also increase with age. Thus, the lower rates of colon cancer that we see among the athletes may have nothing to do with physical activity but instead reflect the fact that the athletes are younger. This phenomenon is termed *confounding*. Statistical adjustment can be made in data analyses to factor in the age differences. If we perform this adjustment and continue to observe lower colon cancer rates among the athletes, it is reasonable to conclude that physical activity is associated with lower colon cancer rates, provided that there are no other confounders.

dose response—For definition, see page 182.

epidemiology—For definition, see page 50.

randomized clinical trial (RCT)—A type of epidemiologic study design in which participants are grouped on the basis of the investigator-assigned exposure of interest, in this case, physical activity. For example, among a group of eligible participants, investigators may randomly assign them to exercise at three levels: no activity, moderate activity, and vigorous activity. These participants are then followed over time to assess the outcome of interest, such as change in abdominal fat. This is often considered the "gold standard" of epidemiologic study designs. However, because of the cost and issues regarding compliance with an assigned activity level, it may not always be feasible, or even desirable, to conduct RCTs.

relative risk—For definition, see page 183.

Study Questions

1. List the leading causes of death in the United States. Where does cancer rank among the causes of death?

2. What are the most common fatal cancers among U.S. men and women? What are the most common newly occurring cancers among them?

3. Describe some of the mechanisms that might explain why physically active persons have lower rates of cancer than inactive persons.

4. What are some of the key differences in the following epidemiologic design strategies: randomized clinical trial, cohort study, case–control study?

5. Explain the meaning of the terms *relative risk* and *confidence interval*.

6. Explain the meaning of the term *confounding*. Why should we be concerned about this in studies of physical activity and cancer?

7. Which cancers have been shown to occur less frequently among physically active compared with inactive persons?

8. How much of an effect does physical activity have in reducing the risk of developing colon cancer?

9. What do the available data show regarding how much physical activity may be needed to reduce the risk of developing colon cancer?

10. Discuss what epidemiologic studies have shown regarding the relationship between physical activity and the risk of developing breast cancer.

References

Byers, T., M. Nestle, A. McTiernan, C. Doyle, A. Currie-Williams, T. Gansler, and M. Thun. 2002. American Cancer Society guidelines on nutrition and physical activity for cancer prevention: Reducing the risk of cancer with healthy food choices and physical activity. *CA: A Cancer Journal for Clinicians* 52:92-119.

Church, T.S., C.P. Earnest, A.M. Thompson, E.L. Priest, R.Q. Rodarte, T. Sanders, R. Ross, and S.N. Blair. 2010. Exercise without weight loss does not reduce C-reactive protein: The INFLAME Study. *Medicine and Science in Sports and Exercise* 42:708-716.

Giovannucci, E., M. Leitzmann, D. Spiegelman, E.B. Rimm, G.A. Colditz, M.J. Stampfer, and W.C. Willett. 1998. A prospective study of physical activity and prostate cancer in male health professionals. *Cancer Research* 58:5117-5122.

International Agency for Research on Cancer (IARC). 2002. *Weight control and physical activity. IARC Working Group on the Evaluation of Cancer-Preventive Agents.* Lyon: IARC.

Irwin, M.L., Y. Yasui, C.M. Ulrich, D. Bowen, R.E. Rudolph, R.S. Schwartz, M. Yukawa, E. Aiello, J.D. Potter, and A. McTiernan. 2003. Effect of exercise on total and intra-abdominal body fat in postmenopausal women: A randomized controlled trial. *Journal of the American Medical Association* 289:323-330.

Jemal, A., R. Siegel, J. Xu, and E. Ward. 2010. Cancer statistics, 2010. *CA: A Cancer Journal for Clinicians* 60: 277-300.

Lee, I. M., R.S. Paffenbarger Jr., and C.C. Hsieh. 1991. Physical activity and risk of developing colorectal cancer among college alumni. *Journal of the National Cancer Institute* 83:1324-1329.

Lee, I. M., R.S. Paffenbarger Jr., and C.C. Hsieh. 1992. Physical activity and risk of prostatic cancer among college alumni. *American Journal of Epidemiology* 135:169-179.

Lee, I. M., K.M. Rexrode, N.R. Cook, C.H. Hennekens, and J.E. Buring. 2001. Physical activity and breast cancer risk: The Women's Health Study (United States). *Cancer Causes and Control* 12:137-145.

Lee, I. M., H.D. Sesso, and R.S. Paffenbarger Jr. 1999. Physical activity and risk of lung cancer. *International Journal of Epidemiology* 28:620-625.

Lee, I. M., H.D. Sesso, and R.S. Paffenbarger Jr. 2001. A prospective cohort study of physical activity and body size in relation to prostate cancer risk (United States). *Cancer Causes and Control* 12:187-193.

Mao, Y., S. Pan, S.W. Wen, and K.C. Johnson. 2003. Physical activity and the risk of lung cancer in Canada. *American Journal of Epidemiology* 158:564-575.

McTiernan, A., C. Kooperberg, E. White, S. Wilcox, R. Coates, L.L. Adams-Campbell, N. Woods, and J. Ockene. 2003. Recreational physical activity and the risk of breast cancer in postmenopausal women: The Women's Health Initiative Cohort Study. *Journal of the American Medical Association* 290:1331-1336.

Monninkhof, E.M., S.G. Elias, F.A. Vlems, I. van der Tweel, A.J. Schuit, D.W. Voskuil, F.E. van Leeuwen; TFPAC. 2007. Physical activity and breast cancer: A systematic review. *Epidemiology* 18:137-157.

Physical Activity Guidelines Advisory Committee. 2008. *Physical Activity Guidelines Advisory Committee report, 2008.* Washington, DC: U.S. Department of Health and Human Services.

Puntoni, M., D. Marra, S. Zanardi, and A. Decensi. 2008. Inflammation and cancer prevention. *Annals of Oncology* 19:vii225-vii229.

Schottenfeld, D., and J. Beebe-Dimmer. 2006. Chronic inflammation: A common and important factor in the pathogenesis of neoplasia. *CA: A Cancer Journal for Clinicians* 56:69-83.

Sternfeld, B., and I.M. Lee. 2009. Physical activity and cancer: The evidence, the issues, and the challenges. In *Epidemiologic methods in physical activity studies*, ed. I.M. Lee, S.N. Blair, J.E. Manson, and R.S. Paffenbarger. New York: Oxford University Press.

Strong, K., C. Mathers, S. Leeder, and R. Beaglehole. 2005. Preventing chronic diseases: How many lives can we save? *Lancet* 366:1578-1582.

Tardon, A., W.J. Lee, M. Delgado-Rodriguez, M. Dosemeci, D. Albanes, R. Hoover, and A. Blair. 2005. Leisure-time physical activity and lung cancer: A meta-analysis. *Cancer Causes and Control* 16:389-397.

Thune, I., T. Brenn, E. Lund, and M. Gaard. 1997. Physical activity and the risk of breast cancer. *New England Journal of Medicine* 336:1269-1275.

Wolin, K.Y., I.M. Lee, G.A. Colditz, R.J. Glynn, C. Fuchs, and E. Giovannucci. 2007. Leisure-time physical activity patterns and risk of colon cancer in women. *International Journal of Cancer* 121:2776-2781.

Wolin, K.Y., Y. Yan, G.A. Colditz, and I.M. Lee. 2009. Physical activity and colon cancer prevention: A meta-analysis. *British Journal of Cancer* 100:611-616.

World Cancer Research Fund/American Institute for Cancer Research. 2007. *Food, nutrition, physical activity, and the prevention of cancer: A global perspective*. Washington, DC: American Institute for Cancer Research.

15

Physical Activity, Fitness, and Joint and Bone Health

Jennifer M. Hootman, PhD, ATC, FACSM, FNATA

CHAPTER OUTLINE

As you learned in chapter 1, the human body is engineered for movement. The neurological, muscular, and skeletal systems all work together to produce human movement, with forces being produced and transmitted within the musculoskeletal system. Most forces are generated for required daily functioning activities and are well tolerated by the human body. However, unexpected, excessive, or accumulated forces can have unwanted effects. This chapter discusses the prevalence of select bone and joint conditions; the ways in which different levels of activity affect the musculoskeletal system; and the relationships among physical activity, functional impairment, and disability prevention.

Scientific Evidence

In this section we review the state of the evidence on musculoskeletal diseases, including their relationships with physical activity as both a treatment and a prevention strategy.

Burden of Selected Musculoskeletal Diseases in the Population

Acute and chronic musculoskeletal conditions, including trauma-related injuries, low back pain, and **arthritis and other rheumatic conditions**, are among the most frequently occurring medical conditions that have substantial impact on the general health of the population and the health care system.

In the United States and worldwide, unintentional injury is a leading cause of death (>124,000 deaths in the United States, 3.9 million globally) and disability (National Center for Injury Prevention and Control 2007; Chandran, Hyder, and Peek-Asa 2010). Globally, 50% of children treated in hospitals for an unintentional injury will have some long-term disability (Chandran, Hyder, and Peek-Asa 2010). Arthritis and rheumatism are the most common causes of disability among persons aged 15 years and older in the United States (Centers for Disease Control and Prevention 2009). In fact, as shown in figure 15.1, 3 of the 10 leading causes of disability are musculoskeletal conditions: arthritis and rheumatism, back and spine problems, and stiffness/deformity of limbs. These three conditions account for 39.4% of all disabilities. Almost 15 million U.S. adults report limitations in activities of daily living due to musculoskeletal conditions, and these conditions account for over 50% of the total lost workdays (437.6 million days) for U.S. adults age 18 and older (United States Bone and Joint Decade 2008).

Osteoarthritis

Osteoarthritis, or degenerative joint disease, affects more than 27 million people in the United States and an estimated 8.6% of men and 18.0% of women globally (Lawrence et al. 2008; Woolf and Pfleger 2003). It most commonly affects the large weight-bearing joints, such as the knees, hips, and spine. Osteoarthritis is more common among women, Caucasians, older persons, overweight or obese persons, and those with a history of significant joint injuries

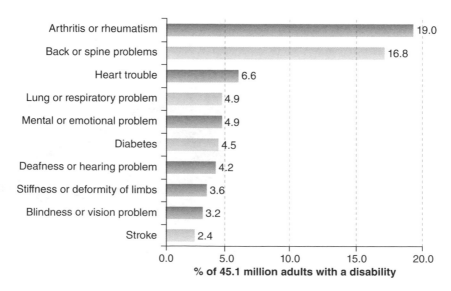

FIGURE 15.1 Top 10 most common causes of disability among persons aged 15 years and older, United States, 2005.
Centers for Disease Control and Prevention 2009.

(Felson et al. 2000; Lawrence et al. 2008; Woolf and Pfleger 2003). Osteoarthritis is a disease of the articular cartilage and the subchondral bone. Clinical diagnosis is based on specific radiographic features including joint space narrowing (indicating cartilage degeneration), **sclerosis** of subchondral bone, and **osteophyte** formation. Physical signs and symptoms include pain, swelling, **crepitus**, and restricted range of motion (Arthritis Foundation 2001).

Limitations attributable to osteoarthritis are common. This condition accounts for the majority of reported difficulties with climbing stairs and walking reported by individuals, as well as lost work time and early retirement. Over 1 million joint replacement surgeries are performed every year, most due to osteoarthritis. Total knee replacements make up more than half of all joint replacement surgeries and have increased threefold in recent years (United States Bone and Joint Decade 2008). These procedures are costly, and many patients require substantial long-term postoperative care. Because of the aging of the population, as well as high rates of obesity (a primary risk factor for osteoarthritis), rates of osteoarthritis and associated disabilities are expected to increase exponentially in the next two decades.

Osteoporosis

Osteoporosis is a disease of the skeletal system that is characterized by low bone mass and fragility. Prevalence estimates of osteoporosis range from 23% to 30% globally, with most cases occurring among post-menopausal white women (National Osteoporosis Foundation 2002; Arthritis Foundation 2001; Kahn et al. 2001; Woolf and Pfleger 2003). In the United States in 2010, approximately 12 million persons had osteoporosis, and another 40 million were at risk because of low bone mass (National Osteoporosis Foundation 2002). Osteoporosis is largely a silent disease, producing no symptoms until it is manifested clinically in the form of overt fractures; common fracture sites include the hip, spine, and wrist. More than 1.3 million osteoporosis-related fractures are treated annually in the United States, resulting in 432,000 hospitalizations and 3.4 million outpatient medical care visits and costing more than $13.8 billion (National Center for Injury Prevention and Control 2005). Approximately 9 million osteoporosis-related fractures occur globally each year, and fracture rates vary by geographic region, with the highest rates seen in Scandinavia, Europe, and North America and lower rates in Africa and Asia (Johnell and Kanis 2006). The lifetime risk of a hip, vertebral, or wrist fracture is 40% for white women; but men, smokers, persons with physical disabilities, and sedentary persons are also at risk.

Osteoporosis-related hip fractures are of special concern among elderly persons. Up to 20% of patients die in the first year after a hip fracture, and two-thirds do not return to their preinjury level of functioning (Arthritis Foundation 2001). Several recent public health intervention studies examined ways to reduce the morbidity and mortality associated

Translating Research to Practice

Tai chi has been shown to be effective in preventing falls among older adults. Despite evidence of effectiveness from randomized controlled trials, widespread adoption and reach of this intervention in local communities are low. In 2005, the Centers for Disease Control and Prevention funded researchers at Oregon Research Institute to translate the research-based "Tai Chi—Moving for Better Balance" program for use in community senior service organizations. Easy-to-use program materials include standardized instructor's manuals and supplementary materials, a videotape/ DVD, and a participant guide. The 12-week program was implemented in senior centers. Participants attended a group class two days per week for 1 h and were instructed to do the exercises at least one additional day per week at home. The program had 100% adoption by all senior centers and reached 87% of the target population. At 12 weeks, participants reported significant improvements in physical function (e.g., gait speed) and quality of life. Attendance rates were >85%, and attrition was acceptable (<25%). Participants felt that the class was enjoyable, safe, and beneficial; said they would recommend it to others; and planned on continuing tai chi at home or in a class if one was offered. Disseminating evidence-based programs in the community is feasible and sustainable. Increasing the reach of these programs can be an effective and low-cost approach to reducing the occurrence and complications of falls among older adults.

Based on Li et al. 2008.

with fall-related injuries among older adults. Promising interventions to prevent falls and hip fractures include programs that integrate improving balance and mobility through exercises such as tai chi with patient education programs that address modification of medical (visual impairment, medications, etc.), as well as environmental factors (trip hazards, poor lighting, etc.) that increase the potential for falling (National Center for Injury Prevention and Control 2006). (See "Translating Research to Practice.")

Total bone strength is determined by both material and structural properties (Kahn et al. 2001). Figure 15.2 depicts the material properties, which are independent of bone size, and the structural features that contribute to skeletal strength. Of these, bone density is the feature most commonly studied in relation to physical activity and exercise; however, all features may respond to changes in activity level.

In addition to nutritional, environmental, and genetic factors, physical inactivity is an established risk factor for low bone density and osteoporosis. Bone becomes brittle and porous without bone-loading stress from weight-bearing activity and muscle contraction (Kahn et al. 2001; Vuori 2001). Bone cells respond to mechanical loading stresses (axial compression and bending forces) generated during weight-bearing activity by temporarily deforming, which in turn stimulates bone formation. This process is described by the **mechanostat theory**, which suggests that a minimum effective strain is needed to initiate and maintain osteogenesis (Kahn et al. 2001). The adaptation of bone to mechanical stress and strain is a complex process, but the three basic elements are as follows:

1. An effective stimulus involves dynamic versus static loading.

2. Relatively few loading cycles of short duration are needed to stimulate bone tissue response.

3. Over time, bone cells cease to accommodate to routine loads, and subsequently a periodic increase in load is necessary to continue to effect bone growth (Arthritis Foundation 2001; Kahn et al. 2001).

Dose–Response Relationships Among Physical Activity, Osteoarthritis, and Osteoporosis

The total amount and type of physical activity play an important role in the development of a variety of musculoskeletal conditions. This section focuses on the relationships among the dose (frequency, intensity, duration, and type) of physical activity, osteoarthritis, and osteoporosis.

Physical Activity and the Development of Osteoarthritis

The incubation period for the development of degenerative joint disease (osteoarthritis) may be upward of 30 years; therefore, it is difficult to connect participation in various activities throughout a lifetime and the long-term risk of osteoarthritis. The important consideration with regard to physical activity is how the specific activity loads the joint surfaces. Three main features likely contribute to the development of osteoarthritis: rate and magnitude of contact forces, extent of joint torsion and twisting, and total dose of exposure to different physical and occupational activities over a lifetime. Sudden, single, high-impact torsional loadings can significantly damage articular cartilage and subchondral bone (Vuori 2001). Loads that are applied more slowly are tolerated much better because muscles can absorb and disperse the load more efficiently and articular cartilage can deform slowly, thereby safely transmitting forces across the joint.

Articular cartilage fractures in response to joint contact forces in the range of 25 MPa (megapascals, newtons per square meter). In activities such as running, jumping, and throwing, peak articular contact forces range from 4 to 9 MPa, which is far below the maximum level associated with cartilage injury (Vuori 2001). In animal models, activity-related joint contact stresses are not associated with cartilage damage in healthy joints. In fact, moderate physical activity produces many structural and functional beneficial alterations within the joint. However,

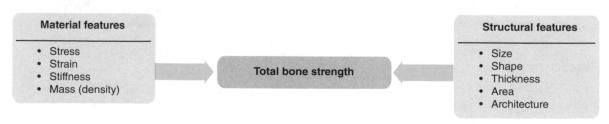

FIGURE 15.2 Structural and material properties that contribute to total bone strength.

in joints with abnormal anatomy or biomechanics related to trauma or injury, even normal joint stress can cause further articular cartilage damage and eventually osteoarthritis. Also, chronic or repetitive articular forces below 25 MPa may eventually lead to articular surface damage, initiating the development of osteoarthritis through a cumulative effect (Vuori 2001). In addition to trauma-related joint injury, the normal physiological changes that occur within the musculoskeletal system attributable to aging may perpetuate the progression of superficial surface damage, leading to disease progression.

Observational data suggest that select types of activities may contribute to the development of osteoarthritis, especially in the knee but also in other large weight-bearing joints such as the hip, ankle, and spine. Activities with high-load joint stresses, especially twisting and torsional types of stress, have been associated with degenerative joint conditions. Soccer, football, tennis, and certain track and field events involve twisting and repetitive impact compression forces. As proposed by the **wear and tear theory** of degenerative joint disease, these forces, over time, may wear excessively on the hyaline cartilage of the joint, resulting in osteoarthritis (Buckwalter 2003; Felson et al. 2000). Occupational activity can also contribute to joint stress. Persons employed in occupations requiring a lot of knee bending, carrying of heavy loads, and twisting, such as farming, warehouse work, and carpet laying, are at high risk for knee and hip osteoarthritis (Felson et al. 2000). These high-load activities, whether from sports, exercise, or occupational exposure, also carry a higher risk of major joint injuries, and joint injuries are a known risk factor for osteoarthritis.

No experimental studies examining the dose–response relationship between various activity levels and the development of osteoarthritis have been published. However, data from observational epidemiological studies may help us understand this relationship. When different categories of athletes are compared, former elite athletes are at higher risk of hip and knee osteoarthritis than are recreational athletes and nonathletes. Also, in terms of total exposure to sports, whether measured in total hours of sport participation or frequency of participation over a lifetime, those in the highest exposure category are two to four times more likely to develop knee or hip osteoarthritis than persons in the lowest exposure groups (Vuori 2001). The combination of high-load occupational activities and high-load sport participation multiplies the risk of osteoarthritis. The dose associated with a significantly increased risk of knee osteoarthritis is approximately 4 h per day of high-load occupational and leisure-time activity.

Participation in moderate types and amounts of activity, such as running, walking, or cycling, poses low, if any, risk of osteoarthritis (Physical Activity Guidelines Advisory Committee [PAGAC] 2008). In fact, we know that some level of activity is necessary for optimal joint health. Loss of joint motion, either from immobilization postinjury or postsurgery or from spinal cord transection, is associated with muscle atrophy, declines in bone density, slowed tissue metabolism, connective tissue stiffness, and other deleterious effects on the musculoskeletal system.

Several factors warrant consideration when we attempt to quantify the dose of activity that may help protect against osteoarthritis. The primary focus should be on the amplitude and rate of force loading, as well as the frequency (number of repetitions and accumulation over a lifetime) and duration of each force loading. It is also necessary to consider the **torsional joint loading** of the activity because torsion or twisting forces produce the highest joint forces and tend to be concentrated in small areas of the joint surface (Buckwalter 2003). For a variety of reasons, most researchers have found it impossible to quantitatively define a dose–response relationship between activity level and osteoarthritis that incorporates all these measurements. However, by incorporating data from animal, clinical, and observational studies, we can hypothesize that the relationship is likely nonlinear and is probably J-shaped. Figure 15.3 illustrates this theoretical relationship. There is no direct

Sport Participation and Osteoarthritis

Lifelong participation in sports that cause minimal twisting and torsional joint forces (e.g., walking, cycling, swimming) does not increase an individual's risk of developing osteoarthritis. High-impact sports such as football and soccer carry higher inherent risks of major joint injuries and are associated with the subsequent development of posttraumatic osteoarthritis. Even in the absence of major joint injury, elite and professional athletes who play high-impact sports for many years likely have a higher risk of developing osteoarthritis compared to nonathletes.

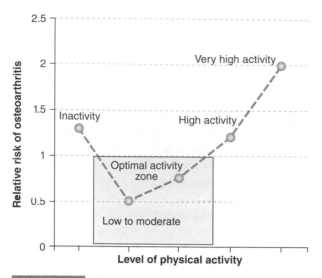

FIGURE 15.3 Theoretical relationship between physical activity level and osteoarthritis.

evidence to show that physical activity in the optimal zone prevents the development of osteoarthritis. However, weak evidence from longitudinal cohort studies suggests that participation in low-impact, health-enhancing physical activity such as walking may provide protection against the development of hip and knee osteoarthritis (PAGAC 2008).

Exercise as a Treatment for Osteoarthritis

The previous section addressed the fact that some types of vigorous sport participation may contribute to an increased risk of osteoarthritis. However, once an individual has been diagnosed with osteoarthritis, physical activity can also ameliorate the effects of the condition. Exercise, both aerobic and resistance, has been consistently shown to decrease pain; improve physical function, mental health, and overall quality of life; and delay disability among adults with knee osteoarthritis (PAGAC 2008; Vuori 2001). For example, persons with knee osteoarthritis typically present clinically with severe quadriceps atrophy and weakness and poor cardiorespiratory fitness. Exercise that strengthens the anterior thigh musculature can vastly improve gait performance, mobility, and balance and helps reduce pain in the long term. Stronger muscles help disperse the joint stresses associated with daily movement. Range of motion and flexibility exercises promote optimal joint lubrication and connective tissue extensibility. Appropriate aerobic exercise improves fitness levels and emotional well-being and is integral to weight loss and maintenance.

What type and amount of exercise are most beneficial for persons with osteoarthritis? The published studies are relatively homogenous with regard to dose of activity. The standard exercise program involves moderate types of exercise such as walking, cycling, or swimming three or four times per week for 30 to 60 min. United States physical activity recommendations suggest that adults with arthritis engage in low-impact, moderate-intensity activity at least 150 min per week and engage in muscle-strengthening exercises at least two days per week (U.S. Department of Health and Human Services [DHHS] 2008). The sessions can be broken into 10-min bouts. Few if any adverse events have been reported with these doses of

Proper exercise technique may reduce the damage to joints that can lead to osteoarthritis.

exercise, and they are generally regarded as safe. However, a small subgroup of persons with severe knee osteoarthritis who also have joint deformity (varus or valgus angulation) and excessive joint laxity should consult a health professional and undergo supervised exercise programs, as some types of quadriceps exercises may worsen symptoms. More research is needed to identify both the minimal dose of activity that has disease-specific benefits and the maximal dose that may exacerbate symptoms or cause injury.

Physical Activity and Bone Development and Maintenance

Physical activity is crucial to total bone health and has the potential to affect bone density at three critical time periods during a lifetime (PAGAC 2008; Kahn et al. 2001; Vuori 2001):

1. Weight-bearing activity in childhood and adolescence helps to develop peak bone mass. Puberty is the most critical time period because the rate of bone mass accrual is highest during this period. In most adolescents, bone mass peaks around age 18. Children who participate in sports that involve running, jumping, and twisting have higher peak bone mass than more inactive children when they reach young adulthood. Targeted bone-loading, school-based interventions have also been successful at increasing bone density as much as 10% above levels in the control group.

2. During the second critical time period, the second through fifth decades, physical activity helps maintain peak bone mass. Longitudinal intervention studies have shown that bone density may be augmented about 1% to 3% in active young adults compared with controls.

3. In later adult life, the third critical time period, physical activity slows the rate of decline in bone mass as much as 1% per year. Because of the hormonal changes associated with menopause and their relation to bone health, postmenopausal women are a high-risk population with an accelerated risk of osteoporosis. However, evidence suggests that physical activity can slow bone loss and even facilitate some gain in bone density among postmenopausal women who already have osteoporosis and are taking hormone replacement therapy. Higher levels of physical activity are associated with lower rates of fracture, particularly of the proximal femur (PAGAC 2008).

Bone responds variably depending on the extent of strain applied. Loads below 50 Φ (microstrain) have no appreciable effect on bone. Loads in the 50 to 200 Φ range are physiological in effect and promote healthy remodeling of bone. Loads ranging from 2,000 Φ to 4,000 Φ overload bone and stimulate new bone formation. Loads in excess of 4,000 Φ are pathological, resulting in microdamage and initiating the production of unorganized bone cells to assist in the repair process. Cortical bone fractures at compressive loads of about 25,000 Φ. Fortunately, loads generated during normal activities, including high-intensity, vigorous sport and exercise, rarely exceed 1,500 to 3,000 Φ (Kahn et al. 2001).

No evidence exists to define a clear dose–response relationship, in terms of frequency, intensity, and duration, between physical activity and bone health. Despite this, research in both animals and humans suggests that several components of exercise provide the appropriate stimulus to bone (Kahn et al. 2001; Vuori 2001). Bone-loading activities should impose forceful and fast mechanical loads that may be generated from either muscular contraction or weight-bearing forces and that load the bone at various angles. Specific bones must be targeted for the best localized effect. For instance, to increase bone density in the tibia, activities must incorporate rhythmic contractions of the large muscles of the lower leg, as well as compressive forces on the foot, ankle, and lower leg. The number of repetitions or loading cycles need not be high; strain magnitude, rate, and distribution play the most important role in stimulating bone growth. Also, because bone accommodates to customary loads over time, the magnitude of loads applied needs to increase periodically to continually stimulate bone formation. Exercise programs that incorporate high-impact jump training (50-200 jumps per day, variable height 7.6-25 cm [3-10 in.]), high-intensity aerobic training (walking, jogging, stair climbing; 50-min sessions four times per week at 55-75% $\dot{V}O_2$max), or resistance training (12 different exercises, three sets of 8-12 repetitions, 70-80% of one-repetition maximum [1RM]) have been successful at maintaining or increasing bone density.

The Role of Physical Activity in Preserving Function

Evidence suggests that physical activity levels are inversely associated with the development of functional and role limitations in middle-age and older adults. In fact, several longitudinal studies show that physically active persons have an approximately 30% lower risk of developing moderate or severe functional limitations compared to sedentary persons (Singh 2002; PAGAC 2008). A recent longitudinal study showed that physically active overweight older adults had significantly lower disability scores over 13 years than their normal-weight sedentary peers (Bruce, Fries, and Hubert 2008). There are a variety of mechanisms by which physical activity and exercise can help prevent incident function loss and disability during the aging process. Directly, exercise may affect physiologic capacity (cardiorespiratory function, muscle strength, flexibility, balance). Indirectly, physical activity may affect psychosocial factors such as low self-efficacy and depression, which in turn can influence activity levels. In the context of a general disability prevention framework, four constructs help elucidate the complex relationship between physical activity level and disability (Singh 2002; Spirduso and Cronin 2001):

1. Physical activity may delay various physiological processes associated with aging.

2. Physical activity has been shown to modify risk factors (hypertension, high blood glucose, hypercholesterolemia) for common chronic and disabling diseases.

3. Physical activity may affect the course and sequelae of diseases already present.

4. Physical activity can affect other contributors to disability, such as depression, low self-efficacy, and lack of social support.

To illustrate these concepts, let's look at muscular function and its relationship to the onset of disability. Sedentary individuals lose about 10% of muscle mass every decade of their adult life. This age-related loss of muscle mass is called **sarcopenia**, which is an established risk factor for disability (Singh 2002). Engaging only in aerobic activities does not provide enough stimulus to preserve muscle mass as we age: Muscle must be properly "loaded" through resistance training activities. Progressive strength training programs have demonstrated increases in muscle strength of 40% to 150%; these strength and subsequent functional gains have been reported for adults as old as 80 to 90 years, suggesting that it is never too late to start a strengthening program. Women may particularly benefit from strength training because they tend to have lower baseline reserves of muscle mass and progress to functional disability an average of 10 years earlier than men (Spirduso and Cronin 2001).

Relatively few studies have investigated primary prevention of disability through exercise interven-tions, and these published studies have been small in size, have been of short duration, and have used exercise interventions of very low intensity. No significant differences were noted in the rate of disability between exercise and comparison groups even though strength gains and other beneficial effects were reported (Spirduso and Cronin 2001). There is a significant need for additional research in this area.

On the other hand, the evidence is much stronger regarding secondary and tertiary prevention of disability among adults with existing functional impairments. Physical inactivity is an independent predictor of functional decline and resulting disability. Moderate-intensity activity such as walking seems to have the same favorable effects on function as more strenuous activity programs. Observational studies suggest a dose–response relationship between activity level and secondary and tertiary prevention of functional disability; however, the ideal dose, and especially the minimum intensity, have not been identified (PAGAC 2008; Singh 2002; Spirduso and Cronin 2001).

Despite the lack of tested exercise prescriptions designed to ward off age- and disease-related functional loss and disability, some recommendations have been published. A comprehensive disability prevention activity prescription should, at minimum, contain

- a functional aerobic component (such as walking activities) to promote cardiorespiratory function as well as muscle mass and strength development,

- static and dynamic balance activities to improve movement self-efficacy and reduce fall risk, and

- resistance training to strengthen all major muscle groups (Singh 2002).

Strengths and Limitations of the Evidence

The health effects of physical activity on the musculoskeletal system are strikingly consistent across a variety of conditions (osteoporosis, osteoarthritis) and health domains (physical function, disability). Although activity-related injuries can happen, many can be prevented if people follow simple steps such as choosing appropriate activities and venues, using safety equipment, and adhering to the rules of the sport. Physical activity can also help people maintain or lose excess body weight, a major risk factor for osteoarthritis, which may help in the primary prevention of this leading cause of disability. Future research should focus on (1) identifying the optimal dose of physical activity that may prevent incident musculoskeletal disease, as well as the dose needed for management of chronic musculoskeletal conditions; (2) clarifying the relationship between physical activity level and the prevention of incident disability; and (3) refining the methods of defining and measuring physical activity as related to the accumulated effects on the musculoskeletal system.

Clarifying the Role of Physical Activity in Disability Prevention

The benefits of physical activity on functional health and disability have been consistently shown in observational studies (PAGAC 2008). Despite this, many gaps in our understanding still exist. One important issue is the fact that no randomized controlled trials (RCTs) have been conducted to prospectively examine the effect of physical activity on the prevention of functional impairment and disability outcomes. Part of the reason is that such studies would require large numbers of subjects to be followed for long periods of time, which suggests that RCTs are not feasible because of the time and cost involved. Also, some experts would argue that randomizing people to a nonexercise control group is not ethical given the wide range of known health benefits of physical activity. However, RCTs are considered the best study design to control for other risk factors for disability, as well as to investigate the interactions between risk factors. A second issue is related to varying definitions of disability and the lack of standardized measures and operational definitions across studies. Disability is a complex concept, and accurately and completely capturing this concept is difficult.

Challenges in Defining Exposure Data for Musculoskeletal Outcomes

It is difficult to determine with certainty the exact relationship between physical activity and various musculoskeletal outcomes for two primary reasons. The first is that historically, physical activity has been measured in epidemiological studies using self-report survey instruments. These surveys tend to have problems with recall bias, especially when participants are asked to recall activity levels over an entire lifetime. With regard to musculoskeletal

outcomes, the lifetime accumulation and patterns of activity are the most important exposure elements to capture accurately. Also, because most surveys have been developed for studies of cardiovascular health or other nonmusculoskeletal disease outcomes, the types of activities that people are asked about to ascertain physical activity level tend to be based on their cardiorespiratory effect and not necessarily their musculoskeletal effect. For example, in terms of cardiorespiratory effect, swimming and running are similar; however, these two activities are very different in terms of their net loading effect on the musculoskeletal system. Many surveys also fail to ask questions regarding light-intensity activities that significantly stress bones or joints (e.g., repetitive squatting and kneeling during household tasks) but do not sufficiently stress the cardiorespiratory system.

The second reason it is difficult to determine the relationship between physical activity and musculoskeletal outcomes is that standardized scoring systems do not assess the dose of activity that is actually delivered to the musculoskeletal system. The most common scoring method uses a standardized set of absolute MET (metabolic equivalent) values for each activity and then applies this information in an equation that includes frequency and duration of activity to calculate a summary dose of activity. However, MET-based intensity values primarily represent the energy cost of a given activity. Thus, neither the design of the self-report surveys nor the scoring algorithms used in analysis reflect the dose of activity that is delivered to the musculoskeletal system. There is a critical need to develop a standardized scoring system that can be used to weight self-reported physical activity data according to the musculoskeletal loading aspects of the activity.

Summary

This chapter discusses physical activity as an essential part of musculoskeletal health. Physical activity plays an important role in preventing fall-related fractures and loss of physical function, as well as in managing arthritis and osteoporosis. Although some sports may be associated with high rates of activity-related injuries, participation in the lifestyle activities typically recommended for health benefits carries little risk of injury. Future research should focus on identifying the optimal dose and type of activity needed for developing and maintaining a healthy musculoskeletal system and to prevent disability.

Key Concepts

arthritis and other rheumatic conditions—A set of more than 120 medical conditions or diseases that primarily affect the musculoskeletal system. The most common conditions are osteoarthritis, rheumatoid arthritis, gout, fibromyalgia, systemic lupus erythematosus, arthritis, polymyalgia rheumatica, and psoriatic arthritis.

crepitus—A grating, crunching, or crackling sensation under the skin, in the lungs, or around the joints; can be felt and sometimes heard on palpation of a joint during movement.

mechanostat theory—Theory of the process by which bone formation is initiated in response to mechanical loading. It posits that a minimum effective strain must be present to promote bone cell proliferation.

osteoarthritis—Condition of the hyaline cartilage and underlying bone in which cartilage fractures and disintegrates, resulting in joint swelling, pain, and deformity; also called degenerative joint disease. Diagnosis is based on specific radiographic features (joint space narrowing, osteophytes, and subchondral bone cysts) and clinical symptoms (pain, swelling, and reduced motion).

osteophyte—Excess bone formation, usually caused by abnormal stress or pressure at joint margins or sites of ligament or tendon attachment.

osteoporosis—Loss of bone mineral density and the inadequate formation of bone, which can lead to an increase in bone fragility and a risk of fracture.

sarcopenia—Loss of lean muscle mass attributable to factors such as aging, poor nutrition, and lack of exercise, often leading to weakness, decreased metabolic energy needs, fatigue, and eventually functional limitation and loss of independence.

sclerosis—Stiffening of bone directly beneath the hyaline cartilage of a joint; caused by accelerated bone turnover in response to superficial cartilage degeneration (cracks and fissures).

torsional joint loading—Stress imposed by activities that involve weight-bearing and twisting motions, such as tennis, soccer, and basketball. These types of activities can transmit high-impact forces to the large weight-bearing joints and may contribute to the development of osteoarthritis.

wear and tear theory—Theory that explains one mechanism by which osteoarthritis may develop. Suggests that accumulation of joint stresses from high-load activities over a lifetime initiates and promotes the degradation of hyaline cartilage in the joints.

Study Questions

1. What three leading causes account for almost 40% of all disabilities?

2. What is the leading cause of death among persons aged 1 to 34 years?

3. Clinical trial research suggests that fall and hip fracture rates can be reduced through exercise programs that focus on which of the following?

   a. Cardiorespiratory fitness

   b. Flexibility

   c. Balance and mobility

   d. Strength and function

4. What is the lifetime risk of osteoporotic fracture among white women?

5. What are the three basic principles of mechanical bone loading?

6. What are four ways in which physical activity can affect the development of disability?

References

Arthritis Foundation. 2001. *Primer on the rheumatic diseases,* 12th ed., ed. J.H. Klippel. Atlanta: Arthritis Foundation.

Bruce, B., J.F. Fries, and H. Hubert. 2008. Regular vigorous physical activity and disability development in healthy overweight and normal-weight seniors: A 13-year study. *American Journal of Public Health* 98(7):1294-1299.

Buckwalter, J.A. 2003. Sports, joint injury, and post-traumatic osteoarthritis. *Journal of Orthopaedic and Sports Physical Therapy* 33:578-588.

Centers for Disease Control and Prevention. 2009. Prevalence and most common causes of disabilities among adults, United States, 2005. *Morbidity and Mortality Weekly Report* 58:421-426.

Chandran, A., A.A. Hyder, and C. Peek-Asa. 2010. The global burden of unintentional injuries and an agenda for progress. *Epidemiologic Reviews* 32:110-120.

Felson, D.T., R.C. Lawrence, P.A. Diepp, et al. 2000. Osteoarthritis: New insights part I: The disease and its risk factors. *Annals of Internal Medicine* 133(8):635-646.

Johnell, O., and J.A. Kanis. 2006. An estimate of the worldwide prevalence and disability associated with osteoporotic fracture. *Osteoporosis International* 17:1726-1733.

Khan, K., H. McKay, P. Kannus, et al. 2001. *Physical activity and bone health.* Champaign, IL: Human Kinetics.

Lawrence, R.C., D.T. Felson, C.G. Helmick, et al. 2008. Estimates of the prevalence of arthritis and other rheumatic conditions in the United States. Part II. *Arthritis and Rheumatism* 58(1):26-35.

Li, F., P. Harmer, R. Glasgow, et al. 2008. Translation of an effective tai chi intervention into a community-based falls-prevention program. *American Journal of Public Health* 98:1195-1198.

National Center for Injury Prevention and Control. 2005. *CDC's unintentional injury activities – 2004.* Atlanta: Centers for Disease Control and Prevention, National Center for Injury Prevention and Control.

National Center for Injury Prevention and Control. 2006. *CDC injury fact book.* Atlanta: Centers for Disease Control and Prevention, National Center for Injury Prevention and Control.

National Center for Injury Prevention and Control. 2007. Web-based injury statistics query and reporting system (WISQARS). www.cdc.gov/injury/wisqars/index.html. Accessed August 3, 2010.

National Osteoporosis Foundation. 2002. *America's bone health: The state of osteoporosis and low bone mass in our nation.* Washington, DC: National Osteoporosis Foundation.

Physical Activity Guidelines Advisory Committee. 2008. *Physical Activity Guidelines Advisory Committee report, 2008.* Washington, DC: U.S. Department of Health and Human Services.

Singh, M.A.F. 2002. Exercise to prevent and treat functional disability. *Clinics in Geriatric Medicine* 18:431-462.

Spirduso, W.W., and D.L. Cronin. 2001. Exercise dose-response effects on quality of life and independent living in older adults. *Medicine and Science in Sport and Exercise* 33(6 Suppl):S598-S608.

United States Bone and Joint Decade. 2008. *The burden of musculoskeletal diseases in the United States.* Rosemont, IL: American Academy of Orthopedic Surgeons.

U.S. Department of Health and Human Services. 2008. *2008 Physical activity guidelines for Americans.* Washington, DC: U.S. Department of Health and Human Services.

Vuori, I.M. 2001. Dose-response of physical activity and low pack pain, osteoarthritis and osteoporosis. *Medicine and Science in Sports and Exercise* 33(6 Suppl):S551-586.

Woolf, A.D., and B. Pfleger. 2003. Burden of major musculoskeletal conditions. *Bulletin of the World Health Organization* 81(9):646-656.

16

Physical Activity, Muscular Fitness, and Health

Neil McCartney, PhD; and Stuart M. Phillips, PhD

CHAPTER OUTLINE

History of Resistance Training and Its Role in Health

Fundamental Aspects of Resistance Training

Effects of Resistance Training

Dose–Response Effects

Turnover of Muscle Proteins

Sex-Based Differences in Hypertrophic Gains

Importance of Skeletal Muscle Mass and Quality

Resistance Training Throughout the Life Span

Resistance Training in Children and Adolescents

Resistance Training in Middle-Aged and Elderly People

Resistance Training in Disease and Disability

Resistance Training, Weight Control, and Obesity

Resistance Training in Coronary Artery Disease

Resistance Training in Chronic Heart Failure and After Heart Transplantation

Resistance Training in Arthritis

Resistance Training in Osteoporosis

Summary

Review Materials

©Anton

This chapter considers muscular fitness in terms of muscular strength and power and the properties of muscle that contribute to its mass and quality. The relationships between muscular fitness and health throughout the life span are reviewed, both in healthy individuals and among groups with diseases and disabilities. We demonstrate that muscular fitness is perhaps as important as aerobic fitness in improving health and maintaining good health. Because progressive overload **resistance training (RT)** is the method of choice to increase muscular fitness, this literature is reviewed.

History of Resistance Training and Its Role in Health

In this chapter we define RT as a technique whereby external weights are used to provide progressive overload to skeletal muscles to strengthen and enlarge them. In any given resistance exercise, it is typical to set the load as a percentage of the individual's **one-repetition maximum (1RM)**, the maximum load that can be lifted only once throughout a complete range of motion. The dose of RT is described by the load lifted, the number of repetitions, and the number of sets of repetitions. It is typically believed that three to five sets using high loads ($\geq 80\%$ 1RM) and low repetitions (three to six) are best for increasing muscle strength and mass, whereas lower loads (50-70%) and higher repetitions (10-20) are best for increasing muscular endurance and muscular power (the ability to move an external load rapidly). As discussed later, the evidence for these assertions is not overwhelming.

The formal origins of RT are found in the 1945 publication by Captain Thomas De Lorme, showing that heavy RT restored muscular strength and power in physically disabled war veterans. De Lorme's strategy of using heavy loads and low repetitions to build strength versus low loads and high repetitions to build muscular endurance is still the accepted approach to this day. After that original publication, RT was often recommended as part of a balanced exercise program until the burgeoning literature on the cardioprotective effects of aerobic exercise tended to make this form of exercise preeminent. The 1968 book *Aerobics*, by Kenneth Cooper, catalyzed a decade of publications on the health and fitness benefits of activities such as running, but RT received little attention. Indeed, the first position statement by the American College of Sports Medicine (ACSM)

in 1978, "The Recommended Quantity and Quality of Exercise for Developing and Maintaining Cardiorespiratory and Muscular Fitness in Healthy Adults," did not include any RT guidelines. By the mid-1980s, publications began to appear demonstrating the benefits of RT in various clinical populations, such as those with coronary artery disease (CAD), hypertension, and low bone mineral density (BMD); and in 1990, the ACSM revised its position statement to include guidelines for RT. This was followed by recognition of the importance of RT by the American Heart Association and the American Association of Cardiovascular and Pulmonary Rehabilitation. Resistance training is now considered an important component of a balanced exercise program for healthy individuals and many patient groups.

> The well-documented cardioprotective effects of aerobic exercise previously overshadowed resistance exercise and training; however, RT is now recognized as a critically important part of any formal exercise program.

Fundamental Aspects of Resistance Training

Muscle tissue is very malleable, changing in response to external stimuli. One such stimulus is resistance exercise, which involves performance of high-force but brief contractions. The usual adaptation to this stimulus is that the muscle grows larger or undergoes **hypertrophy**, which means that the volume of muscle increases so long as the stimulus is maintained and is frequent enough. The volume of high-force contractions necessary to elicit this response is still not completely clear, but enough data are now available on which to base some general guidelines.

We do know that for hypertrophy to occur there must be a shift in the continual turnover of muscle proteins toward a state of net protein synthesis. In this section, we introduce the concept of skeletal **muscle protein turnover** and show how this system is affected by resistance exercise and feeding. We also examine other potentially important health effects of the increase in skeletal muscle mass and introduce the concept of skeletal muscle "metabolic quality." In addition, we discuss differences between men and women with regard to the impact of sex on the ability to accrue muscle proteins.

Effects of Resistance Training

Resistance training is fundamentally anabolic. Thus, when regularly performed, RT stimulates skeletal muscle to synthesize new muscle proteins and to preserve existing proteins. Although RT can damage muscle by disrupting muscle ultrastructure and releasing myocellular constituents, such damage is repaired and the muscle becomes stronger and remarkably resistant to subsequent damage. Resistance training can lead to an accumulation of muscle protein, or muscular hypertrophy, when performed regularly and with sufficient intensity, usually greater than 50% to 60% of a person's 1RM (figure 16.1), and sufficient volume (figure 16.2). In fact, there may be a trade-off between these two variables in terms of promoting hypertrophy. That is, if a person performs RT at a lower intensity but employs a higher volume, this may have effects similar to those of low-volume, high-intensity RT. Muscular **atrophy**, or loss of muscle protein, results from acute (i.e., immobilization, exposure to microgravity) or prolonged and persistent (i.e., aging, wasting diseases) withdrawal of a sufficiently intense contractile stimulus. Hence, as one might imagine, RT is an attractive therapeutic tool for persons who are losing skeletal muscle mass.

> **Resistance training promotes anabolism and thus promotes muscle growth and retention of muscle protein. These characteristics make RT a highly valuable clinical tool in treating declining skeletal muscle mass.**

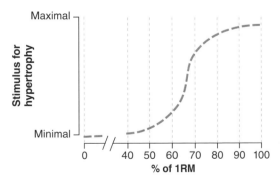

FIGURE 16.1 Theoretical curve showing the relationship of resistance training intensity to a relative stimulus for hypertrophy. Note that at intensities of ~60% to 80% of one-repetition maximum, the stimulus for hypertrophy increases, but it plateaus at higher intensities.

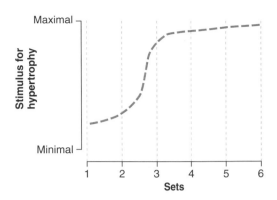

FIGURE 16.2 Theoretical curve showing the relationship between resistance training volume and stimulus for hypertrophy. It appears that beyond three sets, so long as voluntary fatigue has been induced, the stimulus for muscle mass, as well as strength gains, is relatively minimal.

Data from Krieger 2009.

Dose–Response Effects

The effective dose of RT that induces a beneficial outcome depends greatly on the outcome desired (strength, hypertrophy, insulin sensitivity) and the target population (Phillips and Winett 2010). Where strength is concerned, one can see changes in as few as one or two sessions of RT, reflecting neural adaptations rather than muscle protein accumulation (i.e., hypertrophy). However, the optimal dose or volume of exercise required to maximize strength gains is an understudied area in which information is truly lacking. Some attempts to reach consensus in this area have been made, but we believe that the evidence behind such recommendations is inadequate (Feigenbaum and Pollock 1999). We have attempted to summarize several published studies in which hypertrophy, not strength gains, was the measured outcome and have put those results in schematic form in figures 16.1 and 16.2. These are still theoretical constructs, and much work remains to establish true dose–response effects concerning resistance exercise and hypertrophy. For reasons outlined subsequently, we believe that viewing resistance exercise from the standpoint of its ability to increase muscle mass is more beneficial than considering only strength, which almost always increases to a varying degree. Most organizations such as ACSM have aimed to develop both muscular strength and power in relatively sedentary, diseased, or aged individuals; their recommendations can be summarized as one or two sets of 8 to 12 repetitions per set, with 8 to 10 exercises per session and no more than two or three sessions per week

(Feigenbaum and Pollock 1999). However, when viewed from the perspective of "effort" per se, it becomes apparent that a strenuous effort, with any number of sets and intensities, results in very similar adaptations. In other words, there is substantial redundancy in many of the variables that are considered important by certain organizations (Phillips and Winett 2010).

Turnover of Muscle Proteins

The process of muscle turnover, as shown in figure 16.3, illustrates the interplay of protein synthesis and breakdown that occurs simultaneously and continuously in skeletal muscle and all body tissues. In figure 16.3, the rate of protein breakdown (B) exceeds that of protein synthesis (S), and the net protein balance (equal to S minus B) is negative (net loss of protein). This is the situation observed in the fasted state, which begins approximately 4 to 5 h after food consumption. Note that there is a net loss of amino acids from muscle (i.e., the outward flux of amino acids is greater than the inward). Consumption of protein results in a fed-state hyperaminoacidemia (i.e., increase in blood amino acids), promoting inward transport of amino acids from the blood into the skeletal muscle pool of free amino acids. Protein synthesis is then stimulated to yield a net positive protein balance (figure 16.4). The stimulation of muscle protein synthesis appears to be mediated by the amino acids themselves and relies on the presence of a small amount of insulin. As figure 16.5 indicates, the magnitude of the negative net balance with fast-

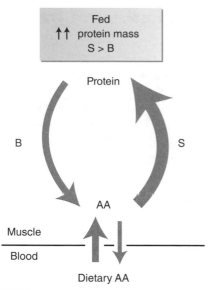

FIGURE 16.4 Schematic of muscle protein turnover in the fed state. Muscle protein synthesis (S) is stimulated by amino acid influx into the cell; breakdown (B) may be suppressed somewhat, but net balance is positive and equal in magnitude to the fasted losses seen in figure 16.3.

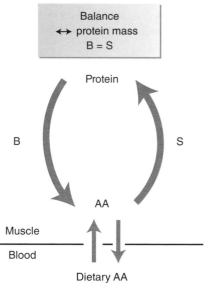

FIGURE 16.5 If adequate energy and sufficient protein (i.e., containing all essential amino acids) are consumed daily, then muscle protein balance is maintained and muscle mass remains relatively unchanged.

ing is equal to that of the positive net balance with feeding; hence, skeletal muscle mass is maintained by feeding. In fact, this happens daily in people who perform no RT but simply eat sufficient protein containing adequate quantities of essential amino acids.

As previously noted, RT is anabolic; it shifts the balance of protein turnover toward net anabolism by increasing both the rate of protein synthesis and the rate of protein breakdown. In a fasted state, performance of RT stimulates protein synthesis, but

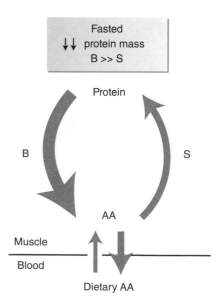

FIGURE 16.3 Schematic of muscle protein turnover in the fasted state. Muscle protein breakdown (B) exceeds synthesis (S) so that net muscle balance (S minus B) is negative, resulting in a net loss of muscle protein.

insufficient amino acids are available in the circulation to be transported into the muscle to shift a negative net protein balance to a state of net protein gain. Net balance is less negative following RT but is not positive (figure 16.6). Only when protein is consumed following RT does a maximal stimulation of protein synthesis yield a net positive muscle protein balance (figure 16.7). The concepts presented in figures 16.3 through 16.7 are shown dynamically in figure 16.8. The curves in this figure show that swings in net protein balance occur with feeding and fasting and that RT results in a reduced loss of protein in the fasted state and a higher gain in the fed state. Although the curves in figure 16.8 are perfectly shaped and are sinusoidal, they are generalized.

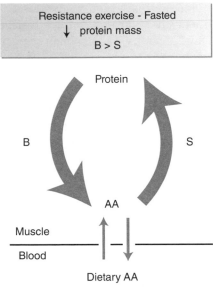

FIGURE 16.6 Resistance exercise is a potent stimulator of synthesis (S) and also of breakdown (B), to a much smaller degree. Net muscle protein balance is improved (i.e., less negative than depicted in figure 16.3), which is attributable to improved intracellular recycling of free amino acids, but is not positive.

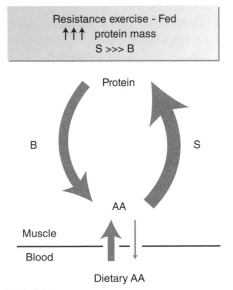

FIGURE 16.7 After resistance exercise, in the presence of amino acids, synthesis (S) is additively stimulated by feeding and resistance exercise. The response yields a greater net balance than feeding or resistance exercise alone, resulting in muscle protein accretion.

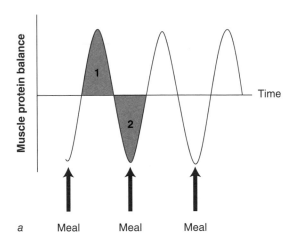

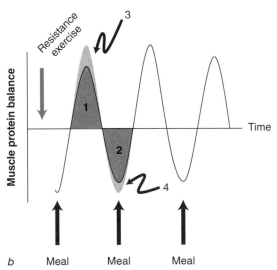

FIGURE 16.8 (a) Normal fed-state gains and fasted-state losses in skeletal muscle protein balance (synthesis minus breakdown). Note that the area under the curve in the fed state (indicated by 1) would be equivalent to the fasted loss area under the curve (indicated by 2); hence, skeletal muscle mass is maintained by feeding. (b) Fed-state gains and fasted-state losses in skeletal muscle protein balance with performance of resistance training. In this scenario, fasted-state gains are enhanced by an amount equivalent to the stimulation of protein synthesis brought about by exercise (indicated by 3). Additionally, fasted-state losses appear to be less (indicated by 4), which is attributable to persistent stimulation of protein synthesis in the fasted state.

Reprinted from *Nutrition*, Vol. 20, S.M. Phillips, "Protein requirements and supplementation in strength sports," pp. 689-95, copyright 2004, with permission from Elsevier.

Greater meal frequency or longer periods of fasting would affect the temporal nature of the response of muscle protein net balance and possibly the rate of rise and fall of the response. It is also likely, but has not yet been shown, that the intensity and volume of exercise affect the amplitude and duration of the anabolic response (Burd et al. 2009).

■ **The balance of muscle protein turnover, which includes both muscle protein synthesis (S) and muscle protein breakdown (B), determines muscle protein gain or loss. When S chronically exceeds B, then muscle protein accretion—hypertrophy—occurs; in the opposite situation (B > S), wasting or atrophy occurs. Both feeding of protein and RT stimulate S and thus shift the balance of muscle protein turnover toward gain of protein mass.**

Sex-Based Differences in Hypertrophic Gains

One school of thought is that a large portion of the acute exercise-induced muscle mass gain (see figure 16.8) is attributable to the postexercise increase in anabolic hormones such as growth hormone (GH), insulin-like growth factor 1, and testosterone. What is now evident is that pharmacological doses of GH do not have a marked effect in increasing muscle mass or strength in persons with normal pituitary function, whereas pharmacological doses of testosterone do. Hence, the acute postexercise increase in testosterone has been implicated as an important factor in determining hypertrophy. The rise in testosterone following intense RT is, however, small (relative to the daily circadian release of the hormone) and remarkably transient (lasting only 30-60 min); it occurs in the absence of a significant increase in luteinizing hormone (LH) and tracks closely with shifts in hematocrit. The latter points highlight the fact that the postexercise increase in testosterone concentration is simply attributable to increased hemoconcentration of the hormone from plasma shifts and not increased production, which would be accompanied by higher levels of LH. Hence, it is questionable whether the acute increase in testosterone (or of GH) after exercise is important in determining the hypertrophic response.

Women have a 10-fold lower concentration of testosterone than men, yet several studies have shown that women who follow an intense, progressive RT program have significant strength gains and muscle hypertrophy. In fact, it has been concluded that abso-lute and relative maximal dynamic strength increase in a similar manner for both sexes. Moreover, data suggest that skeletal muscle adaptations that may contribute to strength gains of the lower extremity are similar for men and women during the early phase of RT. Such RT-induced hypertrophic responses are not related to either transient or chronic changes in circulating testosterone, except possibly in elderly women. Even in elderly women, the change in systemic testosterone cannot explain more than 20% to 25% of the variance in strength gains with RT. Instead, local muscle factors potentially acting in an autocrine or a paracrine manner are more important than circulating systemic hormones in determining exercise-induced hypertrophic gains (West et al. 2010).

Importance of Skeletal Muscle Mass and Quality

Aside from the obvious role that skeletal muscle mass plays in locomotion, why is it important to maintain skeletal muscle mass? Because skeletal muscle has a large working range of adenosine triphosphate turnover rates, it has tremendous potential to consume energy and hence is important in weight maintenance or loss, as discussed later. Because of its mass, skeletal muscle is highly important thermogenic tissue and an important contributor to basal metabolic rate (BMR), which for most people is the largest single contributor to daily energy expenditure; again, this fact highlights the importance of maintaining muscle mass. Because of its oxidative capacity (i.e., mitochondrial content), skeletal muscle is also a large site of lipid oxidation, potentially playing a role in maintaining balance in lipoprotein and triglyceride homeostasis. Skeletal muscle is also, mostly because of its mass, the primary site of glucose disposal in the postprandial state. Hence, maintaining skeletal muscle mass would also reduce the risk for development of type 2 diabetes mellitus. Finally, the decline in maximal aerobic capacity with age and with muscular wasting conditions has been found to result largely from a decline in skeletal muscle mass (figure 16.9).

Although skeletal muscle mass per se is important, the quality of skeletal muscle must be maintained. The term **muscle quality** has usually been used in the context of the ability of the skeletal muscle to generate force. In most cases, the potential for force generation of a skeletal muscle is directly proportional to its cross-sectional area. However, with aging and with other neurological and metabolic disorders, force per cross-sectional area declines, which may be a manifestation of reduced skeletal muscle quality. In addition, a relatively inactive skeletal muscle mass is low in metabolic quality because of a lower oxidative

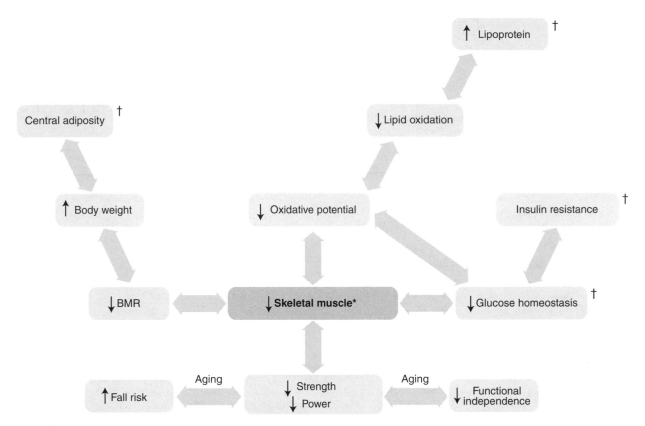

FIGURE 16.9 Interplay of skeletal muscle mass and its relative metabolic activity with certain factors important in the development of various chronic diseases—heart disease, metabolic syndrome, and obesity. In particular, the decline in skeletal muscle mass and the metabolic quality of skeletal muscle can lead to derangements that are considered risk factors for metabolic syndrome. The light blue boxes indicate either a process that negatively affects skeletal muscle mass or a negative outcome of reduced skeletal muscle mass. †Metabolic syndrome abnormalities. *The reduction in skeletal muscle refers to mass, metabolic quality, or both (see text for explanation).

capacity, capillary supply, and transport capacity for fatty acids and glucose. Although aerobic exercise is associated with increased mitochondrial capacity, capillarization, and transporters for glucose and fatty acids, most moderately intense RT programs, particularly in sedentary people, would probably enhance many of these variables also (Phillips 2007).

Beyond the obvious important role that skeletal muscle plays in locomotion, it is also a critically important tissue in maintaining health. Skeletal muscle is active tissue. It burns a tremendous amount of lipid, stores the majority of ingested glucose, and is a significant contributor to BMR. Thus one can easily appreciate how important it is to maintain a large and metabolically active skeletal muscle mass. Doing so likely decreases the risk for numerous health problems including obesity, diabetes, and frailty in the elderly.

Resistance Training Throughout the Life Span

There is no doubt that RT can benefit health, but what about effects on different age groups? Are they similar among young and aged individuals, or is there an age-related loss in the capacity for strength or hypertrophic gain? Should volume and intensity be different? We attempt to answer these questions in the following sections.

Resistance Training in Children and Adolescents

Resistance training programs have proven to be a safe and effective method of "conditioning" for children. A caveat concerning this statement is that exercise guidelines need to be rigorously followed. In fact, participation in well-designed and appropriately supervised programs of strength and conditioning by children and adolescents is supported by the American Academy of Pediatrics and the ACSM. These organizations' position stands are in accordance with such reports as the Surgeon General's *Physical Activity and Health,* which aims to increase the number of children who participate regularly in physical activities that enhance and maintain muscular strength and muscular endurance. Many of the same health-related benefits from RT seen in adults hold true for this age group as well (i.e., improvements in aerobic fitness, reduced risk of osteoporosis and increased bone mass, prevention of obesity and hypertension, increased high-density lipoprotein cholesterol, and improved psychological health indexes). Also, data support the thesis that physically active children are more likely to become physically active adults. Hence, encouraging children who already enjoy free play to participate in RT programs would be advantageous and beneficial.

Although some early reports cast doubt on whether children or adolescents could actually gain strength as a result of RT, subsequent reports clearly showed that RT does result in strength gains, enhanced motor performance skills, and possible benefits to selected anatomic characteristics such as bone density and architecture and lean mass. Furthermore, RT enhances self-efficacy for exercise and other tasks. Significantly, participating in programs of RT appears to reduce injuries in sports and recreational activities. For example, strength-trained athletes (13-19 years old) had a lower injury rate and required less time for rehabilitation than their teammates who did not participate in RT. Resistance training has been shown to reduce the incidence of shoulder pain in 13- to 18-year-old swimmers and to decrease the number and severity of knee injuries in high school football players (Faigenbaum and Myer 2010).

The most efficacious training frequency and intensity rates for children are difficult issues to resolve, because excessive frequency and intensity of resistance activities could lessen adherence and enthusiasm for participation and potentially be injurious. It appears, from the few studies in this area, that a training frequency of two sessions per week results in improved strength gains versus just one session per week. In addition, with respect to enhancing at least upper-body strength and local muscle endurance of untrained children, the prescription of higher-

repetition (13-15 repetitions per set) training protocols, at least during the initial adaptation period, would be optimal.

Resistance Training in Middle-Aged and Elderly People

Beginning in midlife, aging is associated with a time-dependent loss of muscle, known as **sarcopenia**. This loss is commonly thought of as a consequence of old age itself, although chronic illness, poor diet, and inactivity all accelerate its progression. Sarcopenia, because of the associated loss of muscle strength and stamina, is a major cause of disability, frailty, loss of independence, and increased risk of falling and fractures in the elderly. Muscle wasting in the elderly is associated with a 50% loss of muscle fibers between 20 and 80 years and a loss of fiber area from those fibers remaining, particularly type II. The increase in longevity has forced physiologists to reconsider exactly where middle age actually lies in a person's lifetime. It is clear that sarcopenia is not a precipitous phenomenon, because muscle mass is progressively lost at a rate of ~1% to 2% per annum beginning at ~40 years of age. Hence, unlike osteoporosis, sarcopenia is insipid and is not associated with a sudden change in hormonal status. The decline in muscle mass must come about because of an imbalance between muscle protein synthesis and breakdown (figure 16.3); however, exactly which process—synthesis, breakdown, or a combination of these—alters with age to bring about sarcopenia is a matter of great debate. What is clear, however, is that older individuals can accrue muscle mass as a result of RT in the same way as younger subjects. The best evidence suggests that muscle maintains its plasticity and capacity to hypertrophy even into the 10th decade of life.

Projections are that life expectancy of North Americans will have increased by three to four years from 1980 levels by the year 2020, but for more than 40% of people, the years of added life will be spent in a full-time care facility. Data show that 43% of males and 37% of females between the ages of 65 and 74 have a disability or disease that limits the kind or amount of activity they engage in at home or work or in other activities such as travel, sport, or leisure. The most common disabling condition for all elderly persons is arthritis (44% males and 56% females) and the coincident muscle weakness, which may be etiologically linked. Resistance training is an effective intervention to increase total muscle mass, muscle quality (force per cross-sectional area), strength, and power. Because of the lowered functional status of the majority of elderly people, probably no other

It is a common misconception that properly designed RT programs harm children or adolescents. In fact, numerous reports show that RT can increase BMD, enhance skeletal muscle growth, and reduce the incidence of sport-related injuries in children and adolescents.

population segment would benefit more from RT to combat age-related muscle wasting.

The consequences of the age-related reduction in skeletal muscle mass are numerous, including reduced muscle strength and power (Doherty 2001), reduced BMR, reduced capacity for lipid oxidation, increased abdominal adiposity and insulin resistance, and increased risk for falls (figure 16.9). Ultimately, given the decline in skeletal muscle mass and reduction in strength and power in elderly people, sarcopenia contributes substantially to morbidity and reduced mobility in the elderly and thus to significant health care costs (Tseng et al. 1995). According to Tseng and colleagues (1995), sarcopenia is "a progressive neuromuscular syndrome that will lower the quality of life in the elderly by (1) decreasing the ability to lift loads (progressing to difficulty arising from a chair), and (2) decreasing endurance (leading to an inability to perform the activities of daily living, which increases health care costs)" (p. 113). Numerous studies have shown that RT programs for the elderly can increase muscle mass, strength (maximal force-generating capacity), and power (Campbell et al. 1999). It has also been shown that RT programs in elderly men and women can cause substantial fiber hypertrophy (Macaluso and De Vito 2004). All of these RT-induced changes are potent countermeasures to sarcopenia; thus RT is highly effective in reducing disability in the elderly.

Clearly, RT can directly affect sarcopenia, but what about other chronic diseases? As discussed elegantly by Fiatarone Singh (2001), RT presents a unique opportunity to treat **disease clusters** associated with advancing age. For example, it would not be uncommon for a 75-year-old man to be overweight and to have hypertension, vascular disease, hyperlipemia, impaired fasting glucose tolerance (a precursor to diabetes), and some degree of movement impairment. We propose that, for a number of reasons, a regular program of RT might be a useful adjunct to any pharmacological therapy such an individual might be receiving. It is likely that this man would take several different medications each day, but regular RT may substantially reduce his reliance on pharmacological therapy. Evidence is now emerging that RT may also have important spillover effects, aside from those directly related to prevention of disability and movement impairment, such as reducing insomnia, improving poor appetite, improving postural instability, improving muscle weakness, possibly lessening the risk of depression, and delaying the threshold for dependency by promoting increased muscle strength and the ability to perform activities of daily living (Phillips 2007) (figure 16.9).

Given the low functional status and often multitudinous health problems of many elderly persons, there is likely no other population segment that could benefit more from a program of RT. In an elderly person suffering from sarcopenia and health conditions such as central adiposity, vascular disease, impaired glucose tolerance, and osteoarthritis, a program of RT could address all of these diseases directly and not simply treat symptoms as many medicines would.

Resistance Training in Disease and Disability

Numerous studies have demonstrated the effectiveness of RT in rehabilitating patients after myocardial infarction. Substantial evidence is now also accumulating to indicate that RT is effective in controlling weight, treating depression, improving glucose tolerance, and altering blood lipid profile. It can be used not only in a rehabilitative role, but also as a primary treatment for numerous diseases.

Resistance Training, Weight Control, and Obesity

Although it is clear that aerobic exercise is associated with much greater energy expenditure during the exercise bout than RT, numerous studies have shown that regular RT combined with dietary energy restriction is remarkably effective in promoting weight loss. However, in certain populations, such as obese or elderly persons, a program of aerobic training may have only a mild effect in favorably altering body weight or body composition. The reason is that the low aerobic fitness levels prevent significant energy expenditures associated with even longer periods of endurance-based exercise. For example, a 55 kg (121 lb) elderly woman with a $\dot{V}O_2$max of 24 ml · kg^{-1} · min^{-1}, working at 50% of her peak for 30 min, would expend only ~4,200 kJ (~100 kcal), likely with very little residual increase in resting postexercise energy expenditure. In addition, an obese person who restricts energy to lose weight and also performs aerobic exercise would not have an anabolic stimulus to preserve lean mass from the aerobic exercise. Because it is anabolic, RT aids in retaining skeletal muscle mass during energy-restricted periods; although people may lose less weight when they are energy restricted and performing RT, more of the weight they lose is fat mass and not skeletal muscle. Because skeletal muscle is so metabolically active, this pattern of weight loss would be highly advantageous.

Adults who underwent a 12-week program of RT showed an increase in their requirements (~15%) for energy to maintain body weight. The increase was not accounted for solely by the energy cost of the exercise itself, which is relatively low, but instead was attributable to an increase in BMR. In addition, participating in a program of RT promoted an increase in free living physical activity in the elderly. These findings are important because elderly and obese persons do not expend much energy during endurance exercise due to their low fitness levels; hence, RT with its ability to increase BMR and promote greater free-living physical activity is an attractive alternative or a critical adjunct to weight loss programs.

Many of these same arguments can be applied to people with type 2 diabetes mellitus, who are frequently overweight. Another potential advantage of having diabetic patients perform RT along with restricting energy intake is that RT, like aerobic exercise, has been shown to increase the amount and insulin responsiveness of the insulin- and contraction-sensitive muscle glucose transporter, GLUT4. Because muscle accounts for more than 80% of the disposal of an oral glucose load, any weight loss strategy that promotes increased skeletal muscle GLUT4 would be of greater benefit than a strategy simply promoting weight loss alone (i.e., dieting). Hence, it is possible that RT can delay the need for insulin injections or lower the dosage of insulin required for glucose homeostasis.

In 1997, the World Health Organization recognized obesity as a worldwide disease that poses a global challenge to public health and health care systems. Overweight or obese people have a substantially increased risk for morbidity and mortality from numerous chronic disorders, such as diabetes, hypertension, and cardiovascular disease; in fact, excess fat deposited in the abdominal region is a strong predictor of cardiovascular disease and type 2 diabetes mellitus. This may be partially explained by excess accumulation of visceral fat, an independent correlate of insulin resistance and poor blood lipid profile. These observations highlight the need to identify appropriate treatment strategies to prevent and reduce obesity; they also suggest that the effectiveness of these treatments would be enhanced if abdominal obesity, particularly visceral fat, were substantially reduced.

Interventions to treat obesity usually include a combination of diet and pharmacological strategies such as thermogenic agents, appetite suppressants,

gastrointestinal lipase blockers, or synthetic fat derivatives. In addition, several studies have looked at the inclusion of exercise, either alone or combined with diet or drugs, as a strategy to combat obesity. Because of its large energy expenditure, aerobic exercise has often been the intervention of choice for obese persons. Several studies, however, have used RT as an adjunctive therapy in obese persons trying to lose weight. This research showed that in direct head-to-head comparisons of hypocaloric diets with either RT or aerobic exercise, obese men and women lose the same amount of weight. This is despite an almost four- to fivefold difference in exercising energy expenditure between the two exercise modes (Rice et al. 1999). Given the relatively low aerobic fitness of many obese persons, programs of aerobic exercise may not be the best exercise intervention for weight loss for this group. Hence, a general recommendation for obese persons might be that exercise-based interventions to promote weight loss, or at least a beneficial shift of lean:fat mass, incorporate RT as the main form of exercise, or at least that RT accompany aerobic and dietary interventions.

> Despite the observation that aerobic exercise is associated with a larger energy expenditure than resistance exercise, RT has been shown to be an effective tool in promoting weight loss in obese persons. In addition, RT promotes the retention of skeletal muscle mass during periods of energy restriction, making it the logical choice to prevent loss of metabolically valuable skeletal muscle mass during caloric restriction in obese persons undergoing weight loss.

Resistance Training in Coronary Artery Disease

Now considered a standard of rehabilitation for cardiac patients, RT offers this patient group substantial benefits. In this section, we review evidence in this area.

Safety

The first reports of the use of RT in treating CAD appeared in the literature during the mid-1980s (McCartney 1999). Up until that time, RT was contraindicated in this patient group because of untoward symptoms, electrocardiographic changes, and left ventricular wall motion abnormalities that had been observed in patients (with active myocardial ischemia) during sustained isometric handgrip exercise.

Despite these findings with isometric exercise, a later series of studies using intra-arterial catheterization to measure the arterial pressures of patients during RT demonstrated that the responses were no greater than during aerobic exercise and were within clinically acceptable levels (McCartney 1999). Other investigations using two-dimensional echocardiography to assess left ventricular function during RT with intra-arterial pressure measurements reported good preservation of stroke volume and enhanced myocardial contractility, even in patients with documented heart failure and in those with orthotopic heart transplantation. Several investigators reported that RT did not provoke signs or symptoms of myocardial ischemia, most likely because the heart rate was lower and the diastolic pressure was higher than during aerobic exercise, which are conditions favoring coronary artery perfusion. Moreover, myocardial supply-to-demand balance was demonstrated to be more favorable during RT exercises with progressively heavier loads than it was during incremental exercise testing. Another important observation after a period of RT in seniors was that the heart rate and systolic blood pressure product was reduced considerably during lifting of the same absolute heavy loads (McCartney et al. 1993). On the basis of such findings, it is tempting to speculate that RT may reduce myocardial oxygen demands and thus provide some degree of cardioprotection during strenuous isodynamic activities such as lifting, gardening, and snow shoveling. Resistance training could no longer be contraindicated based on the earlier studies of isometric handgrip exercise in patients with active myocardial ischemia.

Efficacy

In addition to research on the acute circulatory and left ventricular responses, a proliferation of studies in the past three decades have examined the efficacy of RT in patients with CAD. Most studies evaluated patients with relatively good left ventricular function who were already participants in community-based maintenance programs, and these studies usually included an aerobic training control group. Training interventions were typically 12 weeks or less and used one or two sets of upper- and lower-body exercises with 10 to 15 repetitions per set. Weightlifting intensities ranged from 30% to 80% of the 1RM, with most studies using moderate loads of 60% or less, and the frequency was most often three times per week. All studies demonstrated increases in the 1RM ranging from 3% to greater than 50%. In one study, patients did as many lifts as possible with their pretraining 1RM to yield a measure of lifting endurance. The 1RM measured at baseline was lifted an average of 14 times before fatigue, indicating that strength-related activities of daily living requiring almost maximum effort at baseline could be done with ease after training and were potentially safer.

In addition to showing improvements in dynamic strength, patients have demonstrated increased endurance of 12% during a standard Bruce treadmill test and a 15% gain in maximum power output during progressive cycle ergometry. One notable observation during cycle ergometry testing was a marked reduction in perceived leg effort at power outputs above 50% of the pretraining maximum, perhaps suggesting that the improved power and endurance were linked to the reduced symptoms of effort from stronger leg musculature. No such increases in cycling power occurred among control participants who had undergone additional aerobic training, further suggesting that the locus of improvement was in the stronger leg muscles. A recent study demonstrated that combining RT two days per week with aerobic training three days per week may yield larger increases in $\dot{V}O_2$max than a regimen of aerobic training five days per week (Marzolini et al. 2008). Studies have shown that improvements in muscular strength correlate strongly with the ability to perform household activities such as climbing stairs, making beds, carrying groceries, and vacuuming (Brochu et al. 2002).

Many patients who have experienced a myocardial infarction are limited more by their own perception that they cannot do certain activities than by any real physical limitation, so an intervention that can favorably alter this perception may significantly affect psychological well-being and quality of life. Such effects may be possible with RT. One study demonstrated that 10 weeks of RT resulted in increased self-efficacy for tasks demanding significant arm or leg strength (self-efficacy defined as one's level of certainty that one can successfully complete a given task or assume a given behavior), whereas there was no change in aerobically trained control participants. In another study, 38 patients added high-intensity RT (up to 80% of the 1RM) to their usual exercise prescription and demonstrated similar improvements in self-efficacy for strength-related tasks and also for jogging. Another interesting finding was the improvement in quality of life parameters such as total mood disturbance, depression or dejection, fatigue or inertia, and emotional health domains scores (McCartney 1998). Although these findings suggest that RT may improve the quality of life of patients with CAD, more research in this area seems warranted.

A number of investigations have focused on the effects of RT on coronary risk factors, but because the findings are largely equivocal (McCartney 1998), we review them only briefly here. Although some reports show decreases in low-density lipoprotein cholesterol and increases in high-density lipoprotein cholesterol after short periods of RT, there are many confounders. These include lack of a control group, only a single blood sample before and after the training, no regulation of diet or account of changes in body mass and composition, and normal preintervention lipid profiles.

The results of individual studies on blood pressure have been variable, with some demonstrating decreases and others showing no change. A recent meta-analysis of 320 normotensive and hypertensive males and females demonstrated a small but significant reduction in resting systolic and diastolic pressures of 3 mmHg. This reduction was also clinically significant because it would theoretically reduce CAD by 5% to 9% (Pescatello et al. 2004).

As mentioned previously, the decrease in muscle mass with advancing age may significantly contribute to the impaired glucose tolerance that is so prevalent among middle-aged and elderly people. It is possible that this impairment may be at least partially reversed by increasing muscle mass with RT, because there are encouraging reports of increased glucose tolerance and insulin sensitivity independent of changes in body fat or aerobic capacity. It is likely that more studies in this area will be forthcoming.

Evidence is accumulating that RT has positive effects on emerging risk factors for CAD, such as hemoglobin A1C, lipid peroxidation, and plasma homocysteine. Moreover, RT may reduce inflammatory markers such as C-reactive protein, interleukin-6, and tumor necrosis factor-alpha. There is also a report that a bout of RT promotes improved fibrinolytic potential and may therefore reduce the risk of acute thrombogenesis (deJong et al. 2006).

Resistance Training in Chronic Heart Failure and After Heart Transplantation

Patients with chronic heart failure (CHF) and heart transplantation (HT) suffer from muscular atrophy and weakness and could theoretically benefit from RT. In CHF, there is a relatively modest association between maximum exercise capacity and left ventricular dysfunction as measured by ejection fraction. Much of the reduced exercise capacity is seemingly attributable to intrinsic abnormalities within peripheral muscles, independent of any reductions in peripheral blood flow. These abnormalities include selective atrophy of fatigue-resistant oxidative muscle fibers (type I), reduced mitochondrial volume and density, and decreased concentration and activity of mitochondrial oxidative enzymes. The overall cross-sectional area of the thigh muscles may decrease by 15% or more, resulting in muscular weakness that in principle could be partially reversed by RT. Nevertheless, despite the hypothetical basis for RT in CHF, there are few reports of RT in this patient group and consequently no published guidelines by health and exercise agencies.

The largest study done so far was the Exercise Rehabilitation Trial (EXERT) from Canada (McKelvie et al. 2002), and the results were inconclusive. A total of 181 patients participated in a 12-month randomized controlled single-blind trial of supervised (three months) and home-based (nine months) aerobic and RT exercise, with blinded evaluation of patients on a range of clinically useful outcomes. After the initial period of supervised exercise training, the exercise group demonstrated significant increases in $\dot{V}O_2$max and 1RM for arm and leg strength compared with control participants, but these differences diminished and became nonsignificant after the nine months of home-based training. The reduced adherence during home-based training may have been a confounder in the study. One encouraging finding was that there were no adverse effects on cardiac function or any greater number of clinical events among the exercising patients. More studies of RT in CHF are needed before evidence-based recommendations can be formulated.

Heart transplantation may ameliorate many of the symptoms of CHF, but HT recipients typically have markedly reduced exercise capacity and a $\dot{V}O_2$max approximately 55% to 60% of predicted values. Similar relative reductions are seen in quadriceps muscle strength, with a strong correlation between quadriceps strength and $\dot{V}O_2$max in HT patients. Moreover, individuals with CHF also manifest peripheral muscle myopathy and osteoporosis as a consequence of immunosuppression with glucocorticoids. Because trabecular bone is lost more rapidly than cortical bone, HT patients are particularly susceptible to loss of bone mineral from the lumbar vertebrae and suffer a very high incidence of vertebral compression fractures. Because RT is inherently anabolic, it might be a useful intervention to help prevent or reverse these musculoskeletal changes.

Published information on the effects of RT on muscle myopathy after HT is limited. Nevertheless, one study (Braith 1998) showed that a six-month program of RT successfully reversed the glucocorticoid-induced muscle atrophy in seven exercising patients versus seven control participants. Again, more work in this area is warranted before definite conclusions can be drawn.

A growing body of literature demonstrates that RT may stabilize the loss of BMD in fracture-prone populations, but the evidence for increasing BMD is more equivocal. Braith (1998) conducted a randomized controlled trial (RCT) to evaluate the effects of RT on the losses of BMD that occur after HT and noted that just two months after the surgical procedure, the control and RT groups had lost 12.2% and 14.9% of lumbar BMD, respectively. After the two-month measurement, the experimental group began RT on two days per week using a single-set, 10- to 15-repetition program, and the control group took part in usual activities. After six months of RT, the BMD of the training group was similar to pre-HT levels, whereas there was no meaningful recovery of BMD among control participants. This preliminary evidence indicates that RT could be a useful intervention to ameliorate losses of BMD after HT, but more studies are needed in this area.

Inclusion and Exclusion Criteria

Recent updated guidelines from the American Heart Association (Williams et al. 2007) define the following contraindications to RT: unstable angina, uncontrolled hypertension (systolic pressure ≥160 mmHg or diastolic pressure ≥100 mmHg), uncontrolled dysrhythmias, a recent history of congestive heart failure without evaluation and effective treatment, severe stenotic or regurgitant valvular disease, and hypertrophic cardiomyopathy. Preferred inclusion criteria are moderate to good left ventricular function and an exercise capacity of >5 METs.

Exercise Prescription

Patients should take part in two to four weeks of aerobic training in a supervised setting before doing RT. Pretraining instruction should emphasize correct weightlifting and breathing techniques. Resistance training should take place twice weekly and include one set of 10 to 15 repetitions of 8 to 10 exercises designed to train all major muscle groups. If the 1RM is determined, patients should begin training with loads equivalent to 30% to 40% of the 1RM for upper-body exercises and 50% to 60% of the 1RM for lower-body exercises. Older and frail individuals may start training at lower intensities and progress more slowly. Determination of the 1RM is not strictly necessary; patients can begin using light loads that result in moderate levels of fatigue by the end of a set of lifting. Once patients can complete their final lift with ease, the weights can be increased; added

loads of 2 to 5 lb (0.9-2.3 kg) per week for the arms and 5 to 10 lb (2.3-4.5 kg) per week for the legs are adequate in most cases. Slower progression may be necessary in older patients. There is no need to rush; patients should determine their own pace based on their levels of fatigue and perceptions of effort.

> **Resistance training was once thought to be too intense and dangerous for patients with CAD and for those who had undergone HT. However, the substantial benefits of RT, including increased strength and endurance, improved self-efficacy, lowered blood pressure, and reduced incidence of depression, are now well documented and recognized. Resistance training can be used safely in patients with CAD, CHF, and HT provided that safety measures are followed.**

Resistance Training in Arthritis

The more than 100 different types of arthritis can be divided into three broad classifications: osteoarthritis (the most common form), inflammatory conditions, and rheumatism. Arthritis is the leading cause of disability among Americans, affecting 50 million individuals of all ages, not just older people. The common symptom of arthritis is pain in joints or soft tissues, leading to restricted joint range of motion, sedentary behavior, and concomitant reductions in physical fitness and muscular strength. Indeed, the reductions in fitness and strength may be a leading cause of disability among arthritis sufferers, so RT may be a useful strategy to improve function.

The past two decades witnessed a proliferation of studies that evaluated the role of RT in arthritis (Callahan 2009). The weight of evidence suggests that RT improves muscle strength, balance, and coordination; reduces pain; and increases functional capacity and health-related quality of life for people with arthritis. These adaptations should decrease disability and dependency and also improve an individual's risk profile for diseases such as CAD. The mechanisms responsible for the improvements with RT are not fully understood but are likely a combination of physiological (e.g., improved joint stability resulting from increased strength) and psychological (e.g., greater self-efficacy, mastery, control) factors.

No evidence-based guidelines exist for RT in arthritis, so caregivers who prescribe exercise must consider the individual differences and the comorbid conditions associated with different types of arthritis. For example, inflammatory conditions of soft tissue are volatile, and RT should be avoided during times of flare-up. Increased symptoms of general fatigue may signal exacerbation of comorbid conditions, and ankylosis conditions make the back vulnerable to forced flexion, extension, and rotation. Because of these and other considerations, the exercise therapist prescribing RT must be aware of each patient's condition and develop an appropriate exercise program.

More research on RT and arthritis is warranted, but the majority of evidence thus far suggests that RT may be an important therapy in this population.

Resistance Training in Osteoporosis

The National Institutes of Health (NIH) Consensus Conference has modified the original definition of osteoporosis to include "a skeletal disorder characterized by compromised bone strength, predisposing a person to an increased risk of fracture. Bone strength reflects the integration of two main features, bone density and bone quality" (Hellekson 2002, p. 161). This definition acknowledges that a decrease in BMD is not the only pathological feature of osteoporosis, and the term *bone quality* refers to microarchitectural elements that contribute to bone strength. Nevertheless, the conventional diagnosis of osteoporosis is a BMD that is 2.5 standard deviations (SD) or more below the level for healthy young adults of the same gender. It is estimated that 10 million Americans over the age of 65 have osteoporosis, with another 18 million exhibiting osteopenia, or low bone mass. At least 700,000 fractures of the spine, 300,000 hip fractures, and 250,000 fractures of the wrist each year in the United States are attributable to osteoporosis. The health costs associated with these fractures are staggering, yet osteoporosis is viewed as largely preventable.

The theoretical basis for RT in osteoporosis is the concept of "minimal essential strain" (subjection of a bone to forces representing at least 10% of the level that would fracture it), which is believed to be the threshold level required for new bone formation. If the muscular contractions associated with RT can repeatedly load the skeleton above the threshold level, then bone mass should increase. Animal studies have demonstrated that within two or three months of external loading at appropriate levels, osteoblasts deposit collagen in the bone matrix, and mineralization follows over the next three months. This evidence indicates that RT programs longer than six months

would be needed to produce any measurable effect on BMD.

The literature on RT and BMD could be viewed as equivocal, because some researchers reported improvements and others reported no change. Nevertheless, differences between the two groups of studies may explain the conflicting results. Most of the investigations yielding positive results were RCTs that focused on adult females up to 75 years of age and lasted for a year or more. Resistance training was performed two or three times each week and included three sets of up to 12 different exercises, using eight repetitions of high-intensity loading up to 85% of the 1RM. The increases in BMD in these studies were significant but generally less than 3%. Moreover, increases were most evident in the axial skeleton, which has more trabecular bone than the appendicular skeleton. Studies that showed no increases in BMD over a one-year period differed from the other investigations in one important aspect: The intensity of loading during RT was moderate. Although the gains in 1RM were comparable, the data suggest that bone loading during RT may be the most important variable for increasing BMD.

Resistance training has been shown to be an effective therapeutic intervention in arthritis and osteoporosis. Not surprisingly, RT has been shown to be as effective as many drug interventions in these diseases.

In summary, RT has a sound theoretical basis with regard to the prevention and treatment of osteoporosis and should be administered as a supplement to conventional treatment, not as a stand-alone modality. It appears that RT programs longer than one year using high-intensity loading are required to increase BMD, but more research in this area is needed.

Summary

Resistance training has been used by athletes successfully for many years to improve performance, to increase muscle strength and mass, and to reduce injury; but the benefits of RT for the general population and those with disease and disability have only recently been appreciated. Research has demonstrated that RT promotes the following adaptations that foster and maintain good health: increases in muscle mass and quality, large increases in dynamic strength and endurance, enhanced exercise and functional capacity, improved balance, decreased falls, reductions in body fat, small but significant reductions in systemic arterial pressure, improved blood lipid and lipoprotein profile, improved disposal of blood glucose, increases in BMD, and increases in self-efficacy and health-related quality of life. Resistance training has been used successfully with obese people, frail elderly people, and various patient groups including those with CAD, arthritis, osteoporosis, and type 2 diabetes mellitus. Limited information indicates that RT may be useful in other cohorts, such as those with neuromuscular disorders, multiple sclerosis, amyotrophic lateral sclerosis, fibromyalgia, kidney disease, chronic obstructive pulmonary disease, and low back pain; but more research is needed. Clearly, RT should be an integral part of a well-rounded exercise program to develop and maintain good health.

Key Concepts

atrophy—Decrease in the cross-sectional area of skeletal muscle fibers and eventually the muscle itself, occurring when muscle protein breakdown exceeds synthesis.

disease cluster—Linked series of adverse health conditions present in a single individual or group of individuals, such as type 2 diabetes mellitus, high blood pressure, and obesity in the metabolic syndrome.

hypertrophy—Increase in the cross-sectional area of skeletal muscle fibers and eventually the muscle itself. For this to occur, muscle protein synthesis must exceed breakdown.

muscle protein turnover—Rates of both muscle protein synthesis and breakdown and ultimately the net flux between these two processes.

muscle quality—Traditionally, the capacity to generate muscle force as a ratio of the muscle's cross-sectional area; here, however, the term refers to the metabolic quality of the muscle, that is, the muscle's capacity for oxidative metabolism.

one-repetition maximum (1RM)—Single highest load that a person can lift once (i.e., that results in instant fatigue).

resistance training (RT)—Training that uses either mechanical or free-moving loads to create a systematic condition of progressive overload on the skeletal muscle. Loads typically range from 50% to 90% of the 1RM.

sarcopenia—For definition, see page 254.

Study Questions

1. Describe how feeding and resistance exercise interact to increase skeletal muscle fiber size (i.e., hypertrophy).

2. Name five reasons why maintenance of muscle mass is important to long-term health in elderly people.

3. Describe why RT would be as effective as or potentially more effective than aerobically based exercise in the treatment of persons who are overweight or obese.

4. Define a disease cluster and give an example of how RT might be able to treat such a cluster; explain why this might be beneficial as opposed to pharmacological treatment of the same disease cluster.

5. Is RT contraindicated in children and adolescents? What benefits might RT offer children and adolescents?

6. Aside from the expected increases in strength with RT in persons with CAD, CHF, and HT, what other benefits have been seen in these patients as a result of engagement in a program of RT?

7. What are the main inclusion and exclusion criteria for participation in a program of RT in patients with CAD, CHF, or HT? What might a typical exercise prescription for the same group of patients look like?

8. Describe a limitation to the use of RT in the treatment of arthritis.

9. Can RT effectively reduce the risk of osteoporosis? Why or why not?

References

Braith, R.W. 1998. Exercise training in patients with CHF and heart transplant recipients. *Medicine and Science in Sports and Exercise* 30:S367-S378.

Brochu, M., P. Savage, M. Lee, J. Dee, M.E. Cress, E.T. Poehlman, M. Tischler, and P.A. Ades. 2002. Effects of resistance training on physical function in older disabled women with coronary heart disease. *Journal of Applied Physiology* 92:672-678.

Burd, N.A., J.E. Tang, D.R. Moore, and S.M. Phillips. 2009. Exercise training and protein metabolism: Influences of contraction, protein intake, and sex-based differences. *Journal of Applied Physiology* 106:1692-1701.

Callahan, L.F. 2009. Physical activity programs for chronic arthritis. *Current Opinion in Rheumatology* 21:177-182.

Campbell, W.W., M.L. Barton Jr., D. Cyr-Campbell, S.L. Davey, J.L. Beard, G. Parise, and W.J. Evans. 1999. Effects of an omnivorous diet compared with a lactoovovegetarian diet on resistance-training-induced changes in body composition and skeletal muscle in older men. *American Journal of Clinical Nutrition* 70:1032-1039.

deJong, A.T., C.J. Womack, J.A. Perrine, and B.A. Franklin. 2006. Hemostatic responses to resistance training in patients with coronary artery disease. *Journal of Cardiopulmonary Rehabilitation* 26:80-83.

Doherty, T.J. 2001. The influence of aging and sex on skeletal muscle mass and strength. *Current Opinion in Clinical Nutrition and Metabolic Care* 4:503 508.

Faigenbaum, A.D., and G.D. Myer. 2010. Resistance training among young athletes: safety, efficacy and injury prevention effects. *British Journal of Sports Medicine* 44(1):56-63.

Feigenbaum, M.S., and M.L. Pollock. 1999. Prescription of resistance training for health and disease. *Medicine and Science in Sports and Exercise* 31:38-45.

Fiatarone Singh, M.A. 2001. Elderly patients and frailty. In *Resistance training for health and rehabilitation*, ed. J.E. Graves and B.A. Franklin (pp. 181-213). Champaign, IL: Human Kinetics.

Hellekson, K.L. 2002. NIH releases statement on osteoporosis prevention, diagnosis, and therapy. *American Family Physician* 66:161-162.

Krieger, J.W. 2009. Single versus multiple sets of resistance exercise: a meta-regression. *Journal of Strength and Conditioning Research* 23:1890-1901.

Macaluso, A., and G. De Vito. 2004. Muscle strength, power and adaptations to resistance training in older people. *European Journal of Applied Physiology* 91:450-472.

Marzolini, S., P.I. Oh, S.G. Thomas, and J.M. Goodman. 2008. Aerobic and resistance training in coronary disease: Single versus multiple sets. *Medicine and Science in Sports and Exercise* 40:1557-1564.

McCartney, N. 1998. Role of resistance training in heart disease. *Medicine and Science in Sports and Exercise* 30:S396-S402.

McCartney, N. 1999. Acute responses to resistance training and safety. *Medicine and Science in Sports and Exercise* 31:31-37.

McCartney, N., R.S. McKelvie, J. Martin, D.G. Sale, and J.D. MacDougall. 1993. Weight-training-induced attenuation of the circulatory response of older males to weight lifting. *Journal of Applied Physiology* 74:1056-1060.

McKelvie, R.S., K.K. Teo, R. Roberts, N. McCartney, D. Humen, T. Montague, K. Hendrican, and S. Yusuf. 2002. Effects of exercise training in patients with heart failure: The Exercise Rehabilitation Trial (EXERT). *American Heart Journal* 14:423-430.

Pescatello, L.S., B.A. Franklin, R. Fagard, W.B. Farquhar, G.A. Kelley, and C.A. Ray. 2004. American College of Sports Medicine position stand. Exercise and hypertension. *Medicine and Science in Sports and Exercise* 36533-36553.

Phillips, S.M. 2004. Protein requirements and supplementation in strength sports. *Nutrition* 20.689-695.

Phillips, S.M. 2007. Resistance exercise: Good for more than just Grandma and Grandpa's muscles. *Applied Physiology, Nutrition, and Metabolism* 32:1198-1205.

Phillips, S.M., and R.A. Winett. 2010. Uncomplicated resistance training and health-related outcomes: Evidence for a public health mandate. *Current Sports Medicine Reports* 9:208-213.

Rice, B., I. Janssen, R. Hudson, and R. Ross. 1999. Effects of aerobic or resistance exercise and/or diet on glucose tolerance and plasma insulin levels in obese men. *Diabetes Care* 22:684-691.

Tseng, B.S., D.R. Marsh, M.T. Hamilton, and F.W. Booth. 1995. Strength and aerobic training attenuate muscle wasting and improve resistance to the development of disability with aging. *Journals of Gerontology: Series A, Biological Sciences and Medical Sciences* 50:113-119.

West, D.W., N.A. Burd, J.E. Tang, D.R. Moore, A.W. Staples, A.M. Holwerda, S.K. Baker, and S.M. Phillips. 2010. Elevations in ostensibly anabolic hormones with resistance exercise enhance neither training-induced muscle hypertrophy nor strength of the elbow flexors. *Journal of Applied Physiology* 108:60-67.

Williams, M.A., W.L. Haskell, P.A. Ades, E.A. Amsterdam, V. Bittner, B.A. Franklin, M. Gulanick, S.T. Laing, K.J. Stewart; American Heart Association Council on Clinical Cardiology; American Heart Association Council on Nutrition, Physical Activity, and Metabolism. 2007. Resistance exercise in individuals with and without cardiovascular disease: 2007 update: A scientific statement from the American Heart Association Council on Clinical Cardiology and Council on Nutrition, Physical Activity, and Metabolism. *Circulation* 116:572-584.

17

Physical Activity, Fitness, and Children

Thomas Rowland, MD

Sedentary habits are dangerous to one's health. Indeed, the major causes of mortality and morbidity in developed populations—atherosclerotic vascular disease, hypertension, obesity, type 2 diabetes, osteoporosis—can be linked to lack of regular exercise habits. The potential of physical activity to improve the health and well-being of the population is thereby well accepted and serves as a basic tenet of preventive health strategies.

But this is a chapter about youth, who don't suffer myocardial infarction, stroke, hypertensive renal disease, or femoral fractures. And children are the most active segment of our society. Is there reason to support exercise habits in young populations? Are there unique features about children that warrant particular attention in activity promotion? The answer to both questions is yes, and initiatives to promote habits of regular exercise in young persons have become a major preventive health focus. Such efforts have been fueled particularly by concerns about (1) the pervasive negative influences of automobile transportation, television time, and computer use; (2) thwarted opportunities for exercise by reductions in physical education class time and lack of environmental opportunities; and (3) dramatic increases in levels of childhood obesity.

The rationale for promoting health through optimizing childhood physical activity, although difficult to prove experimentally, is a compelling one. Here is the story:

A solid base of scientific data in adults supports the benefit of regular exercise habits and diminished risk for chronic disease outcomes whose *clinical outcomes* (heart attacks, stroke, renal failure, bone thinning and fractures) occur in the older years. It is necessary to recognize, however, that these outcomes often reflect pathophysiologic processes that have their genesis during the pediatric years. Early evidence of vascular lesions of atherosclerosis is commonly observed before the end of adolescence. Essential hypertension and obesity frequently have their antecedents during childhood. Osteoporosis and bone fractures in elderly women are outcomes of inadequate bone deposition early in the life span.

The obvious conclusion from this information is that promotion of exercise habits early in life, during childhood and adolescence, should be a particularly effective means of reducing risk for adverse disease outcomes later on in the adult years (figure 17.1). Maintaining high levels of physical activity during the growing years might be expected to impair the natural progression of these pathologic processes and delay or diminish the impact of their later

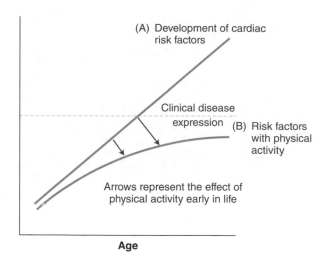

FIGURE 17.1 The expected development of cardiovascular risk factors (A) may be retarded by physical activity in the early years of life, (B) thereby limiting clinical disease expression in the adult years.

Reprinted, by permission, from O. Bar-Or and T.W. Rowland, 2004, *Pediatric exercise medicine* (Champaign, IL: Human Kinetics), 122.

clinical expression. Indeed, the promotion of physical activity in youth is founded on this "window of opportunity" idea.

The methodological challenges surrounding any investigation testing the validity of this premise are highly daunting. It might be presumed that a longitudinal study examining any link of activity habits in childhood—or the effect of exercise interventions in youth—with disease outcomes 40 years later in life will probably never be at hand. Still, the "pediatric rationale" for early promotion of physical activity for physical health is intellectually sound and sufficiently compelling to serve as a major public health stratagem.

Proponents of this approach would point out the potential magnitude of the impact on the population's health of successfully maintaining high levels of physical activity in youth that persist to the adult years. Given (a) the recognized ability of exercise to reduce risks of chronic disease processes of multiple pathophysiologies in adults and (b) the greater time that would be afforded activity to reduce these processes, the possibility that promotion of activity in youth would serve as an effective means of diminishing mortality and morbidity can hardly be overstated.

There exist, of course, other reasons for promoting exercise in children and adolescents—enhancing enjoyment of life, developing social skills, and improving psychological values, for example. In youth, as in adults, there are data suggesting the efficacy of exercise interventions in those with emotional disorders. Exercise has a role as a therapeutic tool in children with chronic diseases as well, including cardiac, pulmonary, musculoskeletal, and metabolic disorders. In this chapter, though, the discussion generally centers on the effect of physical activity and fitness on the physical health of the overall pediatric population.

These comments can be construed, then, as the "good news." On closer inspection, though, exercise scientists face a considerable number of serious challenges in understanding the link of regular physical activity in children with their future health, as well as the means of translating this information into effective methods of exercise promotion. While considerable progress has been achieved in recent years, a number of difficult hurdles remain. In this chapter, we examine these challenges.

Understanding the Exercise–Health Link in Children

In an ideal research world, one would perform a randomized trial in which a cohort of 10-year-olds would be engaged in a program of regular exercise and their health outcomes 40 years later would be compared to those of a nonexercise group, while controlling for factors such as diet, motivation, gender, geographical area, subject dropout, and change of testing methodology. Even better, one might include several interventional groups with different amounts of exercise to quantitate the requisite activity levels. Clearly, as noted previously, an investigation that would establish the pediatric rationale for exercise promotion (or not) is not practically feasible. Consequently, the premise that childhood exercise can blunt the course of lifelong pathologic processes has been tested via examination of the relationships of physical activity to surrogate, or indirect, measures of adult chronic disease. For the most part, these have included disease risk factors identified in adult populations as having predictive value for clinical outcomes. In some cases, such information has clearly supported the pediatric rationale. In others, surprisingly, the evidence has been found lacking.

Cardiovascular Disease

Studies in adults in the 1970s and 1980s clearly implicated a set of modifiable **cardiovascular risk factors**—abnormal serum lipid profile, obesity, hypertension, and cigarette smoking—that identified those at risk for heart attack, stroke, and peripheral vascular disease. Subsequent investigations added a sedentary lifestyle to this list, and, importantly, indicated that physical activity and fitness both were effective in reducing the levels of cardiovascular risk factors as well as clinical disease and mortality. These data have served as the central basis for the promotion of regular exercise in adult populations.

Lacking specific disease outcomes, it was obvious to pediatric researchers that relationships might be examined between physical activity and fitness in children and these risk factors, assumed to serve as indicators of future cardiovascular disease. The findings in such studies, cross-sectional, longitudinal, and interventional, have been less than impressive and have not, in general, mimicked those relationships observed in adults.

Serum Lipids

Adults who have elevated total serum cholesterol levels (TC) associated with increases in the low-density lipoprotein (LDL) fraction (LDL) carry an increased risk for cardiovascular disease. High-density lipoprotein (HDL) cholesterol, on the other hand, confers a protective effect, as does the HDL:TC ratio. Evidence for atherogenic risk associated with increases in serum levels of very low-density lipoprotein cholesterol and serum triglycerides is not clear-cut. Regular exercise habits and level of physical fitness in adults have been shown to improve one's lipid profile, principally via increases in HDL, particularly in males.

In children and adolescents, limited autopsy information suggests an association between these lipid markers and early atherosclerotic lesions. However, in young persons, physical activity and **aerobic fitness** seem to have no important effect on the serum lipid profile (Twisk and Ferreira 2008). Most cross-sectional studies and interventional aerobic training studies (of 4 to 20 weeks' duration) have revealed no alterations in serum levels of TC, HDL cholesterol, or LDL cholesterol. Noninterventional longitudinal studies generally have borne this out. In the Amsterdam Growth and Health Longitudinal Study, which examined subjects serially from 15 years into young adulthood, changes in physical activity were linked to those in HDL, but not TC or HDL:TC. Another long-term investigation, the Cardiovascular Risk in Young Finns Study, showed no relationship between activity and HDL or TC values between ages 6 and 18 years. HDL:TC was related to activity during that time period in boys but not girls. In the Muscatine Study, neither physical activity nor aerobic fitness predicted lipid levels as children aged from 10 to 15 years.

Blood Pressure

Regular aerobic exercise in both normotensive and hypertensive adults effects small but significant reductions in systemic blood pressure. In youth with normal blood pressure, on the other hand, cross-sectional studies generally have shown no relationship between blood pressure and activity or fitness once the confounding influence of body fat content is removed. Similarly, no decreases in blood pressure have been observed in training studies in normotensive children and adolescents. In the long-term longitudinal studies described in the previous section, no relationship was shown between changes in daily activity and those in systolic or diastolic blood pressure.

On the other hand, exercise training has been demonstrated to lower blood pressure levels in youth with mild essential hypertension. This effect has been observed with both aerobic and resistance training. Once training was stopped, improvements in blood pressure levels were lost.

Obesity

Assessing the influence of exercise on obesity in youth is confusing since excessive body fat, in itself, can be expected to reduce both physical fitness and level of habitual activity. Thus, when cross-sectional studies consistently demonstrate that the obese child is habitually less active and less fit than his peers, the direction of the arrow of causality is unclear. Is the overweight girl less active because of the burden of her extra body weight? Or is limited participation in physical activity etiologic—or at least contributory— to her adiposity? Confounding this conundrum is the fact that energy expenditure required to perform a given physical task can be expected to be greater in the obese subject. Thus, investigators in such studies need to distinguish whether they are examining physical activity (amount of movement) or caloric expenditure.

Aerobic training studies in nonobese youth have failed to reveal any significant changes in body fat content. It could be concluded, then, for children with normal body fat content, that the added caloric expenditure during such short-term exercise training is compensated for by alterations in other components affecting energy balance and body composition. The Amsterdam Growth and Health Longitudinal Study indicated that fitness and long-term activity were significantly related to skinfold thickness in adults, but the amount of the variance explained was small.

In obese youth, though, exercise interventions have been shown to be an effective means of reducing body fat via increases in energy expenditure. The magnitude of this effect is apparently not large, approximating 5% to 10%, indicating that successful weight loss programs for obese youth should be multidimensional (diet, exercise, psychological support).

Type 2 Diabetes Mellitus

Insulin resistance—a decrease in tissue sensitivity to the actions of insulin—is the hallmark of type 2 diabetes mellitus and is recognized as a strong predictor of cardiovascular disease in adults. As insulin resistance and hyperglycemia are associated with excessive body fat, a sharp rise in the incidence of type 2 diabetes in children and adolescents has been observed concomitantly with the rise in pediatric obesity. Among adults with type 2 diabetes, physical activity improves insulin sensitivity and glycemic control, presumably due to the insulin-like action of exercise and increases in muscle mass. The question whether youth can experience similar salutary effects of regular physical activity, independent of an effect of reducing body fat, is just beginning to be explored.

In summary, most evidence indicates little or no effect of regular exercise on the traditional cardiovascular risk factors in healthy children and adolescents with "normal" levels of such factors. Whether activity or fitness in youth might affect newer risk factors identified in adults, such as apolipoproteins, homocysteine, and C-reactive protein (CRP), has not yet been adequately studied. This picture differs from that in adults, in whom reductions of cardiovascular risk factors are typically witnessed in healthy, normotensive, nonobese individuals with normal serum lipid profiles.

A number of explanations have been offered for this discrepancy. Inability to identify any important relationships between physical activity and traditional cardiovascular risk factors in youth could reflect one or more of the following: (a) differences in the nature of childhood versus adult activity, (b) inability to accurately measure activity in children, or (c) a relatively stronger genetic influence on risk factor expression in youth. It remains possible, as well, that the multiple mechanisms by which physical exercise improves risk profiles are less operant in children and adolescents compared to mature individuals.

It has long been recognized that certain cardiovascular risk factors tend to cluster together in children, particularly obesity, systemic hypertension, and abnormal serum lipid profile. More recently, such clusters have been linked to insulin resistance and defined as the "metabolic syndrome." Whether this term is applicable to factor clustering in children remains uncertain, largely because of difficulties in defining "normal" cut points for individual risk factors in this age group. Some cross-sectional studies in youth (such as the CASPIAN Study and AVENA Study) have indicated that a negative relationship exists between such factors, considered as a group, and level of physical activity. For instance, information in the European Youth Heart Study revealed a graded inverse relationship between **accelerometer**-measured activity and clustered cardiovascular risk factors in 1,700 youth (Andersen et al. 2006). The strongest link with lowered risk factors was observed in those who were involved in daily moderate and vigorous physical activity of 116 min (9-year-olds) and 88 min (15-year-olds). On the other hand, studies such as the Danish Youth and Sports Study have failed to detect an association between clustered risk and self-reported activity levels in adolescents.

Long-term studies examining the link between activity and fitness and clustered risk factors in childhood and adolescence have provided conflicting results. In the Amsterdam Growth and Health Longitudinal Study, for example, subjects who did and did not demonstrate the metabolic syndrome at age 36 years showed no differences in physical activity or fitness during adolescence. However, daily activity was significantly related to clustering of TC:HDL ratio, blood pressure, skinfold sum, and maximal oxygen uptake ($\dot{V}O_2$max) between ages 13 and 27 years.

As noted previously, excessive body fat is closely associated with elevated blood pressure, insulin resistance, glucose intolerance, and abnormal lipid profiles. Not unexpectedly, adiposity is thus often observed to be a strong mediator between physical activity and risk factors. Once body fat is considered in such analyses, cross-sectional associations between these risk factors and level of physical activity often disappear.

It has been pointed out that increases in physical activity in youth do, in fact, diminish risk factors in those youth who have abnormal levels to start with. That is, as noted previously, exercise can lower blood pressure in children with essential hypertension, improve body composition and lipid profiles in those who are obese, and enhance insulin sensitivity in patients with insulin resistance or type 2 diabetes. Physical activity and fitness appear to be effective in reducing abnormal levels of risk factors in youth. Perhaps, then, regular exercise might serve to prevent the *development* of these factors into abnormal ranges over time during the growing years.

> The relationship between physical activity levels and risk factors for cardiovascular disease appears to be less strong among children than in adult populations.

Endothelial Function

A recent focus on the functional health of the peripheral vascular endothelium may provide new biomarkers of cardiovascular risk in children and permit direct ultrasound visualization of early vascular alterations. In turn, this information promises to provide new insights into the link between physical activity and fitness and the early phases of the atherosclerotic process (Slyper 2004).

This approach is based on the observation that the vascular endothelium, rather than serving as simply a protective inner lining of blood vessels, is a highly active tissue with multiple functions, particularly regulation of vascular tone and thus control of blood flow. In response to wall stress from blood flow, it produces vasoactive agents (particularly nitric oxide) that in turn trigger dilation of the medial wall. When the endothelium becomes dysfunctional, as occurs early in the atherosclerotic process, the bioavailability of nitric oxide is reduced, and vascular reactivity becomes limited. The degree of this dysfunction, which is linked to vascular inflammation, has been considered an indicator of early atherosclerotic change.

The use of **endothelial dysfunction** as a surrogate indicator of early atherosclerosis is made appealing by the fact that high-resolution ultrasound techniques (measuring flow-mediated dilation [FMD], carotid **intima-medial thickness [IMT]**, and arterial compliance) are now available to identify and quantitate dysfunction noninvasively. In addition, the process of endothelial dysfunction is associated with several biochemical markers of inflammation measurable by blood assay that can serve as biomarkers of the degree of dysfunction and, by extension, the extent of early atherosclerosis. Of these, CRP, a traditional acute-phase indicator measured in clinical laboratories, has received the most research attention.

Fernhall and Agiovlasitis (2008) conducted a review of ultrasound studies in youth. These studies indicate that FMD in obese children is approximately one-half that of healthy subjects, and similar reductions in vascular reactivity are observed in those with heterozygous familial lipidemia and type 1 diabetes mellitus. Parallel increases in IMT are observed in the same groups. Several reports indicate that blood CRP levels are linked to body fat content in obese youth. Few studies have examined these factors in the general pediatric population, but early information suggests that CRP levels may be predictive of both FMD and IMT in large groups of healthy young subjects.

In adults, a period of exercise training has been shown to ameliorate endothelial dysfunction in patients with coronary artery disease, congestive heart failure, and systemic hypertension. This has been explained as a manifestation of the stimulation of increases in nitric oxide release with repeated wall stress (from increased blood flow with exercise). Reports indicate that exercise training in obese youth can effect similar improvements in FMD and IMT, although whether these reflect a direct effect of exercise on the endothelium or changes in body fat remains uncertain. Initial studies have suggested that regular physical activity may beneficially affect endothelial function in the general childhood population. Abbott and colleagues (2002) reported that habitual physical activity (as measured by the doubly labeled water technique) was associated with brachial artery FMD in 47 healthy youth ages 5 to 10 years, and this relationship remained after differences in subject body fat content were accounted for.

The assessment of these new diagnostic tools is in its infancy, and their importance as surrogate markers of endothelial dysfunction and atherosclerosis is yet to be determined. If valid and feasible, they could provide important insights into the natural history of atherosclerosis and the role that physical activity might serve in modifying this process. Moreover, such approaches might prove useful in the clinical management of patients at particular risk (i.e., familial hyperlipidemias).

Gene–Environment Interaction

The recognized genetic influence on phenotypic expression of certain risk factors (particularly serum lipoproteins and body fat) raises the question whether specific genotypes might influence the relationships among physical activity, fitness, and cardiovascular risk (Franks and Looker 2008). Investigators are just beginning to examine this issue. In adults, early studies have suggested a **gene-environment interaction** of the angiotensin I converting enzyme 1 (ACE) I/D genotype and habitual physical activity with blood pressure. In the only study that examined this question in children, however, no such relationship was observed. Sarzynski and colleagues (2010) showed no evidence of association between blood pressure and physical activity in 132 children ages 3 to 12 years. The findings indicated that the correlations between blood pressure and physical activity did not differ across genotypes, suggesting a lack of gene–physical activity interaction on blood pressure. In the Cardiovascular Risk in Young Finns Study, a cross-sectional examination of 1,498 males and females ages 9 to 24 years, genotype for apolipoprotein E (APOE) explained some but not all of the associations between physical activity and TC and LDL cholesterol levels.

Bone Health

Calcium deposition in bone increases throughout the growing years, and bone mineral density rises to reach a peak between the second and third decades of life. The early adolescent period is particularly important in this development of bone mass, as 35% of the total rise in bone mineral density occurs at the time of the pubertal growth spurt. After age 20 to 25 years, bone mineral density steadily declines. This progressive thinning of bone can eventually result in the clinical condition of osteoporosis, which puts elderly persons, particularly females, at risk for crippling bone fractures.

Increasing mechanical stress on bones through short-burst, explosive exercise (skipping, stair climbing, jumping) is an effective means of optimizing increases in bone mass. It follows that by augmenting the development of bone mass during the first portions of life, one should be able to shift the life curve of bone density to a greater peak in early adulthood and subsequently an elevated level during its decline into the older years (figure 17.2). Based on this construct, promotion of exercise that stresses bones during childhood and adolescence should pay dividends in reducing risk of osteoporosis and bone fractures in the elderly years.

Again, the definitive study to verify this assumption has yet to be performed. But in this case, a growing body of supportive research findings exists. Randomized controlled trials (RCTs) in both children and adults reveal significant improvements in bone mineral density in response to exercise interventions. In the pediatric age group, this has been most consistently documented in pre- and early-pubertal subjects. Prepubertal subjects among six such studies showed increases in bone mass by 1.1% to 5.5%; in early-pubescent youth, the increase in nine studies over six months averaged 0.9% to 4.9%. Studies on effects of exercise in postpubertal adolescents have provided conflicting results (Kemper 2008; Hind and Burrows 2007).

Findings in the Amsterdam Growth and Health Longitudinal Study revealed that weight-bearing physical activity between ages 13 and 27 years was a significant predictor of bone mineral density at 27

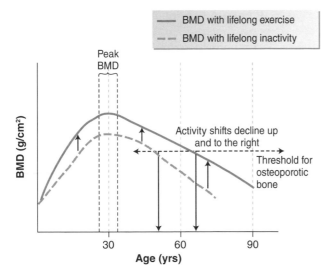

FIGURE 17.2 Weight-bearing exercise during childhood and adolescence may increase development of bone mineral density (BMD), shifting the curve of bone growth and limiting osteoporosis at older ages.

Reprinted, by permission, from H. Kemper, 2000, "Skeletal development during childhood and adolescence," *Pediatric Exercise Science* 12:198-216.

years. In that analysis, the effect of strain-inducing, weight-bearing activities was more closely related to bone density than were estimates of total daily energy expenditure.

There is suggestive evidence, then, that minimizing risk for osteoporosis in the elderly can best be achieved by (1) promoting bone accrual via weight-bearing physical activities in the young and (2) ensuring that adults sustain a physical active lifestyle. Additional research is needed to determine the specific forms and amounts of such activities for the promotion of bone health in the pediatric years.

> Augmentation of bone mineral density through weight-bearing physical activities in the early portions of life may forestall the development of osteoporosis and bone fractures in the elderly.

Measurement of Physical Activity in Youth

The ability to accurately assess the volume, intensity, and pattern of daily physical activity in youth is a critical element in the activity-for-health paradigm. Without such information, dose–response curves to health outcome markers cannot be established, and it will not be possible to formulate appropriate thresholds for "effective" activity behavior. Epidemiologic studies defining activity levels in young populations cannot be performed, and identifying hypoactive youth at risk is impossible without information on daily physical activity. Similarly, without valid measurement tools, the success of activity behavior interventions in children and adolescents cannot be ascertained.

Given the central importance of activity measurement, a great deal of research effort in recent years has focused on the optimal means of capturing activity data in young subjects. A number of both objective and subjective measurement tools are available, each with its advantages and drawbacks. And while considerable progress has been achieved in activity measurement in children, major challenges remain, and the "ideal" measurement technique has yet to be identified.

A key difficulty lies in the observation that a distinct negative relationship exists between the feasibility, ease, and affordability of a measurement technique and its accuracy in capturing activity behavior or energy expenditure. That is, questionnaires and activity diaries are inexpensive and simple to administer but, particularly in children, typically provide limited insights into activity behavior. On the other hand, objective measures such as indirect calorimetry and doubly labeled water can accurately determine energy expenditure but are costly and cannot identify activity parameters in real-world settings.

The following is a brief summary of how the value of each of these techniques is currently viewed with regard to assessing physical activity in children. Readers seeking in-depth reviews should consult other sources (Marshall and Welk 2008; Rowlands 2007; Corder and Ekelund 2008).

Indirect Calorimetry

Measurement of oxygen consumption during physical activity provides an accurate assessment of energy expenditure. Previous use of this technique required that subjects be housed in a metabolic chamber, which hardly mimicked real-world conditions. More recently, lightweight portable oxygen monitors have permitted measurement of VO_2 during free-living activities. This approach requires expensive equipment that is still cumbersome to use in many normal activities. Its use in other than small-scale studies is therefore usually prohibitive. Moreover, it provides no information regarding patterns of activity. Portable VO_2 monitors may be particularly useful, though, in defining caloric consumption of particular activities, information that can then be used for assessment of energy expenditure by other techniques (questionnaire, direct observation, etc.).

Doubly Labeled Water

Doubly labeled water is a safe, noninvasive technique that accurately measures energy expenditure unobtrusively over a period of time. Subjects ingest isotopic tracers (deuterium and oxygen-18), which when eliminated from the body in urine permit a determination of total carbon dioxide production and energy expenditure. The major drawbacks are (1) high cost; (2) inability to determine the relative contributions of activity, basal metabolic rate, and diet-induced thermogenesis to energy expenditure; and (3) inability to provide insight into patterns of activity.

Direct Observation

Grading systems have been developed and used for estimating physical activity of children based on direct observation. These are particular useful in defining the context of physical activity. The direct observation technique can provide accurate information on activity in youth but is clearly labor intensive and limited to use by highly trained personnel in small-scale studies.

Pedometers

Small portable devices that measure the number of steps taken over a given period have become increasingly popular as a means of measuring levels of activity in both individuals and group studies. While inexpensive and easy to use, these devices are limited to assessing step counts. Thus, they do not provide insight into activities that involve the upper extremity, give no information regarding intensity of activity, are influenced by leg length, and do not assess energy expenditure. Still, daily step counts may serve as a gross measure of these nonmeasured variables and provide a general marker of activity level. In fact, recent reports indicate fairly high correlations of energy expenditure and pedometer counts (r = 0.62-0.92). Pedometers may have particular utility in assessing changes in activity in response to exercise interventions.

Accelerometers

These devices are recognized currently as the best means of measuring physical activity in field-based studies in children. Rowlands (2007) has noted that the number of publications related to physical activity and accelerometry in children increased sevenfold between the years 2000 and 2005. Usually worn on the hip, the devices measure acceleration of body motion in one to three planes and record "counts" of activity over a particular time period. Such counts then reflect both volume and intensity of movement. Accurate equations by which counts can be translated to actual energy expenditure, however, remain to be well defined. Proper use of accelerometers requires an appreciation for measurement duration (generally at least seven days) and the sampling interval, or epoch length (less than 10 s is usually optimal). Cut points for defining different levels of physical activity in different age groups in children remain debatable.

Heart Rate Monitors

Heart rate is a sound marker of physiologic stress and is related to intensity of effort during physical activity. The difficulty is that myriad other factors, including age, physical fitness, and psychological stress, also affect heart rate and thereby confound its ability to assess physical activity in children. In small-scale investigations, calibration of heart rate to VO_2 in each individual subject during a laboratory exercise test can sharpen the accuracy of heart rate monitoring for subsequent estimation of energy expenditure in the field. Recent innovations in devices that combine heart rate and accelerometry data may prove particularly useful.

Self-Report Questionnaires and Diaries

In large-scale studies, information derived from diaries and questionnaires completed by subjects (or sometimes by parents or teachers) serves as the only practical means of defining levels of physical activity. The validity of this approach obviously depends on the ability of the children to accurately both estimate and remember particular activities. Children, especially at younger ages, may lack the cognitive ability to use these tools and are often unclear regarding the amount and intensity of exercise they perform. These problems seriously limit accuracy, yet these tools may be the only means of estimating activity levels in large populations of subjects.

Global Positioning Systems

Information from global positioning systems or GPS (satellites) has been used to track the location, as well as the direction and speed, of human subjects. Researchers have also used a combination of GPS and accelerometer data to analyze activity patterns in both children and adults. The role of this approach is just beginning to be examined.

In summary, no ideal means exists at present for measuring physical activity in youth. Each method offers its own particular advantages and weaknesses, and the choice of technique typically involves a compromise between practicality and accuracy. Decisions regarding monitoring methods need to be tailored to the purposes of the investigation, the number of subjects involved, the availability of equipment, and the skills and financial resources of the investigators.

■ Continued efforts to identify accurate measuring tools for physical activity in children are pivotal for understanding and promoting regular exercise for health in the pediatric age group.

Defining the Kinds and Amount of Physical Activities for Health

As an ultimate goal, researchers would like to define the specific forms and amount of exercise required in youth for long-term health outcomes (or at least their surrogate markers). Such information would permit recommendations for thresholds of health-related activity in young people, as well as identify particular individuals or groups at risk. As outlined previously, progress toward this objective is contingent upon the development of accurate activity measurement techniques as well as insights into the specific links—and their dose–effect relationships—between activity and markers of health outcomes. As is clear from the evidence reviewed in this section, headway has been made, but many gaps in the resolution of these challenges remain.

Physical Activity Guidelines

The perceived importance of physical activity to health in children begs identification of particular levels of activity required for health benefits. A number of expert groups in various countries have been convened to construct such guidelines based on the available research data regarding an exercise–health link in young people (Twisk 2001). Initial recommendations of 30 min per day of moderate activity were soon felt to be inadequate, since most children appeared to be meeting this requirement while levels of childhood obesity continued to increase. Subsequently, most guidelines have called for 60 min of daily moderate activity in youth as a threshold for obtaining health benefits. In Canada, recommendations call for increasing activity above current levels by an additional 30 min a day. These guidelines suggest that children should optimally aim for 90 min a day, which is a volume consistent with that seen in children with lower clustered risk factors in the European Youth Heart Study described previously.

Organized games can help children meet the daily recommended amount of physical activity.

However, considering the very limited basis outlined earlier for linking activity in children with future health, it is evident that such guidelines are not based on any firm scientific foundation. Specifically, relationships between activity in healthy children and recognized surrogates of adult disease are weak or nonexistent. Those associations that have been verified are limited to children with preexisting abnormal risk factors; and even in such situations, dose–response relationships have not been established. Consequently, current activity recommendations for youth are best described as only "informed judgments," since "there is no real scientific rationale for these guidelines" (Twisk and Ferreira 2008). Indeed, some have concluded that "the amount and type of physical activity during childhood which is appropriate for optimal health is probably impossible to ascertain" (Cumming and Riddoch 2008). Apart from the problem of delayed expression of disease, such recommendations of amount and types of activity are undoubtedly influenced by age, gender, socioeconomic factors, environmental variables, and—most particularly—type of disease outcome

Physical Activity Versus Physical Fitness

The foregoing discussions have focused on the role of physical activity levels in youth and health outcomes, with little attention to physical fitness. The two are entirely different constructs; activity indicates the amount of body movement, or energy expenditure, in a given period, while fitness relates to the ability of an individual to perform certain motor tasks (running a mile, lifting a weight). In adults, a significant relationship is observed between activity and fitness, a not unexpected finding given that a causal relationship with an arrow in both directions could be expected. That is, more fit people might be expected to enjoy and engage more frequently in exercise, and being more regularly active might result in improvements in physical fitness.

In children, on the other hand, studies have failed to reveal any consistent relationship between amount of daily physical activity and fitness (most commonly assessed as aerobic fitness, or $\dot{V}O_2$max). It has been suggested that this disparity might be explained by the lack of a precise measurement of daily physical activity in youth. In fact, Rowlands and colleagues (1999) found that the relationship between activity and $\dot{V}O_2$max in 8- to 10-year-old children differed depending on whether heart rate, pedometer, or accelerometer data were used.

Other information, however, supports the idea that physical activity and fitness are only weakly linked, if at all, in the pediatric age group. This is evidenced by the limited influence of even extremes of activity on aerobic fitness as observed in youth. A period of high-intensity aerobic training in groups of prepubertal children typically results in average increases of $\dot{V}O_2$max of only about 5% to 6% (compared to 15% to 20% in adults). A period of complete bed rest for nine weeks in five 7- to 11-year-old children in one study caused a fall in $\dot{V}O_2$max estimated to be just 13% (while treadmill endurance time was halved). In the Amsterdam Growth and Health Longitudinal Study, the association between the development of aerobic fitness ($\dot{V}O_2$max by treadmill testing) and daily physical activity (by interview) was examined over a 23-year period beginning at age 13 years. The two were positively related, but the magnitude was small: A 10% difference in activity score over two decades translated into a 0.3% difference in $\dot{V}O_2$max (Kemper and Koppes 2006).

A difficulty faced by those evaluating an association or causal relationship between aerobic fitness and cardiovascular risk factors is the pervasive effect of body fat on the measure of fitness itself. $\dot{V}O_2$max is typically expressed relative to body mass, which is strongly influenced by body fat content. Similarly, in weight-bearing field measures of cardiorespiratory fitness, such as the 1-mile or shuttle run, time is expected to be strongly influenced by adiposity (i.e., the extra inert weight that must be transported during the test). In any accurate assessment of a link between aerobic fitness and a health risk factor, then, accounting for body fat in the analysis is essential.

■ **Excess body fat serves as a strong mediator for relationships of physical activity and fitness with health risk factors such as elevated blood pressure and abnormal serum lipid profile.**

In studies in which aerobic fitness is expressed as $\dot{V}O_2$max per kilogram body mass, significant cross-sectional relationships with health risk factors have generally disappeared when values are adjusted for body fat content. In certain areas, however, limited information does suggest a beneficial role of physical fitness in youth. Those studies that have revealed a salutary effect of exercise training on blood pressure reduction in adolescents with mild essential hypertension have involved both aerobic and resistance training. This observation is consistent with the concept that reductions in resting sympathetic tone via exercise training might prove beneficial in lowering blood pressure.

In adults, a number of mechanistic schemata have been proposed to explain the observation that aerobic exercise training can effect improvements in insulin resistance independent of its influence on body fat content. This issue is just beginning to be examined in the pediatric age group. In one study of obese youth (Allen et al. 2007), $\dot{V}O_2$max expressed relative to lean body mass was observed to be a stronger predictor of insulin levels than body fat content. Another report (Benson, Torode, and Singh 2006) indicated that both muscle strength (supine bench press) and treadmill-measured $\dot{V}O_2$max served as independent predictors of insulin resistance in 10- to 15-year-old nonobese subjects.

This information leads to two important observations. First, adiposity has consistently been shown to play a central role in the link between both physical activity and fitness and health outcome markers among children and adolescents. This observation provides further impetus for identifying the optimal means of preventing and treating childhood obesity. It also points to the need for investigations of the influence of activity and fitness on health outcomes to take subject body composition into account.

Second, it is apparent that the forms of exercise in children that might be expected to improve health are multiple, and that optimal interventions should consider the goals desired. Low-grade or moderately intense activities that increase caloric expenditure might be most appropriate for children who are obese; improvements in aerobic physical fitness may be more important for vascular health; and weight-bearing activities are probably optimal for promoting bone health.

Forms of exercise that relate to particular health outcomes in youth are varied. These need to be taken into account in the design of activity interventions.

Optimal Intervention Strategies

Regular physical activity during the growing years is expected to retard the development of pathologic processes (atherosclerosis, obesity) that will eventuate in clinical disease in the adult years. If so, one approach to intervention would be to identify a particular threshold amount of daily activity "sufficient for health." This would permit the establishment of quantitative guidelines for designing activity programs for youth as well as for identifying sedentary at-risk individuals or groups.

As noted previously, some feel that such a strategy is unreasonable since identifying cut points of activity for different ages, levels of fitness, and a variety of health outcomes becomes quite unfeasible. An alternate strategy has thus been proposed, that of establishing regular exercise habits in youth rather than focusing on achieving a particular volume of activity. According to this concept, a foundation for a lifestyle that includes regular activity would be built early on, and healthy activity behaviors would then persist into adulthood and lead to positive health outcomes.

While this concept is intuitively attractive (indeed, all kinds of lifelong behaviors are often established in childhood), the extent to which activity habits can be established in youth, and the extent to which these can be expected to persist over a lifetime, have not been determined (Rowland 2008). At the present time, no experimental evidence exists one way or the other to indicate the validity of this strategy. To start with, developmental psychologists remain uncertain regarding just how habits are established in children (Pavlovian conditioning? Imitation of parents? Imprinting, as observed in animals?). In addition, the role of genetic influences in the development of human behavior needs to be taken into account.

The "lifestyle habit" strategy would be supported if physical activity behavior during childhood could be documented to "track," or persist, into the adult years. While current cross-sectional and longitudinal studies have failed to demonstrate any impressive degree of such **tracking of physical activity**, these studies are weakened by inadequate measures of activity, focusing only on sport activities and involving adolescents rather than younger children (Malina 2001). Moreover, no study has addressed the most critical question: Can an *increase* in a child's activity level from an early intervention be expected to persist into adulthood?

A more immediate strategy for decreasing the time children spend in sedentary behavior has drawn increasing attention. Studies reveal that youth devote large blocks of time to viewing television, playing video games, and using computers. It has been suggested that limiting such "screen time" will lead to its replacement by more active pursuits. Specific guidelines have been suggested, such as limiting time spent using electronic media for entertainment to not more than 2 h a day. The value of this approach has been supported by evidence that sedentary behavior is more likely to track through life than is physical activity per se.

That sedentary time is related in an important way to physical activity and health markers (particularly obesity), however, has not always been observed in cross-sectional studies. Some have viewed activity and sedentary behavior as separate constructs; indeed, it can be argued that even large amounts of screen time do not preclude participation in sport and high levels of daily physical activity. Moreover, one cannot assume that screen time will not be replaced by other sedentary pursuits. One must also keep in mind that a good deal of inactive time can be spent on laudable pursuits, such as doing homework, reading, creating art, and listening to or playing music.

Biological Effects on Physical Activity in Youth

Over the course of childhood, daily physical activity, measured either as body movement or as energy expenditure per body size, steadily declines. This decrease is more than minor. Between the ages of 6 and 16 years, daily caloric expenditure expressed as kilocalories per kilogram body mass decreases in both males and females by approximately 40% (figure 17.3). This decline has been viewed by some health scientists with dismay, yet a developmental fall in physical activity level is a biological, not behavioral, phenomenon, observed at all levels of the animal kingdom (Sallis 2000).

Studies documenting a significant genetic influence on physical activity levels in youth are consistent with this biological effect. Recent investigations have described the control by specific gene loci of activity in adults. Such possibilities have yet to be explored in children.

It is important, then, for those seeking to promote exercise for health in children to appreciate that their

goal is to define means of "bending" this normal developmental curve of activity. More particularly, it is necessary to assess the extent to which biological versus environmental and psychological factors define a child's level of activity. Such influences might seem to be age related, for example. Biological influences would clearly appear to dominate in a 1-year-old child, while the 16-year-old, with cognitive options for activity behaviors, might be more influenced by extrinsic social and environmental and factors.

Summary

The rationale for promoting physical activity in youth as a means of improving health outcomes in the adult years is intellectually sound. Indeed, it is difficult to visualize a preventive health strategy with more important potential for ensuring the well-being of the population. At the same time, those seeking to translate this concept into feasible, effective means of establishing adequate levels of activity in children and adolescents face a number of difficult challenges. Most particularly, strategies must counter strong contemporary cultural influences that promote a sedentary lifestyle. Nonetheless, considerable progress has been made toward this goal.

Key Concepts

accelerometer—For definition, see page 66.

aerobic fitness—Ability to perform sustained muscular exercise such as running and cycling in conjunction with support of the cardiovascular system.

cardiovascular risk factors—Variables such as abnormal levels of blood lipids, excessive body fat, sedentary lifestyle, and smoking that are predictive of future adverse circulatory outcomes (heart attacks, stroke, high blood pressure from kidney disease, etc.).

endothelial dysfunction—Impaired ability of the lining of blood vessels (the endothelium) to secrete substances such as nitric oxide, which normally cause vessel dilation in response to increased blood flow.

gene–environment interaction—Influence of genetic factors on the response to a change in behavior or to changing environmental conditions.

intima-medial thickness (IMT)—Greater thickness of the inner and middle layers of the blood vessels, an early indicator of atherosclerotic change.

tracking of physical activity—Persistence of exercise habits from childhood through the adult years.

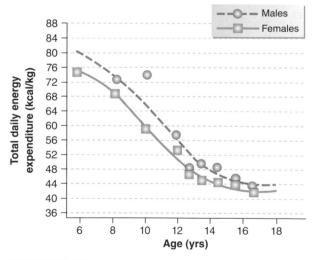

FIGURE 17.3 Physical activity levels decline with age during childhood.

Reprinted, by permission, from T.W. Rowland, 1990, *Exercise and children's health* (Champaign, IL: Human Kinetics), 35.

Study Questions

1. You wish to perform a study to determine the level of habitual physical activity in a cohort of 970 boys and girls. What method of activity measurement would you choose? What would be the advantages and disadvantages of your choice? What if, instead, you wanted to determine the caloric expenditure of normal daily activities (digging in the garden, walking flights of stairs) in a group of 20 youngsters?

2. A study has revealed that maximal oxygen uptake ($\dot{V}O_2max$) expressed per kilogram body mass is closely related to percent body fat in a group of adolescent boys. The authors have concluded, therefore, that obesity limits cardiovascular fitness. How might you argue that this conclusion is inappropriate?

3. In studying longitudinally a group of young females, you find that their level of habitual physical activity declines by 25% between the ages of 10 and 16 years. What factors do you think might be most important in explaining this decline?

4. Promotion of physical activity in children might occur in several settings, including home, school physical education classes, community programs, organized sport settings, and physicians' offices. What are the advantages and disadvantages of each setting? Which would you choose as most feasible and potentially efficacious?

5. Your local school board is planning to eliminate physical education classes in the middle and elementary schools. You decide to make a presentation in front of the board, urging the continuation of physical education at all grade levels. What would you say?

References

Abbott, R.A., M.A. Harkness, and P.S.W. Davies. 2002. Correlation of habitual activity levels with flow-mediated dilation of the brachial artery in 5-10 year old children. *Atherosclerosis* 160:233-239.

Allen, D.B., B.A. Nemeth, R.R. Clark, S.E. Peterson, J. Eickhoff, and A.L. Carrel. 2007. Fitness is a stronger predictor of fasting insulin levels than fatness in overweight male middle-school children. *Journal of Pediatrics* 150:383-387.

Andersen, L.B., M. Harro, L.B. Sardinha, K. Froberg, U. Ekelund, S. Brage, and S.A. Anderssen. 2006. Physical activity and clustered cardiovascular risk in children: A cross-sectional study (The European Youth Heart Study). *Lancet* 368:299-304.

Benson, A.C., M.E. Torode, and M.A.F. Singh. 2006. Muscular strength and cardiorespiratory fitness is associated with higher insulin sensitivity in children and adolescents. *International Journal of Pediatric Obesity* 1:222-231.

Corder, K., and U. Ekelund. 2008. Physical activity. In *Paediatric exercise science and medicine* (2nd ed.), ed. N. Armstrong and W. van Mechelen (pp. 129-144). Oxford: Oxford University Press.

Cumming, S., and C. Riddoch. 2008. Physical activity, physical fitness, and health: Current concepts. In *Paediatric exercise science and medicine* (2nd ed.), ed. N. Armstrong and W. van Mechelen (pp. 327-338). Oxford: Oxford University Press.

Fernhall, B., and S. Agiovlasitis. 2008. Arterial function in youth: Window into cardiovascular risk. *Journal of Applied Physiology* 105:325-333.

Franks, P.W., and H.C. Looker. 2008. Gene-physical activity interactions and their role in determining cardiovascular and metabolic health. In *Paediatric exercise science and medicine* 2nd ed.), ed. N. Armstrong and W. van Mechelen (pp. 353-364). Oxford: Oxford University Press.

Hind, K., and M. Burrows. 2007. Weight-bearing exercise and bone mineral accrual in children and adolescents: A review of controlled trials. *Bone* 40:14-27.

Kemper, H.C.G. 2008. Physical activity, physical fitness, and bone health. In *Paediatric exercise science and medicine* (2nd ed.), ed. N. Armstrong and W. van Mechelen (pp. 365-374). Oxford: Oxford University Press.

Kemper, H.C.G., and L.L.J. Koppes. 2006. Linking physical activity and aerobic fitness: Are we active because we are fit, or are we fit because we are active? *Pediatric Exercise Science* 18:173-181.

Malina, R.M. 2001. Physical activity and fitness: Pathways from childhood to adulthood. *American Journal of Human Biology* 13:162-172.

Marshall, S.J., and G.J. Welk. 2008. Definitions and measurements. In *Youth physical activity and sedentary behavior. Challenges and solutions*, ed. A.L. Smith and S.J.H. Biddle (pp. 3-30). Champaign, IL: Human Kinetics.

Rowland, T. 2008. On Lamarck, lima beans, and learning of physical activity habits. *Pediatric Exercise Science* 20:1-4.

Rowlands, A.V. 2007. Accelerometer assessment of physical activity in children: An update. *Pediatric Exercise Science* 19:252-266.

Rowlands, A.V., R.G. Eston, and D.K. Ingledew. 1999. Relationship between activity levels, aerobic fitness, and body fat in 8- to 10-yr old children. *Journal of Applied Physiology* 86:1428-1435.

Sallis, J.F. 2000. Age-related decline in physical activity: A synthesis of human and animal studies. *Medicine and Science in Sports and Exercise* 32:1598-6000.

Sarzynski, M.A., J.C. Eisenmann, G.J. Welk, J. Tucker, K. Glenn, M. Rothschild, and K. Heelan. 2010. ACE I/D genotype, habitual physical activity, and blood pressure in children. *Pediatric Exercise Science* 22:301-313.

Slyper, A.H. 2004. What vascular ultrasound testing has revealed about pediatric atherogenesis, and a potential clinical role for ultrasound in pediatric risk assessment. *Journal of Clinical Endocrinology and Metabolism* 89:3089-3095.

Twisk, J.W. 2001. Physical activity guidelines for children and adolescents: A critical review. *Sports Medicine* 31:617-627.

Twisk, J.W., and I. Ferreira. 2008. Physical activity, physical fitness, and cardiovascular health. In *Paediatric exercise science and medicine* (2nd ed.), ed. N. Armstrong and W. van Mechelen (pp. 339-352). Oxford: Oxford University Press.

18

Risks of Physical Activity

Evert A.L.M. Verhagen, PhD, FECSS; Esther M.F. van Sluijs, PhD; and Willem van Mechelen, MD, PhD, FACSM, FECSS

CHAPTER OUTLINE

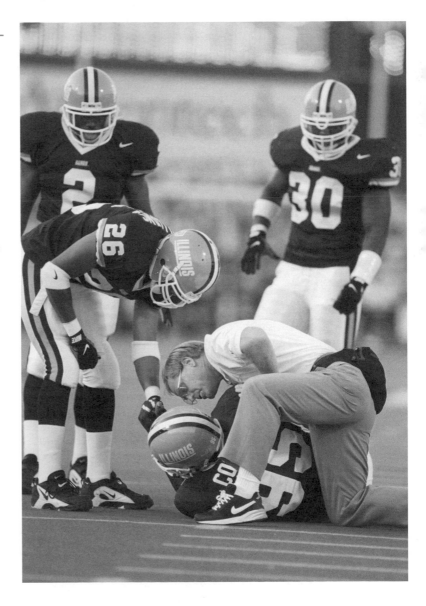

Including daily physical activity in one's life provides a number of health benefits, but such participation, especially in vigorous exercise or sport, can result in injury, disability, or death. This chapter focuses on the risk of sudden cardiac death (SCD) and the female triad, which are important risks in terms of consequences, and on musculoskeletal injuries and asthma, which are important in terms of frequency of occurrence. After discussing the risks and causes, we consider strategies for preventing these risks. After all, the benefits of physical activity need to outweigh the risks in order for people to maintain joyful and healthy physical activity behavior.

Studies performed over the past 25 years confirm the health benefits of regular physical activity, a concept with foundations in antiquity. The effects of physical activity on certain individual health conditions, the precise dose (intensity or amount) of activity that is required for specific benefits, and the biological pathways through which physical activity contributes to health are topics discussed elsewhere in this text. Although numerous details of these topics remain to be clarified by research, it is now clear that regular physical activity reduces the risk of morbidity and mortality from several chronic diseases. It also increases physical fitness, which improves function, physical independence, and quality of life.

In addition to enhancing health, participation in physical activity and, more precisely, vigorous exercise or sport also carries significant risks. These risks can be biomechanical (e.g., injury to various tissues or organs), cardiovascular (ranging from discomfort or pain such as angina pectoris to transient risk of SCD), respiratory (asthma or **anaphylaxis**), heat related (heatstroke), or combined (e.g., the female triad, which is the interrelationship among eating disorders, **amenorrhea**, and **osteoporosis** in the female athlete). The occurrence of any serious health problem is very low in the general population who exercise at moderate intensity and in amounts intended to improve health and physical performance.

In a population of athletes or others who exercise very vigorously, the chances of an injury increase with increasing intensity and amount of exercise. Sudden cardiac death, for instance, has a low incidence of approximately 1 cardiac arrest per 50,000 to 100,000 exercisers per year in the general population (Maron 2007; Pigozzi and Rizzo 2008). However, the risk of SCD during vigorous exercise is 5 to 56 times greater than during usual activities of daily living (Siscovick et al. 1984). The actual prevalence of the female triad is unknown in both the general and the athletic populations. Yet data on eating disorders in the female athlete population suggest the existence of a significant medical problem. As an example, the prevalence of eating disorders has been reported to

be up to sixfold higher in an athletic population in "thin-build" sports as opposed to a control population, 31% versus 5.5% (Nattiv et al. 2007).

Taking the risks of physical activity into account, individuals might adopt a "decision-balance" approach in deciding whether it is worthwhile to continue with the same activity level or to become less or more active. Risk is therefore a perception that may partially guide an individual's physical activity behavior.

> The female triad is a combination of three interrelated conditions that can be associated with athletic training and competition and can result in a significant health risk to female athletes: disordered eating, amenorrhea, and osteoporosis.

Risks of Physical Activity and Sport Participation

In the new public health move toward greater physical activity, low- to moderate-intensity physical activities are promoted to reduce the health risks in otherwise sedentary people. Just as the health benefits from physical activity seem to increase with an increase in physical activity amount and intensity, so do the risks, as depicted in figure 18.1. When talking about physical activity and health benefits, we often mean moderate-intensity daily activities (e.g., gardening, brisk walking, or cycling). It should be obvious that the risks of such low- to moderate-intensity physical activities are relatively low. The more vigorous physical activities and, in particular, sport participation present a greater risk whether the participant is an elite athlete or a recreational athlete. It is important to note, however, that relative injury risk has been recently argued to be highest in the least active part of the population. This means that individuals who start engaging in physical activities for their health may have a higher injury risk than previously thought. Whether the same holds true for the other risks discussed in this chapter is unknown.

Sudden Cardiac Arrest

One of the most serious hazards of vigorous exercise is the transient risk of SCD, which raises concerns regarding the safety of vigorous exercise for participants of all ages. It is fascinating that SCD usually occurs during warm-up, after training, or during a relatively inactive period of a game, and not at peak performance when oxygen demand is the highest

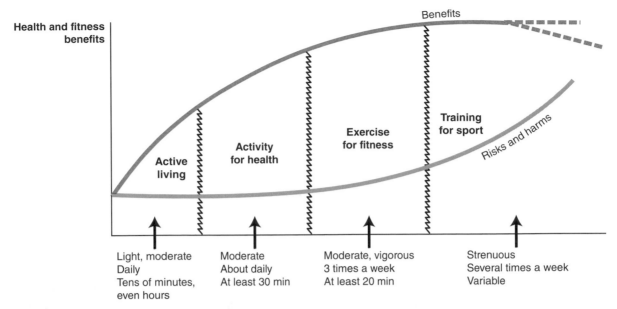

FIGURE 18.1 Benefits and risks related to levels of physical activity.

From van Sluijs, Verhagen, van der Beek, van Poppel, Vuori, 2003, Risks of physical activity. In *Perspectives on health and exercise*, edited by J. McKenna and C. Riddoch (United Kingdom: Palgrave MacMillan), 112. Reproduced with permission from Palgrave MacMillan. Used with permission from Willem van Mechelen.

in the myocardium (Varró and Baczkó 2010). The number of sudden deaths during moderate-intensity exercise is small—0.35% to 0.5% of all sudden deaths in autopsy materials and less than 1 death per 1 million exercise hours in middle-aged men (Vuori 1995). However, it has been argued that SCD is two to four times more frequent in young athletes compared to age-matched persons who do not engage in sport activities. Corrado and colleagues (2007) estimated the incidence of SCD in young competitive athletes at 0.61 per 100,000 person-years. Nevertheless, figures indicate that the absolute risk is low among apparently healthy sport participants. But any death occurring during exercise or sport participation is one too many. Thus, this risk associated with physical activity should be decreased whenever possible.

Sudden cardiac death is usually caused by an abrupt loss of electrical stability of the heart, causing it to beat rapidly and inefficiently or stop beating altogether (cardiac arrest). Or, a myocardial infarction can be attributable to a sudden reduction in coronary blood flow that causes rapid death of heart muscle and, in some cases, sudden death.

The age-specific mortality rate is lower among physically active people than among inactive people. At the same time, there is an increased risk for SCD or myocardial infarction during physical activity com-pared with inactivity. The mechanisms behind this counterintuitive phenomenon are complex and not fully known. The cause of sudden cardiac arrest during exercise cannot be attributed to just one mechanism (Maron et al. 2009; Papadakis and Sharma 2009).

Pathophysiological evidence suggests that exercise, by increasing the oxygen consumption of the heart muscle and at the same time shortening diastole and coronary perfusion time, may evoke a transient oxygen deficiency at the subendocardial level, which can be worsened by abrupt cessation of activity. A shortage of blood in the heart muscle (myocardial **ischemia**) can alter depolarization, repolarization, and conduction velocity, triggering serious ventricular arrhythmia, which in extreme cases may be the forerunner of ventricular tachycardia or fibrillation. Another cause of sudden death may be the rupture of an atherosclerotic plaque located in a coronary artery, causing a localized blood clot to form and block blood flow to the myocardium. The reasons why strenuous activity may cause such a plaque rupture are not well understood (Burke et al. 1999). The biological mechanisms responsible for exercise-related SCD differ with the age of the athlete. The majority of deaths in young athletes (<35 years) have congenital and cardiovascular origins, the most common being hypertrophic cardiomyopathy (46%), followed by coronary artery anomalies (19%). In contrast, the majority of sudden deaths during exercise in older athletes or nonathletes (>35 years) are attributable to myocardial infarction as a result of underlying coronary artery disease (Papadakis and Sharma 2009).

Plaque rupture occurs when an atherosclerotic plaque forms in an artery wall located in the intima just below the endothelium (a one-cell-thick inner lining of the artery). The plaque remains separated from the blood flowing through the artery by the endothelium and additional cells and material that slowly form a "cap" over the plaque. If this cap breaks open (ruptures), material in the plaque can cause the blood in the artery to clot and block blood flow to the heart.

When considering the risk of vigorous physical activity, one must be aware of the intensity and duration of the activity and the health status of the person engaged in it. To determine whether vigorous exercise is worth the risk, Siscovick and colleagues (1984) studied the incidence of sudden death during vigorous exercise, paying special attention to the initial level of habitual physical activity. They showed that the relative risk of cardiac arrest among men with low levels of habitual activity was 56 times greater during vigorous exercise (mainly jogging) compared with other times in their lives. Among men with the highest levels of habitual physical activity, this risk was also elevated but only by a factor of 5. Siscovick and colleagues (1984) (figure 18.2) also studied the overall risk of sudden death (during and not during vigorous physical activity) and showed that men with

high levels of habitual physical activity had a risk of SCD only 40% that for habitually sedentary men. These results support the hypothesis that physical activity both protects against and provokes cardiovascular events: Over the short term, it can provoke clinical cardiac events in those with underlying disease, whereas over the long term, it provides protection.

Female Triad

Competitive athletes are frequently under intense pressure to perform and succeed. Many female athletes experience pressure from their coaches, peers, family, and the public to have a low percentage of body fat. They believe that a low level of body fat increases performance and improves their appearance. This is especially true in dancers, distance runners, and gymnasts. Some athletes are driven people, willing to make extreme personal sacrifices to accomplish their goals. But this willingness can drive the athlete to unhealthy eating and exercise behaviors. The results of such behaviors can lead to what is referred to as the female triad. The female triad is a combination of three interrelated conditions that can be associated with athletic training and competition: disordered eating, amenorrhea, and osteoporosis. Although this phenomenon has been reported properly only in high-level athletes, similar symptoms can be found in a general active population, as a large portion of young female participants engage in sport and physical activities to improve their physique.

The term *disordered eating* refers to a wide spectrum of abnormal patterns of eating that range in severity and include restriction of food intake; use of diet pills, diuretics, or laxatives; periods of binge eating and purging; and anorexia nervosa or bulimia nervosa at the extreme end of the spectrum (Greydanus, Omar, and Pratt 2010). The athlete may start simply by monitoring her food intake and then progress to restricting foods such as fats or red meats, limiting food (and calorie) intake, and finally engaging in voluntary starvation. Although several studies on the prevalence of disordered eating in athletes are available, most have yielded unreliable results because of methodological issues (e.g., nonstandard diagnostic procedures, small sample sizes, lack of control groups) (Byrne and McLean 2001). Two large, well-controlled studies showed that the prevalence of eating disorders in elite female athletes may lie between 25% and 31% (Byrne and McLean 2002; Sundgot-Borgen and Torstveit 2004), compared to a prevalence of between 5% and 9% in the general nonathletic population.

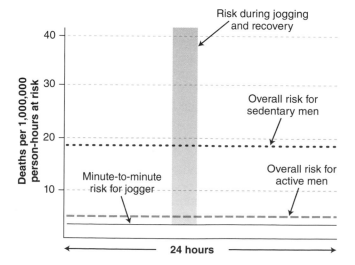

FIGURE 18.2 The relationship of vigorous activity to sudden cardiac arrest.

Data from Siscovick et al. 1984.

Disordered eating consists of a wide spectrum of abnormal patterns of eating, ranging in severity from restriction of food intake to use of diet pills, diuretics, or laxatives; periods of binge eating and purging; and development of anorexia nervosa or bulimia nervosa.

The second aspect of the female athlete triad, amenorrhea, refers to the delayed onset or the absence of menstrual bleeding. The inability to initiate a menses (menarche) before the age of 16 years is called primary amenorrhea, whereas the cessation of the menstrual cycle function after menarche is termed secondary amenorrhea (Greydanus, Omar, and Pratt 2010). Both primary and secondary amenorrhea have a higher prevalence in female athletes than in the general female population. The prevalence in the general population ranges from 2% to 5%, whereas in the female athlete population, prevalence rates of up to 65% have been reported in long-distance runners (Nattiv et al. 2007). However, all female athletes training at high intensities and under mental stress are at risk. The pathophysiology of exercise-induced amenorrhea is complex, with varied contributions of a lowered percentage of body fat, body weight loss, and emotional and physical stress. Although amenorrhea is more prevalent in a population with a coexisting eating disorder, Loucks and Horvath (1985) showed that there is no specific body fat percentage below which regular menses ceases. Some athletes with amenorrhea regain their menses after a period of rest, even without regaining body weight or body fat. These findings suggest that amenorrhea is not caused solely by low body weight or body fat and that other important factors must be considered.

Osteoporosis is the final component of the female athlete triad. Osteoporosis is defined as the loss of bone mineral density (BMD) and the inadequate formation of bone, which can lead to increased bone fragility and an increased risk of fracture (Greydanus, Omar, and Pratt 2010). Premature osteoporosis puts the female athlete at risk of stress fractures as well as more devastating fractures of the hip or vertebral column. A low BMD has been associated with disordered eating even in eumenorrheic athletes, but it is generally lower in amenorrheic athletes than in eumenorrheic athletes (Nattiv et al. 2007). Studies on the prevalence of osteopenia and osteoporosis in female athletes show prevalences of osteopenia ranging from 22% to 50% and prevalences of osteoporosis ranging from 0% to 13% in female athletes (Khan et al. 2002); the same prevalences in a general population are 12% and 2.3%, respectively. A young athlete may find herself with the bone mass of a 60-year-old, with a consequent threefold risk of stress fractures. This accelerated bone loss is the result of estrogen deficiency and subsequent bone resorption. The concern is that the bone loss during early age is partly irreversible, although research shows that regaining the menses can result in increases in BMD. A significant increase in BMD is found in women who decrease their training intensity and regain their menses. However, these athletes are at risk of reaching a BMD level that is far below normal for their age and may never reach a normal level again. A low peak BMD in early life is a major risk factor for osteoporosis and increased bone fractures in old age.

High-Risk Sports for Development of the Female Triad

According to the American College of Sports Medicine, participation in the following sports puts female athletes at high risk for developing the female triad, or components thereof:

- Sports in which performance is subjectively scored (e.g., dance, figure skating, gymnastics)
- Endurance sports favoring participants with a low body weight (e.g., distance running, cycling, cross-country skiing)
- Sports in which body contour–revealing clothing is worn for competition (e.g., volleyball, swimming, diving, running)
- Sports using weight categories for participation (e.g., horse racing, martial arts, rowing)
- Sports in which prepubertal body habitus favors success (e.g., figure skating, gymnastics, diving)

Adapted from Otis et al. 1997.

Injuries Related to Physical Activity and Sport

As opposed to SCD and the female triad, musculoskeletal injury is a risk for all who engage in physical activity, regardless of its level and type. The risk of injury associated with many of the recommended health-promoting physical activities has not been systematically evaluated. Although it is well established that there is an increased risk of injury at the higher end of the physical activity intensity scale, the prevalence of injuries during low-level physical activities (e.g., gardening, walking) is highly variable and not well established. In general, the risks for injury in such low-level physical activities are considered equal to the risks with the activities required for daily living. For instance, walking for half an hour a day carries little or no increased risk for acute or chronic musculoskeletal problems.

One way of looking at the injury problem is to examine the absolute number of physical activity– and sport-related injuries that occur in a specific population over a defined period of time. According to the most recent count in the Netherlands, in a population of about 11 million active participants, a total of 3.6 million sport-related injuries occur annually (Schoots et al. 2009). Of all these injuries, 1.4 million require some form of medical attention. Activity-specific numbers are given in table 18.1. Absolute numbers, however, do not precisely represent the injury risk for a person performing a specific activity. Given the popularity of soccer in the Netherlands and the large number of participants in this sport, it is not surprising to find that in absolute numbers soccer is the most "dangerous" activity. However, a better way of viewing injuries is to look at the number of injuries per 1,000 hours of participation, that is, the injury incidence. Injury incidence numbers give a more precise estimate of the actual risk of engaging in a particular activity. The injury incidence numbers in table 18.1 show that indoor soccer is the activity with the highest injury risk in the Netherlands. The table further shows that martial arts, with an absolute number of only 93,000 injuries, carry an injury risk higher than soccer. Both calisthenics and running/jogging are generally associated with a healthy, physically active lifestyle and are very popular; but whereas the injury risk is relatively low for calisthenics, the risk associated with running/jogging is high. Therefore, from a public health perspective, these sports are of particular interest.

Each sport or activity has its own injury types and causes. Thus, injury prevention in each activity should focus on the risk factors and injuries inherent to that activity (Conn, Annest, and Gilchrist 2003). For instance, soccer players make hard cuts, sharp turns off a planted foot, and intense contact with the ball and other players. This makes them more vulnerable to acute lower-extremity injuries, especially to the knee and ankle. Acute lower-extremity injuries are also the most common injuries in volleyball. There is general agreement that these acute volleyball injuries result from frequent jumping and landing as well as from striking the leg on the floor during defensive maneuvers. The upper extremity is particularly susceptible to injury (acute and chronic) in tennis because of the use of the racket and the stress it places on the dominant arm and shoulder. Finally,

TABLE 18.1 Sport-Specific Injury Numbers

Type of sport	Number of participants	Absolute number of injuries	Number of injuries/1,000 h
Total	11,000,000	3,600,000	1.9
Soccer	1,400,000	580,000	3.7
Volleyball	540,000	170,000	4.6
Outdoor tennis	1,100,000	210,000	2.8
Indoor soccer	340,000	99,000	9.7
Field hockey	270,000	130,000	4.5
Calisthenics	3,100,000	370,000	1.4
Martial arts	340,000	93,000	5.5
Running/jogging	1,400,000	410,000	5.6

Data from Dutch Consumer Safety Institute 2010

running- and jogging-related injuries are primarily chronic injuries to the lower extremities (e.g., stress fractures). It is beyond the scope of this chapter to discuss risk factors and injury types for each activity in detail (for comprehensive information, refer to Bahr and Engebretsen 2009). Table 18.2 presents the most common risk factors for selected activities and sports.

Asthma and Airway Hyperresponsiveness

A high prevalence of asthma and **airway hyperresponsiveness (AHR)** has been reported in athlete populations. In the general population, the prevalence of asthma is about 5% to 10%, whereas in elite endurance athletes, the prevalence ranges from 10% to 50% (Carlsen et al. 2008). The prevalence of asthma and AHR is particularly high among swimmers and athletes exercising in cold air environments. In contrast to elite athletes, amateur endurance athletes do not seem to have an elevated risk for asthma or AHR.

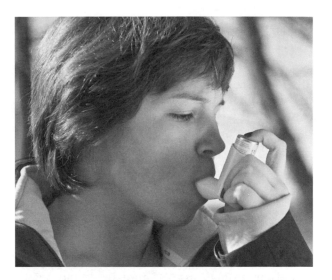

Factors such as cold air, specific activities, and high-intensity physical activity may contribute to asthma.

TABLE 18.2 Various Sports and Their Main Injury Risk Factors

Sport	Risk factors
Soccer	Previous injury
	Player's position
	Playing surface
Volleyball	Jumping technique
	Position
	Playing surface
Tennis	Shoulder strength
	Equipment
	Playing surface
	Muscle imbalance
Hockey	Physical characteristics
	Aggressive play
	Equipment
Gymnastics	Previous injury
	High lumbar curvature
	Protection
Martial arts	Physical characteristics
	Technique
	Equipment
	Opponent
	Skill level

According to a summary by Langdeau and Boulet (2001), although moderate exercise has been shown to be beneficial for patients with asthma, repeated and high-intensity exercise could contribute to the development of asthma. High-intensity physical activity can trigger asthma symptoms in athletes who already have asthma. But high-level exercise performed on a regular basis might also contribute to the development of asthma in previously unaffected athletes. In the general population, the development of asthma is of multifactorial origin. A genetic component can be recognized, in addition to environmental factors such as exposure to inflammatory substances. Among athletes, it has been suggested that both prolonged hyperventilation and the quality of the inhaled air during exercise could be contributing factors. Athletes may have an increased exposure to allergens and pollutants attributable to prolonged hyperventilation during and following intensive exercise. This increased exposure could lead to an inflammatory process that might contribute to the development of AHR. The air temperature may be a factor in that exposure to cold air could induce a bronchoconstrictive response. Whether this is the effect of the low temperature or the low water content in cold air is uncertain. Also, cold air could cause epithelial damage and inflammation and thereby influence airway function. Another factor that may explain the high prevalence of asthma and AHR among athletes is some degree of immunosuppression. Athletes are at increased risk for upper respiratory infection (e.g., common cold, sore throat) during periods of intense training, which may increase their susceptibility for developing asthma.

Minimizing Risk and Maximizing Benefits

Knowing the risks associated with moderate- and vigorous-intensity exercise and, in particular, sport participation raises the question whether performing vigorous exercise or participating in sport is as healthy as many scientists want the public to believe. It is; but to lead a healthy, physically active life, one must minimize the risks in order to maximize the benefits. The next sections present strategies to minimize the risks of SCD, the female triad, and musculoskeletal injury. Prevention of asthma and AHR is not discussed here. Preventive measures against asthma and AHR depend mostly on the environment that the person exercises in and vary from the use of ventilated indoor swimming pools to the use of medications.

Preventing Sudden Cardiac Death During Exercise

Although the absolute numbers of SCDs during exercise are low, an exercise-related SCD is a highly emotional event given the relative youth of the individual and the potential number of life years lost. The counterintuitivity of SCD also astounds communities around the globe because physically active individuals are considered to represent the healthiest segment of our society (Papadakis and Sharma 2009).

People of all ages with congenital, acquired, or degenerative heart disease can be identified through preparticipation evaluations consisting of history and clinical examination. Although known heart defects can be important predictors for sudden death during exercise, about 80% of all SCDs occur in people with latent or subclinical heart disease. Unfortunately, many of these conditions are not detected by the typical medical evaluation (Papadakis and Sharma 2009). A preexercise examination that also includes a 12-lead electrocardiogram (ECG), however, is effective in reducing SCD from cardiomyopathies and electrical disorders of the heart at the expense of a small number of false-positive tests, about 2% (Corrado et al. 2005; Papadakis and Sharma 2009).

Providing information about the nature of exercise-related risks and the safest way to exercise is of great importance. A key is to detect and point out the importance of effort-related symptoms, unexplained tiredness, and febrile infections. Furthermore, attention needs to be paid to the intensity of the activity. Two-thirds of sudden deaths with exercise occur during vigorous physical activity, even though most population data indicate that only a minority of exercising people engage in vigorous physical activity.

As described earlier, the risk of sudden death during vigorous exercise such as jogging is 10 times greater for subjects with a low level of fitness than for subjects with a high level of fitness (Siscovick et al. 1984). Participating in vigorous-intensity exercise when one is not used to it substantially increases the risk for SCD. In health promotion activities, people should be cautioned to exercise in moderation in relation to their own exercise capacity and health status. They should be reminded that engaging in moderate-intensity physical activity provides numerous health benefits while keeping risk low. Sedentary people who are changing their physical activity behavior should be discouraged from starting with vigorous-intensity physical activity but rather should initiate a moderate-intensity program and slowly increase intensity as they become more fit.

Preventing the Female Triad

It is not uncommon for female athletes to train at very high intensities, to have an unhealthy diet, and to perform under a great deal of mental and physical stress. A lack of education about the risks for female athletes from such compulsive behavior might help explain why this unhealthy behavior continues even when competition is finished. Therefore, to identify and prevent the female triad, it is crucial for an athlete's caretaker to provide adequate information about its causes and consequences and to detect the triad early in its evolution using a multidisciplinary approach (Nattiv et al. 2007).

When providing education on this topic, one must realize that educating only the athlete will not solve the problem. Education also needs to be directed at the coach and, especially among adolescents, at the parents. Education should include dispelling myths regarding body weight and body fat (e.g., "Thinner is better"; "Every sport has an ideal body weight") and their relationship to performance; providing nutritional information (e.g., the need to consume adequate calories from healthy foods to meet the energy requirement); and dealing with other issues of personal wellness (e.g., issues related to sexuality, time and stress management, drug and alcohol use). Female athletes need to be made aware of the long-term consequences of the female triad, such as the possible health consequences regarding fertility and osteoporosis. An effort must be made to persuade female athletes to change unhealthy behaviors. This is not an easy task, and little is known about the specific approaches that are effective in this population.

Early detection of the female triad (or separate parts of it) may prevent athletes from experiencing irreversible consequences of their behavior. In

North America, the preparticipation physical evaluation may be the ideal time to screen for the triad, specifically for disordered eating and amenorrhea. However, in Europe, such physical exams are not mandatory for athletes, and the triad may therefore remain undetected for a long time. A sports physician or family physician may also screen during office visits for injuries, weight change, amenorrhea, or disordered eating. The physician should screen for signs of disordered eating by asking the athlete about her past eating habits and asking for a list of "forbidden foods." The patient's highest and lowest body weights should be ascertained, as well as whether she is happy with her current body weight. A history of amenorrhea is another easy way to detect the triad in its earliest stages. The physician should be aware that there is no specific body fat percentage below which regular menses cease. Also, physicians should not discount amenorrhea as a benign consequence of intensive athletic training. This will only reinforce the athlete in her unhealthy behavior.

Important aspects of treating the female triad include decreasing training intensity, increasing periods of rest, and regaining body weight. Although the athlete may be unwilling to cooperate, the physician needs to convince her of the importance of these lifestyle changes. In athletes with signs of the female triad, attention should be paid to the possibility of osteoporosis. Physicians should educate an amenorrheic patient about the possible consequences of long-term amenorrhea and the risks of irreplaceable bone loss. To obtain accurate information about her BMD, they should consider a dual-energy X-ray absorptiometry (DEXA) evaluation. They can use this information in considering whether to start hormone therapy to reduce the decrease in BMD.

Preventing Injuries Related to Vigorous Exercise and Sport Participation

Epidemiological data should serve as a basis for prevention programs designed to reduce the injury risk associated with vigorous exercise and sport participation. As postulated by van Mechelen (1992), measures to prevent injuries during vigorous exercise or sport participation do not stand by themselves; they form part of what might be called a sequence of prevention (figure 18.3):

1. The first step in the sequence of prevention is to define the sport injury problem in terms of incidence and severity. This descriptive information provides insight into the magnitude of the problem. It also shows which types

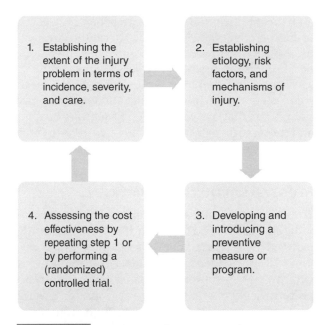

FIGURE 18.3 Sequence of prevention of sport injuries.

Reprinted, by permission, from H. Hlobil, W. van Mechelen, and H.C.G. Kemper, 1987, *How can sports injuries be prevented?* (Netherlands: National Institute for Sports Health Care), 1-134.

of injury are the most common across all sports or in a certain type of sport, as well as which sports are more risky in terms of injury frequency and severity. Furthermore, information on the severity of sport injuries can help to focus on specific preventive measures.

2. Once it is known where preventive measures are warranted, the etiological risk factors and mechanisms underlying the occurrence of the targeted injury need to be identified.

3. With this information regarding underlying risk factors, preventive measures that are likely to work can be developed and introduced.

4. Finally, one should evaluate the cost-effectiveness of these preventive measures by repeating the first step (time–trend analysis) or preferably by performing a randomized controlled trial (RCT). The actual methodology to use in this final step depends on the research question at hand.

As seen in table 18.2, numerous factors that contribute to injuries during vigorous exercise and sport participation have been identified. However, discussing the third and fourth steps of the sequence of prevention is more difficult. A review by Parkkari and colleagues (2001) indicated that only 16 RCTs on the prevention of sport injuries had been conducted over the preceding three decades. A large number of trials have been conducted in more recent years, and this specific field of research has progressed immensely. Nevertheless, evidence on the effectiveness of specific preventive measures remains scarce. Because it is impossible to discuss the prevention of each injury in detail here, we limit our discussion to some general principles of acute and chronic injury prevention that are applicable to vigorous exercise and sport participation. These are divided into three main categories: athlete, sport or exercise, and risk behavior.

Preparation by the Athlete

The use of stretching and warm-up exercises to promote suppleness and flexibility has been historically believed to prevent strain injuries to muscles and tendons. However, there still is much controversy about stretching as a preventive measure. For example, no preventive effect of stretching was found in several studies (Pope et al. 2000; Shrier 2000; Thacker et al. 2004). General and sport-specific conditioning programs, preferably incorporated into the regular warm-up and cool-down, are necessary for athletes to attain successful performance and avoid injury. Avoiding overuse injuries is a critical part of designing and monitoring the conditioning program, particularly in endurance sports such as running. A common rule to help avoid overuse injuries is not to increase more than one exercise parameter (i.e., intensity, frequency, or duration) at a time. Braces and taping are widely used to stabilize weak or unstable joints during the rehabilitation or return-to-play phase. In addition to this secondary preventive application, tape and braces also have a primary preventive function. However, significant controversy exists about the benefit of such procedures. Taping and bracing may function more by improving **proprioception** and thereby stimulating earlier recruitment of supportive muscles than through the actual mechanical restriction or support of a joint. Neither tape nor brace applications can guarantee protection from new injury or from exacerbation of a preexisting trauma.

Exercise- or Sport-Specific Measures

The equipment used in sport has the potential for contributing to injury prevention. For instance, facial and head injuries in American football may be prevented through the use of properly fitted helmets and padded chin straps, which eliminate helmet rotation. The effect of headgear in decreasing injuries has also been shown in wrestling and ice hockey. Also, the use of proper cleats or running shoes reduces lower-extremity injuries. Preventive equipment can be individually applied depending on sport, position,

and bodily dimensions. However, equipment should always fit the conditions for which it is designed; for example, basketball shoes should not be used for long-distance running.

In team sports, an important role in injury prevention lies with the referee. The fact that only 8% of all ice hockey injuries are associated with a penalty suggests that referees may allow dangerous play. Many injuries in soccer, for instance, occur during tackling and contact with the opposing player. A good referee needs to keep the game under tight control and try to prevent dangerous behavior by the participants. Coaching within the spirit as well as within the letter of sport rules and regulations should be emphasized, because aggressive actions often lead to injuries in team sports. For instance, only half of teenage ice hockey players understand the seriousness of checking another player from behind. The coach has the power to emphasize the serious magnitude of the injuries that this move can inflict.

Risk Behavior

In contact sports, a debate has arisen about the introduction of preventive measures because changes in injury patterns and mechanisms have occurred hand in hand with the introduction of protective equipment. An example of this phenomenon is seen in ice hockey, where the introduction of mandatory helmet use reduced the incidence and severity of head injuries. However, evidence suggests that neck injuries may have become more frequent in ice hockey since the introduction of mandatory facial and head protection. In the past, shoulder padding in American football was viewed primarily as "an offensive weapon" and not within the "spirit" of the game. Shoulder padding is a preventive measure—if it is used correctly. It is noticeable that players in a variety of contact sports become more reckless after the introduction of protective measures. This phenomenon is described in the literature as "risk homeostasis."

The theory of risk homeostasis states that individuals maintain their risk behavior at a level they perceive as acceptable and safe. In this viewpoint, individuals adapt their behavior to a certain level of risk that they consider acceptable ("target level of risk"). This explains why the manipulation of risk factors and the introduction of preventive measures frequently do not eliminate or even significantly reduce the risk of injuries. In other words, people compensate for preventive measures by adjusting their risk-taking behavior. Motivating individuals to decrease their target level of risk is the only measure that can decrease the risk for injuries.

> Risk homeostasis is a theory stating that athletes maintain their risk behavior at a level they perceive as acceptable and safe. Such individuals adapt their behavior regarding risk to achieve a "target level of risk."

Recommendations for Future Research

Although a great deal is known about the risks associated with exercise and sport participation, a variety of issues are still under investigation or in need of further clarification. Some of the key issues are summarized here.

Sudden Cardiac Death

To reduce the number of SCDs during vigorous exercise or sport participation, we can make a major advance by increasing our understanding of the pathophysiological mechanisms that underlie such events. Questions to be addressed include "Which persons are at greatest risk?" and "When during their training or competition is their risk the highest?" Such data would enable caregivers to more effectively screen and accurately inform people about the risks associated with exercising vigorously. However, a significant improvement in this effort may be limited by the very low incidence of sudden death during exercise.

Female Triad

The prevalence of the female triad in athletes is substantially higher than in the general population, but estimations for various athlete populations are highly diverse. Researchers should aim to obtain more accurate information about the prevalence of the female triad and its components. They should consider improving methods for diagnosing and screening for the triad and enlarging knowledge about the irreversibility of its detrimental effects. Attention should be given to determining safe training volumes and intensities that will not decrease performance during athletic competition. Finally, theory-based health behavior change programs need to be developed and evaluated to assist these competitors in changing their unhealthy behaviors.

Musculoskeletal Injuries

Numerous questions remain regarding how to more effectively prevent musculoskeletal injuries during exercise of all types and intensities. Procedures to prevent musculoskeletal injuries during exercise do not stand by themselves because they form part of the sequence of prevention. Unfortunately, only a limited number of effect studies on injury prevention have been conducted. Therefore, well-designed randomized studies are needed on preventive measures and devices in common use, such as warming up, proprioceptive training, protective equipment, and education interventions. After evidence of the effectiveness of interventions has been obtained, preventive measures should be implemented. Whereas the number of effective interventions is limited, even less is known about the implementation of such measures in practice. This is a great gap that needs to be closed in order to affect public health.

Summary

An increase in physical activity, especially vigorous-intensity exercise or sport participation, carries a variety of risks as well as a number of well-documented health benefits. Serious health risks associated with increases in moderate-intensity activities, the core of most public health recommendations, are quite rare but still require consideration in the implementation of a physical activity plan designed to maximize benefit while minimizing risk. Major examples of risks associated with vigorous exercise include SCD, the female triad, asthma and related pulmonary disorders, and musculoskeletal injuries.

Factors contributing to the likelihood of an exercise-induced medical complication include the medical status of the participant (underlying disease, prior injury, nutritional deficiency, obesity); current physical activity status (sedentary or active, fit or unfit); type (weight bearing, contact sport), intensity (relative to the person's capacity), duration, and frequency of activity; approach to the exercise session (stretching, warm-up, cool-down), use of appropriate equipment (shoes, protective gear); and the environment (type of surface, air pollution, temperature, humidity). One should consider each of these factors whether implementing a physical activity plan to enhance health or an exercise training program in preparation for athletic competition.

Comprehensive programs to minimize risk include medical screening when indicated, education of participants about their risks during various activities and about how to minimize these risks, professional guidance regarding exercise selection and performance in high-risk persons, and the provision of safe environments for exercise. Additional research is needed on the pathobiological basis for some health risks, the most effective medical screening procedures for selected populations, and intervention programs designed to minimize risk; but broad-scale implementation of risk reduction components based on current knowledge can keep overall risks low while maximizing positive health and performance outcomes.

Key Concepts

airway hyperresponsiveness (AHR)—An abnormal condition in which the airways (especially the bronchi in the lungs) respond to a stimulus such as cold air during exercise by narrowing and restricting airflow.

amenorrhea—Delayed onset or cessation of menstruation for six or more months, associated with high-level exercise training or athletic competition.

anaphylaxis—A sudden, severe, potentially fatal systemic allergic reaction that can involve various areas of the body, such as the skin, respiratory tract, gastrointestinal tract, and cardiovascular system. Anaphylactic reactions can be mild to life threatening. The annual incidence of anaphylactic reactions is about 30 per 100,000 persons. Individuals with asthma, eczema, or hay fever are at greater relative risk of experiencing anaphylaxis at rest or during exercise.

ischemia—Condition in which the oxygen-rich blood flow to a part of the body is not adequate to meet oxygen demands. Cardiac or myocardial ischemia refers to lack of blood flow and oxygen to the heart muscle and happens when sudden exercise is performed without warm-up or when an artery becomes narrowed or blocked for a short time, preventing oxygen-rich blood from reaching the heart. If ischemia is severe or lasts too long, it can cause a heart attack (myocardial infarction) and can lead to heart tissue death. A temporary blood shortage to the heart can cause pressure or pain (angina pectoris). In some cases (40-50%) there is no pain; this condition is called *silent ischemia*.

osteoporosis—For definition, see page 254.

proprioception—Process by which the body can vary muscle contraction in immediate response to incoming information regarding external forces by using stretch receptors in the muscles to keep track of the joint position in the body.

Study Questions

1. Briefly explain how a bout of vigorous exercise might trigger a sudden cardiac death in a 55-year-old man. What are some personal characteristics that might contribute to his risk of having such an event?

2. What are the three major components of the female athlete triad? Briefly explain how a coach, athletic trainer, or physician might identify these features in a 22-year-old highly competitive distance runner.

3. What are the main cardiovascular abnormalities or pathologies that contribute to the risk of sudden cardiac death during vigorous exercise in men under age 35 compared with men over age 35?

4. What is meant by the "sequence of prevention"? Briefly explain the four components of this sequence.

5. If you were advising a group of novice coaches from a youth soccer league about how to reduce the risk of injury in their players, what are four or five issues you would address?

6. A 58-year-old with a body mass index of 33 who has not been physically active since he was 36 years old wants to begin a physical activity program to increase his cardiovascular endurance and muscle strength. How would you advise him to reduce his risk of musculoskeletal injury?

7. Explain how it is possible that even though the risk of sudden cardiac death is substantially increased during vigorous exercise in middle-aged and older adults, compared with periods when they are inactive, these physically active people have an overall lower risk of cardiac death than their sedentary counterparts.

8. What advice would you give a mother regarding participation in cross-country skiing by her 13-year-old son who has exercise-induced airway hyperresponsiveness?

References

Bahr, R., and L. Engebretsen (eds.). 2009. *Sports injury prevention: Olympic handbook of sports medicine.* Oxford: Blackwell.

Burke, A.P., A. Farb, G.T. Malcom, Y-H. Liang, J.E. Smialek, and R. Virmani. 1999. Plaque rupture and sudden death related to exercise in men with coronary artery disease. *Journal of the American Medical Association* 281:921-926.

Byrne, S., and N. McLean. 2001. Eating disorders in athletes: A review of the literature. *Journal of Science and Medicine in Sport* 4:145-159.

Byrne, S., and N. McLean. 2002. Elite athletes: Effects of the pressure to be thin. *Journal of Science and Medicine in Sport* 5:80-94.

Carlsen, K.H., S.D. Anderson, L. Bjermer, S. Bonini, V. Brusasco, W. Canonica, J. Cummiskey, L. Delgado, S.R. Del Giacco, F. Drobnic, T. Haahtela, K. Larsson, P. Palange, T. Popov, P. van Cauwenberge; European Respiratory Society; European Academy of Allergy and Clinical Immunology. 2008. Exercise-induced asthma, respiratory and allergic disorders in elite athletes: Epidemiology, mechanisms and diagnosis: Part I of the report from the Joint Task Force of the European Respiratory Society (ERS) and the European Academy of Allergy and Clinical Immunology (EAACI) in cooperation with GA2LEN. *Allergy* 63:387-403.

Conn, J.M., J.L. Annest, and J. Gilchrist. 2003. Sports and recreation related injury episodes in the US population, 1997-99. *Injury Prevention* 9:117-123.

Corrado, D., P. Michieli, C. Basso, M. Schiavon, and G. Thiene. 2007. How to screen athletes for cardiovascular diseases. *Cardiology Clinics* 25:391-397.

Corrado, D., A. Pelliccia, H.H. Bjørnstad, L. Vanhees, A. Biffi, M. Borjesson, N. Panhuyzen-Goedkoop, A. Deligiannis, E. Solberg, D. Dugmore, K.P. Mellwig, D. Assanelli, P. Delise, F. van-Buuren, A. Anastasakis, H. Heidbuchel, E. Hoffmann, R. Fagard, S.G. Priori, C. Basso, E. Arbustini, C. Blomstrom-Lundqvist, W.J. McKenna, G. Thiene; Study Group of Sport Cardiology of the Working Group of Cardiac Rehabilitation and Exercise Physiology and the Working Group of Myocardial and Pericardial Diseases of the European Society of Cardiology. 2005. Cardiovascular pre-participation screening of young competitive athletes for prevention of sudden death: Proposal for a common European protocol. Consensus statement of the Study Group of Sport Cardiology of the Working Group of Cardiac Rehabilitation and Exercise Physiology and the Working Group of Myocardial and Pericardial Diseases of the European Society of Cardiology. *European Heart Journal* 26:516-524.

Greydanus, D.E., H. Omar, and H.D. Pratt. 2010. The adolescent female athlete: Current concepts and conundrums. *Pediatric Clinics of North America* 57:697-718.

Khan, K.M., T. Liu-Ambrose, M.M. Sran, M.C. Ashe, M.G. Donaldson, and J.D. Wark. 2002. New criteria for female athlete triad syndrome? As osteoporosis is rare, should osteopenia be among the criteria for defining the female athlete triad syndrome? *British Journal of Sports Medicine* 36:10-13.

Langdeau, J.B., and L.P. Boulet. 2001. Prevalence and mechanisms of development of asthma and airway hyperresponsiveness in athletes. *Sports Medicine* 31:601-616.

Loucks, A.B., and S.M. Horvath. 1985. Athletic amenorrhoea: A review. *Medicine and Science in Sports and Exercise* 17:54-72.

Maron, B.J. 2007. Hypertrophic cardiomyopathy and other causes of sudden cardiac death in young competitive athletes, with considerations for preparticipation screening and criteria for disqualification. *Cardiology Clinics* 25:399-414.

Maron, B.J., J.J. Doerer, T.S. Haas, D.M. Tierney, and F.O. Mueller. 2009. Sudden deaths in young competitive athletes: Analysis of 1866 deaths in the United States, 1980-2006. *Circulation* 119:1085-1092.

Nattiv, A., A.B. Loucks, M.M. Manore, C.F. Sanborn, J. Sundgot-Borgen, M.P. Warren; American College of Sports Medicine. 2007. The female athlete triad. *Medicine and Science in Sports and Exercise* 39:1867-1882.

Otis, C.L., B. Drinkwater, M. Johnson, A. Loucks, and J. Wilmore. 1997. The female athlete triad. *Medicine and Science in Sports and Exercise* 29:i-ix.

Papadakis, M., and S. Sharma. 2009. Electrocardiographic screening in athletes: The time is now for universal screening. *British Journal of Sports Medicine* 43:663-668.

Parkkari, J., U.M. Kujala, and P. Kannus. 2001. Is it possible to prevent sports injuries? Review of controlled clinical trials and recommendations for future work. *Sports Medicine* 31:985-995.

Pigozzi, F., and M. Rizzo. 2008. Sudden death in competitive athletes. *Clinical Sports Medicine* 27:153-181.

Pope, R.P., R.D. Herbert, J.D. Kirwan, and B.J. Graham. 2000. A randomized trial of preexercise stretching for prevention of lower-limb injury. *Medicine and Science in Sports and Exercise* 32:271-277.

Schoots, W., I. Vriend, C. Stam, and S. Kloet. 2009. Sports injuries in the Netherlands, the current state of affairs. *Sport and Geneeskunde* 2:16-23.

Shrier, I. 2000. Stretching before exercise: An evidence based approach. *British Journal of Sports Medicine* 34:324-325.

Siscovick, D.S., N.S. Weiss, R.H. Fletcher, and T. Lasky. 1984. The incidence of primary cardiac arrest during vigorous exercise. *New England Journal of Medicine* 311:874-877.

Sundgot-Borgen, J., and M.K. Torstveit. 2004. Prevalence of eating disorders in elite athletes is higher than in the general population. *Clinical Journal of Sports Medicine* 14:25-32.

Thacker, S.B., J. Gilchrist, D.F. Stroup, and C.D. Kimsey. 2004. The impact of stretching on sports injury risk: A systematic review of the literature. *Medicine and Science in Sports and Exercise* 36:371-378.

van Mechelen, W. 1992. *Aetiology and prevention of running injuries* [dissertation]. Amsterdam: Free University of Amsterdam.

Varró, A., and I. Baczkó. 2010. Possible mechanisms of sudden cardiac death in top athletes: A basic cardiac electrophysiological point of view. *Pflügers Archiv: European Journal of Physiology* 460:31-40.

Vuori, I. 1995. Reducing the number of sudden deaths in exercise. *Scandinavian Journal of Medicine and Science in Sports* 5:267-268.

PART IV

Physical Activity, Fitness, Aging, and Brain Functions

Part IV was designed with the goal of bringing together critical information on aging, brain functions, and mental health. It includes three chapters. Chapter 19 focuses on physical activity, fitness, and aging. Physical activity and brain structures and functions are discussed in chapter 20. Finally, the potential roles of regular exercise in alleviating mental health problems are discussed in chapter 21. We feel that these three chapters are of particular importance as they are supported by a young but rapidly expanding body of research.

Physical Activity, Fitness, and Aging

Loretta DiPietro, PhD, MPH

CHAPTER OUTLINE

©Glenda Powers

Older age traditionally has been viewed as a time of inevitable disease and frailty. However, the current view of aging distinguishes true aging-related decline in function from decline that is secondary to other factors known to decline in older age—especially physical activity. Ample data now exist demonstrating that even the frailest members of the older population can respond favorably to exercise. Therefore, physical activity and fitness remain vitally important in older age with regard to maintaining a functional and independent lifestyle. Understanding the role of physical activity and fitness in modifying aging-related changes in health and function has important public health implications for meeting the needs of the ever-growing population of older adults.

The Aging Process

As advances in public health (sanitation, immunizations, improved nutrition) and health care are keeping people alive longer, the population worldwide is growing older. This aging trend has substantial political, social, medical, and economic implications. Therefore, we need to understand the many ways in which the aging process alters human health and function and to distinguish between alterations in function that are reversible and those that are not.

Demographics of Aging

The population aged 65 years and older living in the United States numbers about 39 million and comprises approximately 13% of the population. Because of decreased mortality in older age groups, **demographic trends** in aging will continue, with the population of older people expected to approach 72.1 million (19% of the total) by the year 2030—a twofold increase over the older adult population in 2000 (Administration on Aging, U.S. Department of Health

and Human Services [DHHS] 2008). Perhaps of greatest interest in aging research is the increase in the "oldest old" segment of the population—those people 85 years and older. Since 1930, this oldest segment of the U.S. population has doubled in number every 30 years, and it is projected to be the fastest-growing sector of the older population well into this century. For example, in 2002 there were approximately 4.6 million persons aged 85 years or older living in the United States, and this number will increase to approximately 8.7 million by the year 2030 and then to 19 million by the year 2050 (Administration on Aging, DHHS 2008). The impact of these demographic changes on public health is substantial, particularly because emphasis has begun to shift from tertiary care toward health promotion and disease prevention.

Accompanying the demographic trends in aging is the increasing prevalence of chronic disease and consequent functional impairment. Indeed, more than 80% of older people have at least one chronic health problem such as cardiovascular disease, cancer, diabetes, osteoporosis, **sarcopenia**, or arthritis. Chronic health problems in older adults exact a markedly disproportionate toll on the U.S. economy. For example, despite accounting for about 13% of the population, Americans over age 65 years account for more than 30% of health care expenditures (Federal Interagency Forum on Aging Related Statistics 2004). Clearly, the public health benefits would be enormous if the onset of disease and functional limitations could be postponed or eliminated altogether. With a delay in the onset of these chronic conditions, the maintenance of physical function could be extended to a time closer to the life expectancy. This compression of morbidity will undoubtedly improve quality of life and preserve autonomy for older people, as well as reduce health care costs to the individual and society. Table 19.1 presents a list of prevalent chronic diseases and their risk factors among older persons that can be ameliorated with a regular exercise program.

TABLE 19.1 Chronic Diseases and Risk Factors in Older People That Can Be Ameliorated With Physical Activity

Chronic disease	Risk factor
Cardiovascular disease	Hypertension, dyslipidemia, obesity
Type 2 diabetes	Insulin resistance, glucose intolerance, dyslipidemia, obesity
Cancer	Obesity, bowel immotility, sex hormone profile
Osteoporosis	Low bone density
Physical disability	Sarcopenia, musculoskeletal weakness, poor balance, neuromuscular defects, arthritis

The compression of morbidity refers to the delay of chronic disease and frailty until the end of life or as close to the end of life as possible. This postponement of chronic disease would maintain physical function, autonomy, and quality of life among older people for a greater proportion of their life span.

Mandatory Versus Facultative Aging

To further accelerate the compression of morbidity, we need to improve our ability to distinguish between the aging-related decline that is mandatory and decline that is facultative. **Mandatory aging** is that over which we have no control. In the absence of disease or injury, biological cells, systems, and organs undergo a process of irreversible decline. The underlying basis for mandatory (i.e., biological) aging has been debated for decades, and there are two general classes of hypotheses that attempt to define it, as shown in figure 19.1. Briefly, the first of these hypotheses proposes that random environmental events such as oxygen free radical damage, somatic cell gene mutation, or cross-linkage among macromolecules render the cell incapable of functioning normally. Normal function would require the cell to transfer information from DNA to RNA to the synthesis of protein, thereby allowing the cell to contribute to tissue homeostasis. The consequence of this error buildup is that the genetic

foundation of the cell is altered and the expression of essential protein either is limited or cannot proceed at all. Individual cell loss is not catastrophic to tissue or organ function until a significant complement of cells in the tissue or organ fails. Aging is therefore the consequence of a progressive accumulation of errors in the makeup of the cell attributable to the inability of the cell's repair processes to keep up. The second of the hypotheses proposes that the aging process is actively programmed by the cell's genetic machinery. In the case of programmed cell death, there is some valid evidence of the death of certain cell lines during development and maturation; however, the relevance of this process to the aging of the organism has not been established.

Facultative aging, on the other hand, is that over which we do have control and comprises factors at the community (e.g., quality of health care) as well as the individual (e.g., lifestyle) level. Physiological function and resiliency decline with aging, even among the most robust sectors of the older adult population. The degree to which this decline is attributable to true biological aging and the degree to which it is related to changes in social or lifestyle factors that also accompany older age—particularly physical activity or **disuse**—is a primary focus of this chapter.

In sum, the aging process traditionally has been viewed as an inevitable decline in health and function. Although many physiological functions are known to decline with age, the emerging view of the aging process distinguishes the decline in function and resiliency attributable to biological aging from

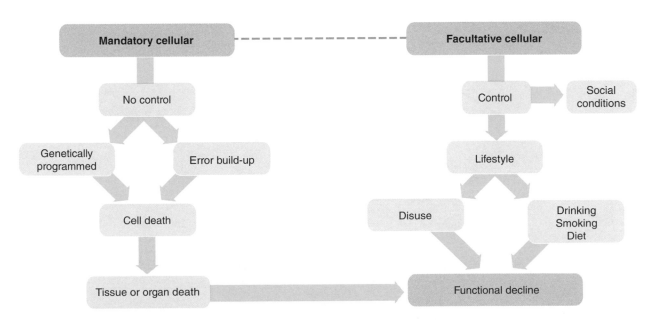

FIGURE 19.1 Hypothesized models of aging. Mandatory aging is that over which we have no control. In the absence of disease or injury, biological cells, systems, and organs undergo a process of irreversible decline. In contrast, facultative aging is that over which we do have control and comprises factors at the community, as well as the individual, level.

that attributable to disuse. Bortz (1982) was an early proponent of the theory that inactivity causes much of the functional loss attributed to aging, from the cellular and molecular to tissue and organ systems. He noted that many of the physiological changes commonly ascribed to aging are similar to those induced by enforced inactivity, as during prolonged bed rest or during spaceflight. He also proposed that the decline in function attributable to disuse could be attenuated, and perhaps reversed, by exercise and stated that this prospect holds much promise for what we now term successful aging. The attenuation in functional decline with exercise in older age is extremely important, because closing the "fitness gap" between active and inactive older people can prolong the time to the disability threshold—often independent of the actual improvements made in muscle strength, balance, or bone strength (figure 19.2).

Successful aging refers to a resilient, disease-free, and highly functional state in older age. Bortz proposed that much of the functional decline commonly attributed to aging per se could be attenuated and even reversed by the reinstatement of regular physical activity.

Physiological Changes Occurring With Aging

As mentioned previously, normal physiological function begins to decline in advancing age as various systems become less pliant and less resilient to environmental stressors. One of the most noticeable and clinically relevant changes occurring with aging is the loss of muscle mass. Longitudinal evidence suggests that during older age, muscle mass decreases about 3% to 6% per decade. This loss of muscle mass, with the accompanying increase in the proportion of body fat, has negative consequences for maintaining resting metabolic rate (RMR) and metabolic resiliency, as well as for maintaining reaction time, strength, flexibility, and balance—all important variables for maintaining an active and independent lifestyle in older age. Another important functional change accompanying older age is the loss of cardiovascular **plasticity**, resulting in a decline in maximal heart rate, stroke volume, cardiac output, and arteriovenous oxygen difference (American College of Sports Medicine [ACSM] 2009; Lakatta 1993; Heckman and McKelvie 2008). These decrements in cardiorespiratory function (along with the loss of lean body mass) contribute substantially to the age-associated decline in maximal aerobic capacity—an important physiological indicator of functional capacity in older age. Indeed, maximal oxygen consumption ($\dot{V}O_2$max) declines by approximately 10% per decade after age 25 (ACSM 2009; Heckman and McKelvie 2008).

Accompanying the decline in the aforementioned physical capabilities is a decline in sensory function. Defects in vision and hearing may increase the psychological stress associated with common daily interactions with the physical and social environment (e.g., walking to the store, going to the senior center). As older people feel less confident in their abilities to

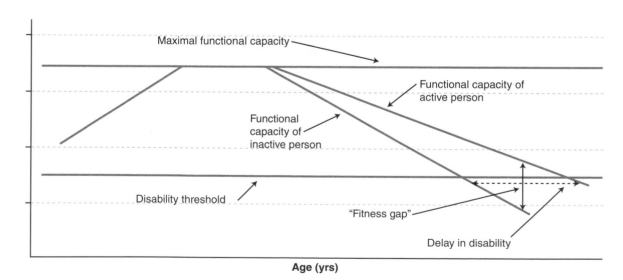

FIGURE 19.2 The fitness gap and disability threshold. The attenuation in functional decline with exercise in older age would contribute to closing the fitness gap between active and inactive older people, thereby prolonging the time to the disability threshold.

venture safely outside the home, these interactions with the environment decrease, thereby compounding the cycle of disuse and functional decline.

Although a genetic predisposition toward the loss of muscle tissue and cardiovascular plasticity is important, disuse is also a significant contributor. For instance, many of these physiological changes are similar to transient responses observed after long periods of bed rest or spaceflight. The common element among these three conditions (aging, bed rest, and spaceflight) is microgravity, and disuse (sedentary behavior) is a form of microgravity, if not hypogravity. However, bed rest and spaceflight are of short duration relative to the life span of humans, and the associated decrements in function are usually reversible within short periods of time. This is not the case with older people who may have been inactive for the majority of their middle age and older age. Indeed, the effect of disuse on function varies by age, with greater impairment in older age attributable to less resiliency and longer duration of disuse. This has important implications with regard to the doses of exercise necessary to reverse existing functional defects (e.g., insulin resistance) in older people.

Nonetheless, ample data have demonstrated the benefits of physical activity in delaying and substantially attenuating aging-related functional decline, even among the least robust members of the older population. Available evidence (see ACSM 2009; Heckman and McKelvie 2008; Vogel et al. 2009; Hollman 2007 for reviews) indicates that older people (even those in their 9th and 10th decades) can respond favorably to both endurance and strength training. Endurance training can maintain and even improve several indicators of cardiorespiratory function—most notably, submaximal work performance, which accounts for the majority of day-to-day physical activity. Strength training is very effective in modulating the loss of muscle mass and its accompanying decline in muscle strength and metabolic function. In addition, both endurance and strength training have demonstrated their effectiveness in improving bone health, postural stability, flexibility, and, in some cases, depressive symptoms and cognitive function in older people (ACSM 2009). More recent evidence points to the benefits of exercise in counteracting age-related neuronal cellular loss and synapse hypotrophy and in improving neurogenesis and capillarization in the brain (Hollman et al. 2007). Such changes have important consequences for delaying or reducing the risk associated with heart disease, diabetes, osteoporosis, dementia, and falling, as well as for increasing life expectancy and quality of life in older age. Thus, it appears that it is never too late in life to achieve the benefits of increased physical activity.

Methodological Considerations in Aging Research

The ability to observe an etiological relationship between physical activity and human function in aging is dependent on our ability to measure these factors with **accuracy**. This section describes several of the methodological problems and concerns inherent in studying physical activity behaviors in older populations.

Assessment Issues

As already noted, exercise and other forms of physical activity are known to provide myriad physiological and psychosocial benefits specific to older people. Although data from intervention studies demonstrate the effect of more moderate and vigorous aerobic or strength training on improvements in physiological function in older people, the benefits of lower-intensity activity, such as that performed as part of an active lifestyle, are less clear. This is attributable, in part, to the difficulty inherent in assessing habitual activity in older people. Physical activity in older age tends to be unstructured, of low intensity, and highly variable. In addition, the issue of recall of such activity patterns in older people leads to less than accurate estimates. Problems in the definition and measurement of physical activity limit the ability to assess it properly and therefore to determine the health consequences associated with an active lifestyle.

Because many activity behaviors common to very old people are the same behaviors used to assess mobility (walking, housework, climbing stairs), often the assessments of behavior and performance are confused. For instance, the question "How often *do* you walk half a mile?" can become confused with "*Can* you walk half a mile?" Because these activities relate to ordinary life, they are measured in units of activities of daily living (ADLs), instrumental activities of daily living (IADLs), and advanced activities of daily living (AADLs). Measuring performance specifically using the standard fitness tests designed for younger populations (e.g., $\dot{V}O_2max$) may not be appropriate in the very old; so, more recently, objective, performance-based measures have been developed that are very effective in discriminating among a wide range of physical function abilities in older people. The combination of these objective *performance-based* measures with self-reported measures of functional ability (i.e., ADLs, IADLs, and mobility) and self-reported *activity behaviors* (typical of the older person's lifestyle) offers the best opportunity to capture the physical activity behaviors that best relate to health and function in aging (figure 19.3).

Functional ability refers to the ability to perform activities of daily living (ADLs: bathing, grooming, dressing, eating), instrumental activities of daily living (IADLs: shopping, cooking, housework), advanced activities of daily living (AADLs: volunteer work, recreational activity), and mobility (stair climbing, walking). In large population-based studies, functional ability typically is assessed by self-report.

Future physical activity assessment for older people needs to focus on separate age categories within older age—that is, age groupings of 65 to 75 years, 75 to 85 years, and the fastest-growing segment of the U.S. population, those people ≥85 years. Given the high prevalence of walking and other even less structured lower-intensity activities in older people, we need better validation measures for establishing the internal validity of physical activity assessment survey questions. The combination of field-based and laboratory-based validation measures is necessary to advance epidemiological research in this area. Also necessary in all physical activity assessments in older populations is the inclusion of physical function items, both self-reported measures of functional ability (e.g., ADLs, IADLs, AADLs, and mobility) and objective, performance-based tasks (timed walk, chair stand, balance). Finally, multiple assessments of physical activity, as well as the health indicator of interest, are necessary to determine the true etiological relationship between being active and being healthy over the life span. Newer, longitudinal statistical techniques are now available to model, for example, the influence of changes in physical activity patterns on the trajectory of change in physical function or body weight through older age.

Study Design Issues

Both cross-sectional and longitudinal experimental approaches have been used to study the influence of aging on human function at rest and in response to acute and chronic exercise. Cross-sectional approaches, which are the most common, involve comparisons of different physiological variables among different age groups (say, 20-25 years vs. 40-45 years vs. 65+ years). Longitudinal studies, which make serial measurements within the same subjects over time (e.g., 10-20 years), are less common but give more valuable information with regard to aging-related changes. For instance, in cross-sectional studies, physiological variables among older trained (master) athletes are often compared with those of older sedentary control subjects, whereas in longitudinal designs, functional adaptations to a period of exercise training are determined in older sedentary subjects. In general, the observed positive associations between exercise and physiological function are much stronger in cross-sectional than in longitudinal studies. Indeed, previously sedentary older people who undergo training are unable to achieve the same high level of physiological adaptation as their athletic counterparts. This is presumably attributable to the genetic advantages of the older athlete, as well as the athlete's ability to exercise regularly at much greater intensities and durations than is possible for the untrained older person, who has a much lower physical working capacity. Thus, data from cross-sectional

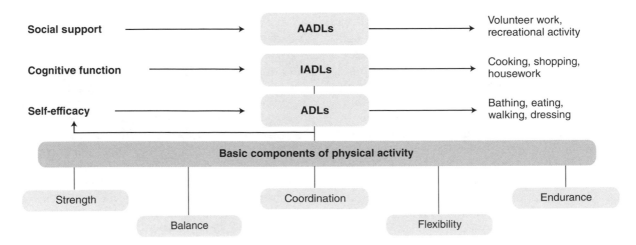

FIGURE 19.3 Relationship between physical activity and physical function.

studies should be viewed cautiously because the influence of regular activity on health and function may be exaggerated based on such data.

Finally, selective survival may influence the variability of adaptation to exercise training both within older populations and between younger and older adults. The individuals who are most susceptible to putative risk factors and who may benefit most from training die at earlier ages. Thus, research results may be biased because they are derived from studies of the effects of exercise training on the more robust surviving cohort.

Selective survival refers to a select sample of older people who have survived the putative effects of various risk factors in middle age and thus present a more robust physiological profile compared with the general population. Issues of selective survival often affect our ability to generalize results from specific aging studies to the general population of older people or to the population at large.

Demographics of Physical Activity Among Older Adults

Physical activity behavior is not uniform across all age sectors of the population, nor is it particularly stable over time. This section describes the

- activity patterns and choices of older people,
- age differences in activity patterns between younger and older people,
- cross-sectional trends in activity patterns, and
- longitudinal trends in physical activity patterns.

Cross-Sectional Patterns

The *Healthy People 2010* objectives state the goal "to increase to at least 30% the proportion of people aged 19 and older who engage in moderate physical activity for at least 30 min/day." Approximately 15% of the U.S. adult population met this goal in 1999. Cross-sectional studies consistently show that the prevalence of reported inactivity increases with age and is especially evident among older women. Behavioral Risk Factor Surveillance System (BRFSS) survey data suggest that more than 40% of U.S. women aged 65 years and older reported no leisure-time physical

activity in 1992 (Centers for Disease Control and Prevention 1995). Although older women reporting inactivity rose to 51% in 1999, recent data suggest that the percentage dropped to 26% in 2007 (Physical Activity Guidelines Advisory Committee [PAGAC] 2008). Most population-based studies in the United States as well as in other countries report an inverse relationship between weekly energy expenditure from physical activity and age.

Interestingly, some national surveys in the United States show a higher prevalence of vigorous activity among older age groups. This presumably is attributable to the way in which vigorous activity is defined. The BRFSS survey defines vigorous activity as activity that is performed at >60% of maximal aerobic capacity. Because $\dot{V}O_2$max decreases with age, activities of moderate intensity for middle-aged persons (say, brisk walking) may be performed at a greater percentage of $\dot{V}O_2$max in older people and therefore meet the criterion for "vigorous." Differences in physical activity classification across the life span need to be carefully considered when results from different surveys are compared.

Walking is the most prevalent activity reported among adults of all socioeconomic strata in the United States, Canada, Latin America, and Europe (figure 19.4). The most prevalent activities among older people tend to be lower-intensity but sustained activities, such as walking, gardening or yard work, bicycling, and golf. Data from the 1991-1992 National Health Interview Survey–Health Promotion Disease Prevention Supplement (NHIS–HPDP) showed cross-sectional patterns in strengthening and stretching activity by age group (Caspersen, Pereira, and Curran 2000). Although the prevalence of strengthening activity was markedly lower among older people compared with the youngest age group, the prevalence of stretching activities remained fairly stable across age groups. These data have important clinical relevance for older people, among whom the loss of muscle strength and flexibility contributes significantly to the decline in physical functioning and mobility.

Cross-Sectional Trends

Cross-sectional trend data rely on information from different representative populations over two or more time intervals. As such, they provide valuable information on temporal trends in physical activity patterns. These data are often referred to as **surveillance data**. Surveillance data from the United States suggest that the prevalence of inactivity (i.e., no reported leisure-time physical activities in the past month) decreased between 1986 and 1990, although this decline was not apparent among nonwhite adults

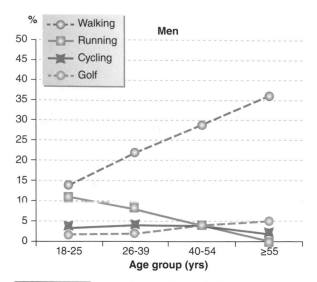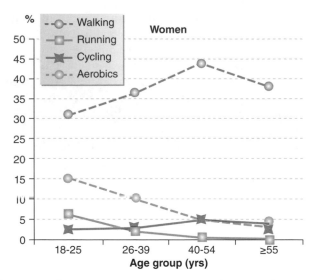

FIGURE 19.4 Prevalence of popular leisure-time activities by age and sex among persons trying to lose weight.

From Centers for Disease Control and Prevention 2003.

or adults of lower educational attainment. When the data were stratified further by age groups, surprisingly they showed that this overall decline in reported inactivity was explained primarily by reduced inactivity among people more than 55 years old. Older data from the NHIS–HPDP between 1985 and 1995 show a general increase in reported walking over these 10 years and stable trends in reported gardening activity (figure 19.5). More recent data from the same survey show a small yet consistent increase in

reported physical activity of any kind among older persons living in the United States between 1998 and 2009 (National Center for Health Statistics 2008). In fact, among people aged 65 to 74 years, reports of "some physical activity" increased from 5.9% in 1998 to 33% in 2009. Similarly, among those 75 years and older, the percentage reporting physical activity increased from 15.4% in 1998 to 17.5% in 2009. Although these surveillance data are important for understanding temporal changes, they do not give information about aging-related changes among a single population over time. Nonetheless, multiple assessments in the same population of older people over many years have added tremendously to the study of successful aging.

Longitudinal Trends

Longitudinal trend data reflect serial measurements on a population and thus can provide information on aging- and time-related changes in activity patterns, as well as differences in these trends by birth cohort. Longitudinal data from several countries suggest that although total activity and common activities of gardening and bicycling decline with aging among older people, time spent walking remains relatively stable. For instance, in the Zutphen Elderly Study, walking became an increasing proportion of total activity over 10 years, and this was especially evident in the youngest birth cohort (born between 1916 and 1920) between 1990 and 1995 (Bijnen et al. 1998) (figure 19.6).

Retirement is an important time with regard to potential changes in the physical activity pattern.

FIGURE 19.5 Participation in activities among older persons living in the United States. Data are from the National Health Interview Survey–Health Promotion Disease Prevention Supplement, 1985-1995.

From Centers for Disease Control and Prevention 2003.

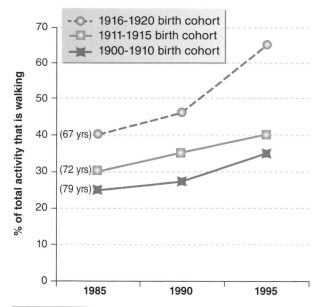

FIGURE 19.6 Age and cohort trends in walking among older men: The Zutphen Elderly Study, 1985-1995.

Reprinted from F.C.H. Bijnen et al., 1998, "Age, period, and cohort effects on physical activity among elderly men during 10 years of follow-up: The Zutphen Elderly Study," *Journal of Gerontology - Series A: Biological Science and Medicine Sciences* 53: M235-M241. Copyright © The Gerontological Society of America.

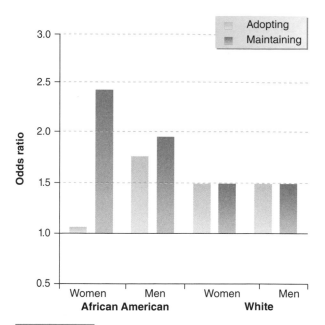

FIGURE 19.7 The six-year odds of maintaining or adopting activity among workers who had retired at follow-up: The Atherosclerosis Risk in Communities Study.

Adapted from K.R. Evenson et al., 2002, "Influence of retirement on leisure-time physical activity," *American Journal of Epidemiology* 155: 692-699, by permission of Oxford University Press.

Not surprisingly, data from the Atherosclerosis Risk in Communities Study suggested that the six-year odds of adopting or maintaining physical activity were significantly higher among older people who had retired at follow-up relative to those who had continued working (Evenson et al. 2002), although there was significant effect modification in these odds with regard to race and sex. Older black persons who were already active had a higher odds of maintaining that activity compared with older white persons, and this was especially so among older black women. Older black men had the highest odds of adopting an activity on retirement, whereas older black women had the lowest odds, which were no different from those for their counterparts who were still working (figure 19.7). Favorite activities adopted on retirement among older people of both races and sexes were walking briskly, walking for pleasure, gardening or yard work, floor exercises, and exercise cycling. In contrast, recent prospective data from the Netherlands suggest that retirement resulted in a reduction in physical activity from work-related transportation that was not compensated for by an increase in sport activity or nonsport leisure-time physical activity (Slingerland et al. 2007).

In sum, there is encouraging although limited evidence from cross-sectional surveillance data that the prevalence of reported inactivity is decreasing over time among some sectors (namely, older adults) of the general U.S. population, with an increased prevalence of regular leisure-time activity. However, leisure-time physical activity constitutes only a portion of total daily activity. Other components involve work or household activity and transportation. Although few surveillance data are available for these other components, one can reasonably assume that energy spent in work and household tasks, as well as in transportation, has progressively declined over the years with increasing automation and labor-saving devices and increased use of the Internet. This time-related decline in daily energy expenditure is further accelerated by aging-related declines in physical activity. It therefore is possible that overall total daily energy expenditure from physical activity has declined substantially among our older population despite small increases in their leisure-time activity. Nonetheless, both cross-sectional and longitudinal trend data demonstrate that even though daily energy expenditure from physical activity declines with aging, walking remains an important activity in older age. Because walking is weight bearing, uses large muscle groups, can be sustained, and, if performed regularly, can improve maximal and submaximal physical functioning, its merits should be promoted among older populations.

Dimensions of Physical Activity and Their Relationship to Health and Function in Aging

The relation between physical activity and health is well established. It is important to understand, however, the different aspects of physical activity (and of disuse) that relate in a specific manner to health and function and how the relative importance of these characteristics changes with age.

Consequences of Sedentary Behavior

Disuse and a sedentary lifestyle are especially detrimental to older people. For example, low levels of physical activity and cardiorespiratory fitness are primary determinants of the decline in metabolic and functional reserve observed in older age. Sedentary behavior has consistently demonstrated a relationship with the development of chronic disease, premature mortality, poor quality of life, and loss of function and independence with aging. Sedentary behavior accumulated over a lifetime may be the risk condition with the biggest global public health impact that is faced by older people. Indeed, population risk estimates for heart disease attributed to low activity or low fitness levels in middle-aged and older people are comparable to, or even greater than, those attributed to other well-known predictors of morbidity and mortality, such as hypertension, smoking, dyslipidemia, and obesity.

Given the high prevalence of sedentary behavior, especially among older individuals, and the strong association of low activity or low fitness to many chronic diseases of aging, the direct costs of a lifetime of inactivity among Americans may be as high as $24 billion per year (Colditz 1999). This figure, however, does not account for the economic costs associated with frailty and loss of physical function among older people, so it actually may underestimate the true economic burden of a sedentary lifestyle.

Physical activity and fitness have been associated with as much as a 40% reduction in morbidity and mortality from a number of major chronic diseases affecting older people (Vogel et al. 2009), namely, coronary heart disease, cancer, and type 2 diabetes, as well as with a lower incidence of most of the risk factors preceding these diseases, such as hypertension, dyslipidemia, and insulin resistance. The psychological benefits of regular physical activity have also been well documented (ACSM 2009), as have the

significant protective effects on the risk of bone loss, hip fracture, factors associated with falls, and the rate of functional decline so common with aging (Ceceli et al. 2009). There is evidence to suggest that recent or current activity is more protective than past activity; however, cumulative, lifetime activity patterns may be a more influential factor for most of these diseases, especially those with a long developmental (latency) period, such as cancer, bone integrity, or weight maintenance. These etiological relationships are discussed in greater detail in other chapters.

The inverse association between physical activity and disease or functional decline is consistently strong, graded, and independent and is biologically plausible and specific, thereby meeting most criteria for inferring a causal relationship. The challenge often encountered in epidemiological research in aging—especially when it relies on self-reported measures of physical activity—is that of establishing temporal sequencing (i.e., does sedentary behavior truly precede the onset of functional decline?).

Characteristics of Physical Activity Affecting Health

It is difficult to determine the characteristics of physical activity related most specifically to different aspects of health because there are several dimensions of physical activity behaviors. These dimensions, which are exclusive neither to any one type of activity nor to each other, include exercise intensity, energy expenditure, weight bearing, flexibility, and muscle strength. Certainly, these dimensions are interrelated because, for example, activities that increase aerobic capacity also require energy expenditure. Any sustained weight-bearing activity also expends energy and if done vigorously enough increases aerobic capacity.

The relative importance and specificity of each of these dimensions to health alter with age. For example, among adolescent girls and younger women, dimensions of physical activity related to muscle and bone growth may be of primary interest, whereas among middle-aged people, the influence of energy expenditure on weight regulation or of exercise intensity on cardiovascular health becomes more important. Among older adults, the influence of weight-bearing, strength, and flexibility aspects of activity on bone and lean mass preservation and balance assumes highest priority with regard to maintaining functional ability and independence.

In the past, it was proposed that exercise of sufficient frequency, intensity, and duration to improve physical fitness was necessary to promote resistance to disease. Although the effects of physical activity on health status may be mediated primarily by the physiological changes that accompany increased fitness, recent data suggest that physical activity as such may have an independent positive impact on several health indicators in aging. A growing epidemiological literature shows significant relationships between low- and moderate-intensity activity and overall health and longevity in aging (see ACSM 2009; Heckman and McKelvie 2008; Vogel et al. 2009; Hollman et al. 2007). Of most recent interest are data demonstrating the importance of short but frequent bouts of low-intensity walking and standing with regard to their ability to break up sedentary time (Healy et al. 2008) and interrupt the molecular dysfunction associated with prolonged sitting (Hamilton, Hamilton, and Zderic 2007). This pattern of physical activity may be particularly beneficial for older people, especially when repeated several times over the course of the day.

The exercise prescription or plan (i.e., frequency, duration, intensity) optimal in achieving one type of health outcome (e.g., weight loss) may be quite different from that necessary to achieve another (e.g., increased bone mineral content or muscular strength). Again, age may be an important variable in modifying the effects of a given exercise stimulus on health and functioning. That is, the amount of exercise related to disease-specific morbidity in middle age may be very different from that related to successful aging and overall longevity. Disease status may also be an important effect modifier in that the volume (frequency × duration × intensity) of physical activity necessary to prevent functional decline or to maintain health and function may be lower than the amount needed to reverse an already established condition such as sarcopenia or insulin resistance (table 19.2). Thus, physical activity may be far more effective with regard to prevention than with regard to therapy, and this concept is emphasized repeatedly in the 2008 federal physical activity guidelines (PAGAC 2008). Nonetheless, physical activity and exercise have demonstrated effectiveness in improving function across the aging spectrum—from the very healthy to the most frail—and should be encouraged among all older people.

In summary, limited information exists on the amount of physical activity needed to promote optimal health and function in older age, although research is now focusing more on the specificity of the exercise prescription—especially with regard to the timing of the exercise bout (e.g., postmeal or later in the day). Recommending levels of exercise necessary to achieve aerobic fitness may no longer be appropriate in older age; rather, public health policy should focus on sufficient levels when promoting physical activity among more sedentary older or frailer sectors of the general population. Given that health benefits may accrue independently of the fitness effects achieved through sustained vigorous activity, a new lifestyle approach should be emphasized from childhood through older adulthood. This approach calls for incorporating at least 30 min of any activity of moderate intensity (sustained or accumulated) into the daily schedule (PAGAC 2008). Moreover, recent ACSM guidelines for older adults recommend a regular program of resistance, flexibility, and balance training on two days per week (ACSM 2009). Thus, regular participation in activities of moderate intensity (such as walking, climbing stairs, biking, yard work or gardening, or walking the dog), which increase accumulated daily or weekly energy expenditure and maintain muscular strength but may not be of sufficient intensity for improving aerobic fitness, should be encouraged in the community. Because the greatest gains in protection from disease and functional decline continue to be observed with increases in moderate-intensity activity or fitness levels, public health efforts geared toward the most sedentary members of the middle-aged community would most likely offer the most benefits with regard to the prevention of premature morbidity and mortality in later years.

TABLE 19.2 Public Health Strategy for Preventing Functional and Metabolic Decline

Level	State	Intervention
Primary	Normal function	30 min or more of moderate-intensity activity on most days
Secondary	Compromised function	?
Tertiary	Diseased state	?

? = unknown.

Reprinted, by permission, from J. Dziura and L. DiPietro, 2003, The importance of body weight management in successful aging. In *Obesity, etiology, assessment, treatment and prevention*, edited by R.E. Anderson (Champaign, IL: Human Kinetics), 141-153.

Programmatic Issues in Promoting Physical Activity in Older Populations

Translating scientific knowledge into practical interventions that benefit the health of the community is difficult. A number of factors must be identified and managed before the public health potential of physical activity for older persons can be fulfilled.

Determinants

Older people report a variety of reasons for not exercising, including lack of time, fear of injury, caregiving duties, lack of energy, lack of a safe place to exercise, and concerns about appearance. The often cited factors may also be secondary to a general lack of motivation or readiness to change or both. Thus, an array of physiological, psychosocial, and environmental factors may determine physical activity behavior throughout the life span, and these factors become even more important in older age. Physiological factors include aerobic capacity, speed, strength, balance, and flexibility. As these physiological attributes tend to decline with aging, they become strong determinants of activity level and choice of activity, because older people participate in activities at which they feel competent and that they feel safe performing. Psychosocial determinants include personality, knowledge and beliefs about the health effects of exercise, and social support from family and friends. Safety and accessibility are the two most important environmental factors affecting activity participation across the life span. Although these environmental factors have not been studied extensively, they are becoming quite important as public health efforts are shifting toward modifying the built environment.

Many of these determinants, particularly some of the psychosocial and environmental factors, are particularly amenable to change and should be the focus of community intervention efforts. Strategies for increasing physical activity among the older sectors of the community include

- increased public education about the health effects of regular low- and moderate-intensity physical activity;
- increased senior center and community center programs that are supervised and that provide social support and other incentives for exercise; and
- increased community availability and accessibility of safe physical activity and recreational facilities such as hiking, biking, and fitness trails; public swimming pools; and acres of park space.

Setting

Preferences for where to exercise may vary in older people, but there is evidence that individual home-based programs are preferable to group- or class-based formats. Indeed, data from the Stanford–Sunnyvale Health Improvement Project showed significantly greater adherence over two years to a higher-intensity home-based program compared with a lower-intensity home-based and a higher-intensity group-based program (King et al. 1997). Although differences in adherence between the two home-based programs became apparent only in the second year, adherence was substantially higher in both home-based programs than in the group-based program from the start of the project. These data suggest that programs that individuals can perform on their own, whenever they want, may be the most effective types of exercise programs to promote for older people. However, social support for the home-based groups in the Stanford–Sunnyvale project was provided through telephone calls from the research staff, and it is not clear how well home-based exercise works for older people who are not part of a formal research study and don't receive such support.

Programs offered through community settings can be appealing to older people who want to "get out" but may not be willing to leave their familiar surroundings to participate in activity or who may live in environments that are not safe for exercise. Churches, shopping malls, neighborhood recreation or senior centers, and retirement homes are examples of community settings in which physical activity programs can be implemented. Data from community-based interventions suggest that walking programs are the most popular with older adult populations; however, tai chi, yoga, dance, and strength programs are gaining in popularity. Given the multiple components involved in maintaining physical function and mobility in older age, programs that combine the basic elements of aerobic endurance, strength, flexibility, balance, and coordination through a variety of activities (e.g., walking, tai chi, resistance bands, juggling, yoga, dance) may be the most efficacious programs for promoting and maintaining the level of activity both necessary and sufficient for achieving the health benefits related to mobility and functional ability in older age. Research that integrates basic and applied physiological science with the social sciences is necessary to advance understanding of how improvements in basic components of function (e.g., strength, coordination, balance, endurance, flexibility) translate into higher-order functions of

IADLs (shopping, cooking, housework) and AADLs (volunteer work and recreational activity).

Environmental Versus Individual Approaches to Promoting Physical Activity

Environmental interventions that remove barriers to activity at the community level have a greater public health impact than attempts to change barriers at the individual level. Many of the most innovative and successful health interventions have little to do with health per se but rather with altering the barriers that prevent people from gaining control over the determinants of their own health. Public policy that ensures equal and safe access to public spaces for walking, biking, and other forms of recreation is an example of such an environmental strategy. As with any environmental action, an informed and active public takes the lead in influencing the environmental structural changes. Similarly, community agencies can work together to ensure a health-promoting environment for older people.

In addition to involving physicians and other allied health personnel, the fitness industry, and medical agencies like the visiting nurses associations, collaboration can come from governmental bodies such as state school boards and local health departments, parks and recreation departments, and such nontraditional partners as departments of urban planning and transportation. These groups should cooperate in setting physical activity standards for the entire community (figure 19.8).

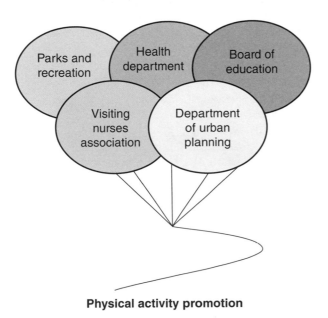

Physical activity promotion

FIGURE 19.8 Interagency involvement in physical activity promotion.

Environmental interventions are those performed at the community level rather than at the individual level. These types of interventions have the ability to affect everyone equally and target risk conditions rather than individual risk factors or behaviors.

Summary

Overall physiological function is known to decline with aging; however, the emerging view of the aging process distinguishes the decline in function and resiliency attributable to biological aging from that attributable to disuse. Low levels of physical activity and fitness are primary determinants of the decline in metabolic and functional reserve observed in older age, and therefore accumulated lifetime sedentary behavior among older adults is a risk condition of substantial public health importance. There is now substantial evidence demonstrating the benefits of physical activity in attenuating aging-related functional decline, even among older people with established chronic disease and frailty. Therefore, because health benefits can accrue independently of the fitness effects achieved through sustained vigorous activity, a lifestyle approach of incorporating at least 30 min of any moderate-intensity activity into the daily schedule should be encouraged throughout the life span. Population-based data from several countries demonstrate that walking is an important activity through older age, and therefore its merits should be promoted among older people. Public policy that incorporates education and access to safe opportunities for exercise at home and in the community should be a priority among community agencies working together to ensure a healthy lifestyle for all.

Key Concepts

accuracy—Combination of validity (hitting the mark) and precision (low variability) in measuring a study variable.

demographic trend—A consistent pattern in a given population characteristic.

disuse—Chronically low levels of physical activity and high levels of sedentary behavior.

facultative aging—Components of the aging process that we have control over and that are potentially reversible.

mandatory aging—Components of the aging process over which we do not have control.

plasticity—Flexibility in function.

sarcopenia—For definition, see page 254.

surveillance data—Regularly collected data that reflect temporal changes in population characteristics.

Study Questions

1. Explain several differences between mandatory and facultative aging.

2. Plot a curve describing an age-associated decline in maximal oxygen consumption, and discuss briefly the implications of this decline for health and function in older age.

3. List three problems in assessing physical activity accurately in older people.

4. Describe how selective survival in an older study sample may limit the ability to observe large improvements in function with exercise training.

5. Differentiate between temporal trends and aging-related trends in walking behaviors.

6. Describe the dimensions of physical activity and their relative importance in an older population.

7. List several determinants of physical activity among older people, and identify those that are most amenable to change at the individual or community level.

8. Define and describe an environmental intervention to promote physical activity in older people.

References

Administration on Aging, U.S. Department of Health and Human Services. 2008. *A profile of older Americans.* Washington, DC: U.S. Department of Health and Human Services.

American College of Sports Medicine. 2009. Exercise and physical activity for older adults. *Medicine and Science in Sports and Exercise* 41:1510-1530.

Bijnen, F.C.H., E.J.M. Feskens, C.J. Caspersen, W.L. Mosterd, and D. Kromhout. 1998. Age, period, and cohort effects on physical activity among elderly men during 10 years of follow-up: The Zutphen Elderly Study. *Journal of Gerontology: Medical Sciences* 53A:M235-M241.

Bortz, W.M. 1982. Disuse and aging. *Journal of the American Medical Association* 248:1203-1208.

Caspersen, C.J., M.A. Pereira, and K.M. Curran. 2000. Changes in physical activity patterns in the United States, by sex and cross-sectional age. *Medicine and Science in Sports and Exercise* 32:1601-1609.

Ceceli, E., F. Gokoglu, M. Koybasi, O. Cicek, and R. Yorgancioglu. 2009. The comparison of balance, functional activity, and flexibility between active and sedentary elderly. *Topics in Geriatric Rehabilitation* 25:198-202.

Centers for Disease Control and Prevention. 1995. Prevalence of recommended levels of physical activity among women – Behavioral Risk Factor Surveillance System, 1992. *Morbidity and Mortality Weekly Report* 44:105-113.

Colditz, G.A. 1999. Economic costs of obesity. *Medicine and Science in Sports and Exercise* 31:S663-S667.

Evenson, K.R., W.D. Rosamond, J. Cai, A.V. Diez-Roux, and F.L. Brancati. 2002. Influence of retirement on leisure-time physical activity. The Atherosclerosis Risk in Communities Study. *American Journal of Epidemiology* 155:692-699.

Federal Interagency Forum on Aging Related Statistics. 2004. *Older Americans 2004: Key indicators of well-being.* Washington, DC: Federal Interagency Forum on Aging Related Statistics.

Hamilton, M.T., D.G. Hamilton, and T.W. Zderic. 2007. The role of low energy expenditure and sitting on obesity, metabolic syndrome, type 2 diabetes, and cardiovascular disease. *Diabetes* 56:2655-2667.

Healy, G.N., D.W. Dunstan, J. Salmon, E. Cerin, J.E. Shaw, P.Z. Zimmet, and N. Owen. 2008. Breaks in sedentary time: Beneficial associations with metabolic risk. *Diabetes Care* 31:661-666.

Heckman, G.A., and R.S. McKelvie. 2008. Cardiovascular aging and exercise in healthy older adults. *Clinical Journal of Sports Medicine* 18:479-485.

Hollmann, W., H.K. Strüder, C.V. Tagarakis, and G. King. 2007. Physical activity and the elderly. *European Journal of Cardiovascular Prevention and Rehabilitation* 14:730-739.

King, A.C., M. Kiernan, R.F. Oman, H.C. Kraemer, M. Hull, and D. Ahn. 1997. Can we identify who will adhere to long-term physical activity? Signal detection methodology as a potential aid to clinical decision making. *Health Psychology* 16:380-389.

Lakatta, E.G. 1993. Cardiovascular regulatory mechanisms in advanced age. *Physiological Reviews* 78:413-467.

National Center for Health Statistics. 2008. www.agingstats.gov. Accessed December 1, 2010.

Physical Activity Guidelines Advisory Committee. 2008. *Physical Activity Guidelines Advisory Committee report, 2008.* Washington, DC: U.S. Department of Health and Human Services.

Slingerland, A.S., F.J. van Lenthe, J.W. Jukema, C.B.M. Kamphuis, C. Looman, K. Giskes, M. Huisman, K.M. Narayan, J.P. Mackenbach, and J. Brug. 2007. Aging, retirement, and changes in physical activity: Prospective cohort findings from the GLOBE study. *American Journal of Epidemiology* 165:1356-1363.

Vogel, T., P.H. Brechat, P.M. Leprêtre, G. Kaltenbach, M. Berthel, and J. Lonsdorfer. 2009. Health benefits of physical activity in older patients: A review. *International Journal of Clinical Practice* 63:303-320.

20

Physical Activity and Brain Functions

Kirk I. Erickson, PhD

CHAPTER OUTLINE

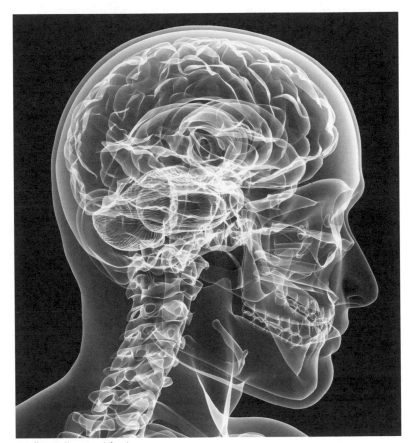

©Jeffrey Collingwood/fotolia.com

The term *brain exercises* is often used to refer to intellectual activities such as crossword puzzles and word or math games that are meant to challenge one's cognitive skills and to "flex" the brain with the hope of improving **cognition**. Such improvement is analogous to the way in which physical activity and aerobic exercise enhance physical and motor function. Indeed, in lay terms, the brain is often referred to as a "muscle" that should be consistently worked and flexed like any other muscle in the body. The argument, and the analogy with muscle, continues with the statement that without consistent intellectual stimulation one's mind will weaken and atrophy, similar to the way in which muscles weaken and atrophy if they are not consistently used. Accordingly, it is natural to believe that mental activity would be the most effective method for exercising the brain, enhancing cognitive potential, and perhaps reducing susceptibility to psychiatric or neurologic illness. All the same, what is usually referred to as intellectual activity is clearly different from physical activity. Categorical distinctions are often made between "nerds" and "jocks," those who stress intellectual activity over physical activity and those who stress physical activity over intellectual activity, respectively. However, such categorical distinctions are, at best, overly simple and, at worst, blatantly wrong. Realistically, physically active individuals often outperform their more sedentary counterparts on many different measures of cognitive performance and academic achievement. Thus, the Platonic claim that maintaining a healthy body is essential for maintaining a healthy mind is not far off from current scientific viewpoints.

While it is clear that intellectual stimulation such as brain exercises can be effective at enhancing cognition, it is also now clear that physical activity has similar effects. In fact, as described in this chapter, research in both humans and animals suggests that physical activity can have a direct and stimulating effect on the brain and cognition and that physical activity might have both broader and more robust effects on cognition than brain exercises. Although the relative contribution of physical activity versus mental activity on enhancing cognitive function remains a matter of debate, it is evident that physical activity has an effect on both the body and the mind.

This chapter is organized according to four main types of questions that are meant to provide some overarching principles and structure to the field (see table 20.1). These types of questions can be broadly categorized as descriptive, mechanistic, applied, and moderating. The next sections briefly discuss each of these categories of questions, their relative importance, and the scientific evidence gathered in an attempt to answer them.

TABLE 20.1 Questions on Physical Activity and Brain Function

Category	Sample questions
Descriptive	What types of physical activity are most effective at enhancing cognitive and brain function?
	How much physical activity is necessary for detection of an effect on the brain and cognition?
Mechanistic	What are the molecular pathways involved in the improvements in cognitive and brain function?
	What are the changes at the systems or regional level of brain function that contribute to enhanced cognition?
	Are there any social or emotional factors that mediate the effect of physical activity on cognition?
Applied	Which populations show the greatest benefit of physical activity on the brain and cognition?
	Is physical activity a feasible treatment for people with neurologic or psychiatric illness?
	Can physical activity protect against the development of neurologic or psychiatric illness?
Moderating	Are there other lifestyle factors or habits that either attenuate or potentiate the effect of physical activity on the brain and cognition?
	Are there genetic factors that moderate the effect that physical activity has on cognitive function?

Descriptive Questions

Descriptive questions deal with the effect of physical activity on the brain and cognition without considering the possible mechanisms or applicability of the effect. For example, questions related to the *types* of physical activity that may be most efficacious for the brain and cognition may seek to determine the specific effect of physical activity but would not address the mechanisms or applicability of the effect on the brain and cognition. Other questions that could be categorized under the umbrella of descriptive questions might include the following: How much physical activity is necessary to enable detection of an effect on the brain or cognition? How quickly do the cognitive benefits accrue through physical activity or dissipate when physical activity patterns are interrupted or discontinued?

Although scientists often dismiss descriptive questions as rudimentary, they are not simple to empirically address or to answer. In fact, to address these questions, complex study designs with large sample sizes are often needed, with separate groups participating in different activities for variable amounts of time. Further, the intensity of the exercise and skill levels across activities must be rigorously controlled. Nonetheless, the answers to these questions have practical significance. For example, if physical activity is going to be prescribed by physicians and health care workers either as prevention or as treatment for cognitive and brain diseases, then patients need to know how much and what types of physical activity to aim for. Without this explicit information, patients will be unlikely to adopt physical activity as a nonpharmaceutical method to enhance or protect brain function.

In this section, we briefly review the structure and function of the brain as well as the cognitive domains that will be relevant for understanding the material in later sections. Then we consider the empirical data gathered in an attempt to address two descriptive questions: (1) Which cognitive and brain functions are enhanced by physical activity, and (2) how much physical activity is necessary before an effect on cognitive and brain function can be detected?

Overview of the Brain and Cognition

The brain is composed of several different types of cells, but the neuron is the main type supporting cognitive function. The neuron is composed of a cell body; dendrites, which are the prolific branches that receive signals from other neurons; and a single axon, which sends signals from the active cell to neighboring cells. Neurons communicate with one another by sending an electrochemical signal down the length of the axon to the synaptic terminals, where neurotransmitters, neuromodulators, or neurotrophins are released into the cleft and bind to receptors located on the dendrites of the neighboring cells. The binding of the molecule to the receptor results in either an excitatory or an inhibitory transmission, depending on the type of neurotransmitter secreted from the presynaptic cell. In this fashion, neurons rapidly communicate with one another and transfer information encoded in one cell to other cells in different brain regions. Changes in receptor density, neurotransmitter concentration, or the efficacy with which (or the rate at which) neurotransmitters are catabolized could influence the incidence or prevalence of psychiatric or neurologic diseases; but these changes are also involved in learning, memory, attention, emotion, and all other aspects of brain function.

The brain is largely divided into separate lobes including the occipital, frontal, parietal, and temporal, all with separate functions but often working as a network to produce complex cognitive function (see figure 20.1). For example, the frontal lobe, especially the dorsolateral regions of the prefrontal cortex, is involved in higher levels of cognition, including planning, coordinating multiple tasks, selectively attending to certain things while ignoring other distracting items, planning and sequencing, and working memory. Other regions, such as the parietal lobe, are also involved in some aspects of higher-level cognitive function but also control spatial planning and spatial processing, including sensorimotor function and the distribution of attention across space. In addition to the four major lobes, brain tissue is often divided into gray matter and white matter, consisting of cell bodies and dendrites (gray matter) and axons (white matter). Further, several subcortical structures

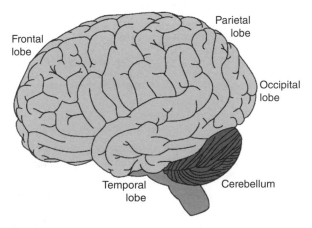

FIGURE 20.1 The four lobes of the brain and the cerebellum.

function in learning and memory, such as the medial temporal lobe, of which the **hippocampus** is the major structure involved in memory formation. Another set of structures sitting deeper in the brain is the basal ganglia, which are involved in procedural types of memory, reward, and habit formation.

Alzheimer's disease is most notably associated with early and profound deterioration of the hippocampal formation and surrounding entorhinal cortex tissue, whereas **Parkinson's disease** is associated with neuronal loss and dysfunction of the basal ganglia. The presumed causes of Alzheimer's disease include the accumulation of plaques (amyloid beta) that disrupt the communication and survival of neurons and the accumulation of intra-axonal protein (tau) that disrupts the structure and function of the axon. On the other hand, Parkinson's disease is associated with the loss of dopamine-producing cells located in a small midbrain structure called the substantia nigra. The cells of the substantia nigra project to the basal ganglia where dopamine is released in several regions. The loss of dopamine cells in the substantia nigra gives rise to the deficits that characterize Parkinson's disease including tremors, rigidity, and **cognitive impairment**.

Rodent models are often used to examine the molecular processes and cascades responsible for learning, memory, and cognition. However, the expansive differences in the cognitive complexity between humans and rodents are an inherent limitation to rodent models. Therefore, human studies are frequently conducted to examine the cognitive architecture and processing streams associated with higher-level cognition. The advent of sophisticated and noninvasive **neuroimaging** techniques such as **magnetic resonance imaging** (MRI) over the past two decades has ushered in avenues of research that are capable of investigating the network of brain regions responsible for higher-level cognitive function.

Cognitive and Brain Functions Enhanced by Physical Activity

The most thorough and comprehensive investigations of physical activity on cognitive and brain function have been conducted in older adults (over age 65) who are free of dementia but are experiencing cognitive decline. Unfortunately, cognitive decline remains one of the most ubiquitous characteristics of late adulthood, yet decline does not occur uniformly across all cognitive domains. Cognitive processes such as working memory, episodic memory, control of directed attention, maintenance of goals, and switching between multiple ongoing tasks tend

to show disproportionate age-related losses. These cognitive functions are often categorized under the umbrella term "executive function," which largely refers to controlled processes that are supported by frontal and parietal brain regions. Whereas executive functions often show decline, other cognitive processes such as recognition memory, object recognition, semantic memory, and emotional regulation show little decline and sometimes even improve across the life span. Accordingly, regional variation in the time course and magnitude of cortical decay mirrors the decline in cognitive processes, such that frontal and parietal regions that support executive functions show the earliest and greatest amount of decay while visual cortical regions show the least amount.

The hippocampus and caudate nucleus (a structure that is part of the basal ganglia) also show substantial decay in late adulthood. Several studies have demonstrated that the hippocampus decays in a nonlinear manner, with relative stability until the fifth decade of life followed by a volume loss of 1% to 2% per year. However, in people with dementia, this annual rate of loss increases to nearly 5%. The caudate nucleus also decays in late adulthood at an annual rate of about 1%. Changes in volume for both the hippocampus and the caudate nucleus are accompanied by changes in cognitive functions that are dependent on these structures, including episodic memory formation and switching rapidly between ongoing tasks.

Interestingly, studies on physical activity have demonstrated that the cognitive functions and brain regions showing the most rapid decay in late adulthood are the same neurocognitive processes and regions that are amenable to improvements. For instance, in one study, 124 sedentary older adults aged 60 to 75 were randomly assigned to either a walking group or a stretching control group (Kramer et al. 1999). The walking group reported to a track three times per week and walked at a moderate intensity for 45 min each day. The stretching control group also attended physical activity sessions, but instead of walking, they participated in low-intensity stretching exercises for the same amount of time as the walking group. A comprehensive neuropsychological evaluation was conducted both before and after the six-month intervention (see figure 20.2). The authors reported that the walking group became more aerobically fit throughout the intervention but also showed enhanced cognitive function compared to the stretching control group. Importantly, the exercise intervention was most effective at enhancing performance on tasks that contained an executive component and was less effective at enhancing performance on other tasks. These results suggested that (1) six months of

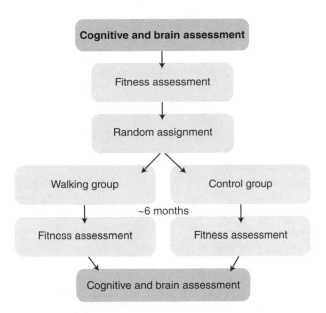

Cognitive and brain assessment

Fitness assessment

Random assignment

Walking group | Control group

~6 months

Fitness assessment | Fitness assessment

Cognitive and brain assessment

FIGURE 20.2 A figural display of the design used by Kramer and colleagues (1999) and Colcombe and colleagues (2004, 2006).

frequent walking could improve cognitive function; (2) the benefits of walking on cognitive function are greater for tasks of executive function, suggesting that this type of exercise has some specificity with respect to brain function; and (3) older adults retain their capacity for plasticity and even executive processes, or those processes that show the greatest rate of decline remain tractable.

Consistent with this conclusion, a meta-analysis of 18 randomized exercise trials conducted in non-demented older adults showed that the effect of exercise on cognition is both general and specific (Colcombe and Kramer 2003). The effect of physical activity is *general* in the sense that most cognitive functions improve with physical activity but *specific* in the sense that executive functions improve more than other cognitive domains.

The specific effect that physical activity has on cognition led to the hypothesis that it would also have a regionally specific effect, such as improvement in those regions supporting executive function. To test this, older adults were randomly assigned to either a moderate-intensity walking intervention for six months or a stretching–toning control group (similar to the design of Kramer et al. 1999, described previously). High-resolution brain scans were obtained both before and after the intervention. With use of a semiautomated approach to examine brain volume, the stretching–toning control group showed a slight decline in volume over the six-month interval, while the exercising group showed a significant increase in volume over the same period (Colcombe et al.

2006). The increased volume was localized to the gray matter in the prefrontal and temporal cortices as well as the anterior white matter tracts. Importantly, the increased volume of tissue corresponded to the same regions that support executive function—the cognitive processes that are enhanced by physical activity.

The evidence that fitness and exercise are associated with greater tissue volume in late adulthood is quite convincing. However, volumetric studies of the brain have failed to link cognitive function with brain function. To examine whether physical activity results in improved brain function specific to regions supporting executive function, functional magnetic resonance imaging (fMRI) was used in a group of 41 participants between 58 and 77 years of age. Using an **executive control** task that required participants to inhibit irrelevant information and selectively attend to a target, this study examined both the cross-sectional and longitudinal effects of fitness and exercise on the brain networks involved with task performance. In the cross-sectional analysis, older adults with higher fitness levels demonstrated increased brain activity in prefrontal and parietal regions compared to less fit adults and superior performance on the task (Colcombe et al. 2004). A second study, a randomized trial over six months, demonstrated that moderate-intensity walking increased activity in prefrontal and parietal brain areas during the same selective attention paradigm. These results provide converging evidence that the prefrontal and parietal brain regions make up a network that is influenced by moderate-intensity aerobic exercise.

The morphology and function of the hippocampus also appear to be affected by physical activity. Much of this research has been in rodents, and it is evident that, at least in rodents, physical activity, especially in the form of aerobic activity, unequivocally affects the hippocampus and related memory function. In humans, the association between aerobic fitness and hippocampal morphology has only recently been examined. In 165 older adults between 59 and 81 years of age, cardiorespiratory fitness levels were quantified using $\dot{V}O_2$peak, and the volume of the hippocampus was acquired using MRI and an algorithm to segment the hippocampal formation from surrounding tissue. After age, sex, years of education, and intracranial volume were controlled for, the results showed that older adults with higher fitness levels had larger hippocampal structures than their less fit peers (Erickson et al. 2009). Further, the more fit individuals had better memory performance compared to less fit individuals, and memory performance was positively associated with hippocampal volume. Therefore, consistent with findings from rodent research, physical activity is associated with enhanced hippocampal morphology and memory function in humans (see figure 20.3).

In sum, physical activity influences both the brain and cognition, but it does so in a relatively specified fashion. Physical activity improves most cognitive functions, but the greatest effects appear to occur on tasks that measure executive function. Likewise, several brain regions are affected by physical activity, with the largest effects occurring in the prefrontal, parietal, and temporal cortices, including the hippocampus.

Contrary to popular belief, older adult brains retain the capacity for plasticity and adaptation. Physical activity is one way to take advantage of this plasticity to enhance cognitive and brain function in late adulthood.

The effect of physical activity on cognitive function is both general and specific—general in the sense that most cognitive domains are improved after physical activity, but specific in the sense that executive functions improve more than other cognitive domains.

Sufficient Amounts of Physical Activity

Another major question concerns the effect of physical activity on the brain and cognition. Inherent within this question is whether there is a threshold, or minimum amount, of physical activity that needs to be reached before physical activity can exert any beneficial effect on cognition. As noted earlier, this question is critical if health care professionals are to

prescribe physical activity to prevent or treat brain disorders. Unfortunately, we have yet to see properly designed experiments in which the intensity and duration of physical activity are systematically manipulated so that dose–response information on the brain and cognition can be identified.

Randomized trials have demonstrated that cognitive improvements can be detected in as little as one month of consistent moderate-intensity physical activity. However, other studies, including meta-analyses of randomized trials, have indicated that longer periods of physical activity are more consistently associated with improved cognitive function (Erickson and Kramer 2009). In the series of studies described previously (Kramer et al. 1999; Colcombe et al. 2004, 2006), six months of frequent walking was sufficient for improving both executive function and brain function in a sample of healthy older adults.

It appears that six months of moderate-intensity physical activity several days per week is also effective at enhancing cognitive function in older adults who have cognitive impairment (Baker et al. 2010). For example, in one study, 138 participants over the age of 50 with self-reported memory problems were randomly assigned to either a home-based exercise regimen or an educational control group (Lautenschlager et al. 2008). Participants in the intervention group demonstrated marked improvements on the Alzheimer's Disease Assessment Scale, while the educational control group experienced a slight decline on this measure over the same period.

The results from randomized trials are compatible with those from longitudinal observational studies that have examined the effect of physical activity on cognitive function. These studies have demonstrated that greater amounts of physical activity are associated with maintenance of cognitive function. Several studies have shown an inverse linear association

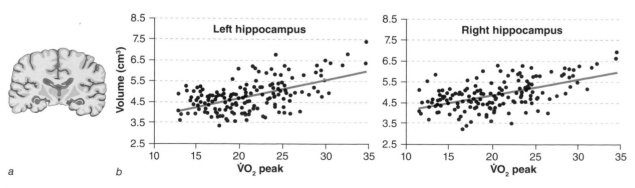

FIGURE 20.3 (a) A coronal view of the hippocampus in the medial temporal lobe. (b) Both the left and right hippocampal volumes are positively correlated with aerobic fitness levels.

Adapted from Erickson et al. 2009.

between energy expenditure and cognitive decline, such that greater amounts of physical activity are associated with a reduced risk of cognitive impairment (e.g., Podewils et al. 2005). These results support the claim that moderate-intensity physical activity is sufficient for enhancing cognitive function. However, these studies were observational and could have been confounded by factors that covary both with risk for cognitive decline and with physical activity patterns. Nonetheless, the overarching theme in both randomized trials and longitudinal studies is that moderate-intensity exercise on about three days per week over a six-month period enhances cognitive function and reduces the risk for developing dementia.

It is important to note that the research described here focused on older adults. The amount of exercise that is needed to enhance cognitive and brain function in other populations remains unclear. In young adults, six weeks of exercise was effective at enhancing visuospatial memory and affect (Stroth et al. 2009). In children, 15 weeks of high-dose (40 min/day) physical activity effectively enhanced executive functioning skills (Davis et al. 2007). In short, the amount of physical activity necessary to improve cognitive and brain function may be dependent on the sample being examined, the starting level of cognitive functioning of the sample, and the types of cognitive domains tested.

> ■ Six months of moderate-intensity exercise is sufficient for enhancing cognitive and brain function.

Mechanistic Questions

It is clear from the previous sections that moderate-intensity physical activity is effective at improving the brain and cognition. However, an important question to ask is, What are the mechanisms by which physical activity improves cognitive and brain function? From a biological perspective, it is customary to define the term *mechanism* as the molecular processes and cascades that give rise to the systems-level changes influencing cognitive functions. However, mechanisms can be thought of on several levels besides the molecular. For example, using cognitive neuroscience tools, we could ask how the different brain regions affected by physical activity work together to enhance cognitive function. With this question we are asking whether physical activity enhances cognition by changing or improving the connectivity of particular brain regions. Finally, an alternative way of thinking about mechanisms is to look at changes in socio-emotional function that mediate the improvements in the brain and cognition. In this mechanistic scenario, we might expect physical activity to improve measures of mood and life satisfaction, which in turn improve the brain networks that support complex cognition. Thus, mechanistic hypotheses can occur on at least three different levels: (1) the molecular level, (2) the systems or cognitive neuroscience level, and (3) the socioemotional level.

Molecular Mechanisms

One of the benefits of using animal models to assess the effect of physical activity on brain function is that the molecular mechanisms can be more easily investigated than in humans. Researchers using nonhuman animal models often manipulate exercise either by providing voluntary access to a running wheel in the cage or by forcing exercise on a treadmill. Many studies have used a voluntary exercise model with the argument that forced exercise could inadvertently attenuate cognitive function by increasing stress levels. In addition, it has been argued that voluntary exercise is more similar to exercise behavior in humans than is forced exercise. In any case, animal models have focused on three main possible mechanisms by which exercise enhances cognition: **neurogenesis**, or the proliferation and survival of new neurons; **angiogenesis**, or the proliferation and survival of new vasculature; and elevated levels of neurotransmitters and neurotrophic factors, among which the most frequently studied is **brain-derived neurotrophic factor** (BDNF). This does not mean that there are only three potential molecular pathways by which exercise exerts its effects. Indeed, many paths and molecules are influenced by exercise, many of which are only now becoming recognized as important contributors to the enhancing effect of physical activity on cognition. Therefore, this review should serve only as a guide to understanding some of the molecular pathways that have received the most attention.

Providing animals with voluntary access to a running wheel has been demonstrated to enhance hippocampal-dependent learning and memory. For example, wheel running accelerates learning on the Morris water maze. In the typical Morris water maze, animals learn to swim to a platform just beneath the surface of a pool of opaque water. The platform remains stationary on all trials, but the location where the rodent is submerged is varied so that the animal must learn to associate the location of the platform with the location of the extra-maze cues. In several studies, exercise has been shown to increase task acquisition and promote greater retention of the platform location in both young and aged animals. Other tasks, such as T-maze, radial arm maze, spatial separation, and avoidance tasks, have also shown enhancements in learning rates in exercising animals (see Kramer, Erickson, and Colcombe 2006).

Exercise-associated improvements in learning and memory have been related to the number of newborn neurons produced in the hippocampus, even in aged rodents (van Praag et al. 2005). In fact, an increase in cell proliferation and cell survival in the hippocampus is one of the most consistently observed effects of exercise (e.g., Cotman and Berchtold 2002). For example, in one study, mice were assigned to either a running condition with voluntary access to a running wheel or one of a few control conditions (van Praag et al. 2005). Cell proliferation was measured by bromodeoxyuridine, a marker that labels dividing cells. Exercise was found to increase the total number of newborn and surviving cells in the hippocampus, and the number of new cells generated was positively correlated with performance on the Morris water maze. Importantly, exercise-induced neurogenesis has been observed in both younger and older animals (van Praag et al. 2005), but cell proliferation in older animals appears to be attenuated relative to that in younger animals—an effect that might indicate a limit to the extent to which exercise can reverse age-related decay. Consistent with this hypothesis, Kronenberg and colleagues (2006) found that voluntary exercise prevented an age-related decline in precursor cell activity in the hippocampus but failed to maintain cell activity at the level of younger animals. In sum, exercise has been convincingly shown to increase cell proliferation and survival in the hippocampus for both younger and older animals, and this increase is linked to better performance on several different cognitive paradigms (van Praag et al. 2005).

The proliferation of new cells in the brain is accompanied by an increased need for nutrients. This demand is met by the stimulation of new blood vessels. Exercise increases the production and secretion of molecules involved with the formation of new blood vessels, including insulin-like growth factor and vascular endothelial growth factor. In one study, Black and colleagues (1990) examined the differences between physical training and skill training in rodents. Rats were exposed to either an acrobatic condition with high motor learning demands, a forced exercise condition, a voluntary exercise condition, or an inactive control group. Using stereological estimates and camera lucida drawings, the researchers reported that motor learning in the acrobatic condition caused synaptogenesis, or the development of new synapses between cells, but that forced and voluntary exercise conditions produced new vasculature in the cerebellum (see figure 20.1). New vasculature as a result of exercise has also been reported in the motor cortex, the hippocampus, and the basal ganglia. Because of angiogenesis, physical activity may protect the brain from stroke damage and promote faster recovery after vasculature insult.

One molecule involved in learning, memory, and neurogenesis is brain-derived neurotrophic factor (BDNF). BDNF, a small molecule produced and secreted by neurons, has been related to long-term potentiation—a cellular analog of memory formation—and to cell proliferation in the hippocampus, olfactory bulb, and striatum. Interestingly, BDNF levels are increased as a function of exercise in the hippocampus, frontal cortex, striatum, and cerebellum (Cotman and Berchtold 2002). In one study, blocking the binding of BDNF to its receptor effectively abolished the exercise-related improvements in Morris water maze performance (Vaynman, Ying, and Gomez-Pinilla 2004), suggesting that BDNF mediates the exercise-related memory enhancements. Although exercise influences the concentration of other molecules (e.g., dopamine, serotonin), BDNF has received considerable attention for its putative role in learning, memory, and neurogenesis.

Exercise also seems to offset the behavioral symptoms observed in rodent models of Alzheimer's and Parkinson's diseases. For example, in a mouse model of Alzheimer's disease, exercising animals showed a reduction in amyloid beta deposits, reduced tau formation, and superior learning rates compared with sedentary animals (Parachikova, Nichol, and Cotman 2008). The offsetting effect of exercise on Alzheimer's disease pathophysiology is encouraging and suggests that physical activity might be an effective intervention in individuals at high risk for developing dementia.

In sum, exercise improves cognitive function in both young and aged animals, increases levels of important biomarkers (e.g., BDNF) that may be contributing to exercise-induced neurogenesis and angiogenesis, and offsets the molecular cascades associated with Alzheimer's and Parkinson's diseases. The results from these studies overlap with those from human studies and provide a low-level biological explanation for the effects observed in humans.

Cognitive Neuroscience and Systems-Level Mechanisms

There are several different ways in which the network of brain regions involved with complex cognitive function could be altered or enhanced by physical activity. As described by the field of cognitive neuroscience, some brain regions are responsible for initiating and maintaining certain strategies that can be used to complete a task. For example, if you are studying for an exam, you may try to memorize the material by using either rote memorization or mnemonic strategies, and these two strategies for memorizing material rely on separate brain regions that could be differentially affected by physical activity. Along these lines, it is possible that exercise enhances the functioning of some brain regions (e.g., prefrontal cortex) that allow a more flexible use of particular strategies or a more efficient engagement of those strategies. Thus, more efficient engagement or flexibility in strategy use could result in elevated cognitive performance. Another possibility is that exercise influences the ability of brain regions to communicate effectively with one another. Enhanced communication between regions may result in enhanced cognitive performance. Unfortunately, this avenue of research is still in its infancy, but a few studies now point to several systems-level changes in the brain as a result of physical activity or higher aerobic fitness levels.

In one study, aerobic fitness levels were quantified using $\dot{V}O_2$max, and a network of brain regions active at rest was identified using a resting-state protocol with fMRI (Voss et al. 2010). The resting-state network includes parietal, prefrontal, and hippocampal regions, and the connectivity among these regions declines with advancing age and with Alzheimer's disease. Importantly, older adults with higher fitness levels showed enhanced *connectivity* of the default resting-state network including prefrontal and hippocampal regions. In addition, increased resting-state connectivity mediated a fitness-related improvement in executive function. This result suggests that even at rest, the brains of more fit older adults are working more effectively than those of less fit older adults. It also provides a mechanistic explanation for how physical activity and aerobic fitness are associated with enhanced cognition. In short, this study identifies increased connectivity between regions as a mediating factor in fitness-related cognitive enhancement. Consistent with this explanation, white matter, comprising the axons that allow regions to communicate, increases as a function of physical activity (e.g., Colcombe et al. 2006).

Similar to animal studies, human studies have also begun to focus on the possible role of increased blood flow and vasculature to the brain in mediating cognitive improvements. In fact, several studies have shown that physical activity increases blood flow in several brain regions, including the hippocampus, of humans (Pereira et al. 2007). Interestingly, increased blood flow is associated with both enhanced memory function and neurogenesis. Since fMRI results are inherently based on blood flow, it is possible that improved blood flow and circulation are contributing to the results described earlier. However, based on the molecular mechanisms outlined in the previous section, it is clear that increased vascularization of brain tissue is unlikely to explain all of the benefits associated with exercise. In addition, the fMRI results described previously (Colcombe et al. 2006) are difficult to interpret solely by group differences in the extent to which the brain is vascularized. The logic is that increased vascularization caused by exercise should result in enhanced fMRI signal for all cognitive conditions and not just the more challenging tasks. What was seen, however, was that exercise selectively enhanced brain function during the more challenging task conditions (Colcombe et al. 2004). Global increases in brain vascularization cannot explain this pattern of results. Nonetheless, one could still argue that regional differences in perfusion resulting from exercise could explain several of the volumetric and functional MRI results described. It will be important for future studies to determine the degree to which vasculature changes mediate exercise-related improvements in cognition.

Finally, several studies in older adults suggest that physical activity improves the capacity for the brain to *prepare* to make a response (e.g., Hillman et al. 2006). According to this hypothesis, physical activity improves cognition by enhancing the cortical flexibility to quickly and efficiently prepare a response. Therefore, physical activity would improve cognition by enhancing preparatory-related brain signals.

In sum, although we are far from a clear understanding of the cognitive neuroscience and systems-level mechanisms, there is growing evidence that physical activity alters communication between regions, enhances vasculature, and improves the functioning of brain regions supporting preparation of a response. These changes at the systems, or cognitive neuroscience, level of function provide a mechanistic explanation for how the changes in brain function observed in humans elevate cognitive function. More research on the cognitive neuroscience mechanisms of physical activity on brain function is necessary before strong claims about the specific mediating pathways can be made.

Socioemotional Mechanisms

Unfortunately, the question of socioemotional mechanisms remains unanswered. However, it is clear that emotion and mood influence cognitive function and that exercise influences mood. Many of the brain regions discussed in this chapter, including the prefrontal and parietal regions, have significant roles in regulating emotion and mood. Therefore, it is plausible that changes in emotion and mood precede changes in cognition and mediate the effect of physical activity on cognition. However, alternative hypotheses for this association exist. For example, it may be that exercise directly affects several different brain regions involved with both executive function and emotion and that the improvements in these processes are affected in parallel and do not interact with or mediate one another. In short, the possible mediating role of socioemotional processing on the effect of physical activity on the brain and cognition remains an open and unstudied question.

■ **Physical activity influences the brain and cognition by increasing the proliferation and survival of new neurons and new vasculature.**

■ **Important molecules critical in learning and memory are elevated after exercise.**

■ **In human studies, it appears that improvements in the connectivity of the brain contribute to enhanced cognitive function.**

Applied Questions: Populations Benefiting From Physical Activity

Most of the research examining physical activity's effect on the brain and cognition has been limited to older adult populations without dementia (e.g., Kramer et al. 1999). In these cases, it is clear that physical activity improves cognition, increases brain function, and increases gray and white matter tissue volume in select cortical areas. However, what is not as clear from the preceding sections of this chapter is whether physical activity or exercise influences the brain and cognition in other populations such as children, young adults, or individuals with manifest neurologic or psychiatric problems. Along these same lines, we could ask whether physical activity is a reliable nonpharmaceutical method to prevent the occurrence of neurologic or psychiatric diseases, and if so whether we could prescribe physical activity as a treatment for cognitive impairment or neurologic disease. These types of questions can be classified under the category of "applied" since they do not

address the phenomenology of physical activity and cognition or the mechanisms by which physical activity influences the brain. Instead, these questions address fundamental concerns regarding the ability of exercise to enhance cognitive and brain function across different populations.

As described previously, animal models of Alzheimer's disease have shown that voluntary exercise is effective at preventing the accumulation of amyloid beta deposits and neurofibrillary tangles, two of the putative causes of Alzheimer's disease. These results suggest that physical activity may act to prevent the development of Alzheimer's disease. Consistent with these findings, epidemiological studies have reported that greater amounts of physical activity are often associated with less cognitive decline and a reduced risk of developing dementia later in life. For example, Podewils and associates (2005) studied the relationship between physical activity and dementia in 3,375 individuals over the course of 5.4 years. Physical activity was assessed by a self-report questionnaire in which participants were asked about the frequency and duration of 15 types of physical activities over the past two weeks. The authors found an inverse relationship between Alzheimer's disease and the number of physical activities performed by the participants. Other epidemiological studies have shown similar associations between physical activity and cognitive function in late life.

Interestingly, these findings were recently extended to a series of intervention studies that examined the effect of physical activity on cognitive impairment. In a sample of individuals with mild cognitive impairment, a cognitively labile state that increases the risk for dementia, Baker and colleagues (2010) found that six months of aerobic exercise improved performance on tests of executive function, but this was specific to the women in the sample. The men with mild cognitive impairment showed no improvements with the exercise intervention. In a similar study, described earlier, Lautenschlager and colleagues (2008) found that six months of exercise improved cognitive function in a sample with diagnosed Alzheimer's disease.

Alzheimer's disease and mild cognitive impairment are associated with a rapidly deteriorating brain. For this reason, it is imperative to examine the potential for physical activity to offset or slow the rate of brain deterioration. Unfortunately, randomized trials of physical activity have not yet been conducted to investigate this question. However, several cross-sectional studies have examined the association between aerobic fitness and brain volume. These investigators hypothesized that if physical activity improves cognitive function and delays the onset of Alzheimer's disease, then higher aerobic fitness levels might be associated with the sparing of brain volume. In one study, aerobic fitness was assessed in individuals in the early stages of Alzheimer's disease. Using MRI techniques to examine brain volume, the researchers found that individuals with Alzheimer's disease who were more aerobically fit had greater total gray matter volume than their less fit peers (Burns et al. 2008). This intriguing finding suggested that there might also be regional variation in brain volume as a function of aerobic fitness. In fact, in one recent study, Honea and colleagues (2009) found greater medial temporal lobe volume in aerobically fit individuals with dementia than in their less fit counterparts. Therefore, consistent with meta-analyses of randomized interventions of exercise in cognitively impaired populations and animal models of disease, higher fitness levels are associated with less cortical decay in populations with neurological disease. It will be important for future researchers to design randomized interventions to determine whether increased physical activity increases brain volume in cognitively impaired populations.

In animal models of Parkinson's disease, exercise increases dopamine release in the basal ganglia and protects against parkinsonian symptoms. In humans with Parkinson's disease, walking interventions improve executive function and motor control. However, to date, the effect of physical activity on brain volume or function in Parkinson's patients remains unknown.

Unfortunately, few randomized trials have been conducted with other populations. Several cross-sectional studies have demonstrated an association between aerobic fitness levels and cortical plasticity in multiple sclerosis patients, while others have focused on physical activity's cognition-enhancing properties in children. Further, exercise appears to be a powerful antidepressant, and it may moderate attention-deficit hyperactivity disorder in children. Several studies have even shown that exercise has the potential to enhance cognitive function and hippocampal volume in people with schizophrenia.

In sum, exercise may have far-reaching effects on the brain and cognition in populations ranging from children to old adults, with and without cognitive impairment and with neurologic or psychiatric disease. At this point, there is too little research to allow definitive conclusions about the preventive or treatment potential of physical activity; however, the growing amount of evidence suggests that physical activity could serve as a possible preventive treatment for brain disorders.

Physical activity improves cognitive and brain function in many different populations including normal, healthy individuals and people with a brain or psychiatric disease. This suggests that the effects of exercise on the brain and cognition are relatively broad.

Moderating Questions: Factors Moderating the Effect of Physical Activity

It is clear that not everyone benefits at the same rate or to the same degree from physical activity. This suggests that individual difference factors moderate physical activity's effect on the brain and cognition.

Physical activity does not occur in isolation from other lifestyle habits such as dietary patterns, intellectual stimulation, cigarette smoking, or stress. To what extent do these lifestyles or habits influence the effect of physical activity on the brain and cognition? Unfortunately, very few studies have investigated these associations; but the few that have done so show that dietary patterns, hormone use, and social participation might moderate the rate at which, or the extent to which, physical activity improves the brain and cognition.

Such variables may moderate the effect of physical activity on the brain and cognition in at least two different ways. First, physical activity's effect may be attenuated by the presence or absence of other lifestyle factors. For example, physical activity may improve cognition, but if it is accompanied by an unhealthy diet, the beneficial effect may be significantly attenuated. Second, other lifestyle factors may potentiate the effect of physical activity on the brain and cognition. Thus, the combination of physical activity and an intellectually stimulating environment may promote a healthier brain than either by itself.

Studies in rodents have shown that diet can moderate the effect of physical activity on BDNF. For example, voluntary exercise potentiates the beneficial effects of omega-3 fatty acids on cell membrane proteins involved in synaptic plasticity and cognition. Further, exercise offsets the negative effects of a high-fat diet on BDNF levels in the hippocampus. Unfortunately, we can only speculate about the interactive effect of dietary patterns and physical activity on the brain and cognition in humans.

Other factors are also likely to moderate the effect of physical activity. For example, exercise and estradiol increase the production and secretion of some of the same molecules (e.g., BDNF). The similar effects of exercise and estradiol on BDNF levels suggest that administration of estradiol to female rats may interact with physical activity patterns in such a way as to moderate BDNF levels. Consistent with this hypothesis, the combination of exercise and estradiol administration increased BDNF levels more than either treatment by itself. A study with human postmenopausal women showed a similar interaction between hormone therapy and aerobic fitness levels on executive function and brain volume (Erickson et al. 2007). Social engagement may also moderate physical activity's effects. For example, socially housed rodents showed more rapid increases in BDNF levels than animals that exercised in social isolation. In short, factors such as diet, hormone administration or supplementation, and social engagement all have been hypothesized to moderate the effect of physical activity on the brain and cognition. Although this field is in its infancy, it would seem that it will be important for future research to advance understanding of the ways in which multiple lifestyle factors contribute to a healthy brain.

It is also important to note that lifestyle factors are not the only contributors to variation in cognitive and brain function. Clearly, genetics plays an important role, and it is likely that genetic factors and polymorphisms moderate the influence of physical activity. For example, the *BDNF* gene has a single nucleotide polymorphism that influences the regulated secretion of BDNF from the cell. The *BDNF* polymorphism has been associated with variation in brain volume and cognitive function (Erickson et al. 2008). Given that exercise increases the production and secretion of BDNF, it is likely that the polymorphism moderates the effect of exercise on learning, memory, and brain function. Future research will need to examine the interactions between physical activity and genetic polymorphisms such as the *BDNF* gene.

Summary

We can conclude from this review that physical activity improves cognitive and brain function and does so through several molecular and systems-level mechanisms. It appears that at least several months of physical activity are needed to allow detection of an effect on cognition, but this may vary as a function of the population under study. Executive functioning is the cognitive domain most amenable to a physical activity training program. Consistent with this, the prefrontal and parietal regions, or those regions that largely support executive function, demonstrate the greatest change in both volume and function from physical activity. However, growing evidence indicates that the hippocampal region is also highly plastic and amenable to adaptations due to physical activity. Although it is clear that physical activity is effective at enhancing the brain and cognition, it is also obvious that many unanswered and perplexing questions remain. What types, intensities, and amounts of physical activity are most effective at enhancing cognitive function? What populations might benefit the most from physical activity interventions? What other lifestyle agents moderate the effect of physical activity? And what are the underlying mechanisms that are driving these effects in humans? These and many more questions remain poorly understood. Future research should be directed toward understanding the potential for exercise as an easy and accessible method for enhancing cognitive and brain function.

Key Concepts

Alzheimer's disease—A form of dementia that results in memory and cognitive impairment during late adulthood.

angiogenesis—Growth and production of new capillaries.

brain-derived neurotrophic factor—A small molecule secreted by neurons and other cells that is critical in memory formation and neurogenesis.

cognition—Set of mental processes that contribute to perception, memory, intellect, and action.

cognitive impairment—A general term referring to cognitive dysfunction in late adulthood that includes both mild memory problems and more severe dementia.

executive control—A set of cognitive operations underlying the selection, scheduling, coordination, and monitoring of complex, goal-directed processes involved in perception, memory, and action.

hippocampus—A medial temporal lobe region involved in memory formation. Damage results in amnesia or Alzheimer's disease.

magnetic resonance imaging—A noninvasive medical technique that provides detailed images of soft biological tissues such as the brain.

neurogenesis—Growth and production of new neurons.

neuroimaging—A general term referring to several medical techniques that measure and identify the operations of the living brain.

Parkinson's disease—A disease characterized by loss of dopamine-secreting neurons in the midbrain regions, resulting in loss of motor and cognitive function.

Study Questions

1. Why would physical activity improve executive functions more than other cognitive domains?

2. What are the hypothesized mechanisms for how physical activity influences brain function?

3. Several of the studies discussed in this review employed randomized trials, while others were observational. How could this variability in study design influence the pattern of results and interpretations?

4. The role of BDNF in the brain is linked to physical activity levels. Describe the reasons why BDNF might be a molecule that mediates the effect of physical activity on cognition.

5. Some have argued that physical activity improves brain function solely by influencing brain vasculature. Describe the evidence that supports and refutes this claim.

6. Is starting an exercise regimen in late adulthood futile? Why or why not?

References

Baker, L.D., L.L. Frank, K. Foster-Schubert, P.S. Green, C.W. Wilkinson, A. McTiernan, S.R. Plymate, M.A. Fishel, G.S. Watson, B.A. Cholerton, et al. 2010. Effects of aerobic exercise on mild cognitive impairment: A controlled trial. *Archives of Neurology* 67:71-79.

Black, J.E., K.R. Isaacs, B.J. Anderson, A.A. Alcantara, and W.T. Greenough. 1990. Learning causes synaptogenesis, whereas motor activity causes angiogenesis, in cerebellar cortex of adult rats. *Proceedings of the National Academy of Sciences* 87:5568-5572.

Burns, J.M., B.B. Cronk, H.S. Anderson, J.E. Donnelly, G.P. Thomas, A. Harsha, W.M. Brooks, and R.H. Swerdlow. 2008. Cardiorespiratory fitness and brain atrophy in early Alzheimer's disease. *Neurology* 71:210-216.

Colcombe, S.J., K.I. Erickson, P.E. Scalf, J.S. Kim, R. Prakash, E. McAuley, S. Elavsky, D.X. Marquez, L. Hu, and A.F. Kramer. 2006. Aerobic exercise training increases brain volume in aging humans. *Journals of Gerontology: Series A, Biological Sciences and Medical Sciences* 61:1166-1170.

Colcombe, S.J., and A.F. Kramer. 2003. Fitness effects on the cognitive function of older adults: A meta-analytic study. *Psychological Science* 14:125-130.

Colcombe, S.J., A.F. Kramer, K.I. Erickson, P. Scalf, E. McAuley, N.J. Cohen, A. Webb, G.J. Jerome, D.X. Marquez, and S. Elavsky. 2004. Cardiovascular fitness, cortical plasticity, and aging. *Proceedings of the National Academy of Sciences* 101:3316-3321.

Cotman, C.W., and N.C. Berchtold. 2002. Exercise: A behavioral intervention to enhance brain health and plasticity. *Trends in Neurosciences* 25:295-301.

Davis, C.L., P.D. Tomporowski, C.A. Boyle, J.L. Waller, P.H. Miller, J.A. Naglieri, and M. Gregoski. 2007. Effects of aerobic exercise on overweight children's cognitive functioning: A randomized controlled trial. *Research Quarterly for Exercise and Sport* 78:510-519.

Erickson, K.I., S.J. Colcombe, S. Elavsky, E. McAuley, D.L. Korol, P.E. Scalf, and A.F. Kramer. 2007. Interactive effects of fitness and hormone treatment on brain health in postmenopausal women. *Neurobiology of Aging* 28:179-185.

Erickson, K.I. J.S. Kim, B.L. Suever, M.W. Voss, B.M. Francis, and A.F. Kramer. 2008. Genetic contributions to age-related decline in executive function: A 10-year longitudinal study of COMT and BDNF polymorphisms. *Frontiers in Human Neuroscience* 2:11.

Erickson, K.I., and A.F. Kramer. 2009. Exercise effects on cognitive and neural plasticity in older adults. *British Journal of Sports Medicine* 43:22-24.

Erickson, K.I., R.S. Prakash, M.W. Voss, L. Chaddock, L. Hu, K.S. Morris, S.M. White, T.R. Wojcicki, E. McAuley, and A.F. Kramer. 2009. Aerobic fitness is associated with hippocampal volume in elderly humans. *Hippocampus* 19:1030-1039.

Hillman, C.H., A.F. Kramer, A.V. Belopolsky, and D.P. Smith. 2006. A cross-sectional examination of age and physical activity on performance and event-related brain potentials in a task-switching paradigm. *International Journal of Psychophysiology* 59:30-39.

Honea, R.A., G.P. Thomas, A. Harsha, H.S. Anderson, J.E. Donnelly, W.M. Brooks, and J.M. Burns. 2009. Cardiorespiratory fitness and preserved medial temporal lobe volume in Alzheimer disease. *Alzheimer Disease and Associated Disorders* 23:188-197.

Kramer, A.F., K.I. Erickson, and S.J. Colcombe. 2006. Exercise, cognition, and the aging brain. *Journal of Applied Physiology* 101:1237-1242.

Kramer, A.F., S. Hahn, N.J. Cohen, M.T. Banich, E. McAuley, C.R. Harrison, J. Chason, E. Vakil, L. Bardell, R.A. Boileau, et al. 1999. Ageing, fitness and neurocognitive function. *Nature* 400:418-419.

Kronenberg, G., A. Bick-Sander, E. Bunk, C. Wolf, D. Ehninger, and G. Kempermann. 2006. Physical exercise prevents age-related decline in precursor cell activity in the mouse dentate gyrus. *Neurobiology of Aging* 27:1505-1513.

Lautenschlager, N.T., K.L. Cox, L. Flicker, J.K. Foster, F.M. van Bockxmeer, J. Xiao, K.R. Greenop, and O.P. Almeida. 2008. Effect of physical activity on cognitive function in older adults at risk for Alzheimer disease: A randomized trial. *Journal of the American Medical Association* 300:1027-1037.

Parachikova, A., K.E. Nichol, and C.W. Cotman. 2008. Short-term exercise in aged Tg2576 mice alters neuroinflammation and improves cognition. *Neurobiology of Disease* 30:121-129.

Pereira, A.C., D.E. Huddleston, A.M. Brickman, A.A. Sosunov, R. Hen, G.M. McKhann, R. Sloan, F.H. Gage, T.R. Brown, and S.A. Small. 2007. An in vivo correlate of exercise-induced neurogenesis in the adult dentate gyrus. *Proceedings of the National Academy of Sciences* 104:5638-5643.

Podewils, L.J., E. Guallar, L.H. Kuller, L.P. Fried, O.L. Lopez, M. Carlson, and C.G. Lyketsos. 2005. Physical activity, APOE genotype, and dementia risk: Findings from the Cardiovascular Health Cognition Study. *American Journal of Epidemiology* 161:639-651.

Stroth, S., K. Hille, M. Spitzer, and R. Reinhardt. 2009. Aerobic endurance exercise benefits memory and affect in young adults. *Neuropsychological Rehabilitation* 19:223-243.

van Praag, H., T. Schubert, C. Zhao, and F.H. Gage. 2005. Exercise enhances learning and hippocampal neurogenesis in aged mice. *Journal of Neuroscience* 25:8680-8685.

Vaynman, S., Z. Ying, and F. Gomez-Pinilla. 2004. Hippocampal BDNF mediates the efficacy of exercise on synaptic plasticity and cognition. *European Journal of Neuroscience* 20:2580-2590.

Voss, M.W., K.I. Erickson, R.S. Prakash, L. Chaddock, E. Malkowski, H. Alves, J.S. Kim, K.S. Morris, S.M. White, T.R. Wojcicki, et al. 2010. Functional connectivity: A source of variance in the association between cardiorespiratory fitness and cognition? *Neuropsychologia* 48:1394-1406.

21

Exercise and Its Effects on Mental Health

John S. Raglin, PhD; and Gregory S. Wilson, PED, FACSM

CHAPTER OUTLINE

© Konstantin Sutyagin

Physical activity has long been advocated as a means of attaining optimal health and preventing disease. As the often repeated Latin dictum states, *mens sana in corpore sano* ("a healthy mind in a healthy body"). Today, overwhelming research supports the benefits of exercise for physical health, and empirically based activity regimens have been established for myriad **disorders** across the life span. However, the benefits of exercise are not limited to the impact on aspects of physical health and disease states. Physical activity has also been promoted as a means to enhance various aspects of mental health, including emotional disorders, while improving or preserving aspects of cognitive function, particularly higher-order executive functions (see chapter 20). Unfortunately there remains much less empirical information concerning the mental benefits of exercise in comparison with its physiological consequences. To some extent, this is because exercise has long been regarded as an unorthodox treatment by psychologists and other mental health workers; but in recent years, landmark documents such as the Surgeon General's report on physical activity have done much to change this perspective. Another factor constraining study of the psychological consequences of physical activity has been a lack of interdisciplinary research. Cooperation between exercise physiologists and sport psychologists would serve to greatly advance research on this topic but has proven to be the exception rather than the rule in exercise science research.

Despite these and other challenges, there is compelling scientific support for the efficacy of physical activity in providing significant psychological benefits to persons with emotional illnesses of mild to moderate severity, as well as those with good emotional health. Recent research studies have also provided important information about how specific exercise variables (mode, intensity, duration, frequency, setting) affect psychological outcomes. However, other extant research reveals that exercise is not always associated with beneficial psychological outcomes and in some circumstances may even result in detrimental changes in emotional health.

> Physical activity can improve and maintain mental health, but the optimal mode, intensity, frequency, and duration of exercise have not been precisely determined.

This chapter provides a general summary of the major findings of the exercise and mental health literature, giving particular emphasis to the effects of exercise on the major emotional disorders of anxiety and depression. We present some discussion on the relationship of exercise to schizophrenia and the mechanisms that are believed to be responsible for the psychological benefits of exercise. We also describe findings suggesting that extreme exercise loads can have detrimental consequences on mental health.

Research Paradigms of Exercise and Mental Health Research

Studies of the psychological consequences of exercise have generally focused on either the short-term changes associated with a single bout (i.e., **acute exercise bout**) or long-term changes following participation in physical activity for a period of weeks or longer (i.e., **chronic exercise**). In an attempt to determine the efficacy of a particular "dosage" on either the general population or samples of people with special characteristics (e.g., older adults, sedentary individuals, anxiety patients), research examining the effects of acute exercise on mental health has utilized various modes of activity ranging in duration from bouts as brief as 5 min to 9 h or longer, as in the case of athletes completing a triathlon (Petruzzello et al. 1991).

Acute and Chronic Exercise Research

Investigators typically measure the psychological responses to either acute or chronic exercise by having participants complete self-report questionnaires that measure various aspects of mental health such as anxiety, mood state, or emotion-related feelings such as fatigue and energy. In the case of acute exercise, researchers use psychological questionnaires that assess psychological **states**, which are transitory emotions that can change dramatically across a span of minutes or even seconds. Because of the labile nature of psychological states, it is preferable to assess them several times following an exercise session rather than only once; a single postexercise assessment could be misleading and would not provide information about the duration or nature of changes in psychological states. For example, immediately after the cessation of high-intensity exercise (>70% $\dot{V}O_2$max) there often are transient elevations in anxiety and mood disturbance. A single postexercise assessment taken at this point would lead to the conclusion that high-intensity exercise worsens mood. However, assessments conducted 15 min or later reveal that state anxiety has dropped significantly below baseline and so would present a truer picture of how high-intensity exercise influences mood. In fact, research using multiple postexercise assessments indicates that the psychological benefits of a single exercise bout typically persist for 2 to 4 h or longer before mood state gradually returns to baseline values (Raglin 1997).

Chronic exercise research typically examines more stable psychological measures referred to as **traits**. Traits are stable moods or emotions that reflect how a person feels over a longer time period ("last week," "in general"); consequently, traits are not altered by single bouts of exercise or brief psychological interventions. Accordingly, an adequate test of the psychological effects of exercise on traits or clinical conditions such as major depression and anxiety disorders requires testing programs that last two to three weeks, if not longer.

> Reported psychological benefits of exercise include reductions in anxiety and depression with improvements in self-esteem and overall well-being.

Quantifying the psychological outcomes of acute or chronic exercise has largely been achieved using standardized psychological inventories of mood states (e.g., the Profile of Mood States; McNair, Lorr, and Droppleman 1992) or measures of emotions such as anxiety (e.g., the State-Trait Anxiety Inventory; Spielberger et al. 1983) that were developed and validated for use with various populations across a wide variety of settings (e.g., hospitals, workplaces). In studies in which the participants have been diagnosed with emotional disorders such as depression, it is preferable to employ questionnaires specifically designed for diagnostic purposes (e.g., Beck Depression Inventory). Although specialized questionnaires have been developed to examine psychological responses in the specific context of exercise, many of these instruments have not been adequately validated, and questions remain concerning the degree to which they provide either greater sensitivity or additional information compared with general measures. Additionally, other relevant biological variables such as stress hormones or brain metabolic activity (e.g., by functional magnetic resonance imaging [fMRI]) are sometimes assessed concomitantly with psychological variables to identify the biological processes

or mechanisms that underlie exercise-associated benefits to mental health.

Study Sample Characteristics

One of the fundamental concerns of exercise and mental health researchers has been quantifying the extent to which exercise benefits persons with symptoms of emotional illnesses such as anxiety or depression compared to standard forms of treatment (i.e., medication or therapy). Unfortunately, much of this research has been of limited value because it has utilized psychologically healthy individuals instead of cases with significant symptoms or, most appropriate, samples of people diagnosed with psychological disorders. Access to clinical samples has been hampered by the typical challenges associated with soliciting sedentary adults to participate in physical activity programs, as well the reluctance of many clinicians to subject their patients to non-traditional and unproven remedies. However, in recent years, researchers have worked to circumvent these challenges and have had greater success in testing the effects of physical activity on patients who have been diagnosed with major depression and anxiety.

Other researchers, however, are specifically interested in studying the potential benefits of exercise for persons with average or even above-average mental health. Findings from this work are crucial for determining the efficacy of physical activity in maintaining normal emotional health and preventing the development of mental illness in people who are free of symptoms or who are in remission. Identifying exercise regimens associated with optimal psychological outcomes could also help in getting inactive individuals to start exercising, as well as in keeping them exercising. Barely more than 20% of the adults in the United States exercise at least three times a week for 20 min or more, and approximately one-half quit exercising within six to eight weeks of starting a program.

Although not typically a target population for exercise and mental health research, athletes consistently possess mood state profiles that are superior to those of nonathletes (Morgan 1997). These findings provide another line of evidence that a physically active lifestyle can benefit both physical and mental health. However, this same research has also shown that when athletes undergo periods of physically demanding training, their emotional health worsens as a direct consequence of exercise, a phenomenon that we describe later in the section on detrimental consequences of physical activity.

Exercise and Depression

Depressive disorders will afflict one out of three persons at some point in their lives. Population studies in several countries indicate that at any given time, 4% of men and 8% of women meet the diagnostic criteria for clinical depression. The main subtypes of depression are major depressive disorder, dysthymia, and bipolar disorder (American Psychiatric Association 1994). Unfortunately, estimates suggest that less than one of three depressed individuals will seek treatment, and of those who do, 90% receive less than adequate professional attention (Buckworth and Dishman 2002).

Research on the Efficacy of Exercise as a Treatment for Depression

Over the past four decades, more than 1,200 studies have been published on the effects of exercise for treating depressive symptoms. Cross-sectional and prospective epidemiologic studies consistently report relationships between physical activity and reduced depressive symptoms, and these relationships hold even when adjustments are made for potentially confounding variables (Buckworth and Dishman 2002). Studies have shown similar effects among adults and adolescents, and most existing research has demonstrated similar effects among men and women.

The benefits of exercise for depression are most evident in persons with significant symptoms, since research has shown that persons scoring in the low to average range on standardized measures of depression exhibit little or no reduction in depression scores following chronic exercise. For example, in the first randomized exercise and depression study, Morgan and colleagues (1970) assigned participants to one of several six-week exercise programs and found no significant reduction in depression for any of the programs. However, depression scores decreased significantly among individuals who possessed elevated baseline scores. In work involving patients diagnosed with mild to moderate depression, exercise provides benefits that equal and sometimes exceed those associated with standard forms of psychotherapy. Moreover, available evidence suggests that aerobic and anaerobic forms of activity provide similar benefits (Buckworth and Dishman 2002).

There is only limited research directly comparing the efficacy of exercise to antidepressant medication. Blumenthal and colleagues (1999) randomly assigned 156 older adults clinically diagnosed with moderate depression to 16-week interventions of a walk–run program, antidepressant medication, or both exercise and medication. The reduction ($p < .05$) in symptoms was equal across groups, and the dropout rate for exercise (26.4%) was marginally higher than for medication (16.4%). Notably, medication was reported to alleviate symptoms more rapidly, particularly in patients with the highest levels of depression. A follow-up study conducted six months after the intervention showed that among participants who were in remission at the end of the initial study (N = 83), the relapse rate for depression was far lower for exercise (8%) than for either medication (38%) or medication combined with exercise (31%). The authors reported that for each 50 min of physical activity, there was a corresponding 50% decrease in the likelihood of being classified as depressed at the end of the follow-up period.

One notable limitation in this study was the lack of a placebo condition. This is an important research issue since it is known that up to 30% of depressive patients may show favorable changes in symptoms under convincing placebo interventions, and some researchers have contended that some or perhaps all of the psychological benefits of exercise are due to placebo effects or expectation effects. In a follow-up designed to remedy this concern as well as more definitively examine the benefits of exercise, Blumenthal and colleagues (2007) randomly assigned 202 sedentary adults to 16-week interventions of antidepressant medication, placebo pills, home-based aerobic exercise, or supervised exercise. Improvement was defined as no longer meeting the clinical criteria for depression (i.e., remission). At the end of the intervention period, the remission rates were 45% for supervised exercise, 40% for home-based exercise, and 47% for medication but only 31% for placebo, a difference that neared ($p = .057$) conventional levels of significance. The authors concluded that exercise provided benefits that were comparable to those of medication while acknowledging that the improvement found in the placebo condition was likely driven by participant expectations and additional nonspecific effects.

Dosage

Definitive recommendations for the optimal exercise prescription have yet to be established for depression (Buckworth and Dishman 2002). Moreover, individuals suffering from depression typically possess lower fitness levels when compared with normal populations, which presents challenges for exercise testing and prescription and increases the risk of dropout. Additionally, while it has been widely assumed that the mental health benefits of physical activity are closely tied to increases in cardiorespiratory fitness, the correlation of increased aerobic capacity with reduced symptoms of anxiety and depression has been found to be only modest ($r = .20$-$.30$), indicating that a vigorous regimen does not necessarily enhance psychological health.

The minimal exercise dose necessary to provide effective treatment was examined by Dunn and colleagues (2005), who randomly assigned 80 adults diagnosed with major depressive disorder to 12-week programs of either a minimal activity placebo condition (flexibility training three times a week) or one of four aerobic activity programs that varied by energy expenditure (low: $7.0 \text{ kcal} \cdot \text{kg}^{-1} \cdot \text{week}^{-1}$; moderate: $17.5 \text{ kcal} \cdot \text{kg}^{-1} \cdot \text{week}^{-1}$) and frequency (three or five days a week) based on public health recommendations. The reduction in depression was significantly ($p > .05$) greater for the moderate energy expenditure condition compared with the low expenditure or the minimal activity placebo condition. Interestingly, exercise frequency had no effect on depression. These results led the authors to conclude that major depressive disorder can be effectively treated with an exercise prescription of moderate dosage based on public health recommendations.

Exercise and Anxiety

Anxiety disorders are the most common form of emotional illness. Large-scale studies indicate that in a 12-month period, 17% of adults in the United States will be afflicted by an anxiety disorder; the most common of these is major anxiety disorder. Other anxiety illnesses include phobias, panic disorder, obsessive–compulsive disorder, and posttraumatic stress disorder (American Psychiatric Association 1994). Despite the scope of anxiety disorders, relatively few studies have tested the efficacy of exercise as a potential treatment compared with the scope of research on exercise and depression (O'Connor, Raglin, and Martinsen 2000).

The first meta-analysis examining the influence of exercise on anxiety was conducted by Petruzzello and colleagues (1991), who reported moderate but significant effect sizes for both acute (0.24) and chronic exercise (0.34) interventions. Because of the paucity of studies, results were collapsed across clinical and nonclinical samples, but a recent meta-analysis (Herring, O'Connor, and Dishman 2010) of the exercise literature in patients with chronic illnesses revealed that exercise programs were associated with a comparable overall effect size of 0.29. In statistics, an effect size is a measure of the strength of the relationship between two variables that provides an estimate of the magnitude of the relationship. Hence, while these studies show comparable results, the relationship, while significant, appears to be moderate.

The results of the Petruzzello meta-analysis indicated that the benefits of exercise were greater than those associated with control conditions. While the inclusion of control conditions is a standard aspect of experimental design, exercise research has been criticized for not including conditions with documented mental health benefits. In a pioneering study that contrasted exercise with Bensonian relaxation—a technique with empirically demonstrated antianxiety properties—Bahrke and Morgan (1978) found that each condition resulted in a similar decrement in state anxiety. Unexpectedly, a control treatment consisting of seated rest in a sound-dampened room was associated with an anxiety reduction equal to that in the other conditions. This led to the hypothesis that exercise reduces anxiety and improves mood because of mental diversion rather than because of physiological changes due to exertion, an explanation referred to as the distraction hypothesis.

Dosage

Knowledge of programmatic exercise factors such as mode or intensity is a prerequisite for establishing evidence-based exercise prescriptions for mental health. However, in a comprehensive review of the exercise and mental health literature, Buckworth and Dishman (2002) found that studies conducted before 1993 on the antianxiety effects of exercise rarely quantified exercise intensity; those that did so relied on indirect and inexact methods (i.e., age-predicted exercise heart rate). Studies comparing different exercise modes were even rarer. Definitive information on exercise intensity is also important because it has been contended that vigorous physical activity not only is ineffective for improving mood in anxiety patients but also could worsen symptoms or provoke panic attacks. However, carefully conducted research on this topic does not support these suppositions. Studies in which exercise intensity is accurately regulated and anxiety is measured repeatedly following exercise indicate that although high-intensity physical activity (>70% $\dot{V}O_2$max) often results in elevations in state anxiety immediately postexercise, after another 10 to 15 min in most participants, state anxiety decreases to a value similar to that found with less intense exercise bouts.

The widespread reluctance to subject individuals with anxiety disorders—particularly panic disorder—to exercise regimens was to a large extent a consequence of the widely cited study by Pitts and McClure (1967), who reported that infusions of

Few studies on the effect of exercise on decreasing anxiety have focused on specific exercise prescriptions.

sodium lactate precipitated panic attacks in patients with anxiety disorders. This work has often been interpreted as suggesting that vigorous exercise and the concomitant increase in blood lactate are likely triggers of panic attacks. But in a study on the ability of anxiety patients to tolerate high-intensity exercise, Martinsen and colleagues (1998) had 35 individuals diagnosed with panic disorder undergo supramaximal (~120% $\dot{V}O_2$max) bouts of bicycle ergometry. Only one participant (4%) experienced a panic attack during exercise, and this individual successfully completed the exercise protocol. On the basis of these findings and the results of similar work, the authors concluded that high-intensity exercise resulting in significant elevations in blood lactate (mean = 10.7 mmol/L) is not a trigger for panic attacks.

In perhaps the first published study to directly test the consequences of exercise in clinically diagnosed anxiety patients, Martinsen and colleagues (1998) found that eight weeks of walk–run exercise or a mild stretching program each significantly reduced anxiety symptoms. None of the patients experienced a panic attack during exercise, and the low dropout rate of 11% indicated that the regimens were well tolerated by the participants. Contrary to the authors' expectations, improvements in symptoms were not associated with fitness gains, indicating that a relatively light exercise prescription ineffective for improving aerobic capacity is equal to more vigorous activity in its ability to reduce symptoms.

In contrast, high-intensity exercise may mitigate antianxiety responses in the case of anaerobic exercise. Studies on acute strength training performed at moderate or hard intensities indicate that state anxiety either does not fall below baseline or that reductions are delayed by 90 to 120 min following the cessation of exercise (Raglin 1997) except in those participants who have elevated baseline levels of state anxiety. Mild strength training bouts are more frequently associated with reductions in state anxiety, and strength training across a wide range of intensities has been found to result in increased feelings of energy and reduced fatigue.

The best randomized study involving clinically diagnosed anxiety patients remains a study conducted by Broocks and colleagues (1998). Forty-six patients diagnosed with panic disorder with or without agoraphobia were assigned to 10 weeks of either a moderate walking–running exercise regimen, antianxiety medication (clomipramine), or a placebo pill. Symptoms were assessed using a variety of clinical measures, and it was found that the exercise program reduced anxiety symptoms more effectively than a placebo but did not achieve the same level of benefit as the medication. In addition, exercise took longer to relieve symptoms. No cases of panic attacks were reported during exercise participation, and the 31% dropout rate for exercise compares favorably to the 50% average noted in the general literature.

Chronic Exercise and Anxiety

The literature on the effects of chronic exercise programs on anxiety continues to be sparse compared to the literature on depression and exercise, but reviews of the extant research indicate that long-term participation in physical activity programs is associated with modest reductions in trait anxiety (Petruzzello et al. 1991; Raglin 1997; Herring, O'Connor, and Dishman 2010). Aerobic and anaerobic regimens appear to be associated with similar benefits, although direct comparisons between exercise modes within studies are lacking.

Research has shown that acute aerobic exercise completed at light to hard intensities consistently results in reductions in state anxiety that persist for several hours. Long-term exercise programs involving either aerobic or anaerobic activities have been found to be associated with reductions in symptoms among clinically depressed or anxious individuals that are near or equal to the benefits of medication.

Exercise and Schizophrenia

While the relationships between exercise and improvements in anxiety and depression have received considerable attention, the effects of exercise on individuals suffering from schizophrenia are less well understood. However, the relatively few studies (Adams 1995; Beebe et al. 2005; Faulkner and Biddle 1999; Fogarty and Happell 2005) in this area suggest the possible efficacy of exercise in the treatment of the physical and mental comorbidities of schizophrenia. Preliminary research has shown that exercise is effective in reducing both anxiety and depressive symptoms in schizophrenics. For instance, Beebee and colleagues (2005) reported significant reductions in levels of both anxiety and depression following a 16-week exercise program of treadmill walking. Additionally, a major health concern for schizophrenics is obesity. While effective pharmacological management of schizophrenia is often required, a serious side effect of prescribed medication is significant weight gain, which poses additional negative health consequences (Fogarty and Happell 2005). Hence, an additional benefit of exercise for schizophrenia patients has been the reduction of body fat through individually based aerobic exercise programs (Beebe et al. 2005; Fogarty and Happell 2005).

Although the extant studies in this area do indicate the potential for exercise to serve as an effective adjunct to the successful treatment of schizophrenia, further work is needed before definitive conclusions can be drawn. For example, Beebe and colleagues noted that existing studies may be affected by factors such as the adequate diagnosis of schizophrenia or individualized response to exercise protocols. They cited other factors such as motivation, type of pharmacological intervention, and baseline functioning, all of which may detract from the ability to generalize these studies across populations.

All these research results indicate that exercise provides important benefits to patients suffering from mild to moderate depression or anxiety. In the case of depression, the benefits of exercise appear to equal those associated with medication, whereas available evidence indicates that medication provides somewhat greater relief for anxiety patients with panic disorder. The modest exercise prescription required for alleviating symptoms of moderate anxiety or depression is well tolerated by individuals with these conditions, and there is some evidence that exercise may be particularly effective in reducing the risk of relapse. Finally, the rates of exercise adherence in most studies of clinical samples are actually higher than those observed for the general population.

Putative Mechanisms for the Psychological Benefits of Exercise

While there is consistent evidence indicating that exercise is strongly associated with improvements in mental health, the underlying mechanism for this relationship remains unknown, and thus a causal link cannot yet be established. This section presents a brief overview of the current physiological and psychosocial mechanisms that have been advanced as explanations for the psychological benefits of exercise. A variety of neurobiological, physiological, and cognitive processes have been hypothesized to explain these benefits. Of the neurotransmitters and neurohormones that have been proposed to contribute to exercise-mediated improvements in mood, no hypothesis has captured the attention of the public or scientific community like the endorphin hypothesis. Endorphins, which include β-endorphins and enkephalins, are produced by the brain, pituitary glands, and other bodily tissues and act as natural opiates that produce analgesia by binding to opiate receptor sites involved in perceptions of pain. Endorphins have been implicated in reward mechanisms and positive emotion. Physical stressors such as exercise can stimulate the production of endorphins, and this has led to the belief that endorphins are responsible for the popular but unsubstantiated phenomenon known as the "runner's high." Recent reports that appear to provide support for the involvement of endorphins in the exercise high and general mental health benefits have been compellingly criticized by Dishman and O'Connor (2009) on the basis of both methodological weaknesses and theoretical limitations. Moreover, research also has not shown endorphin levels following exercise to be correlated with mood improvements, and the use of pharmacological blocking agents such as naloxone has provided only mixed support for the mood-altering effects of endorphins. Because of the lack of direct evidence, most investigators now favor alternative hypotheses.

> Both physiological and cognitive explanations have been used in an attempt to explain improvements in mental health as a result of exercise participation. However, the exact mechanism responsible for this relationship is unknown.

The monoamine hypothesis is based on the knowledge that neurotransmitters such as norepinephrine, dopamine, and serotonin (5-HT) play a role in both depression and various forms of schizophrenia. It is believed that physical activity may alter mood state by increasing the production or upregulation of these brain monoamines. However, exercise research on the monoamine hypothesis has been constrained because of the difficulty of accurately measuring levels of hormones such as norepinephrine in the brain. Animal studies have shown increases in central and peripheral levels of monoamine hormones following exercise, and other work has shown alterations in receptor sensitivity of monoamine neurons in the brain; but further research needs to be conducted to validate these changes in humans (Buckworth and Dishman 2002).

Another popular hypothesis revolves around the changes in internal core temperature, which can increase by several degrees Fahrenheit during exercise and may stay at higher levels for some time following the cessation of activity. According to the so-called thermogenic hypothesis, the increase in body temperature that occurs with vigorous exercise can stimulate neurological changes associated with improved mood; but studies examining this hypothesis have not definitively linked temperature to the mood changes that occur following exercise. One continuing challenge to this line of research is the difficulty of attenuating the temperature rise following exercise.

Finally, cognitive and social factors have been proposed to contribute to improved mood following exercise, for example social interaction and support; feelings of achievement, self-mastery, or self-efficacy; and distraction or diversion of attention. The distraction hypothesis posits that exercise imparts psychological benefits because of the simple fact that the exercise setting itself is generally removed from the workplace or other stressful environments, resulting in a form of pleasant diversion. Another popular cognitive explanation is that feelings of achievement or personal mastery resulting from successfully completing an exercise bout, reaching an exercise or sport goal, or adhering to a long-term exercise program may contribute to improved mood and mental health.

To date, tests of these and other hypothetical mechanisms for the psychological benefits of exercise have failed to provide unequivocal support for any single explanation. Given the complexity of physical activity behavior and the various psychological responses to exercise, it is likely that no particular mental health effect of exercise can be adequately reduced to a single process. The psychological responses to exercise may involve complex interactions of or even different mechanisms for acute and chronic activity. While the question of the mechanism or process responsible for exercise-mediated improvements in mental health may seem purely academic, a greater understanding would have important practical consequences for establishing optimal exercise prescriptions for enhancing mental health, either as a primary treatment or as an adjunct to other therapies. Table 21.1 summarizes the proposed mechanisms.

Detrimental Psychological Responses to Exercise: The Overtraining Syndrome

While regular participation in physical activity routinely results in desirable changes in mental health, there is evidence that in specific circumstances it can worsen aspects of emotional health and result in undesirable behavioral changes. Studies of

TABLE 21.1 Proposed Biological and Psychosocial Mechanisms for the Psychological Benefits of Exercise

Mechanism	Hypothesis	Comment
Biological	Thermogenic	This hypothesis proposes that changes in body temperature that occur during exercise are associated with increased central and peripheral neuron activity in the brain, as well as decreased muscle tension. These physiological changes enhance mood state (i.e., reduce state anxiety).
	Monoamine	Neurotransmitters such as norepinephrine, dopamine, and serotonin play a role in both depression and various forms of schizophrenia. Monoamine oxidase inhibitors have been used for several decades to treat depression, although it is not clearly understood how these inhibitors function. Physical activity may alter mood states through its effect on one or all three of these neurotransmitters. Unfortunately, it is impossible to accurately measure levels of these hormones in the brain.
	Endorphins	Endorphins, the most prominent of which are β-endorphins and enkephalins, are chemical substances produced by the brain, pituitary glands, and other bodily tissues. Endorphins act as natural opiates by binding to opiate receptor sites involved in perceptions of pain, and they have been implicated in reward mechanisms and positive emotion. Physical stressors such as exercise can stimulate the production of endorphins, and this has led to the belief that endorphins are responsible for the popular but unsubstantiated phenomenon known as the "runner's high." Research has not shown endorphin levels following exercise to be correlated with mood change, and the use of pharmacological blocking agents such as naloxone has provided only mixed support for the mood-altering effects of endorphins.
Psychosocial	Distraction	Psychological benefits associated with exercise may not be related to underlying physiological mechanisms but rather to the simple fact that participation in physical activity typically occurs in a setting removed from the workplace or other stressful environments. Hence, a person is distracted from potential stressors and provided with a pleasant diversion.
	Mastery	Feelings of achievement or personal mastery result from the successful completion of tasks. In the case of exercise, the successful completion of an acute bout of exercise or a long-term exercise regimen improves mood and mental health through enhancing an individual's self-efficacy.

competitive men and women athletes in sports such as swimming and distance running reveal that the intensive training they must undergo to achieve peak conditioning is associated with mood disturbances, despite the fact that these athletes typically possess superior mental health profiles in the off-season or during routine maintenance training (see figure 21.1) (Raglin and Wilson 2000). Moreover, research indicates that the training load (volume or intensity) is directly associated with the magnitude of mood disturbance in a **dose–response relationship** such that increases in training load result in concomitant elevations in negative moods including anxiety, depression, and anger, as well as a drop in desirable moods such as vigor. At the peak of training, mood disturbances are at their greatest, with scores that typically exceed the population norm. Conversely, reductions in training (i.e., tapers) result in reductions in negative mood states and elevations in vigor, such that when training volumes are at their lowest, the mood profiles of most athletes resemble desirable preseason values.

■ Some athletes who engage in intense physical training for their sport experience a condition known as overtraining syndrome or staleness, which results in a long-term loss of physical capacity. Overtraining syndrome is commonly associated with clinical depression and other mood disturbances.

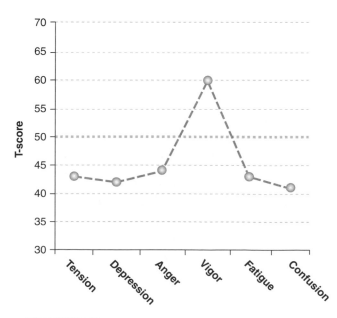

FIGURE 21.1 An example of the "iceberg profile" of mood states typically found in successful athletes.

The majority of athletes can tolerate the physical and emotional stresses of intense training, but between 5% and 10% respond adversely, exhibit more severe mood disturbances, and suffer from substantial performance decrements that persist for weeks or even months. This condition is commonly referred to as overtraining syndrome or staleness, and it has been observed in endurance athletes as well as in competitive nonendurance sports that require high levels of physical conditioning (see figure 21.2). Clinical depression is the most common psychological manifestation of overtraining syndrome, and many athletes are afflicted to a degree that requires appropriate treatment such as counseling or antidepressant medication. It is unclear why some athletes are more prone to developing overtraining syndrome while others who undergo the same training never develop the disorder; but researchers agree that its primary cause is the physical and mental stress of intensive training.

Although detrimental consequences of intensive training and the overtraining syndrome are seldom considered or even acknowledged by researchers who study the psychological benefits of physical activity, this work has potential relevance to the broader literature because it reveals physical activity to be a complex psychobiological phenomenon associated with either beneficial or detrimental psychological responses. Obviously, the psychological outcome is largely dictated by the dosage of physical activity, but it is also evident that a significant factor is individual differences between athletes in their ability to adapt to intensive training.

Summary

A substantial body of research has substantiated long-held beliefs in the beneficial effects of physical activity for mental health. For persons who are free from emotional disorders, the primary psychological benefits occur following acute exercise. Brief bouts of aerobic exercise completed at intensities ranging from mild to intense consistently result in decreases in state anxiety as well as significant mood improvements that last for several hours, whereas for anaerobic activities such as weight training, these benefits may be limited to light to moderate intensities. Studies on the psychological effects of chronic exercise have shown it to be efficacious for cases of mild to moderate depression and anxiety, with benefits that approach or equal those of psychopharmacological medication and therapy. Moreover, the minimal exercise prescription required to yield these benefits has been shown to be moderate, which is an

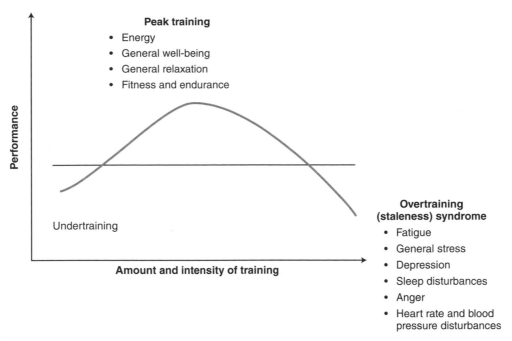

FIGURE 21.2 Although most athletes can cope with high levels of training, a minority of them experience the negative symptoms of the overtraining syndrome if intense training is prolonged.

important consideration given that many depressed or anxious individuals are sedentary and below average in fitness. Additional research, however, is needed to provide more specific information as to what exercise prescriptions are most effective, as well as to test the potential for exercise to benefit other forms of psychological illness. Given evidence that emotional disorders such as anxiety and depression have a significant impact on lifestyle diseases such as coronary heart disease, this line of research has important implications for a modern society that has become increasingly sedentary.

Key Concepts

acute exercise bout—A single exercise session. In the case of acute exercise, transitory psychological emotions known as "states" are assessed.

chronic exercise—Long-term participation in physical activity for a period of weeks or months. In research involving chronic exercise, more stable psychological measures called "traits" are examined.

disorder—A chronic condition versus an episode, which is a single acute instance.

dose–response relationship—The relationship that exists between exercise volume and psychological response. There appears to be a point at which this positive association may become detrimental.

states—Transitory emotions that reflect how a person feels "right now."

traits—Stable emotions that reflect how a person feels "in general."

Study Questions

1. Explain how trait and state anxiety are affected differently by acute and chronic exercise.

2. Discuss the Pitts-McClure hypothesis and explain why these findings have hampered clinical research involving exercise and patients suffering from anxiety disorders.

3. Describe the research examining the effects of exercise as a treatment for depression. Why is it difficult to establish definitive exercise recommendations for individuals suffering from depression?

4. List the three biological and two psychological theories used to explain the affective benefits of exercise. Why must these theories be considered "associational" rather than causative?

5. Explain the dose–response relationship between exercise volume and intensity and mental health. How is staleness related to the dose–response relationship?

References

Adams, L. 1995. How exercise can help people with mental health problems. *Nursing Times* 91:37-39.

American Psychiatric Association. 1994. *Diagnostic and statistical manual of mental disorders.* 4th ed. Washington, DC: American Psychiatric Association.

Bahrke, M.S., and W.P. Morgan. 1978. Anxiety reduction following exercise and meditation. *Cognitive Therapy and Research* 2(4):323-333.

Beebe, L.H., L. Tian, N. Morris, A. Goodwin, S.S. Allen, and I. Kuldau. 2005. Effects of exercise on mental and physical health parameters of persons with schizophrenia. *Issues in Mental Health Nursing* 26:661-676.

Blumenthal, J.A., M.A. Babyak, P.M. Doraiswamy, L. Watkins, B.M. Hoffman, K.A. Barbour, S. Herman, W.E. Craighead, A.L. Brosse, R. Waugh, A. Hinderliter, and A. Sherwood. 2007. Exercise and pharmacotherapy in the treatment of major depressive disorder. *Psychosomatic Medicine* 69:587-596.

Blumenthal, J.A., M.A. Babyak, K.A. Moore, W.E. Craighead, S. Herman, P. Khatri, R. Waugh, M.A. Napolitano, L.M. Forman. M. Appelbaum, P.M. Doraiswamy, and K.R. Krishnan. 1999. Effects of exercise training on older patients with major depression. *Archives of Internal Medicine* 159:2349-2356.

Broocks, A., B. Bandelow, G. Pekrun, A. George, T. Meyer, U. Bartmann, U. Hillmer-Vogel, and E. Rüther. 1998. Comparison of aerobic exercise, clomipramine, and placebo in the treatment of panic disorder. *American Journal of Psychiatry* 155:603-609.

Buckworth, J., and R.K. Dishman. 2002. *Exercise psychology.* Champaign IL: Human Kinetics.

Dishman, R.K., and P.J. O'Connor. 2009. Lessons in exercise neurobiology: The case of endorphins. *Mental Health and Physical Activity* 2:4-9.

Dunn, A.L., M.H. Trivedi, J.B. Kampert, C.G. Clark, and H.O. Chambliss. 2005. Exercise treatment for depression: Efficacy and dose response. *American Journal of Preventive Medicine* 28:1-8.

Faulkner, G., and S. Biddle. 1999. Exercise as an adjunct treatment for schizophrenia: A review of literature. *Journal of Mental Health* 8:441-457.

Fogarty, M., and B. Happell. 2005. Exploring the benefits of an exercise program for people with schizophrenia: A qualitative study. *Issues in Mental Health Nursing* 26:341-351.

Herring, M.P., P.J. O'Connor, and R.K. Dishman. 2010. The effect of exercise training on anxiety symptoms among patients: A systematic review. *Archives of Internal Medicine* 170:321-331.

Martinsen, E.W., J.S. Raglin, A. Hoffart, and S. Friis. 1998. Tolerance to intensive exercise and high levels of lactate in panic disorder. *Journal of Anxiety Disorders* 12(4):333-342.

McNair, D.N., M. Lorr, and L.F. Droppleman. 1992. *Profile of Mood States manual.* San Diego: Educational and Industrial Testing Service.

Morgan, W.P. (ed.). 1997. *Physical activity and mental health.* Washington, DC: Taylor & Francis.

Morgan, W.P., J.A. Roberts, F.R. Brand, and A.D. Feinerman. 1970. Psychological effects of chronic physical activity. *Medicine and Science in Sports and Exercise* 2:213-217.

O'Connor, P.J., J.S. Raglin, and E.W. Martinsen. 2000. Physical activity, anxiety, and anxiety disorders. *International Journal of Sport Psychology* 31:136-155.

Petruzzello, S.J., D.M. Landers, B.D. Hatfield, K.A. Kubitz, and W. Salazar. 1991. A meta-analysis on the anxiety reducing effects of acute and chronic exercise: Outcomes and mechanisms. *Sports Medicine* 11(3):143-182.

Pitts, F.J., and J.J. McClure. 1967. Lactate metabolism in anxiety neurosis. *New England Journal of Medicine* 277:1329-1336.

Raglin, J.S. 1997. Anxiolytic effects of physical activity. In *Physical activity and mental health*, ed. W.P. Morgan. Washington, DC: Taylor & Francis.

Raglin, J.S., and G.S. Wilson. 2000. Overtraining in athletes. In *Emotions in sport*, ed. Y.L. Hanin. Champaign, IL: Human Kinetics.

Spielberger, C.D., R.L. Gorsuch, R.E. Lushene, P.R. Vagg, and G.A. Jacobs. 1983. *Manual for the State-Trait Anxiety Inventory (Form Y).* Palo Alto, CA: Consulting Psychologists Press.

PART V

How Much Is Required and How Do We Get There?

An important question frequently asked by physical activity and health professionals, as well the general public, relates to the amount and type of activity necessary to produce specific health outcomes. Part V of the book deals with a variety of issues raised by this question. A sound scientific basis is needed for recommending a specific activity regimen to improve a person's health. In parts II through IV, various biological and clinical responses to changes in chronic physical activity status were presented. Given the numerous health outcomes influenced by physical activity and the range of possible activities, the optimal amount and type of activity are still under investigation. In chapter 22, evolving concepts regarding exercise dose and health response, referred to as "dose response," are presented, along with a discussion of the interplay between physical fitness and health. Chapter 23 provides some insights into the process used and the challenges encountered in the development of national and international physical activity and health guidelines or recommendations. This is an important topic since the body of knowledge is constantly evolving and continuously informs the process of converting scientific advances into practical and efficacious public health guidelines.

22

Dose–Response Issues in Physical Activity, Fitness, and Health

William L. Haskell, PhD

CHAPTER OUTLINE

There is good documentation that moderate amounts of physical activity added to the routine activities of daily living or moderate levels of physical fitness provide meaningful health benefits. Also, evidence indicates that additional activity and higher fitness levels generally confer somewhat greater benefits. What is still at issue is the more precise dose–response relationship for specific health outcomes, taking into account the activity profile (type, intensity, frequency, duration, and amount), a number of characteristics of the person (gender, age, health history, prior exercise training status, heredity), and the major health outcomes desired. Such data are critical to designing and implementing scientifically based recommendations to achieve desired outcomes safely and cost-effectively in specific populations. This chapter addresses some of the key issues regarding the concept of dose response and presents data that support current physical activity guidelines for disease prevention and health promotion that have been published by major health organizations.

Many people have questions about using physical activity to enhance their health or physical performance. How much activity is enough? How little activity can I get away with? How physically fit do I need to be to be healthy or to help prevent a specific disease? To address such questions, exercise scientists and practitioners have attempted to determine the characteristics of physical activity or level of fitness required to produce meaningful changes in specific health and performance outcomes. *Dose* is the term frequently used to represent the characteristics of the activity (such as type, intensity, frequency, duration, and amount) that are important in producing improvements in health or performance. The terms *response* or **health response** are used to represent the various changes that occur when a specific dose of activity is performed. Thus, for physical activity and health, *dose response* refers to the relationship between the characteristics of the activity performed and the nature of the health-related changes produced. For physical fitness and health, dose response typically refers to the relationship between levels of fitness as defined by a designated measure (e.g., MET capacity on a treadmill or one-repetition maximum [1RM] on bench press) and a specific health outcome, such as the development of arthritis or death attributable to coronary heart disease (CHD).

Principles Guiding the Body's Response to Activity

To better appreciate dose–response issues, it is useful to understand several exercise training principles.

These principles form much of the scientific basis for the development of activity plans or exercise prescriptions to improve health and performance. First is the principle of **overload**, which refers to the fact that when a tissue, organ, or system is stressed by an increase in physical activity, it typically responds by increasing its capacity or efficiency. To detect this response, a person may have to perform the exercise bout for a number of days or weeks. But if the activity is greater in intensity or amount than typically performed by this person, the body slowly adapts. For example, in response to an increase in walking speed or distance, components of the oxygen transport and utilization systems slowly increase their capacity (e.g., increase in stroke volume and skeletal muscle capillary density) or efficiency (decrease in heart rate and systolic blood pressure). This overload principle applies to many of the tissues that are involved in supporting the increase in energy expenditure required for sustained activity or that are exposed to the mechanical forces produced by muscle contractions or gravity.

> The principles of overload, progression, and specificity are major determinants of how the body will respond to a dose of physical activity.

A second training principle that is important in establishing a health-promoting dose of activity is **progression**. For overload to produce health benefits rather than injury, it needs to be increased in small amounts or at a slow rate of progression. If the overload progresses too rapidly, it can produce strain or injury and create a negative health outcome. Thus, people who have been very inactive should begin to increase their activity levels by walking daily at increasing speeds over weeks or months rather than immediately starting a jogging program. However, if the overload is not increased and no progression in intensity or amount occurs, then very little or no improvement will take place.

Progression does not need to continue when the person has reached an appropriate goal of activity or fitness, such as 30 min of brisk walking on most days. Issues involving progression generally are addressed in detailed activity plans but not in the more general recommendations made to the public. Understanding progression scientifically and clinically is necessary to design and implement activity programs that maximize benefit but minimize injury and chronic fatigue.

The last of the three exercise training principles is **specificity**. The specific types of changes that take place in the body in response to an increase in activity depend substantially on the characteristics of the activity. That is, the changes that occur are specific to the nature and degree of the demands produced by the exercise. The most frequently recognized applications of specificity are the prescribing of resistance exercise to increase muscle size and strength and of endurance exercise to increase the capacity of the oxygen transport and utilization systems. However, specificity has other applications as well, including the need to exercise a specific muscle to maximize its strength or endurance (arm vs. leg exercise) or to stretch specific muscles and ligaments to increase the flexibility or range of motion of a particular area of the body.

Components of the Physical Activity Dose

When exercise dose in relation to health-associated outcomes is discussed, the exercise is characterized by its type, intensity, session duration, session frequency, and total amount per day or week. Recently the terms *activity profile*, *activity volume*, and **accumulation** have been included in these discussions. In this section, key features or issues related to each of these characteristics are discussed as they pertain to defining the health-related response to exercise.

Type of Activity

Because muscle contractions have both mechanical and metabolic properties, exercise has been classified by both of these characteristics, a situation that tends to cause some confusion. Mechanical classification stresses whether the muscle contraction produces movement of the limb or not: isometric (same length) or static exercise if no movement of the limb occurs, or isotonic (same tension) or dynamic exercise if movement of the limb occurs. A muscle contraction can be either concentric (shortening the muscle fibers) or eccentric (lengthening the muscle fibers). The metabolic classification primarily involves the immediate availability of oxygen for the contraction process and includes aerobic (oxygen available) or anaerobic (without oxygen) processes. Whether an activity is aerobic or anaerobic depends primarily on its intensity relative to the participant's capacity for that type of exercise. Most activities involve both static and dynamic contractions as well as aerobic and anaerobic metabolism. Thus, activities tend to be classified based on their dominant metabolic or mechanical characteristics. Other classifications of exercise include endurance (aerobic) versus strength (resistance), upper versus lower body (arms vs. legs),

and classification by purpose (occupation, household chores, self-care, transportation, recreation, leisure time, and physical conditioning).

■ **The major components used to define a dose of physical activity include activity type, intensity, session frequency, and session duration. The amount or volume of activity is determined by intensity × frequency × duration.**

A key issue about exercise type in terms of desired changes is the specificity of the response to a given type of exercise. Specificity refers to the fact that the biological changes that occur in response to exercise are dependent on (1) which tissues or systems are activated or stressed by the exercise and (2) the nature of the stress placed on these tissues or systems. If this activation or stress is of appropriate intensity and amount, the tissues or systems respond favorably by increasing their capacity or efficiency (training effect). If the activation or stress is too little (not more than that frequently experienced by the tissues or systems), then there is very little or no training effect or adaptation. If the activation or stress is too great, overuse injury or chronic fatigue can occur. A good example of specificity is the increase in muscle and bone mass seen in the dominant arm versus the nondominant arm of a professional tennis player as a result of playing tennis. The response of the dominant arm to hitting thousands of tennis balls is not transferred to the nondominant arm, which performs limited exercise just tossing the ball into the air. Vigorous endurance-type exercise such as running provides a major stimulus to the oxygen transport system, substrate (glycogen and fat) processing systems, and oxidative processes in leg skeletal muscle fibers, especially in slow-twitch (ST) fibers. These systems increase in efficiency or capacity in response to running and other endurance-type exercise. In contrast are the demands made by heavy resistance exercises, such as powerlifting, that primarily activate fast-twitch (FT) muscle fibers and stress the body's support structures, such as bones and connective tissue. Thus, when we describe the response to a dose of activity, the type of activity needs to be accurately defined.

Physical Activity Intensity

Intensity is a key factor when one considers the dose of physical activity required to achieve specific health outcomes. Not only does an increase in activity intensity play a major role in producing many favorable adaptations; it also has a key role in the various health risks resulting from activity (see chapter 18). The intensity of an activity can be described in both *absolute* and *relative* terms. In absolute terms, intensity is either the magnitude of the increase in energy required to perform the activity (endurance exercise) or the force produced by the muscle contraction (resistance or strength exercise). During endurance exercise, the increase in energy is usually determined by measurement or estimation of the increase in oxygen uptake, which is expressed in units of oxygen (liters or METs) or converted to an equivalent measure of heat produced or energy expended (kilocalories or kilojoules). The intensity can also be expressed as a walking or running speed or the work rate (power) on a cycle ergometer (watts). The force of the muscle contraction is measured by how much weight is moved (dynamic contraction) or the force exerted against an immovable object (isometric contraction) and usually is expressed as "work" in kilograms. If the work performed is expressed per unit of time—such as an oxygen uptake of 1.5 L/min—then the correct term is *power* and not *work*.

> When we consider the intensity component of an activity dose, we need to understand the differences between relative and absolute intensity. Relative intensity takes into consideration the exercise capacity of the person, whereas absolute intensity considers only the demands of the activity.

In relative terms, the intensity of the activity is expressed in relation to the capacity of the person performing the activity. For energy expenditure, the relative intensity usually is expressed as a percentage of the person's aerobic power, also termed maximal oxygen uptake ($\%\dot{V}O_2max$). Because the relationship between the increase in heart rate and the increase in oxygen uptake during dynamic exercise is nearly linear, "percentage of maximum heart rate" or "heart rate reserve" (HRR) (maximum heart rate minus resting heart rate) is also used as an expression of exercise intensity relative to the person's capacity. One advantage of using percentage of maximum heart rate when assigning exercise intensity to a participant is that the recommended value (i.e., 70-85%) can stay the same as the person becomes more fit (but the intensity of the activity will have to increase to reach this heart rate). This is not the case if an absolute intensity is assigned, such as starting at a brisk walk at 4.0 mph. After some weeks at this intensity, the person will have to increase the walking

speed to provide a stimulus of a similar magnitude to the cardiorespiratory system. For muscle force, the relative intensity of the contraction is expressed as a percentage of the maximal force that can be generated for that activity (percentage of maximum voluntary contraction or percentage of a 1RM).

In most experimental studies evaluating the effects of increased activity on various fitness and health outcomes, intensity is expressed relative to each person's capacity (e.g., 60-75% of $\dot{V}O_2max$). However, in most, if not all, large prospective observational studies, exercise intensity is expressed in absolute terms (no adjustment made for each person's exercise capacity). These differences in methodology prevent direct comparison of dose–response data from these two major sources of information. For an activity of a given intensity, such as walking at 3.0 mph or 4.8 km/h (3.3 METs), the relative intensity varies inversely to the aerobic capacity of the individual. As shown in figure 22.1, for highly fit people with an aerobic capacity of 14 METs, walking at 4.8 km/h has a relative intensity of 24% (light intensity); but for people with only a 4-MET capacity, the relative intensity is 83% (hard intensity). The percent of relative capacity required at 4.0 mph or 6.4 km/h (5.0 METs) also is displayed in figure 22.1. An example of activity classification by both absolute and relative intensity is provided in chapter 1, figure 1.4. Standardization of activity intensity classification is essential for accurately establishing the relationship between intensity and health or performance outcomes.

Over the past century, the intensity of numerous physical activities has been defined by the energy expenditure required to perform the activity. Table 22.1 lists the energy requirements for selected activities performed during home and self-care, transportation, occupation, leisure time, and physical conditioning (Ainsworth et al. 2011). There can be substantial interindividual variation in the energy cost of performing some activities, especially when the activity has a large skill component, such as swimming. However, for many activities like walking or jogging, the between-person variation in energy expenditure when adjusted for body weight is quite small, and thus the average values for such activities listed in table 22.1 apply to most adults.

For some health outcomes, it appears that the total amount of exercise performed is more important than the intensity (above some minimal level) at which the exercise is performed (Physical Activity Guidelines Advisory Committee [PAGAC] 2008). Establishing which outcomes can be achieved at lower activity intensities is important when a public health model is used to promote physical activity because risks of orthopedic injury or cardiac arrest

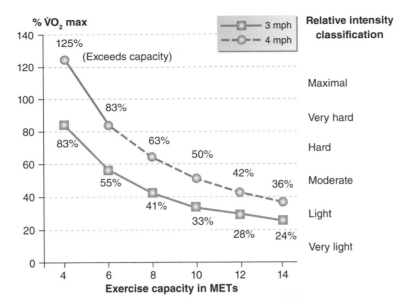

FIGURE 22.1 At a walking speed of 3.0 mph, the average energy cost for an adult is 3.3 METs. Here we see that for highly fit adults (aerobic capacity of 14 METs), walking at this speed on a flat surface requires only 24% of their aerobic capacity (light activity) whereas for adults with an exercise capacity of only 4 METs, such walking requires 83% of their capacity (very hard intensity). Similar data are provided for walking at 4.0 mph (5 METs).

are closely linked to higher intensities of activity and some people prefer to perform only moderate-intensity activity.

Activity Session Duration and Frequency

Discrete recommendations for session duration and frequency traditionally have been provided in the prescription of exercise that focuses on performing a defined bout a designated number of times per week, such as 20 min or more of vigorous exercise three or more times per week. The newer recommendations that focus on adding more activity back into the routine activities of daily living (the lifestyle approach) emphasize activity accumulation throughout the day: accumulating at least 30 min of activity on five or more days of the week. With resistance exercise, the duration or amount usually is not measured in time but by the number of exercises performed, the number of sets per exercise, and the repetitions per set.

When sedentary adults start a program of increased activity, it appears that even quite small amounts of moderate- or vigorous-intensity activity improve both aerobic capacity and strength. For example, a program of jogging 10 min per day three times a week for 12 weeks improved measures of cardiorespiratory fitness in middle-aged men (Wilmore et al. 1970). Church and colleagues (2007) demonstrated a dose response for an increase in aerobic capacity in women who walked and rode a stationary cycle at 50% of $\dot{V}O_2$max for an average of 72, 136, and 192

min per week. The results of other studies evaluating the effects of varying exercise session duration or frequency on health and performance outcomes generally support the concept that the greater the increase in the total amount of exercise (duration of session × frequency of the session × intensity), the greater the improvement.

Data are available to support this dose–response concept as long as the exercise intensity is at least moderate relative to the person's capacity and the total energy expended per session is in the range of 500 to 1,000 MET-minutes per week. On the low side, very limited scientific data support health or performance benefits from performing just light-intensity exercise (<40% of $\dot{V}O_2$max reserve) or expending less than 50 MET-minutes per day during planned activity. On the high end, few data are available on the level of health benefits achieved when people consistently exercise at very hard intensities (≥85% of $\dot{V}O_2$max reserve) or expend 900 MET-minutes a day during vigorous-intensity endurance exercise.

Factors Determining Optimal Activity Dose

The interplay of the many factors that one should consider when determining the optimal activity dose for a specific individual can be extremely complex. In this section, we consider issues that should be taken into account in determining the optimal activity dose for oneself, a client, or a patient.

TABLE 22.1 Estimated Energy Cost (METs) of Selected Physical Activities

Activity	METs
Household and self-care	
Clearing table and washing dishes (with some walking)	2.5
Ironing	1.8
Cleaning house—general	3.5
Vacuuming	3.3
Scrubbing floors on hands and knees	3.5
Sweeping garage, sidewalk, other outdoor locations	4.0
Automobile repair at home	3.3
Carpentry, finishing or refinishing furniture	3.3
Carrying, stacking, or unloading wood	5.5
Mowing lawn—walking with power mower	5.0
Occupational	
Data entry, typing on computer while sitting	1.5
Machine tooling—welding	3.0
Masonry—concrete	4.5
Forestry—general	8.0
Shoveling—heavy, more than 16 lb/min	8.5
Transportation	
Bicycling—leisurely, <16 km/h (<10 mph)	4.0
Bicycling—vigorously, 22.4-25.4 km/h (14.0-15.9 mph)	10.0
Walking briskly—6.8 km/h (3.5 mph)	4.3
Recreational, leisure time, sports	
Playing piano	2.3
Dancing (ballroom, slow—waltz, foxtrot)	3.0
Dancing (ballroom, fast—disco, line dancing, folk)	7.8
Jogging—9.6 km/h (6 mph)	9.8
Running—12.8 km/h (8 mph)	11.8
Tennis—singles	8.0
Skateboarding—general	5.0
Conditioning	
Stretching (hatha yoga)	2.5
Calisthenics (light to moderate effort)	3.5
Calisthenics (heavy, push-ups, sit-ups, jumping jacks)	8.0
Aerobic dancing (with 6- to 8-in. step)	8.5
Weightlifting (light weights)	5.0
Weightlifting (heavy weights—vigorous effort)	6.0
Circuit training with some aerobic activities	8.0

Data from Ainsworth et al. 2011.

Accumulation of Physical Activity Throughout the Day

The concept of accumulating physical activity by performing multiple short bouts (8-10 min each) throughout the day was first included in major guidelines in 1995 (Pate et al. 1995). These recommendations were based on (1) indirect data from prospective observational studies relating amount of activity to CHD, cardiovascular disease, and all-cause mortality; and (2) results of several experimental studies that evaluated the effects of a single longer bout per day versus three or four shorter bouts per day (total exercise performed held constant) on aerobic capacity, plasma high-density lipoprotein (HDL) cholesterol concentration, and body weight. Data from these studies indicated that performing multiple short bouts of moderate- to vigorous-intensity activity produced results not too dissimilar from those produced by the longer bouts. Additional evidence supporting these initial results has been reported (Hardman 2001). Data from experimental studies evaluating bouts of activity less than 8 to 10 min in duration are still limited. However, Macfarlane and colleagues (2006) reported that multiple short bouts (approximately 6 min each) of moderate-intensity endurance exercise spread throughout the day produced increases in cardiorespiratory fitness similar to those with exercise performed in bouts of approximately 30 min each. These types of data are needed so that we can better determine what the minimum duration of

an activity bout needs to be for inclusion in "daily accumulated activity."

Because it now appears that the accumulation concept is valid, we must consider the role of the amount or volume of activity when defining dose response. Volume is the product of exercise intensity, duration, and frequency, usually expressed in some unit of energy expenditure (e.g., kilocalories per day, MET-minutes per day). If the key stimulus for certain health benefits is activity volume above some minimal intensity threshold, then a wide variety of activity profiles can be encouraged for promoting health. Figure 22.2 depicts four different exercise profiles that provide a similar exercise volume expressed in total energy expended (720 MET-minutes per week). These various profiles of activity, if reasonably equal in effectiveness, provide a wide range of opportunities for achieving an adequate volume of activity to maintain or improve health.

Unresolved dose–response research issues regarding the role of session duration and frequency include the following:

- How short can a bout of endurance-type exercise be and still contribute to specific health outcomes (e.g., does stair climbing for 1 min per session 20 times per day provide benefits similar to a single session per day of 20 min)?

- Do sessions of long duration (75 min) on two days per week provide benefits similar to 30-min sessions performed five days per week (150 min of exercise per week in both cases)?

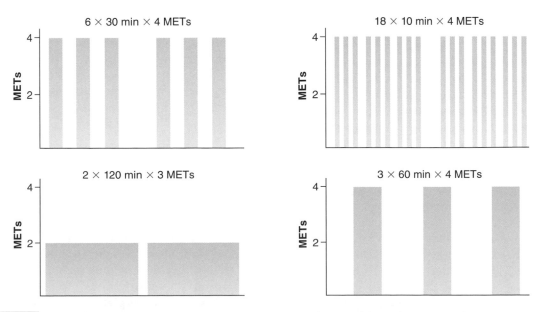

FIGURE 22.2 Examples of different physical activity regimens that would result in expending approximately 720 MET-min per week. The examples include performing three sessions per day (4 METs) for 10 min each, six days per week, and one session per day (3 METs) for 120 min each on just two days per week.

- Do multiple short bouts of exercise spread throughout the day provide more benefit than a single longer bout of equal amount, resulting from repeated acute effects of exercise on specific health outcomes?

Minimal Versus Maximal Dose

Much of the public would like to know the minimal amount or intensity of physical activity needed to produce significant health benefits. From a public health perspective, this presents a problem because most likely the answer depends in a significant way on a person's current level of physical activity or fitness. For example, a person who has been on bed rest for more than several days will benefit significantly by just sitting upright in a chair or standing and moving about slowly for a few minutes several times a day. Also, sedentary older persons appear to benefit from even short bouts of slow walking (<4.8 km/h or 3 mph) or light-intensity activity such as tai chi. Some data from epidemiological studies indicate that there are favorable health outcomes from moderate-intensity activity performed for no more than 10 min per day. Although small increases in activity appear to provide some benefit, especially in the very inactive or unfit, we have very little data on this issue in real-life situations for the general public. See chapter 4 for a discussion of the possible health risks associated with sedentary behavior and why some activity appears to be better than just sitting for most of the day.

Also, at the high end of the **physical activity dose** continuum, we still lack data that define the point where an increase in activity provides no additional health benefits. This would be the point of maximum benefit for a specific health outcome. Documenting such an activity dose would help answer the question, "How much activity do I need to perform to obtain all of the health benefits provided by being physically active?" For some people, this amount of activity appears to be quite high. For example, there is a dose–response relationship between distance run per week and cardiovascular disease risk factors in male runners even up to 80 km (50 mi) per week (Williams 1997). Figure 22.3 shows a positive dose response between distance run per week up to 80 km per week and the percentage of men with clinically defined low levels of HDL cholesterol considered at increased risk for developing CHD. Also, there is a negative dose response between distance run and percentage of men with clinically high levels of HDL considered to be at decreased risk for developing CHD.

Somewhere between the minimal and maximal doses of activity that provide health benefits there is

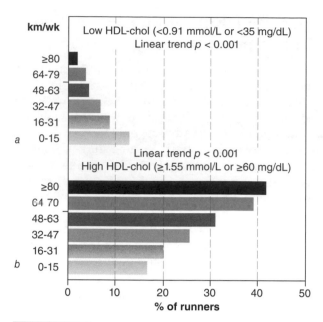

FIGURE 22.3 Dose–response relationships between the distance run per week and the prevalence of (a) low (bad) values and (b) high (good) values of high-density lipoprotein (HDL) cholesterol in men who are regular runners. As the distance run increases, the percentage of men with low levels of HDL decreases whereas the percentage of runners with high levels significantly increases.

likely an optimal dose. This is the dose of activity that would provide the greatest net health benefits for the least amount of time and effort spent in performing the activity. The optimal net health benefit is reached when the difference between positive and negative health outcomes is maximized. Based on what we currently know about dose response, an optimal dose will vary from one health benefit to another and from person to person. This interindividual variation will have a genetic component and could be influenced by factors such as gender, age, exercise capacity, nutritional status, and overall health.

Use of Threshold Values for Dose–Response Recommendations

Studies addressing dose response that have sufficient data to evaluate various amounts of activity generally indicate that the relation between activity and positive health outcomes is more a continuous relationship than a threshold effect (Manson et al. 2002). In some studies, only a few categories of activity amount are provided, leading to the appearance of a threshold effect (Manini et al. 2006). Also, the way in which some guidelines or recommendations have been written makes it appear that a minimum threshold of activity needs to be achieved before any benefit occurs, or that above some amount of activity no further benefit takes place. This situation is not surprising, because it is very similar to the situation with guidelines addressing the relationship between other major chronic disease risk factors (e.g., blood pressure, cholesterol, obesity) and health outcomes. For example, it is well established that the relationship between level of blood pressure and stroke risk is continuous from very low levels of blood pressure to very high levels. Yet in blood pressure guidelines from major health organizations, presentation of a value such as 120/80 mmHg gives the impression that this is a threshold for benefit.

Thus, the use of specific target values, although appropriate for providing easy-to-understand guidelines for the general public, distorts the true relationship between activity or fitness and specific health outcomes. This creates problems when we attempt to develop brief statements on how much activity at what intensity needs to be performed when the range in physical activity habits and exercise capacity in the target population is very wide. More data are needed on the shape of the relationship between the activity dose (amount and intensity) and specific health outcomes. Such research will depend on having very accurate measures of activity as well as of the change in health outcomes.

Multiple Biological Changes May Produce a Specific Clinical Benefit

For some clinical conditions such as CHD, it appears that a number of activity-induced biological changes contribute to a reduction in morbidity and mortality. These changes include, but are not limited to, improvements in the plasma lipoprotein profile, a decrease in insulin resistance, reductions in blood pressure, increases in coronary blood flow, and decreases in myocardial oxygen demand. It is likely that the dose–response relationship varies among these biological changes and that the dose–response relation for CHD clinical events is some integration of the dose–response relation for all of the biological mechanisms. Thus, the dose–response relation for any one biological variable does not necessarily represent the dose–response relation for a reduction in CHD clinical events. It would be valuable to know how the dose–response relations for specific biological risk factors relate to the dose–response relations for clinical outcomes such as myocardial infarction, stroke, type 2 diabetes, and site-specific cancers. All of these major clinical entities are likely to benefit through a variety of activity-induced biological changes.

Different Doses Required for Different Clinical Benefits

When we try to determine what dose is needed for better health, we must take the specific outcome of interest into account because of the specificity of the physical activity effect. Substantial data demonstrate that the dose–response relation varies widely for different health benefits. For example, the stimuli for changes in fat and carbohydrate metabolism (decreasing risk of CHD and type 2 diabetes) are most likely to be responsive to increases in total energy expenditure, whereas changes in bone density are likely attributable primarily to the stress placed on bone by forces working against gravity or by vigorous muscular contraction. To prevent weight gain or to produce weight loss, the most important component of the activity dose is the total amount of activity performed expressed as energy expenditure. To increase aerobic capacity, an increase in intensity appears to be critical; but in the prevention or treatment of hypertension, moderate-intensity exercise appears to be at least as effective as vigorous-intensity activity, if not more so. Current public health physical activity guidelines try to take these differences, as best we know them, into account by promoting a variety of activities.

Activity Dose for Maintaining a Healthy Body Weight

The major physical activity and health recommendations of ≥150 min per week of moderate-intensity activity or 75 min per week of vigorous-intensity activity are based primarily on chronic disease outcomes, such as decreased rates of heart attack, stroke, type 2 diabetes, colon cancer, and osteoporosis. Although this amount of activity performed over months and years is likely to contribute to the maintenance of a healthy body weight, recent recommendations based on data from various sources have indicated that greater amounts of activity may be needed by many people to prevent unhealthy weight gain, lose weight, and prevent weight regain. For example, the International Association for the Study of Obesity concluded that for adults, 45 to 60 min per day is required to prevent the transition to overweight or obesity and that prevention of weight regain in formerly obese individuals requires 60 to 90 min of moderate-intensity activity or smaller amounts of vigorous activity (Saris et al. 2003). Similar recommendations were made in *Dietary Guidelines for Americans 2005* (U.S. Department of Health and Human Services [DHHS] and U.S. Department of Agriculture 2005).

Recent WHO recommendations shift primary focus away from losing weight and toward first meeting the weekly physical activity target.

Whether a person is in energy balance or either gains or loses body mass, especially adipose tissue, is determined both by the number of calories consumed and by the number expended through physical activity. Thus, any physical activity dose recommendations intended to prevent weight gain or regain or produce weight loss need to consider what is happening to caloric intake. *Physical Activity Guidelines for Americans*, published by DHHS in 2008, and *Global Recommendations on Physical Activity for Health*, published by the World Health Organization in 2010, recommend that all adults first achieve the public health target of 150 min per week of moderate-intensity or 75 min per week of vigorous-intensity activity. Once they reach this goal, they should assess their body weight and, if it is not optimal for them, should consider a weight management program that includes a reduction in calorie intake and possibly adds more activity throughout the week. It is very important to remember all the health benefits of regular activity, in addition to its role in helping to achieve and maintain an optimal body weight. The reader is referred to chapter 12 on physical activity, physical fitness, and obesity for further discussion of this topic.

Acute Versus Chronic Training Responses

When we attempt to define the dose response for the health benefits of activity, it is useful to consider that some effects may occur as an acute response to a single or several bouts of exercise, some only as a training response, and some as the result of an interaction between acute and chronic training responses. As an example of an acute response, after older men and women with moderate hypertension performed a 30-min bout of stationary cycling at either 40% or 70% of VO_2max, they showed significant decreases in blood pressure for 8 to 12 h postexercise (Pescatello et al. 1991). Also, it has been demonstrated that in men with high plasma triglyceride concentrations, fasting levels on the morning after a 45-min bout of exercise at approximately 75% of aerobic capacity were lower than when such exercise was not performed (Gyntleberg et al. 1977). Over a five-day period, if exercise was performed every day, fasting triglyceride concentration decreased further on each successive day. Thus, this acute response was augmented by repeated bouts of exercise. The reduced triglyceride concentration was rapidly reversed if the exercise was not performed for several days. This could be called an augmented acute response in that there was no further decrease even after weeks of regular activity. It is likely that as a person's physical working capacity increases and the absolute intensity of activity performed during a session increases (relative intensity stays the same), the acute responses of various biological reactions will be enhanced. A more fit person can expend more energy during a set period of time. This would be true for any benefit directly tied to the magnitude of energy expended during the exercise session. For a more detailed discussion of acute responses to exercise, see chapter 6.

Maximizing Health Benefits While Minimizing Health Risks

With regard to health status, physical activity can present a two-edged sword. As the intensity or amount of activity is increased, the greater is the risk of injury, especially musculoskeletal problems for many individuals and cardiovascular complications in people who already have underlying disease. Of particular concern in any attempt to establish the optimal dose of activity for health outcomes is intensity, because it is the major contributor to activity-induced medical complications. Thus, the dose–response assessment needs to consider not only what dose induces the greatest health benefit but also the risk profile for that dose. It may be that high-intensity activity

■ **To maximize the health outcomes of an activity dose, the benefits need to be optimized while the medical risks are kept to a minimum.**

(running) will provide greater benefit for a specific biological outcome but that a moderate-intensity activity (brisk walking) will provide the best overall health benefit in at-risk populations because of its lower risk profile. It will be valuable to establish the risk profile for various activity regimens in different populations, especially those at an increased risk of injury such as obese and older persons. See chapter 18 for more information on medical risks associated with increased activity and ways to reduce these risks.

Physical Activity and Fitness: Dose for Health Benefits

In addition to the large volume of published data that link higher levels of activity to reduced chronic disease and all-cause mortality rates, a number of studies have investigated the relationship between level of cardiorespiratory fitness and similar clinical outcomes (see chapters 10, 11, and 14). In general, studies show that adults who are less fit have higher morbidity and mortality rates than those who are more fit, with the lowest clinical event rates usually occurring in the most fit men and women. In these studies, cardiorespiratory fitness levels were determined using submaximal as well as maximal exercise tests on a motor-driven treadmill or cycle ergometer. Measures of fitness have included heart rate measured at a standard submaximal workload, aerobic capacity (maximum METs) estimated from peak work performed, and $\dot{V}O_2$max measured through analysis of samples of expired air.

As is discussed elsewhere in this book (chapter 24), a person's fitness and health are determined by a complex set of interactions between her genes and the environment, particularly behavior. The major environmental factor that influences cardiorespiratory fitness of most healthy adults is their habitual performance of endurance-type physical activities. Thus, measures of cardiorespiratory fitness indicate the amount and intensity of endurance-type activity performed by an individual over the past weeks or months. In some prospective observational studies evaluating predictors of chronic disease morbidity and mortality or all-cause mortality, a measure of cardiorespiratory fitness has been used as a surrogate for, or an indicator of, habitual physical activity. In studies with a larger sample size and a substantial

number of clinical events, it has been possible to establish a dose–response relationship between level of fitness and mortality.

Figure 22.4 shows the dose response between maximum METs determined during symptom-limited treadmill testing and all-cause mortality in men with and without evidence of cardiovascular disease at time of testing and followed for an average of 6.2 ± 3.7 years (Myers et al. 2002). Mortality is expressed as "relative risk" of death, with the risk for the most fit quintile expressed as 1.0. These data show a strong inverse dose–response gradient across all levels of fitness, with the least fit men having a risk of dying during the follow-up period that was four times greater than that of their most fit counterparts. The MET capacity at each quintile of fitness was higher for the participants free of cardiovascular disease at baseline compared with those with disease. It is possible that the patients with cardiovascular disease were less active, but their overall lower exercise capacity may have been attributable to their cardiovascular disease.

The data from Myers and colleagues (2002) in figure 22.4 are quite similar to findings from a number of other studies that have evaluated the dose–response relationship between cardiorespiratory fitness and all-cause mortality. All have shown that the least fit men and women fare the worst. Men and women who have the highest levels of fitness have the lowest age-specific all-cause mortality rates.

Data relating all-cause mortality to levels of reported physical activity and to measures of cardiorespiratory fitness from selected prospective observational studies are plotted in figure 22.5. Fitness levels were categorized by the authors in either quartiles (Laukkenen et al. 2001) or quintiles (Blair et al. 1989; Myers et al. 2002), whereas activity levels were categorized in tertiles (Leon et al. 1987), in quintiles (Kujala et al. 1998), or by eight categories of energy expenditure (Paffenbarger et al. 1986). The horizontal axis is an approximate percentile ranking for either fitness or activity for subjects in each study. The relative risk is consistently lower for men in the higher fitness categories compared with the men reporting higher levels of physical activity. Mortality data are presented as relative risks, with the risk of death for the least active or least fit group in each study designated as 1.0. (Note that these data are displayed with different reference groups—in figure 22.4 the most fit group is at the reference risk of 1.0, and in figure 22.5 the reference category is the least fit group.) The pattern of results is similar, but the relative risks are different because of the different reference categories. This figure shows that for these studies, the gradient across fitness levels is steeper than the gradient across physical activity levels. This

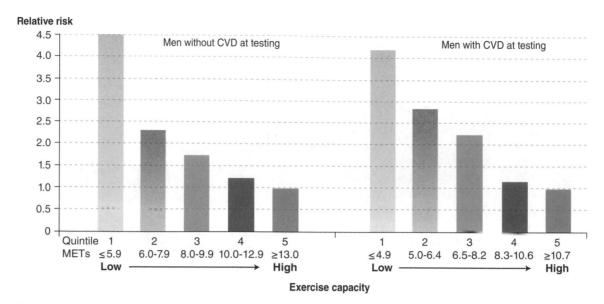

FIGURE 22.4 Age-adjusted relative risk for all-cause mortality according to quintile of exercise capacity (METs) for men with and without cardiovascular disease (CVD) at the time of exercise testing. Average duration of follow-up was 6.2 ± 3.7 years. The reference was set at 1.0 for the highest quintile of fitness.

Adapted, by permission, from J. Myers, M. Prakash, V. Froelicher, D. Do, S. Partington, and J.E. Atwood, 2002, "Exercise capacity and mortality among men referred for exercise testing," *New England Journal of Medicine* 346: 793-801. Copyright ©2002 Massachusetts Medical Society. All rights reserved.

stronger association between fitness and death than between activity and death is likely seen because fitness is a more accurate and reliable measure than physical activity, and fitness is influenced by factors

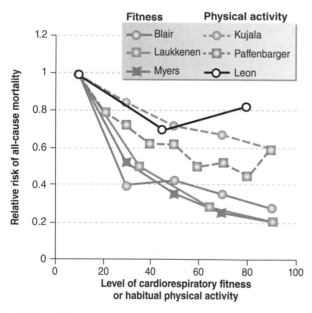

FIGURE 22.5 Physical activity and fitness levels versus all-cause mortality. The figure shows the dose response for all-cause mortality by level of either cardiorespiratory fitness determined by exercise testing or reported habitual physical activity in men from six studies reported in the scientific literature. The relative risk for the least fit or least active men was set at 1.0.

other than physical activity that can affect health and mortality.

Because of the complex nature of habitual physical activity, highly accurate and reliable measures that allow correct activity classification of each person in a large population are difficult to obtain. Whenever participants are misclassified as to activity level (e.g., they should have been in the first quintile but were placed in the second or third quintile because of overreporting), the observed relation between activity and mortality is not as strong as the true relation. If the measure of fitness results in a more accurate classification of individuals than the measure of physical activity, this alone can account for a stronger association between fitness and mortality than for physical activity and mortality. This stronger relation between fitness and mortality may also be attributable, at least in part, to factors other than physical activity that improve fitness and increase longevity. Both cardiorespiratory fitness and all-cause mortality have significant genetic as well environmental determinants (e.g., not smoking, good nutrition); these could contribute additionally to the relationship between fitness and decreased mortality.

Although the magnitude of the dose–response relation between fitness and mortality appears greater than that for activity and mortality, data indicate that levels of activity associated with lower mortality rates also produce meaningful increases in cardiorespiratory fitness. This is consistent with the notion that the lower mortality rates seen in more fit persons are in part attributable to higher levels of activity. For example, in the study by Myers and colleagues (2002), capacity that was a single MET higher was associated with a 12% lower all-cause mortality rate. For many sedentary adults, following the recommendation of adding 150 min per week of moderate-intensity endurance exercise will increase their capacity by at least 2 METs over 6 to 12 months.

> A stronger dose–response relationship exists between measured cardiorespiratory fitness and mortality than between reported physical activity and mortality. This difference is likely attributable to factors other than physical activity that influence fitness and mortality, including heredity, and the more accurate classification of people into fitness categories than into activity categories.

Summary

It is now well established that most people who perform moderate-intensity physical activity for at least 150 min per week, in addition to the routine activities of daily living, significantly reduce their risk of various chronic diseases. Added health benefits frequently accrue at higher doses of activity with increases in either intensity (from moderate to vigorous but not exhausting) or duration. Some benefits, like significant weight loss, may have a higher threshold in terms of activity amount than others, such as reductions in moderately elevated blood pressure or insulin resistance. For selected health-related benefits, people can obtain an adequate dose by accumulating moderate-intensity activity throughout the day in bouts of ≥10 min each. The characteristics of the dose of activity needed to produce a positive health outcome vary from person to person depending on health and fitness status, genetic makeup, and the specific health outcome desired.

Key Concepts

accumulation—Process of adding together the amount of time spent in performing physical activity (usually in short bouts) throughout the day. The concept of accumulating rather than performing activity in a single bout has been added to public health recommendations, but its impact on various health outcomes is still under investigation.

health response—A change (usually an improvement) in a specific measure of health (broadly defined as in chapter 1) or a health-related biological or performance measure. Examples of health-related biological measures include HDL cholesterol, blood pressure, and bone density, whereas examples of health-related performance include changes in aerobic capacity, joint flexibility, and muscle strength.

overload—Increased demands placed on the body by a dose of physical activity characterized by type, intensity, duration, and amount. When tissues, organs, or systems are exposed to an appropriate overload, they typically respond by increasing their capacity or efficiency.

physical activity dose—Characteristics of the physical activity performed by an individual over a defined period of time, which include type, intensity, bout frequency, bout duration, and total amount or volume of activity performed.

progression—Rate at which the physical activity dose (intensity, duration, or amount) is increased over days, weeks, or months. For overload to produce health benefits rather than injury, it needs to be increased in small amounts or at a slow rate of progression.

specificity—Concept that changes in the body attributable to an increase in physical activity are specific to the nature of the demands produced by that activity. Specificity applies to the type of activity (endurance vs. strength or arms vs. legs), intensity, and amount.

Study Questions

1. Briefly describe the three major principles that are the basis for the way the body responds to a habitual increase in physical activity.

2. Name the four main components of a physical activity plan that one needs to consider when defining a dose of physical activity.

3. What is meant by an acute response to a bout of physical activity, and how does this differ from a training response?

4. Which component of the exercise dose is most closely linked to the risk of orthopedic injury or cardiac arrest? How can the activity plan be modified to reduce this risk?

5. Describe the dose–response relationship between cardiorespiratory fitness and all-cause mortality. Explain how the dose–response relationship differs for physical activity versus cardiorespiratory fitness and all-cause mortality.

6. Describe the concept of activity accumulation and the characteristics of the activity that contribute to accumulation.

References

Ainsworth, B.E., W.L. Haskell, S.D. Hermann, N. Meckes, D.R. Bassett Jr., C. Tudor-Locke, J.L. Greer, J. Vezina, M.C. Whitt-Glover, and A.S. Leon. 2011. 2011 Compendium of physical activities: A second update of activity codes and MET intensities. *Medicine and Science in Sports and Exercise* 43:1575-1581.

Blair, S.N., H.W. Kohl III, R.S. Paffenbarger, D.G. Clark, K.H. Cooper, and L.W. Gibbons. 1989. Physical fitness and all-cause mortality: A prospective study of healthy men and women. *Journal of the American Medical Association* 262:2395-2401.

Church, T.S., C.P. Earnest, J.S. Skinner, and S.N. Blair. 2007. Effects of different doses of physical activity on cardiorespiratory fitness among sedentary, overweight or obese postmenopausal women with elevated blood pressure. *Journal of the American Medical Association* 297:2081-2091.

Gyntelberg, F., R. Brennan, J.O. Holloszy, G. Schonfeld, M. Rennie, and S. Weidman. 1977. Plasma triglyceride lowering by exercise despite increased food intake in patients with Type IV hyperlipoproteinemia. *American Journal of Clinical Nutrition* 30:716-720.

Hardman, A.E. 2001. Issues of fractionalization of exercise (short vs long bouts). *Medicine and Science in Sports and Exercise* 33:S421-S427.

Kujala, U.M., J. Kaprio, S. Sarna, and M. Koskenvuo. 1998. Relationship of leisure-time physical activity and mortality: The Finnish twin cohort. *Journal of the American Medical Association* 279:440-444.

Laukkanen, J.A., T.A. Lakka, R. Rauramaa, R. Kuhanen, J.M. Venalainen, R. Salonen, and J.T. Salonen. 2001. Cardiovascular fitness as a predictor of mortality in men. *Archives of Internal Medicine* 161:825-831.

Leon, A.S., J. Connett, D.R. Jacobs, and R. Rauramaa. 1987. Leisure-time physical activity levels and risk of dose response and death: The Multiple Risk Factor Intervention Trial. *Journal of the American Medical Association* 258:2388-2395.

Macfarlane, D.J., L.H. Taylor, and T.F. Cuddihy. 2006. Very short intermittent vs continuous bouts of activity in sedentary adults. *Preventive Medicine* 43:332-336.

Manini, T.M., J.E. Everhart, K.V. Patel, D.A. Schoeller, L.H. Colbert, M. Visser, F. Tylavsky, D.C. Bauer, B.H. Goodpaster, and T.B. Harris. 2006. Daily activity energy expenditure and mortality among older adults. *New England Journal of Medicine* 296:171-179.

Manson, J.E., P. Greenland, A.Z. LaCroix, M.L. Stefanick, C.P. Mouton, A. Oberman, M.G. Perri, D.S. Sheps, M.B. Pettinger, and D.S. Siscovick. 2002. Walking compared with vigorous exercise for the prevention of cardiovascular events in women. *New England Journal of Medicine* 347:716-726.

Myers, J., M. Prakash, V. Froelicher, D. Do, S. Partington, and J.E. Atwood. 2002. Exercise capacity and mortality among men referred for exercise testing. *New England Journal of Medicine* 346:793-801.

Paffenbarger, R.S., R.T. Hyde, A.L. Wing, and C.C. Hsieth. 1986. Physical activity, all-cause mortality, and longevity of college alumni. *New England Journal of Medicine* 314:605-613.

Pate, R., M. Pratt, S.N. Blair, W.L. Haskell, C.A. Macera, C. Bouchard, D. Buchner, W. Ettinger, G.W. Heath, A.C. King, A. Kriska, A.S. Leon, B.H. Marcus, R. Paffenbarger, S.K. Patrick, M.L. Pollock, J.M. Rippe, J. Sallis, and J.H. Wilmore. 1995. Physical activity and public health: A recommendation from the Centers for Disease Control and Prevention and the American College of Sports Medicine. *Journal of the American Medical Association* 273:402-407.

Pescatello, L.S., A.E. Fargo, C.N. Leach, and A.E. Scherzer. 1991. Short-term effect of dynamic exercise on arterial blood pressure. *Circulation* 83:1557-1561.

Physical Activity Guidelines Advisory Committee. 2008. *Physical Activity Guidelines Advisory Committee report, 2008.* Washington, DC: Department of Health and Human Services.

Saris, W.H.M., S.N. Blair, M.A. van Back, E.A. Eaton, P.S.W. Davies, L. Di Pietro, M. Fogelholm, A. Rissanen, D. Schoeller, B. Swinburn, A. Tremblay, K.R. Westerterp, and H. Wyatt. 2003. How much physical activity is enough to prevent unhealthy weight gain? Outcome of the IASO 1st Stock Conference and consensus statement. *Obesity Reviews* 4:101-114.

U.S. Department of Health and Human Services. 2008. *Physical activity guidelines for Americans.* Washington, DC: U.S. Department of Health and Human Services.

U.S. Department of Health and Human Services and U.S. Department of Agriculture. 2005. *Dietary guidelines for Americans 2005.* 6th ed. Washington, DC: U.S. Government Printing Office.

Williams, P. 1997. Relationship of distance run per week to coronary heart disease risk factors in 8283 male runners: The National Runners' Health Study. *Archives of Internal Medicine* 157:191-198.

Wilmore, J.H., J. Royce, R.N. Girandola, F.I. Katch, and V.L. Katch. 1970. Physiological alterations resulting from a 10-week jogging program. *Medicine and Science in Sports* 2:7-14.

World Health Organization. 2010. *Global recommendations on physical activity for health.* Geneva: World Health Organization.

23

From Science to Physical Activity Guidelines

Mark S. Tremblay, PhD; and William L. Haskell, PhD

CHAPTER OUTLINE

©Paylessimages

The volume and strength of scientific evidence documenting the positive relationship between physical activity and health are already quite substantial as described in this text. However, concerns regarding low levels of physical activity remain (World Health Organization 2009), and evidence is accumulating on the detrimental health effects of the long hours spent sitting or simply sedentary. Building on this scientific base, one strategy to promote and monitor population physical activity levels is to establish and disseminate physical activity **guidelines**. Indeed, many countries have recently engaged in a process to develop or revise national physical activity guidelines (Physical Activity Guidelines Advisory Committee [PAGAC] 2008; Tremblay et al. 2010; British Heart Foundation National Centre for Physical Activity and Health 2010; Sims et al. 2006; European Union 2008). Physical activity guidelines provide a basis for public health **messaging**, a standard for population surveillance, a foundation for future research, and a credible target for social marketing efforts to promote physical activity. The purpose of this chapter is to describe a systematic approach to the development, promotion, and evaluation of robust, evidence-informed physical activity guidelines, as well as to identify limitations, challenges, research gaps, and future directions.

Throughout most of this book, the primary focus has been on documenting the numerous health benefits that occur when people are physically active during various periods of their lives and why being sedentary confers so many health risks. As data supporting these relationships have accumulated over the past half century, the role of physical activity in public health and medicine has continued to evolve. In order for physical activity to be accepted as a legitimate component of public health or as a medical modality, it is necessary that knowledgeable scientists, using systematic approaches, objectively review and synthesize the relevant research to inform practice guidelines. The results of such a review then need to be converted to accurate messages for use by health and exercise professionals, the general public, and policy makers. Because of the rapid growth of the number and scope of relevant scientific publications, it is appropriate that physical activity and health guidelines be updated at least every five years or so.

While the results of individual studies are very important in moving the science base forward, no one study or even a limited number of studies provide adequate data for the development of appropriate guidelines pertaining to a broad population of men and women throughout their life span. It is only through a comprehensive and thoughtful review process of numerous studies of varying designs that an overall consensus of the research can be obtained and evidence-informed guidelines prepared. While many readers of this book will not be directly involved in the guideline development process, we believe it is important for physical activity and health professionals to be familiar with the process and appreciate its complexity. Understanding the process will assist researchers in the design and execution of their research, as they will be better able to identify important gaps in knowledge that they can fill. Within the physical activity practitioner community, it is valuable for people to be familiar with the guideline development process so they can appreciate why recommendations are based on a consensus and not the opinion of a few experts. Specific outcomes need to be supported by strong evidence, and updates must be based on the evolving nature of the science, such as that presented in the preceding chapters of this book.

Stages of Physical Activity Guideline Development

Our aim in this section is to describe the sequence of events (stages) recommended in the development, promotion, and evaluation of physical activity guidelines. These stages consolidate the best practices from recent guideline development processes in several countries and provide an indication of the complexity of establishing evidence-informed guidelines. Table 23.1 summarizes the sequence of the stages and estimates of the amount of time required to complete them.

Leadership Team (Stage 1)

The process of initiating the development of physical activity guidelines requires a team of champions committed to leading and resourcing the process. This team may originate from within government, a professional or scientific organization, an academic institution, a lobby group, or a combination of these. The leadership team may be motivated to develop physical activity guidelines for several reasons:

- There is a public health need for and an absence of guidelines.
- Current guidelines are out of date and require revisions.
- Pressure is coming from domestic or international
  - government sectors,
  - lobby groups,

- corporate or industry coalitions,
- media,
- academic or scientific organizations, or
- combinations of the above.

The leadership team needs to develop a strategy to oversee, administer, and resource the guideline development process. Team members need to determine the scope of the guideline development process (e.g., prevention vs. treatment, age ranges, health vs. performance, special population concerns, inclusion of sedentary behavior guidelines). Furthermore, to adhere to rigorous guideline development protocols (Tremblay et al. 2010; Davis, Goldman, and Palda 2007), the leadership team should implement an appraisal methodology (process assessment procedures) prior to the subsequent stages. The work of the leadership team may be facilitated through a guideline development secretariat that provides staff support for the initiative.

Process Assessment Procedures (Stage 2)

The integrity, credibility, and acceptance of guidelines are directly related to the rigor and objectivity of the process used in guideline development. In fact, a variety of valid and reliable instruments have been developed specifically for the assessment and quantification of guideline development processes. Ideally, process assessment methods will be in place before guideline development begins, allowing the process to be assessed from the outset and the assessment procedures to guide and inform the process as it evolves.

The Appraisal of Guidelines, Research and Evaluation (AGREE) was established by the AGREE Collaboration as a generic instrument to assess the methodological rigor, process, and transparency with which practice guidelines are developed. The purpose of the instrument is to assess the quality of reporting in practice guidelines across the spectrum of health, provide direction on future guideline development, and help guideline developers determine what specific information should be included when they are developing new or updating existing guidelines.

The latest version of the instrument is AGREE II, which consists of 23 items, two overall assessment items, and a user's manual (www.agreetrust.org). The following are the six domains included in AGREE II:

Domain 1: Scope and Purpose

1. The overall objective(s) of the guideline is (are) specifically described.

2. The health question(s) covered by the guideline is (are) specifically described.

3. The population (patients, public, etc.) to whom the guideline is meant to apply is specifically described.

Domain 2: Stakeholder Involvement

4. The guideline development group includes individuals from all the relevant professional groups.

5. The views and preferences of the target population (patients, public, etc.) have been sought.

6. The target users of the guideline are clearly defined.

Domain 3: Rigour of Development

7. Systematic methods were used to search for evidence.

8. The criteria for selecting the evidence are clearly described.

9. The strengths and limitations of the body of evidence are clearly described.

10. The methods for formulating the recommendations are clearly described.

11. The health benefits, side effects, and risks have been considered in formulation of the recommendations.

12. There is an explicit link between the recommendations and the supporting evidence.

13. The guideline has been externally reviewed by experts prior to its publication.

14. A procedure for updating the guideline is provided.

Domain 4: Clarity of Presentation

15. The recommendations are specific and unambiguous.

16. The different options for management of the condition or health issue are clearly presented.

17. Key recommendations are easily identifiable.

18. The guideline describes facilitators and barriers to its application.

Domain 5: Applicability

19. The guideline provides advice, tools, or both regarding how the recommendations can be put into practice.

20. The potential resource implications of applying the recommendations have been considered.

21. The guideline presents monitoring or auditing criteria or both.

TABLE 23.1 Physical Activity Guideline Development Time Line

Guideline development stage	Key stakeholders	1	2	3	4
1. Establishing a leadership team	Government, physical activity scientific and professional groups, public health groups, guideline development secretariat	■			
2. Establishing and implementing process assessment procedures	Leadership team, process consultants, systematic review experts		■	■	
3. Forming a guideline development and research committee	Leadership team, stakeholder groups		■	■	
4. International and interjurisdictional harmonization	Leadership team, guideline development and research committee, international partners, stakeholder groups	■	■	■	
5. Literature review	Guideline development and research committee, leadership team, systematic review researchers, guideline development secretariat			■	
6. Interpretation of findings	Guideline development and research committee, independent scientific panel, process consultants				
7. Identification of research gaps	Leadership team, research committee, research granting agencies				
8. Stakeholder consultation, consensus, and endorsement	Government, physical activity scientific and professional groups, public health, stakeholder groups	■	■	■	■
9. Knowledge translation strategy	Leadership team, guideline development and research committee, international partners, stakeholder groups, guideline development secretariat				
10. Language translation					
11. Presentation and messaging					
12. Communication					
13. Dissemination					
14. Evaluation	Leadership team, guideline development and research committee, guideline development secretariat				
15. Planning and cycling of updates and revisions	Leadership team, guideline development secretariat				

Domain 6: Editorial Independence

22. The views of the funding body have not influenced the content of the guideline.

23. Competing interests of guideline development group members have been recorded and addressed.

Overall Guideline Assessment

AGREE II is designed to assess current and proposed guidelines developed by local, regional, national, or international organizations. AGREE is specifically focused for use by the following groups:

- *Health care providers*—to assess a guideline before implementing it in their own practice

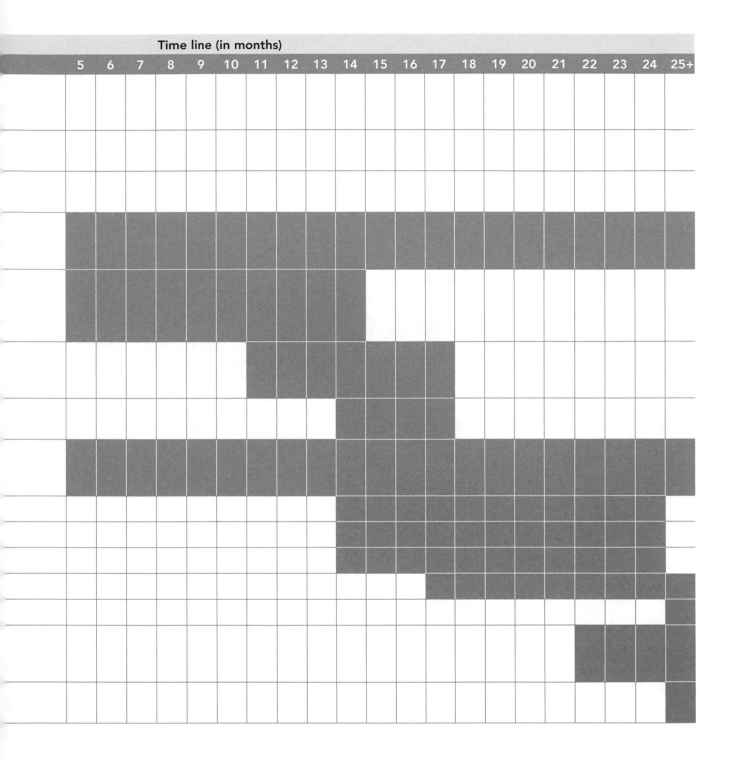

- *Guideline developers*—to construct rigorous methods when proposing work for new guideline development or to assess work put forth by other organizations

- *Policy makers*—to help inform policy decisions or to help make decisions on which guideline to follow

- *Educators*—to help increase critical appraisal skills and to educate on the steps needed for proper guideline development

Ideally, two to four impartial but knowledgeable appraisers (process consultants) will provide a quantitative assessment of the guideline development process. These appraisers or consultants need to observe, have access to, and provide advice on the entire guideline development process. All AGREE II items are graded on a 7-point scale. A score of 1 (strongly disagree) is given when information is not provided, not relevant to the AGREE II item, or poorly reported. A score of 7 (strongly agree) is given when the quality of reporting is exceptional and full criteria and all considerations have been met. A higher domain or overall score indicates a higher quality of guideline development. More details on the AGREE II instrument can be found elsewhere (Brouwers et al. 2010a, 2010b, 2010c).

Specific instruments and procedures also exist for assessing the rigor and objectivity of the background literature search and systematic review. The process of assessment of systematic reviews begins with assessing the quality of evidence available in a particular field of research. This allows researchers to determine the level of confidence they have in the recommendations they are making, and it gives guidance for future research. If the quality of recommendations is deemed appropriate, then guidelines can be developed and disseminated with confidence. Throughout this process, researchers should be transparent with the methods they are using and stay true to their a priori research question to avoid introducing bias (e.g., data dredging or "picking the research"). Before beginning the guideline development process, the leadership team, in consultation with guideline development process experts, should determine what instruments will be employed to assess the literature review and systematic reviews and what procedures will be used to assess and grade the quality of evidence and the strength of the emerging recommendations.

The Assessment of Multiple Systematic Reviews (AMSTAR) instrument was created to assess the methodological quality of systematic reviews. The tool consists of 11 items to which the reviewer can answer "yes," "no," "can't answer," or "not applicable." These are the questions included in AMSTAR:

1. Was an a priori design provided?
2. Was there duplicate study selection and data extraction?
3. Was a comprehensive literature search performed?
4. Was the status of publication (i.e., grey literature) used as an inclusion criterion?
5. Was a list of studies (included and excluded) provided?
6. Were the characteristics of the included studies provided?
7. Was the scientific quality of the included studies used appropriately in formulating conclusions?
8. Were the methods used appropriately in formulating conclusions?
9. Were the methods used to combine the findings of studies appropriate?
10. Was the likelihood of publication bias assessed?
11. Was conflict of interest stated?

AMSTAR can be scored by individual items or by summing the item scores (to give an overall score). All questions were developed to have equal weight. The AMSTAR tool takes 15 to 20 min to complete per review and should be completed by an informed but impartial expert or consultant (Shea et al. 2009, 2007).

The Grading of Recommendations Assessment, Development and Evaluation (GRADE) (table 23.2) is a tool used by guideline developers to rate the quality of evidence and strength of **recommendations from scientific literature** presented in systematic reviews. GRADE is a two-part assessment that allows researchers to go from reviewing available evidence to making recommendations to the general public.

GRADE offers four levels of evidence quality: high, moderate, low, and very low. Randomized controlled trials (RCTs) begin as high-quality evidence and observational studies as low-quality evidence. Quality may be downgraded as a result of limitations in the study design or implementation, imprecision of estimates (wide confidence intervals), variability in results, indirectness of evidence, or publication bias. Quality may be upgraded because of a very large magnitude of effect, a dose–response gradient, or if all plausible biases would reduce an apparent treatment effect.

GRADE has two categories for strength of recommendations: strong and weak. Strength of recommendations is determined by the balance between desirable and undesirable consequences of alternative management strategies, quality of evidence, variability in values and references, and resource use. A recommendation may be strong if the researcher is confident that the desirable effects of adherence to it outweigh the undesirable effects. A weak recommendation may be made if the researcher feels that the desirable effects of adherence to it probably outweigh the undesirable effects but is less confident. Further details on the GRADE instrument are available elsewhere (Guyatt et al. 2008a, 2008b, 2008c, 2008d; Jaeschke et al. 2008).

TABLE 23.2 Grading of Recommendations Assessment, Development and Evaluation (GRADE): Quality of Evidence and Strength of Recommendation Summary

Quality of evidence

Randomized controlled trials are ranked as having the highest quality of evidence and observational studies as the lowest quality of evidence.

Factors in deciding on quality of evidence

Factors that decrease quality	Factors that increase quality
Study limitations Inconsistency of results (variability) Indirectness of evidence Imprecision (wide confidence intervals) Publication bias	Large magnitude of effect Dose–response gradient Plausible confounding that would reduce a demonstrated effect
High quality Further research is very unlikely to change confidence in the estimate of effect.	● ● ● ● or A
Moderate quality Further research is likely to have an important impact on confidence in the estimate of effect and may change the estimate.	● ● ● ○ or B
Low quality Further research is very likely to have an important impact on confidence in the estimate of effect and is likely to change the estimate.	● ● ○ ○ or C
Very low quality Any estimate of effect is very uncertain.	○ ○ ○ ○ or D

Strength of recommendation

Strength of recommendations is determined by:
The balance between desirable and undesirable consequences of alternative management strategies
Quality of evidence
Variability in values and references
Cost (i.e., resource use and allocation)

Weak recommendations mean that patient choices will vary according to their values and preferences and that clinicians must ensure that patients' care is in keeping with their values and preferences.	
Strong recommendation for "Definitely do it" Desirable clearly outweighs undesirable	↑↑ or 1
Weak recommendation for "Probably do it" Desirable probably outweighs undesirable	→ ? or 2
No specific recommendation Trade-offs equally balanced or uncertain	? ? or 0
Weak recommendation against "Probably don't do it" Undesirable probably outweighs desirable	↔ ? or 2
Strong recommendation against "Definitely don't do it" Undesirable clearly outweighs desirable	↓↓ or 1

Other instruments and procedures have been developed and used to evaluate guideline development and assess the level and quality of evidence. The important points are that rigorous, unbiased procedures are put in place *before* the guideline development process begins and that assessments are performed by impartial experts.

Guideline Development and Research Committee (Stage 3)

As indicated in table 23.1, this stage, in which a *guideline development and research committee* is formed, should probably be conducted concurrently with the *process assessment procedures* stage. Whereas the leadership team provides overall leadership, direction, administration (perhaps through a secretariat), and resources, the establishment of a guideline development and research committee (which may also be called or considered the "expert advisory committee") is essential to serve as the engine of the initiative. The committee should be made up of 10 to 20 members, with due attention to gender, geography, sector, and academic representation (e.g., pediatrics, geriatrics, physiology, epidemiology, psychology, medicine, public health, communications). Process consultants may also serve on this committee or provide guidance on its membership complement. The process for forming the committee may vary by jurisdiction and according to who is providing the leadership. For example, the United States committee members were selected through formal government processes involving applications, reviews, and public announcements (PAGAC 2008). Alternatively, in the more informal Canadian guideline development process, the leadership team (housed within the Canadian Society for Exercise Physiology) invited individuals to join the committee (Tremblay et al. 2010).

The guideline development and research committee should meet early on in the guideline development process and regularly thereafter. The committee needs to establish decision-making rules (e.g., consensus, voting) and to have a clear understanding of its responsibilities (e.g., provide recommendations to government for eventual guideline development or the actual establishing of guidelines). Committee members are usually recognized in the published materials emanating from the guideline development (PAGAC 2008; Tremblay et al. 2010; British Heart Foundation National Centre for Physical Activity and Health 2010; Sims et al. 2006).

International and Interjurisdictional Harmonization (Stage 4)

One of the first tasks of the leadership team and guideline development and research committee is to perform a domestic and international scan of similar initiatives under way or recently completed. This serves several purposes, including the following:

- Collating existing guidelines for potential harmonization or even standardization
- Exploring possible consultants, experts, advisors, reviewers, and committee members
- Determining whether recent literature reviews have been completed and can be used or updated, saving valuable time and resources
- Gathering insights and lessons learned from a process perspective

International and interjurisdictional cooperation and harmonization are important not only for reconnaissance purposes, but also to avoid conflicts and confusion in guideline presentation, messaging and wording, and interpretation of the given body of scientific literature. Imagine if the American Heart Association (AHA), the American College of Sports Medicine (ACSM), and the National Association for Sport and Physical Education (NASPE) each came out with different and conflicting physical activity guidelines for children—or if Canada, Australia, and the World Health Organization came out with different adult physical activity guidelines. This would create confusion and competition and raise concerns over the credibility of the guideline development processes, ultimately decreasing the potential public health benefit.

The recent U.K. physical activity guideline revision project is a best practice example of international harmonization. Guideline developers relied completely on recent comprehensive and systematic reviews completed by the United States, Canada, and Australia to inform their guidelines revisions (British Heart Foundation National Centre for Physical Activity and Health 2010). This cooperation reduced the time frame from 24 months to 12 months. The World Health

Organization followed a similar process in the development of its Global Recommendations on Physical Activity for Health (World Health Organization 2010).

Literature Review (Stage 5)

With the initiative leadership, assessment, committee, and international linkages in place, the laborious stage of the process can begin—the systematic review of the available evidence. The leadership team and guideline development and research committee need to be clear on the scope of the review; the following are examples of questions they need to ask:

- Are the guidelines for health promotion (primary prevention in apparently healthy populations) or for treatment of those at high risk (secondary prevention) or those with existing disease (tertiary prevention)?
- What are the target age groups (infants, children, youth, adults, older adults, everyone)?
- Is the focus health promotion, fitness development, or athletic performance?
- Will guidelines include considerations for special populations (e.g., pregnant women, medically underserved populations, people with disabilities, ethnic minorities)?
- How is physical activity defined (e.g., certain intensity threshold, lifestyle activities, leisure time as well as occupation)?
- Will guidelines address both physical activity and sedentary behavior?
- What outcomes will be included in the search (e.g., obesity, heart disease, diabetes, cancer, hypertension, mental health, academic achievement, prosocial behavior, sleep quality)?
- Will both benefits and risks be assessed?
- What types of studies will be accepted for inclusion in the systematic review? Will there be a minimum sample size, length of follow-up, or adherence rate requirement for inclusion?
- How will study quality be assessed?

Obviously, the prework of a systematic review is detailed and complicated but forces decisions on the scope, limitations, and delimitations of the review process. Having a good question a priori is imperative to the success and rigor of the review.

A well-thought-out question allows the review to move forward in a timely manner and reduces risk of bias. The PICO framework, originally designed to focus clinical questions, is a commonly used tool in the initial stages of a systematic review. The PICO framework is as follows:

- P: *Patient or population.* This can refer to a specific patient group (e.g., with a particular condition or health problem) or population (e.g., all Canadian children and youth).
- I: *Intervention.* The "I" represents the intervention of interest. For epidemiological studies, exposure can replace intervention.
- C: *Comparator or control.* This represents the standard to which the intervention or exposure group is being compared. A comparator is not always used, especially in population-based science.
- O: *Outcome.* This is the expected result(s)—the results that will be seen because of the intervention or exposure (e.g., obesity, coronary heart disease mortality).

The following are *optional* but can be useful additions to a PICO strategy.

- S: *Study design.* This helps to narrow the search and clarify what type of information is being sought.
- T: *Time line.* This allows date limits for the review to be set. The time line can be accelerated if previous reviews have recently been published in the area (e.g., is only an update required?).

Once a carefully constructed research question is agreed upon and the PICO elements are determined, decisions on the use of grey literature and key informants are required. Grey literature includes results or documents of potential interest that may not be captured through searches of published manuscripts in traditional peer-reviewed, indexed journals. For example, grey literature could include conference proceedings from relevant international meetings; unpublished doctoral dissertations; websites of key organizations, government departments, or research groups; or general Internet (Google/Wikipedia) searches for lay articles (e.g., technical reports, government reports). Decisions regarding the inclusion and scope of grey literature need to be made early on in the search process, and the decision should be informed by the volume and quality of research found through the main search procedures. Inclusion of grey literature can help reduce publication bias. Similarly, the use of key informants can help to ensure that all key research is captured. Key informants (recognized experts) may be asked to send key manuscripts and reports based on the PICO framework, or they may be sent the list of papers captured through the review process and asked whether any key references are missing.

Finally, a system or instrument for assessing the quality of each paper or research study is required. This is necessary to inform the GRADE assessment (see section on process assessment procedures) and to allow for assessments of findings based on study quality. The GRADE process has a study quality assessment built in, though other tools exist. Many quality assessment instruments are very specific to clinical trials, but the Downs and Black tool is a checklist for the assessment of methodological quality in both randomized and nonrandomized studies (Downs and Black 1998), making it extremely valuable for those conducting epidemiological or population-based searches. The final Downs and Black checklist includes 27 items, with a maximum score of 32. The checklist is broken down into five subscales: reporting (10 items); external validity (3 items); internal validity—bias (7 items); internal validity—confounding or selection bias (6 items); and power (1 item). Not surprisingly, shorter papers take less time to review than longer papers, with an average of 20 to 25 min needed to score each paper. The studies can then be ranked as high quality versus low quality (or some alternate grouping) to deter-

mine whether any difference in effect is due to study design or quality and not exposure.

The literature gathered through the various search processes needs to be screened and abstracted according to standard systematic review procedures (Higgins and Green 2009). The studies meeting all inclusion criteria are then consolidated and synthesized in a summary fashion for interpretation by the guideline development and research committee. When the evidence relating physical activity to health is consolidated, data must be integrated from studies conducted using a diversity of designs and methods. As indicated in figure 23.1, data need to be considered from both observational and experimental studies, studies that evaluate changes in biomarkers thought to be in the causal pathways between activity and clinical outcomes (involving humans and other animals), and studies with a clinical outcome such as diabetes mellitus or all-cause mortality.

Interpretation of Findings (Stage 6)

With the systematic review of the evidence complete and the quality of studies assessed, the guideline development and research committee can interpret the findings objectively based on the system previously agreed on (e.g., GRADE). The quality, consistency, and amount of evidence should be used to develop summary recommendations from the scientific literature, and the strength of or confidence in the recommendations should also be assigned by the guideline development and research committee or, ideally, an independent expert panel

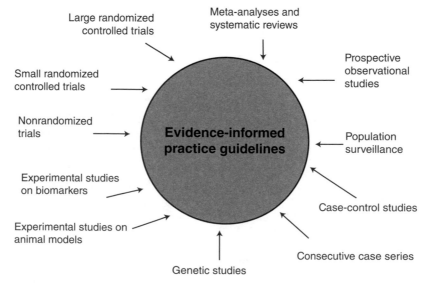

FIGURE 23.1 Sources of data to support evidence-informed physical activity practice guidelines.

(Tremblay et al. 2010; Kesaniemi et al. 2010). These evidence-informed recommendations represent the culmination of the scientific review process and are the starting point for the knowledge translation stage (stage 9) described later.

Identification of Research Gaps (Stage 7)

Through the process of the systematic review of the evidence and the consultations with various stakeholder groups, areas of data and evidence gaps will emerge, creating an evidence-informed opportunity for establishing future research priorities. Data and evidence gaps may be related to certain populations, the relationship between physical activity and specific health conditions, the physical activity and health dose–response relationship, identification of thresholds, prevention versus treatment of health conditions, temporal trends, and so on.

It is important to engage research granting agencies in discussions as the systematic review and research synthesis process approaches completion to ensure that strategic research programs in physical activity, exercise science, and public health focus on these research priorities.

Consensus and Stakeholder Engagement (Stage 8)

As illustrated in table 23.1, it is important to engage stakeholder groups throughout the guideline development process. It is through these groups (professional and scientific organizations, public health and medical associations, fitness leaders and personal trainers, corporate and private sector groups, media, governments) that support or resistance will emerge and dissemination and uptake will be facilitated or opposed. Once scientific recommendations are formed from the systematic review and research synthesis, these stakeholder groups should be formally consulted to provide comments on the scientific recommendations, a rating of support for these statements (e.g., strongly disagree, disagree, neutral, agree, strongly agree), a formal endorsement or rejection, and a list of organizations and individuals that should be added to a network for dissemination.

Knowledge Translation Strategy (Stages 9-13)

The scientific summary recommendations that emerge from the systematic review and literature synthesis process are usually not in language or in a format appropriate for public health messaging to the population. They often contain technical terms or details that may be confusing to the target audience. In the case of young children, the target audience may not even be the population you are trying to affect; instead, it may be parents, caregivers, teachers, and community leaders. Accordingly, knowledge translation processes are required (Brawley and Latimer 2007; Latimer, Brawley, and Bassett 2010).

- *Language translation*: The most obvious translation required is language translation, for example from English to Spanish or French, depending on the linguistic makeup of the jurisdiction in question. Depending on the extent of materials produced to communicate the guidelines (e.g., physical activity guides, posters, website), this process can require considerable time and resources. Some phrases or concepts may not be translatable and may make authentic translation difficult.

- *Presentation*: The physical activity guidelines can be presented in various formats, for example as television or radio sound bites, schematics, images, video vignettes or pictograms, slogans (in print, audio, video), or some combination of these. Each format has advantages and disadvantages and varying cost and reach. The ultimate impact and reach of the guidelines are in large part determined by the success and appeal of the presentation format(s).

- *Communication*: A communication strategy should be created by the leadership team with input from marketing and communication experts. This strategy may include a launch event, determination of presentation format(s), development of promotional merchandise, a social marketing campaign, and linkages with other complementary events, activities, or campaigns. Of importance is sustainability of the guideline promotion in the communication strategy to keep the guidelines fresh and omnipresent. The duration of communication campaigns should align with the timing of guideline revisions.

- *Dissemination*: A dissemination strategy should be prepared as a part of or complementary to the communication strategy. This strategy should draw on the reach of all partner and stakeholder groups involved in the guideline development and the various elements of the communication strategy. Dissemination in most jurisdictions will have media, hard copy, and electronic distribution channels.

Evaluation (Stage 14)

It is essential that the leadership team plan for comprehensive evaluations of the physical activity guideline

369

development process and impact. The guideline development process we have described has process assessment procedures integrated into the various stages. The results of these assessments should be published, disseminated, and used to improve or revise future iterations. Furthermore, the various stakeholder groups should be surveyed to get their evaluation of the process, including an assessment of the appropriateness of the level of their engagement, their satisfaction with the final product, and their activity in disseminating and promoting the guidelines.

Evaluations are also required for the communication and dissemination strategies referred to in discussion of stage 9. Using the cascade of communication effects (Cameron et al. 2007), the actual impact of the guidelines on the target population should be assessed to see the impact on awareness, knowledge, beliefs, intentions, and behaviors. Quantitative (survey) methods can be used, as well as qualitative (interview, focus group) methods, to gather information to inform future revisions to the guidelines themselves and to the knowledge translation strategies.

Updates and Revision Planning (Stage 15)

The level of research currently under way in the physical activity sciences is unprecedented. Consequently, new evidence is constantly emerging. For this reason, and to streamline planning processes, it is recommended that physical activity guidelines be reviewed and updated on a five-year cycle similar to that for food intake guidelines in the United States. International and interjurisdictional cooperation and harmonization in revision planning could lead to resource efficiencies and stronger, more consistent public health messaging.

Strengths, Limitations, and Challenges

The development of physical activity guidelines is complicated and complex, requiring a careful blend of science and art. Strengths of the process described in this chapter include the fact that it is systematic, transparent, consultative, informed by evidence, and robust, as well as the fact that it undergoes evaluation and is a "living" process designed to accommodate ongoing reviews and revisions. However, several important limitations and challenges are associated with physical activity guideline development, including the following:

- The need to reconcile multiple guidelines within the same jurisdiction
- The need to convey **minimal versus optimal** guidelines without leading to confusion—the population has a tendency to quickly accept minimal guidelines as "targets"
- The need to capture the importance of **baseline physical activity** (background- and lifestyle-embedded activity) as activity that is above and beyond the articulated physical activity guidelines
- Different cultural interpretations of physical activity
- The need to devise simple messages from complex data with multiple parameters (frequency, intensity, duration, type, context)
- The reality that there remain many important data gaps; this means that some aspects of physical activity guidelines require expert consensus
- Inconsistent findings in the evidence base
- Knowledge translation—translating research into messages relevant and meaningful to daily life while staying true to the evidence (see table 23.3 for subtle differences in guideline language between jurisdictions)
- Demand for a specific target for physical activity guidelines (e.g., minutes per day or week at a certain intensity) while the evidence clearly indicates a dose–response continuum (with arguably no "clear" threshold for minimal or optimal benefits)
- The need to incorporate the context of physical activity (home, school, work, community, travel) into guidelines and messaging when the context often varies by geography, climate, and culture
- The need to clarify the target populations for the guidelines (e.g., primary, secondary, tertiary prevention guidelines vs. treatment guidelines)
- Age segmentation of guidelines and the target audience for the age segmentation (e.g., individual, parent, caregiver, teacher)
- Cultural segmentation
- Geographical segmentation
- Publication bias (researchers and publishers tend to favor studies with positive findings)
- Cost and expense of guideline development, promotion, dissemination, evaluation, and revision

TABLE 23.3 Examples of Physical Activity Guidelines From Different Jurisdictions

Country	Organization	Date of recommendation and guideline process	Population	Guideline or recommendation
Australia (Sims et al. 2006)	Australian Department of Health and Aging	2005 Scientific Advisory Board, review work, development of guidelines	Children and adolescents (5-18 years)	At least 60 min of moderate- to vigorous-intensity physical activity most days of the week.
			Adults	30 min of moderate-intensity physical activity most days of the week. This can be accumulated in bouts of 10 to 15 min.
			Older adults	Same as adult guidelines, incorporating fitness, strength, balance, and flexibility. Should start at a low level and gradually increase activity.
Canada (Kesaniemi et al. 2010; Janssen and LeBlanc 2010; Warburton et al. 2010; Paterson and Warburton 2010; Tremblay et al. 2011)	Canadian Society for Exercise Physiology	2011 Systematic reviews, critical analysis, summary statements	Children and youth (5-17 years)	For health benefits, at least 60 min of moderate- to vigorous-intensity physical activity daily. This should include: Vigorous-intensity activities at least three days per week Activities that strengthen muscle and bone at least three days per week More daily physical activity provides greater health benefits.
			Adults (18-64 years)	To achieve health benefits, at least 150 min of moderate- to vigorous-intensity aerobic physical activity per week, in bouts of 10 min or more. It is also beneficial to add muscle- and bone-strengthening activities that use major muscle groups, at least two days per week. More physical activity provides greater health benefits.
			Older adults (≥65 years)	To achieve health benefits and improve functional abilities, at least 150 min of moderate- to vigorous-intensity aerobic physical activity per week, in bouts of 10 min or more. It is also beneficial to add muscle- and bone-strengthening activities that use major muscle groups, at least two days per week. Those with poor mobility should perform physical activities to enhance balance and prevent falls. More physical activity provides greater health benefits.

(continued)

TABLE 23.3 *(continued)*

Country	Organization	Date of recommendation and guideline process	Population	Guideline or recommendation
England and Northern Ireland (United Kingdom Department of Health 2004)	A report from the Chief Medical Officer, Department of Health	2004	Children	60 min per day of at least moderate-intensity physical activity. At least twice a week, this should include activities to improve bone health, muscle strength, and flexibility.
			Adults	A total of at least 30 min of at least moderate-intensity physical activity a day on five or more days a week.
			Older adults	Same as adult guidelines.
European Union (Oja et al. 2010; European Union 2008)	European Union and its member states European network for the promotion of health-enhancing physical activity (HEPA Europe)	2008 An expert group came together to create recommendations for guidelines	Children and adolescents	Minimum of 60 min a day of moderate-intensity physical activity.
			Adults and seniors	Minimum of 30 min of daily moderate-intensity physical activity. (Refer to the guidance documents of the World Health Organization regarding obesity and physical activity.)
International (World Health Organization 2010)	World Health Organization World Health Assembly	2010 An expert panel used science base developed for U.S. and Canadian guidelines process plus additional reviews	Children and adolescents (5-17 years)	Accumulate ≥60 min of moderate- and vigorous-intensity activity daily. Most activity should be aerobic, but activities that strengthen muscle and bone should be performed at least three times per week.
			Adults (18-64 years)	At least 150 min of moderate- or 75 min of vigorous-intensity aerobic activity each week, or an equivalent combination of moderate- and vigorous-intensity activity.
			Older adults (≥65 years)	Same as adult guidelines, incorporating activities that promote aerobic power, strength, balance, and flexibility. If limited, people should be as physically active as their abilities and conditions allow.

Country	Organization	Date of recommendation and guideline process	Population	Guideline or recommendation
Scotland (Scotland Physical Activity Task Force 2003)	Scotland Physical Activity Task Force Healthy Living Scotland	2003	Children	60 min of moderate-intensity physical activity on most days of the week.
			Adults	30 min of moderate-intensity physical activity on most days of the week.
			Older adults	Same as adult guidelines, with additional strength and balance activities three times per week.
United Kingdom (Chief medical officers for England, Northern Ireland, Scotland, and Wales 2011)	Chief medical officers for four countries	2011 Technical report, assessment of existing and new reviews, inter-country harmonization	Children and adolescents (5-18 years)	Moderate- to vigorous-intensity physical activity for at least 60 min and up to several hours every day. Vigorous-intensity activities, including those that strengthen muscle and bone, should be incorporated at least three days a week. Minimize the amount of time spent being sedentary (sitting) for extended periods.
			Adults (19-64 years)	Aim to be active daily. Over a week, activity should add up to at least 150 min of moderate-intensity activity in bouts of 10 min or more—one way to approach this is to do 30 min on at least five days per week. Alternatively, comparable benefits can be achieved through 75 min of vigorous-intensity activity spread across the week or a combination of moderate- and vigorous-intensity activity. Undertake physical activity to improve muscle strength on at least two days a week. Minimize the amount of time spent being sedentary (sitting) for extended periods.
			Older adults (≥65 years)	Older adults who participate in any amount of physical activity gain some health benefits, including maintenance of good physical and cognitive function. Some physical activity is better than none, and more physical activity provides greater health benefits. General recommendations are the same as adult guidelines. Those at risk of falls should incorporate physical activity to improve balance and coordination on at least two days a week. Minimize the amount of time spent being sedentary (sitting) for extended periods.

(continued)

TABLE 23.3 *(continued)*

Country	Organization	Date of recommendation and guideline process	Population	Guideline or recommendation
United States (Physical Activity Guidelines Advisory Committee 2008)	U.S. Department of Health and Human Services Physical Activity Guidelines Advisory Committee, Centers for Disease Control and Prevention, American College of Sports Medicine, American Heart Association	2008 Systematic review, critical analysis, and summary statement	Children and adolescents (6-17 years)	60 min of moderate- to vigorous-intensity physical activity per day; should include muscle strengthening and bone strengthening activities.
			Adults (18-65 years)	150 min per week of moderate-intensity or 75 min per week of vigorous-intensity aerobic physical activity. Aerobic activity should be performed in bouts of at least 10 min, spread throughout the week. Additional health benefits are provided with increasing time spent being physically active. Muscle strengthening activities that involve all major muscle groups performed on two or more days per week.
			Older adults (>65 years)	General recommendations are the same as adult guidelines. If this is not possible because of limiting chronic conditions, older adults should be as physically active as their abilities allow. They should avoid inactivity. Older adults should do exercises that maintain or improve balance if they are at risk of falling.
Wales (Welsh Assembly Government 2009)	Welsh Assembly Government	2009	Children	60 min of moderate-intensity physical activity on at least five days of the week.
			Adults	30 min of moderate-intensity physical activity on at least five days a week.

Physical Activity Guideline Differences

As discussed earlier, there are many challenges with developing physical activity guidelines and crafting the accompanying public-facing messages. As described in detail throughout this book, the science and research on physical activity and health have proliferated in recent years, and consequently the knowledge base has evolved rapidly. This rapid change in the evidence upon which guidelines are developed makes it difficult to keep guidelines and messages current and congruent with those of other countries or jurisdictions. Accordingly, guidelines may differ between or even within jurisdictions depending on when the evidence examined to inform the guidelines was compiled. Guidelines may also differ slightly between jurisdictions, even if based on the same evidence, because of other cultural, programmatic, or contextual differences or nuances.

Table 23.3 includes quite recent guidelines only, but a comparison of only the guidelines from Scotland (Scotland Physical Activity Task Force 2003) and those from the World Health Organization (2010) for children reveals the following important differences:

Children

- The World Health Organization uses the word "accumulate" to respect the fundamental movement patterns of children and the growing evidence base showing that such intermittent physical activity is related to positive health outcomes.
- The newer guidelines also say "greater than" or equal to 60 min to make the important point that the dose–response evidence is increasingly clear: More physical activity provides greater health benefits, well beyond those associated with the threshold recommendation of 60 min.
- The Scotland guidelines say "moderate-intensity physical activity" whereas the World Health Organization adds "vigorous" because of the growing literature demonstrating that higher-intensity physical activity provides different or added health benefits or both.
- The World Health Organization guidelines also say "daily" in contrast to "most days of the week" because the evidence suggests that as frequency of activity increases, so too does the health benefit; furthermore, the recommendation of daily is more succinct, easier to remember, and is supportive of the underlying message of making physical activity habitual.
- The World Health Organization states that "most activity should be aerobic, but activities that strengthen muscle and bone should be performed at least three times per week." This additional specificity is an indication of the growing evidence base relating specific activity interventions to health outcomes.
- The World Health Organization guidelines also clarify the age of children and adolescents as 5 to 17 years to align with what "school-aged" means in most countries and the fact that much of the research is done on school-aged children.

Adults

- Whereas the Scotland recommendation refers to "most days of the week" the World Health Organization refers to "each week." This difference reflects the fact that the research is unclear on what combination or sequence of physical activity confers the greatest health benefits; what the current evidence shows is that 150 min per week of moderate-intensity physical activity confers significant health benefits, irrespective of how it is accumulated throughout

The need for guidelines or recommendations to provide a specific target is complicated by the fact that the research evidence supports a dose–response continuum of benefit from physical activity. Guidelines or recommendations frequently provide a specific target such as 150 min per week of moderate-intensity activity. This is necessary in order to provide brief and concise messaging that quickly and precisely delivers a simple message to the target audience. However, statements like this can make it appear that the given amount of activity needs to be achieved before any benefit accrues or that higher amounts will provide no further benefit. Our understanding of the biological changes produced by increasing physical activity is that most changes occur on a continuum, so that every increase in activity intensity or amount provides an additional benefit.

Physical activity refers to all body movement produced by muscle contractions; thus it occurs for many different reasons and in a variety of settings (e.g., work, school, play, transportation, family, community). To understand why people are not sufficiently active or what messaging or strategies might be helpful in increasing their activity, the context of the activity is important in developing guidelines and recommendations.

the week (provided that it is in bouts of 10 min or more). Presenting weekly guidelines has the inherent advantage of additional flexibility to accommodate individual schedules and situations.

- The newer guidelines also make the important point that less duration is required if the activity is of higher intensity. In recent years, many studies have shown that more and different health benefits accrue from vigorous-intensity physical activity compared to moderate-intensity activity. In other words, the dose–response relationship between physical activity and health holds true for frequency, duration, and intensity. By stating guidelines for moderate and vigorous intensity, the World Health Organization guidelines incorporate this growing evidence base.

These differences in the guidelines for children and adults illustrate the importance of regular updates to stay true to the best available evidence and to provide the best possible public health and clinical guidance. Each word in a guideline or physical activity recommendation is important. Differences among physical activity guidelines may reflect their vintage, jurisdictional context, variations in the quality of the development process, underlying biases, or some combination of these. It is hoped that this chapter will assist physical activity leaders and researchers in understanding and assessing the relative contribution of these various factors.

Research Gaps and Future Directions

Research gaps are typically identified through the guideline development process (stage 7). These may include the following:

- Specific populations (e.g., age groups, ethnic groups, geographic groups, pregnant women, those with disabilities)

- The relationship between physical activity and specific health conditions (e.g., mental health, osteoporosis, arthritis, social well-being)

- The physical activity and health dose–response relationship (e.g., evidence at high or low ends of the continuum, differences in dose–response relationships among age groups or health conditions)

- Identification of thresholds (e.g., minimal vs. optimal levels of physical activity for the prevention of certain health conditions) for duration, intensity, frequency, and volume of physical activity

- Prevention versus treatment of health conditions

- Contrast and interaction between physical activity and sedentary behavior dose–response relationships

- Temporal trends of physical activity within different segments of the population

In addition, there is a need for research related to guideline messaging (Brawley and Latimer 2007; Latimer, Brawley, and Basset 2010), guideline impact (Cameron et al. 2007), the efficacy of alternate format

dissemination (e.g., paper vs. electronic, generic vs. customized or personalized), and the importance of adjuncts (e.g., pedometers, social marketing campaigns, merchandise). Future research should also explore the possibilities of creating web-based individualized guidelines and integrated lifestyle guidelines (physical activity guidelines merged with healthy eating, sleeping, and sedentary behavior guidelines).

Summary

The process of going from science to physical activity guidelines can be complicated and complex, as described in this chapter. The process, if done rigorously and comprehensively, generally takes approximately two years to complete, depending on how much previous work has been done, how close the working relationships are among the various researchers and stakeholder groups, how many resources (human and financial) are available, and how much the process is able to benefit from comparable efforts done in other countries or jurisdictions or by other organizations. There are several sequential stages to complete, each requiring the involvement and input of a variety of partners in the guideline development process. The stages of the process described here are applicable to physical activity guideline development for various ages and subgroups and are generally applicable or transferable to development of other guidelines (e.g., sedentary behavior guidelines).

Despite the rigorous nature of the process described in this chapter, there remain important limitations to the research informing physical activity guidelines, and translating often complicated research into meaningful and yet substantiated public health messages is very difficult. Future research should work to address knowledge gaps, provide a better understanding of the most efficacious presentation of physical activity guidelines, and ensure that robust evaluation procedures serve to inform guideline revisions.

Key Concepts

baseline physical activity—Usual type, amount, and intensity of physical activity a person performs throughout the day. Also referred to as "usual daily activity," it includes the activity performed during self-care, household chores, transportation, occupation, and leisure time.

guideline—Summary statement based on the scientific recommendation but translated to be understandable to the general public with due consideration of implementability and uptake. By definition, following a guideline is never mandatory. Guidelines are an essential part of the larger process of governance. Guidelines may be issued by and used by any organization (governmental or private) to increase awareness and provoke action.

messaging—Stage in the guideline development process in which guidelines are brought to life. This may include the use of supplemental images, graphics, stories, or text that helps translate guidelines into meaningful applications in the life of the target audience.

minimal versus optimal—In the development of physical activity guidelines or recommendations, minimal refers to the least amount or lowest intensity of activity that will produce a desired benefit. Optimal in this context refers to the profile of activity (type, intensity, frequency, duration) that results in the greatest benefits, with the costs (time, effort) and risk of injury taken into consideration. An optimal activity program would produce many of the health and fitness benefits of being physically active but would fit well into the person's schedule, and the risk of injury would be small. What is minimal or optimal may vary from person to person.

recommendation from scientific literature—Evidence-informed summary statement related to the particular research synthesis.

Study Questions

1. List and briefly describe the key stages recommended for the development, promotion, and evaluation of physical activity guidelines.

2. What are three reasons a leadership team might be motivated to develop or update physical activity guidelines?

3. Why are international and interjurisdictional cooperation and harmonization important in the development of physical activity guidelines?

4. What are at least four issues that the guideline development and research committee should consider before initiating a systematic review of the literature?

5. Describe the PICO framework in the context of performing a systematic review.

6. What are three knowledge translation processes used to convert the summarization of the science to effective physical activity messaging?

7. Discuss reasons why physical activity guidelines developed from the same evidence base may be different in their final wording.

8. Explain the difference between a physical activity guideline and physical activity messaging.

References

Brawley, L.R., and A.E. Latimer. 2007. Physical activity guides for Canadians: Messaging strategies, realistic expectations for change, and evaluation [in French]. *Applied Physiology, Nutrition, and Metabolism* 32(Suppl):S189-S205.

British Heart Foundation National Centre for Physical Activity and Health. 2010. *Technical report: Physical activity guidelines in the U.K.: Review and recommendations.* Leicestershire, UK: School of Sport, Exercise and Health Sciences, Loughborough University.

Brouwers, M.C., M.E. Kho, G.P. Browman, J.S. Burgers, F. Cluzeau, G. Feder, B. Fervers, I.D. Graham, J. Grimshaw, S.E. Hanna, P. Littlejohns, J. Makarski, and L. Zitzelsberger. 2010a. AGREE II: Advancing guideline development, reporting and evaluation in health care. *Canadian Medical Association Journal* 182: E839-E842.

Brouwers, M.C., M.E. Kho, G.P. Browman, J.S. Burgers, F. Cluzeau, G. Feder, B. Fervers, I.D. Graham, S.E. Hanna, J. Makarski, AGREE Next Steps Consortium. 2010b. Development of the AGREE II, part 1: Performance, usefulness and areas for improvement. *Canadian Medical Association Journal* 182:1045-1052.

Brouwers, M.C., M.E. Kho, G.P. Browman, J.S. Burgers, F. Cluzeau, G. Feder, B. Fervers, I.D. Graham, S.E. Hanna, J. Makarski, AGREE Next Steps Consortium. 2010c. Development of the AGREE II, part 2: Assessment of validity of items and tools to support application. *Canadian Medical Association Journal* 182:E472-E478.

Cameron, C., C.L. Craig, F.C. Bull, and A. Bauman. 2007. Canada's physical activity guides: Has their release had an impact? *Canadian Journal of Public Health* 98(Suppl):S161-S169.

Chief Medical Officers for England, Northern Ireland, Scotland, and Wales. 2011. *Start active, stay active: A report on physical activity for health from the four home countries' chief medical officers.* London: Crown.

Davis, D., J. Goldman, and V.A. Palda. 2007. *Canadian Medical Association handbook on clinical practice guidelines.* Ottawa: Canadian Medical Association.

Downs, S.H., and N. Black. 1998. The feasibility of creating a checklist for the assessment of the methodological quality both of randomised and non-randomised studies of health care interventions. *Journal of Epidemiology and Community Health* 52:377-384.

European Union. 2008. *European Union physical activity guidelines. Recommended policy actions in support of health-enhancing physical activity.* Brussels: European Union.

Guyatt, G.H., A.D. Oxman, R. Kunz, Y. Falck-Ytter, G.E. Vist, A. Liberati, H.J. Schunemann, GRADE Working Group. 2008a. Going from evidence to recommendations. *British Medical Journal* 336:1049-1051.

Guyatt, G.H., A.D. Oxman, R. Kunz, R. Jaeschke, M. Helfand, A. Liberati, G.E. Vist, H.J. Schunemann, GRADE Working Group. 2008b. Incorporating considerations of resources use into grading recommendations. *British Medical Journal* 336:1170-1173.

Guyatt G.H., A.D. Oxman, R. Kunz, G.E. Vist, Y. Falck-Ytter, GRADE Working Group. 2008c. What is "quality of evidence" and why is it important to clinicians? *British Medical Journal* 336:995-998.

Guyatt, G.H., A.D. Oxman, G.E. Vist, R. Kunz, Y. Falck-Vitter, P. Alonso-Coello, H.J. Schunemann, GRADE Working Group. 2008d. GRADE: An emerging consensus on rating quality of evidence and strength of recommendations. *British Medical Journal* 336:924-926.

Higgins, J.P.T., and S. Green (eds.). 2009. *Cochrane handbook for systematic reviews of interventions*. Version 5.0.2. London: Cochrane Collaboration.

Jaeschke, R., G.H. Guyatt, P. Dellinger, H. Schunemann, M.M. Levy, R. Kunz, S. Norris, J. Bion, GRADE Working Group. 2008. Use of GRADE grid to reach decisions on clinical practice guidelines when consensus is elusive. *British Medical Journal* 337:a744.

Janssen, I., and A.G. LeBlanc. 2010. Systematic review of the health benefits of physical activity and fitness in school-aged children and youth. *International Journal of Behavioral Nutrition and Physical Activity* 7:40.

Kesaniemi, A., C.J. Riddoch, B. Reeder, S.N. Blair, and T.I.A. Sorensen. 2010. Advancing the future of physical activity guidelines in Canada: An independent expert panel interpretation of the evidence. *International Journal of Behavioral Nutrition and Physical Activity* 7:41.

Latimer, A.E., L.R. Brawley, and R.L. Bassett. 2010. A systematic review of three approaches for constructing physical activity messages: What messages work and what improvements are needed? *International Journal of Behavioral Nutrition and Physical Activity* 7:36.

Oja, P., F.C. Bull, M. Fogelholm, and B.W. Martin. 2010. Physical activity recommendations for health: What should Europe do? *BMC Public Health* 10:10.

Paterson, D.H., and D.E. Warburton. 2010. Physical activity and functional limitations in older adults: A systematic review related to Canada's Physical Activity Guidelines. *International Journal of Behavioral Nutrition and Physical Activity* 7:38.

Physical Activity Guidelines Advisory Committee. 2008. *Physical Activity Guidelines Advisory Committee report, 2008*. Washington, DC: U.S. Department of Health and Human Services.

Scotland Physical Activity Task Force. 2003. *Let's make Scotland more active: A strategy for physical activity*. Edinburgh: Healthy Living Scotland.

Shea, B.J., J.M. Grimshaw, G.A. Wells, M. Boers, N. Andersson, C. Hamel, A.C. Porter, P. Tugwell, D. Moher, and L.M. Bouter. 2007. Development of AMSTAR: A measurement tool to assess the methodological quality of systematic reviews. *BMC Medical Research Methodology* 7:10.

Shea, B.J., C. Hamel, G.A. Wells, L.M. Bouter, E. Kristjansson, J. Grimshaw, D.A. Henry, and M. Boers. 2009. AMSTAR is a reliable and valid measurement tool to assess the methodological quality of systematic reviews. *Journal of Clinical Epidemiology* 62:1013-1020.

Sims, J., K. Hill, S. Hunt, B. Haralambous, A. Brown, L. Engel, N. Huang, N. Kerse, and M. Ory. 2006. *National physical activity recommendations for older Australians: Discussion document*. Canberra: Australian Government Department of Health and Ageing.

Tremblay, M.S., D.E.R. Warburton, I. Janssen, D.H. Paterson, A.E. Latimer, R.E. Rhodes, M.E. Kho, A. Hicks, A.G. LeBlanc, L. Zehr, K. Murumets, and M. Duggan. 2011. New Canadian Physical Activity Guidelines. *Applied Physiology, Nutrition and Metabolism* 36(1):36-46.

Tremblay, M.S., K.E. Kho, A.C. Tricco, and M. Duggan. 2010. Process description and evaluation of Canadian Physical Activity Guidelines development. *International Journal of Behavioral Nutrition and Physical Activity* 7:42.

United Kingdom Department of Health. 2004. *At least five a week: Evidence on the impact of physical activity and its relationship to health. A report from the Chief Medical Officer*. England: Department of Health.

Warburton, D.E., S. Charlesworth, A. Ivey, L. Nettlefold, and S.S. Bredin. 2010. A systematic review of the evidence for Canada's Physical Activity Guidelines for Adults. *International Journal of Behavioral Nutrition and Physical Activity* 7:39.

Welsh Assembly Government. 2009. *Climbing higher: Creating an active Wales A 5 year strategic action plan consultation document*. Welsh Assembly Government.

World Health Organization. 2009. *Global health risks: Mortality and burden of disease attributable to selected major risks*. Geneva: World Health Organization.

World Health Organization. 2010. *Global recommendations on physical activity for health*. Geneva: World Health Organization.

PART VI

New Challenges and Opportunities

You have now been exposed to the broad areas of sedentary time, physical activity, fitness, and health. You should have a better understanding of how the field has developed over the past several decades. In part II, international experts described our current understanding of how physical activity challenges the human organism and specific physiological systems. The nine chapters in part III contain what is perhaps the core material of the book. A group of distinguished investigators presented material on what we know about how regular physical activity, or in some cases sedentary lifestyles, affect several different health outcomes. You learned that a sedentary and unfit way of life has devastating effects on many health outcomes and decreases longevity. Part IV focused on aging, brain functions, and mental health. Part V included two chapters on what we know about specific amounts, types, and intensities of physical activity in relation to health outcomes. It also incorporated a chapter on the process to be ideally followed in the development of activity- and fitness-related guidelines.

Part VI reviews current knowledge on exercise genomics issues that are relevant to the aim of this book. It illustrates how genetic factors contribute to human heterogeneity in exercise-related traits and how they influence the changes observed in response to regular exercise. Finally, the editors present some concluding thoughts on the integration of data and concepts presented here and what this means for the promotion of physical activity. The chapter concludes with a discussion of selected research areas on which we should focus our attention in the future.

24

Genetic Differences in the Relationships Among Physical Activity, Fitness, and Health

Tuomo Rankinen, PhD; and Claude Bouchard, PhD

CHAPTER OUTLINE

©Imago/Icon SMI

This book provides ample evidence of the benefits of regular physical activity and the advantages of maintaining a reasonable level of health-related fitness. What has not been discussed in any detail thus far is the magnitude of individual differences in the relationship between physical activity or fitness and health outcomes and the role of genetic heterogeneity in accounting for human variation. These topics are the focus of this chapter. After a brief overview of the human genome, the unique features of human **genes** and the regulation of their expression or repression are described. The extent of variation in the genome of *Homo sapiens* is reviewed. Next we discuss how genetic variation affects sedentary people. The magnitude of the individual differences in response to regular exercise is then defined. We discuss the role of genes and DNA sequence variants in the response of blood pressure, lipids and lipoproteins, glucose and insulin, adiposity, and cardiorespiratory endurance. Finally we highlight the implications for public health and individualized preventive medicine approaches.

Basics of Human Genetics

The first section of the chapter is devoted to defining the human genome and its key characteristics. The human gene, gene expression regulation, and the role of **alternative splicing** are then briefly described.

Human Genome

The biology of the gene and the characteristics of the human genome are complex topics about which more is learned every day. The blueprint of the human body is contained in the genetic code specified in the deoxyribonucleic acid (DNA) sequence of chromosomes found in every nucleated cell. Human **genome** is a term that refers to the total genetic information in human cells. It consists of 22 pairs of autosomes and two sex chromosomes in somatic (nonreproductive) cells. Male and female gametes (germ line cells) each contain a nucleus in which 23 chromosomes are normally present. The gametes are haploid, which means they contain a single set of chromosomes. The female gamete has a single copy of each of 22 autosomes and an X chromosome. The male gamete has a similar complement of autosomes and either an X or a Y chromosome. At fertilization, the nuclear contents of the female and male gametes fuse, and the diploid number of chromosomes (23 pairs) is restored with either two X chromosomes (XX, a female zygote) or an X and a Y chromosome (XY, a male zygote).

Chromosomes are composed of long chains of DNA and basic and acidic proteins packed with the DNA. The genetic material in each chromosome is a long string of the four DNA bases: adenine (A), cytosine (C), guanine (G), and thymine (T), joined together via phosphate bonds. The two complementary strands are precisely folded and twisted around one another to form a double helix, with the informative base on the inside. The strands of paired complementary sequences of nucleotides (bases) are held together with relatively weak hydrogen bonds (figure 24.1). C pairs with G, and A pairs with T. The 23 chromosomes contain about 2 m of linear DNA, or about 3 billion pairs of nucleotides. The linear structure of bases in the DNA strands is called the primary structure of the chromosome. The secondary structure of the chromosome arises when the two complementary strands of DNA twist to form a double helix. One turn of the helix is called a pitch and accommodates 10 nucleotides.

The chromosomes are arranged by size and position of the centromere; autosomes are numbered from 1 to 22, and the sex chromosomes are noted as X and Y. The short arm of a chromosome is denoted as p and the long arm as q. Each arm is subdivided into regions numbered consecutively from the centromere to the telomere (tip of the chromosome), and each band within a given region is identified by a number. With this nomenclature, it is possible to specify any chromosomal region by a "cytological address"; for example, 2p25 refers to chromosome 2, p arm, region 2, band 5 (figure 24.2). However, the sequence of almost all the DNA bases of the entire human genome is now available (www.ncbi.nlm.nih.gov), and this is rapidly changing the way an address on a chromosome is defined. Indeed, it is now possible to specify a physical position on a given human chromosome in terms of the exact base number in a sequence ranging from one to millions. Thus, the availability of the DNA sequence for each chromosome makes it possible to go beyond the crude cytological address and specify an actual physical position on a chromosome (in terms of base number).

Each mitochondrion of a cell contains several copies of circular, double-stranded DNA molecules composed of 16,569 base pairs (bp). This is a small number of base pairs compared with nuclear DNA. Mitochondrial DNA (mtDNA) is able to replicate itself independently of nuclear DNA and has its own system of **transcription** and **translation**. Mitochondrial DNA is inherited from the mother through the cytoplasm of the ovum. This small DNA codes for 13 polypeptides associated with the regeneration of adenosine triphosphate (ATP) in the mitochondrion

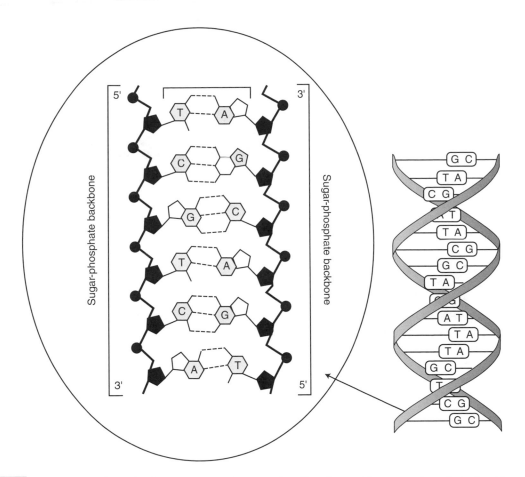

FIGURE 24.1 Two polynucleotide chains running in opposite directions with sugar-phosphate backbones on the outside and the nitrogenous bases inside paired to each other by hydrogen bonds. Adenine (A) pairs with thymine (T), whereas cytosine (C) pairs with guanine (G).

Reprinted from R. Roberts et al.,1992, *A primer of molecular biology* (Heidelberg, Germany: Springer Verlag), 22, with kind permission of Springer Science and Business Media.

and for two ribosomal ribonucleic acid (rRNA) and 22 transfer ribonucleic acid (tRNA) molecules. The remaining genetic information required for the synthesis of the proteins of the mitochondrion originates from nuclear DNA. The importance of mtDNA cannot be overestimated because it has been implicated in several fundamental biological processes associated with the ability of the cell to meet changing energy needs, as well as with aging, a number of anomalies and diseases, and cell death.

The double-stranded DNA of a chromosome is packed very tightly in nucleosomes, which consist of a DNA strand wrapped around histone proteins. The adjacent nucleosomes are linked to each other by a stretch of DNA stabilized by a linker histone. Chemical modifications, such as acetylation, methylation, and phosphorylation, may alter the histone proteins of nucleosomes. Such histone modifications can change the conformation of **chromatin** and thereby potentially change transcription activity. Also, DNA

can be modified via cytosine base methylation in the 5'-position. Methylation occurs when cytosine is followed by guanine, a sequence known as a CpG island. DNA methylation usually reduces transcription, although if it occurs at a repressor element, it may actually increase transcription. Most of the CpG islands in the genome are methylated in the basal, natural state. However, other sites can be methylated, as occurs during development or with the repression of tumor suppressor genes. Also, the inactivation of one X chromosome in the complement of cells of females is based on DNA methylation.

DNA methylation and chemical modification of the histone proteins constitute the **epigenome**. Epigenetic alterations do not involve changes in the DNA sequence. However, they are often transmitted from one cell to daughter cells at mitosis. Some of the epigenetic changes have also been shown to be passed from parents to offspring through the germ lines. Epigenetic modifications can occur at all ages

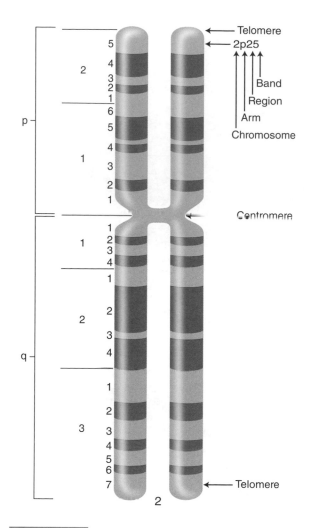

FIGURE 24.2 Numbering system used to specify a cytological address on a chromosome.

Reprinted, by permission, from R.M. Malina, C. Bouchard and O Bar-Or, 2004, *Growth, maturation, and physical activity*, 2nd ed. (Champaign, IL: Human Kinetics), 371.

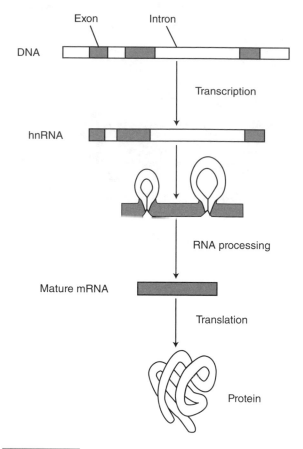

FIGURE 24.3 A gene includes introns and exons that are transcribed into messenger RNA (mRNA). The introns are spliced out as the transcript is processed into heterogeneous RNA (hnRNA) and then into mature mRNA that is subsequently translated into a polypeptide.

Reprinted, by permission, from R.M. Malina, C. Bouchard and O. Bar-Or, 2004, *Growth, maturation, and physical activity*, 2nd ed. (Champaign, IL: Human Kinetics), 373; adapted from Francomano and Kazazian 1986.

Human Gene

A typical gene (figure 24.3) consists of coding sequences (**exons**), noncoding regions (**introns**), and regulatory sequences located both before (5' end; **promoter** region) and after (3' untranslated region [UTR]) the coding regions of the gene. The number of exons is quite heterogeneous, with a range from one (e.g., intronless G-protein–coupled receptor genes) to several hundred (e.g., titin gene with 363 exons). Introns were originally considered to be nonfunctional stretches of DNA between the coding regions, but they may harbor regulatory elements, such as alternative promoters, and splicing enhancers and suppressors. Recent findings also indicate that introns and other noncoding regions of the genome may encode small noncoding RNA genes, such as micro RNAs (miRNAs), that play an important role in the regulation of gene expression.

The number of amino acids in proteins varies from very few to 1,000 and more, with an average of about 100. Thus, the average protein requires the coding information from about 300 DNA base pairs. With at least 3 billion pairs of nucleotides in the haploid human genome, about 20 million genes could be encoded. In contrast, a much lower number of genes has been deduced from the completed human genome sequence. The current estimate is about 20,000 genes. However, other lines of evidence have shown that the human genome encodes a much larger number of proteins than is suggested by the simple sequence data.

as a result of exposure to environmental factors, such as nutrients, cellular insults, and stress. Whether regular exercise or chronic inactivity are stimuli that can trigger epigenetic events remains to be established.

Gene Expression

Gene expression can be simply defined as the process by which genes are activated or repressed in response to biological signals from the internal or external milieu. Gene expression is an area of active research because of its potential significance for understanding the processes of growth, maturation, and aging as well as adaptation to environmental challenges or abnormal cellular growth as in cancer. Gene expression is also a fundamental process in the adaptive mechanisms to exercise.

Every nucleated cell contains a complete copy of the human genome. Some genes are expressed in most or all cells, because the gene product is essential for the function of the cell (housekeeping genes). However, the majority of the genes in the human genome are expressed only in specific organs, tissues, or cells; and some are expressed only at certain stages of a cell cycle or during specific periods of the development of an organism. Moreover, the expression of a gene may be enhanced or depressed in response to external stimuli, such as changes in the metabolic milieu of the cell or in extracellular concentrations of certain ions and nutrients. To accommodate all these specific demands, gene expression must be closely controlled and coordinated. The adaptability and coordination of the gene expression process are achieved through several regulatory mechanisms, such as transcription factors, alternative splicing, alternative promoters, **genomic imprinting**, and gene silencing through epigenetic mechanisms.

Alternative Splicing

One of the most striking surprises generated by the production of the human genome sequence was the low number of genes. The previous estimates based on the number of expressed sequences ranged from 80,000 up to 150,000, whereas the human genome project revealed the presence of only about 20,000 annotated and predicted genes. This finding emphasized the fact that several genes must produce more than one transcript and that the phenomenon of one gene's encoding several gene products is more common than previously thought.

The main factor contributing to the disparity between the number of genes and gene transcripts is alternative splicing. Alternative splicing refers to a situation in which a single gene produces multiple messenger RNAs (mRNAs) through different combinations of exons (figure 24.4). It is estimated that about 75% of the human multiexon genes have alternative splice forms and that at least 70% of splice isoforms change the characteristics of the gene product (Modrek and Lee 2002). Alternative splicing may cause either inclusion or exclusion of one or several exons, or alternative 5' or 3' sites. These changes may induce an in-frame addition or deletion of functional units in the gene product (alternative exons), change the amino terminus of the polypeptide (alternative initiation), or modify the carboxyl terminus of the gene product because of frameshift or alternative termination. The regulation of alternative splicing is still poorly understood. Multiple transcripts may

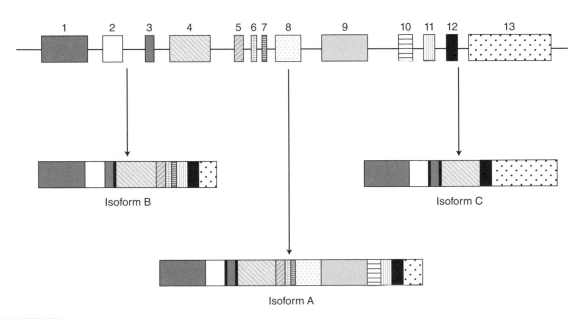

FIGURE 24.4 A gene consisting of 13 exons encodes three distinct messenger RNAs (mRNAs). Isoform A contains all 13 exons, whereas exons 8 through 10 and exons 5 through 11 are spliced off from isoforms B and C, respectively. Isoform C also uses an alternative stop codon.

Reprinted, by permission, from F.C. Mooren and K. Volker, 2005, *Molecular and cellular exercise physiology* (Champaign, IL: Human Kinetics), 42.

Important Features of the Human Genome

- 3 billion base pairs of DNA
- About 20,000 genes
- Alternative promoters
- Alternative splicing for about 75% of multiexon genes
- About 2,000 **transcription factors**
- Some genes with only one exon and no intron
- Most genes with many exons
- Micro RNAs and other small RNAs that contribute to regulation of gene expression

also arise from a single gene as a result of the transcription of both strands of DNA combined with alternative splicing.

Events in Human Genes and Genomes

We know how genes determine the sequences of proteins. In this section, the relationship between the encoded genetic information, proteins, and phenotypes is discussed.

From Genes to Proteins

The role of genetic information is to specify the sequence of amino acids that ultimately form all proteins synthesized by the cellular machinery. The set of all possible DNA triplets and the corresponding amino acids that they encode are called the genetic code. The process that allows the instructions contained in a given gene to be converted to a final gene product may appear to be simple but is in reality highly complex: The DNA sequence is first converted to an RNA sequence, which is then translated to a polypeptide giving rise to the final protein. The first step of the process, transcription, takes place in the nucleus of the cell. Subsequently, the mRNA specifies the primary sequence of a polypeptide, which may undergo posttranslational modifications, such as phosphorylation, methylation, acetylation, carboxylation, or glycosylation. The polypeptides may also be enzymatically cleaved to produce smaller functional products.

Transcription

The initiation of transcription requires the presence of transcription factors that bind to specific DNA sequence elements located in the immediate vicinity of a gene. These sequence elements, which are usually clustered upstream of the coding sequence, form the promoter region of the gene. Once the necessary transcription factors are bound to the promoter, an RNA polymerase binds to the transcription factor complex and is activated to start the synthesis of RNA. The common promoter elements recognized by several transcription factors include a TATA box (usually TATAAA) located about 25 bp before (–25 bp) the transcriptional start site, a CAAT box (–80 bp), and a GC box. In addition to the common promoter elements, enhancers, silencers, and response elements form a group of regulatory sequences that can enhance or inhibit the transcriptional activity of specific genes. These sequence elements are usually located quite far from the transcription initiation site (several thousands of base pairs). They bind gene regulatory proteins, and the DNA strand between the promoter and enhancer–silencer folds in a loop allowing the regulatory proteins to interact with the transcription factors bound to the promoter.

Once the synthesis of an RNA molecule from the DNA template is finished, the primary RNA transcript (i.e., the full copy of the original template DNA) undergoes various posttranscriptional modifications. These include removal of the unwanted internal segments (i.e., intronic sequences), rejoining of the remaining segments (exonic sequences), and capping at the 5' end of the transcript and polyadenylation at the 3' end. The removal of the intronic RNA segments is called RNA splicing. The process is directed by specific nucleotide sequences at the exon–intron boundaries (splice junctions).

Transcription Factors

Gene expression is acutely regulated in response to several external stimuli. These stimuli activate specific transcription factors, which bind to specific regulatory sequences in the promoter region (response elements) of the target genes, leading to their activation. Activation of RNA polymerases I and III, which transcribe housekeeping genes, requires a host of ubiquitous transcription factors. On the other hand, the transcription of polypeptide-encoding genes by RNA polymerase II uses complex sets of general and tissue-specific transcription factors. The complex of RNA polymerase II and general transcription factors is sufficient to initiate gene transcription at a minimum rate, which can be increased or turned off by additional positive or negative regulatory elements. However, the majority of the genes transcribed by RNA polymerase II show tissue-specific expression. The tissue specificity is achieved by interactions between special enhancer and silencer sequence and a variety of promoter sequence elements that are recognized only by tissue-specific transcription factors. The number of transcription factors, rather than the number of genes, has been suggested to be a key determinant of the biological complexity of an organism. For example, the human genome contains more than 2,000 genes for transcription factors, whereas these number about 500 and 700 in the worm (*C. elegans*) and fruit fly genomes, respectively (Szathmary, Jordan, and Pal 2001).

Small RNAs

The latest advance in the understanding of the regulation of gene expression is the discovery of the role played by small RNAs. Micro RNAs (miRNAs) and small interfering RNAs (siRNAs) are short (~22 nucleotides) RNA molecules that are cleaved from longer precursor mRNA sequences by a ribonuclease called Dicer. The precursors for miRNAs are about 70-nucleotide-long, hairpin-shaped RNAs transcribed from small noncoding genes, whereas siRNAs are produced from apparently aberrant double-stranded RNAs. The miRNAs can suppress gene expression either by binding to target mRNA and repressing translation or by degrading the mRNA. The effects of siRNA on gene expression are mediated through its binding on the RNA-induced silencing complex, which gives rise to an endonuclease that cleaves target mRNAs. Recent studies also suggest that small RNAs can regulate gene expression by affecting the form of chromatin. Changes in the compactness of chromatin can alter which genes are expressed without affecting the coding sequence of the genes.

From Proteins to Phenotypes

Translation, or the synthesis of polypeptides from an mRNA template, takes place in the cytoplasm. After mRNA molecules migrate from the nucleus to the cytoplasm, they bind with the ribosomes where translation takes place. Ribosomes are large RNA–protein complexes providing a structural framework for polypeptide synthesis.

Proteins are ubiquitous throughout the body and constitute more than 50% of the dry weight of a typical cell. Proteins are best understood in the context of their functions, which are summarized in table 24.1. Nine types of proteins are indicated in the classification. Several subdivisions could be added, but this classification is sufficient to demonstrate the central role of proteins in mediating the chain of events between genetic specifications and the expression of phenotypic characteristics. For example, enzymes are a very diverse class of proteins that have in common the capacity to increase the rate of biochemical reactions in cells. An enzyme typically has the property of accelerating a specific chemical reaction, although there are several exceptions.

As an example of the concepts presented here, consider an enzyme relevant to physical activity and health, glycogen synthase, which is a regulatory enzyme involved in the glycogen synthesis pathway of the liver and skeletal muscle. The enzyme exists in cells of both tissues in two forms—a less active, phosphorylated form and a more active, nonphosphorylated form. Phosphorylation is the process by which a phosphate molecule is added to a protein to alter its activation state. The phosphorylation of the enzyme is achieved by enzymes from the kinase family; dephosphorylation is brought about by the action of a phosphatase enzyme. This example shows that the effectiveness of a gene product (e.g., glycogen synthase in one tissue) is modulated in part by molecules that are themselves products of other genes. The resulting phenotype (i.e., measurable liver or muscle glycogen concentrations) is thus dependent on several genes. It is also dependent on other mechanisms, such as those related to the availability of glucose precursors and their entry into the glycolytic or glycogenic pathway, which may or may not be dependent on immediate genetic influences as well.

Acute exercise and regular exercise training induce several physiological responses that necessitate increased protein synthesis. This is necessary to meet the needs for higher levels of specific gene products (e.g., enzymes of energy production pathways), to replenish proteins that are catabolized during exercise, or to support the adaptive changes associated with the improved capacity to perform physical activity (e.g., enhanced blood flow in working muscles). It has been argued that the human genome evolved over a long period of time during which high levels of physical activity were necessary for survival. This

TABLE 24.1   A Classification of Proteins by Function

Protein	Example
Structural	Collagen, the human body's most abundant protein, is found in various kinds of connective tissues.
Storage	Ovalbumin is a major source of material and energy during embryonic development.
Transport	Hemoglobin transports oxygen from areas of high concentration in the lungs to areas of lower concentration in the tissues.
Receptor	Insulin receptors are proteins found embedded in the cell membrane and exposed on the surface of the cell. When insulin and the receptor combine, a complex signaling cascade is activated that leads to glucose uptake.
Hormone	Growth hormone, released by the pituitary gland, stimulates growth of most body tissues and has widespread metabolic effects.
Protective	Antibodies are produced in response to the presence of foreign substances, organisms, or tissues in the body.
Contractile	Actin, myosin, and other contractile proteins arranged in orderly arrays in muscle fibers reduce length by sliding past each other in a controlled manner.
Regulatory	Regulatory proteins influence which genes are expressed and when they are expressed. Transcription factors are proteins that bind to DNA sequence and control the expression of specific genes.
Enzymes	Enzymes are the largest and most diverse class of proteins. Creatine kinase, which allows the phosphorylation of ADP into ATP using creatine phosphate as the substrate, is an example.

ADP = adenosine monophosphate; ATP = adenosine triphosphate.

Adapted, by permission, from S. Singer, 1985, *Human genetics: An introduction to the principles of heredity,* 2nd ed. (New York, NY: W.H. Freeman and Company), 60.

would support the view that the link between exercise and the regulation of gene expression is deeply ingrained in our biology.

An example of the effect of exercise on gene expression is muscle contraction–induced skeletal muscle hypertrophy. Muscle contraction has been shown to increase both transcription and translation of myofibrillar proteins, such as α-actin. Data from animal models suggest that a transcription factor called serum response factor and its response element in the promoter of the α-actin gene are involved in muscle contraction–induced α-actin expression and consequently muscular hypertrophy. Muscle overload also increases expression levels of several growth factors, such as insulin-like growth factor 1 and signaling molecules (e.g., calcineurin).

This example illustrates the fact that the acute or chronic adaptation to exercise depends on the expression level of specific genes. However, the net effect of exercise-induced gene expression is determined by the integrated action of multiple genes. Recent advances in microarray technology have opened new opportunities to investigate the expression levels of thousands of genes simultaneously in a single experiment. This allows exploration of the effects of a specific stimulus, such as exercise, on the expression patterns of several gene families, such as transcription factors or genes involved in specific metabolic or physiologic pathways.

For instance, Roth and colleagues (2002) investigated the effects of a nine-week strength training program on the vastus lateralis gene expression profile in 20 sedentary participants. The strength training program consisted of unilateral knee extension exercises of the dominant leg. The participants exercised three times per week, and each training session consisted of four sets of high-volume, heavy-resistance knee extensions. A total of 69 genes showed >1.7-fold difference in expression levels after the training period in the pooled data. Fourteen of these genes were identified in all age-by-sex subgroups, 12 of them showing decreased and two showing increased expression levels after the training program (table 24.2).

Sequence Variation in the Human Genome

Two important events of meiosis contribute to the extraordinary amount of genetic diversity that is characteristic of humans and other sexually reproducing species. The first event is the independent assortment of chromosome pairs during their migration to daughter cells in gametogenesis. With 23 pairs of chromosomes in humans, 2^{23} or 8,388,608 different combinations of paternal and maternal chromosomes can occur in a haploid gamete. The second event that enhances genetic variation is recombination. Before the migration of chromosomes to daughter cells, when homologous chromosomes are paired, crossing-over occurs. Crossing-over refers to the exchange of chromosomal segments between homologous

TABLE 24.2 Summary of the Genes Showing Significant Increase or Decrease* in Expression Levels in Response to a Nine-Week Strength Training Program

Gene name	Gene symbol	Fold change
Downregulated genes		
Four and a half LIM domains 1	FHL1	0.248
Myosin, light polypeptide 2	MYL2	0.262
Cold shock domain protein A	CSDA	0.265
Glyceraldehyde-3-phosphate dehydrogenase	GAPDH	0.297
Actin, α 2	ACTA2	0.405
Myosin, light polypeptide 3	MYL3	0.442
Dynactin	ACTB	0.446
Eukaryotic translation elongation factor 1 γ	EEF1G	0.484
ATP synthase, mitochondrial F1 complex, β polypeptide	ATP5B	0.505
Troponin I	TNNI1	0.508
Actin-related protein 1, centractin α	ACTR1A	0.513
Topoisomerase (DNA) I	TOP1	0.547
Upregulated genes		
Tetraspanin 5	TSPAN5	2.131
TNF receptor–associated factor 6	TRAF6	1.852

LIM = comes from the three transcription factors, Lin-11, Isl-1, and Mec-3, in which the double zinc finger motif (LIM domain) was first identified; ATP = adenosine triphosphate; TNF = tumor necrosis factor.

*Significant increase is defined as >1.7-fold higher expression level, significant decrease as <0.6-fold reduction in expression level.

Data from Roth et al. 2002.

Key Sources of Variation in the Human Genome Among People

- Independent assortment of chromosomes at gametogenesis
- Recombination events during meiosis (crossing-over)
- Base substitutions
- Base deletions
- Base mutations
- Copy number variants

chromosomes and results in the recombination of **alleles** (alternative forms of genes) between the homologous chromosomes of maternal and paternal origins. This is illustrated in figure 24.5. For example, if a pair of chromosomes carries three genes, each existing in two different forms in the population (e.g., A and a, B and b, C and c), a crossing-over taking place between loci B and C will result in two recombinant chromosomes with new gene combinations and two nonrecombinant chromosomes carrying the parental gene combination. It is estimated that about two or three recombination events take place between each pair of homologous chromosomes (i.e., between pairs of chromosomes of maternal and paternal descent) during meiosis. The existing genetic variability in a population is thus amplified by independent assortment and recombination during meiosis to yield a very large number of different unique gametes, ensuring the genetic uniqueness of each individual (except monozygotic [MZ] twins).

In addition, a major source of genetic variation is the variety of heritable changes (mutations) in the

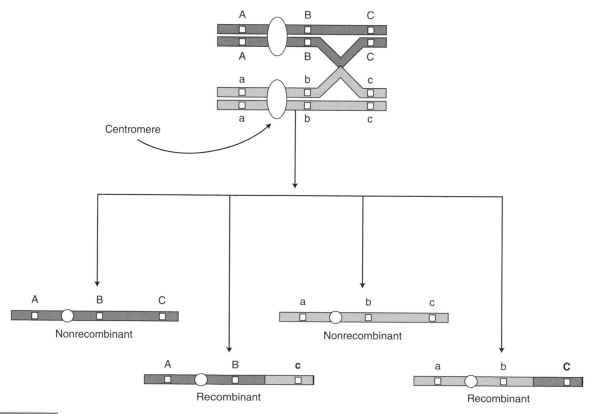

FIGURE 24.5 Recombination between homologous chromosomes.

Reprinted, by permission, from R.M. Malina, C. Bouchard and O. Bar-Or, 2004, *Growth, maturation, and physical activity*, 2nd ed. (Champaign, IL: Human Kinetics), 372.

nucleotide sequence of the DNA. Mutations can be grouped into three classes:

- Base substitutions
- Deletions
- Insertions (figure 24.6)

Base substitutions usually involve the replacement of a single base. Synonymous or silent substitutions, which do not change an amino acid in the final gene product, are the most frequently observed in coding DNA. Nonsynonymous substitutions result in an altered codon that specifies either a different amino acid (a missense mutation) or a termination codon (a nonsense mutation). A missense mutation can induce either a conservative or nonconservative amino acid substitution. A conservative substitution refers to a situation in which the new amino acid is chemically similar to the old amino acid, whereas the amino acid introduced by a nonconservative substitution has different chemical characteristics. Thus, nonconservative substitutions are more likely to change the properties of the gene product than conservative substitutions. Deletions and insertions refer to the removal or addition, respectively, of one or a few nucleotides from the DNA sequence. These variations are relatively common in noncoding DNA. They are less frequent in exons where they may introduce frameshifts, that is, alter the normal translational reading frame of the gene and thereby change the final gene product. Copy number variants (CNVs) represent deletions or duplications of larger genomic regions. Copy number variants usually range in size from a few thousand to several million base pairs, and they could affect entire genes or gene clusters. Recent studies have shown that CNVs play a role in cancers, as well as in developmental and neuropsychiatric disorders. However, it is still unclear whether CNVs are associated with common complex traits, such as type 2 diabetes, cardiovascular disease, and obesity. No studies are available on CNVs and exercise-related traits.

Traditionally, mutations have been considered functionally significant only if they alter the amino acid sequence. However, it is now acknowledged that silent substitutions in exons, as well as mutations in the noncoding sequence, may also have strong effects on gene transcription and on the final gene product. For instance, it has been estimated that as much as 15% of point mutations that cause human diseases affect the splicing process. Mutations in the 5' regulatory region may disrupt a transcription factor binding

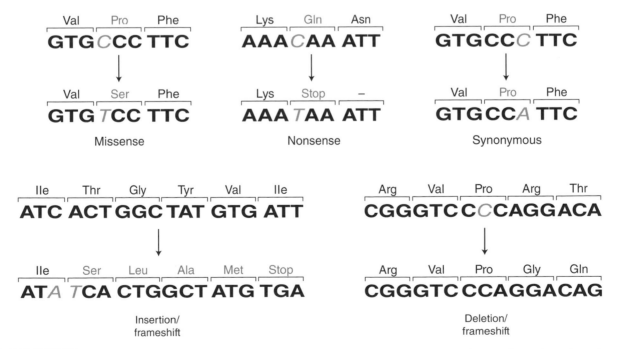

FIGURE 24.6 The top two lines present single-base substitutions, where a change of a single nucleotide induces either a change in amino acid (missense) or a premature stop codon (nonsense) or has no effect on the gene product (synonymous—silent). The lower two lines show examples of small-scale insertion (left) and deletion (right) mutations that alter the translational reading frame of the gene (frameshift). Polymorphic nucleotides and resulting changes in amino acids are indicated with gray font.

Reprinted, by permission, from F.C. Mooren and K. Volker, 2005, *Molecular and cellular exercise physiology* (Champaign, IL: Human Kinetics), 45.

site, a response element, or an enhancer or silencer sequence and thereby affect the rate at which a gene is transcribed. The 3' untranslated region (UTR) harbors several sequence elements that affect nuclear transport, polyadenylation, subcellular targeting, and stability of mRNA. Mutations in these sequences could also potentially influence gene transcription and translation. Both synonymous and nonsynonymous substitutions in the coding sequence may alter splicing sites as well as splicing enhancers and silencers and thereby influence the properties of the mature polypeptide.

A question of concern is the extent of variation in the DNA sequences of human chromosomes. The concept of polymorphism is important when we consider genetic variation. A polymorphism is defined as an alteration in DNA sequence that is present in the population with a frequency of at least 1%. However, recent large-scale sequencing studies have shown that rare DNA sequence variants (frequency less than 1% in the population) contribute significantly to the genetic architecture of complex traits. Quite often, the rare variants are located in the coding regions and regulatory elements of a gene and therefore could introduce major changes in gene function, consequently affecting the trait of interest. These observations have provided new support to the (many) rare variants–common disease hypothesis, which states that the genetic architecture of common diseases (or traits) is strongly influenced by a (large) number of rare variants that, despite their low frequency in the population, have major functional effects on the gene product and function of the gene.

Genetic variation is quite extensive at the DNA sequence level. It is estimated that there is a base variation about every 600 to 1,000 bp in the human genome. A single change in a nucleotide base is known as a **single-nucleotide polymorphism (SNP)**. Because the genome has about 3 billion base pairs of DNA, this would imply that any given individual would carry 3 to 5 million variants compared with the consensus *Homo sapiens* sequence of DNA. Already more than 25 million SNPs have been uncovered, and the number is growing (see 1000 Genomes project

website [www.1000genomes.org] for details). Polymorphic sequences are ubiquitous in introns, flanking regions of genes, and throughout the genome. Variable numbers of nucleotide sequences (e.g., CG or CAG repeats) are found throughout the genome and are highly polymorphic in populations. It is not uncommon to find 10 alleles or more of a given length polymorphism in such tandem repeats. They have become very useful markers of human diversity. In addition, insertions and deletions of from one to many hundreds of nucleotides are also found throughout the genome.

The proportion of genetic variation that is common to all humans and that is specific to a particular population has been the object of discussion for several decades. It is now generally recognized that most genetic variants are shared by the human species and that only about 3% to 5% are unique to specific populations. Genetic differences between populations or ethnic groups are thus relatively small compared with the overall genetic diversity observed in *Homo sapiens*.

Genetic Variation in Exercise Traits Among Sedentary People

The purpose of this chapter is to describe human variation in the response to regular exercise and define the role of genetic differences. All the traits of interest to the physical activity and health paradigm are characterized by a significant genetic component even among totally sedentary people. A few examples will suffice to illustrate this point.

The heritability of cardiorespiratory endurance phenotypes in the sedentary state has been estimated from twin and family studies; the most comprehensive of these is the HERITAGE Family Study (Bouchard et al. 1998). An analysis of variance revealed a clear familial aggregation of $\dot{V}O_2$max in the sedentary state. The variance in $\dot{V}O_2$max (adjusted for age, sex, body mass, and body composition) was 2.7

Extent of Human DNA Sequence Variation

- Each individual carries at least 3 million DNA variants compared with the common *Homo sapiens* sequence.
- Most DNA sequence variants are seen in all ethnic groups, although the allele frequencies may vary considerably by ethnicity.
- Only a small fraction (3-5%) of variants are specific to ethnic groups.

times greater between families than within families (figure 24.7). Maximum likelihood estimation of familial correlations (spouse, four parent–offspring, and three sibling correlations) revealed a maximal heritability of 51% for $\dot{V}O_2$max. However, the significant spouse correlation suggested that the genetic heritability was likely less than 50% (Bouchard et al. 1998). For additional information on the HERITAGE Family Study, visit www.pbrc.edu/HERITAGE.

Research with MZ and dizygotic (DZ) twins, biological brothers, and unrelated individuals suggests that the fiber type composition of a mixed muscle (vastus lateralis), although mediated by the genes, is not completely regulated by genetic mechanisms and may also be influenced by regular exercise and other agents. A summary of the genetic, environmental, and methodological sources of variation in the proportion of type I fibers in human skeletal muscle is illustrated in figure 24.8. The genetic component accounts for about 45% of the variation in the proportion of type I muscle fibers in humans (Simoneau and Bouchard 1995).

The genotype also plays a role in the quantity of key enzymes in skeletal muscle. For example, phosphofructokinase and oxoglutarate dehydrogenase are often considered regulatory enzymes of the glycolytic

and citric acid cycle (CAC) pathways, respectively. The quantity of these two enzymes in muscle fibers is critical for their activities, as well as central to the flow of substrates through the glycolytic and CAC pathways and, in turn, to the replenishment of ATP for the energy needs of the fiber. A study of young adult DZ and MZ twins as well as nontwin brothers suggested that at least 25% and perhaps more of the variation in the muscle content of these two key enzymes is associated with a genetic effect (Bouchard et al. 1986).

Twin and family studies indicate a significant genetic contribution to fat-free mass (FFM). In a cohort of 706 postmenopausal women, including 227 pairs of MZ twins and 126 pairs of DZ twins, FFM was estimated by dual-energy X-ray absorptiometry (DEXA). The results yielded a heritability estimate of 0.52 for FFM (Arden and Spector 1997). In the Quebec Family Study, path analysis of familial correlations computed among various pairs of relatives by descent or adoption indicated a genetic effect accounting for about 30% of the variance for FFM as assessed by underwater weighing (Bouchard et al. 1988). In the HERITAGE Family Study, the heritability estimates for FFM ranged from 40% to 65% depending on the assumptions made

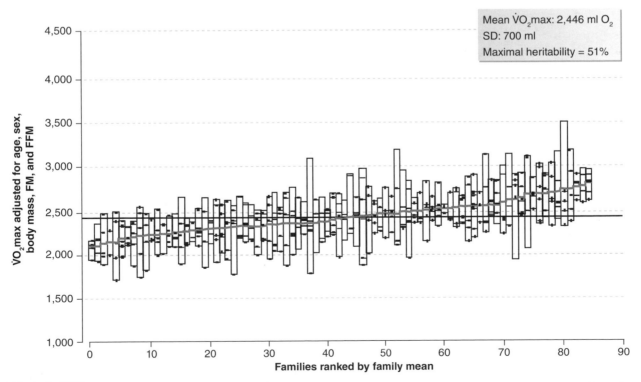

FIGURE 24.7 Family lines with low and high $\dot{V}O_2$max phenotypes in the sedentary state based on data from the HERITAGE Family Study.

Reprinted, by permission, from C. Bouchard et al., 1998, "Familial resemblance for VO₂max in the sedentary state: The HERITAGE Family Study," *Medicine and Science in Sports and Exercise* 30: 252-258.

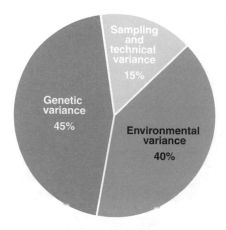

regarding the role of mitochondrial contribution (Rice et al. 1997).

Genetics of Physical Activity Level

Regular physical activity is a central component of current public health recommendations. While psychological, social, and environmental factors contribute significantly to physical activity behavior, it is important to recognize that activity behavior also has a biological basis and that genetic variation could affect individuals' propensity to be physically active or sedentary. Twin studies, as well as studies in nuclear and extended families, have provided maximal heritability estimates ranging from 15% to 60% for total physical activity level as well as for sedentarism, leisure-time activity, and sport participation.

At this point it would be helpful to introduce genome-wide association studies (GWAS). Rapid technical improvements in microarray-based high-throughput genotyping methods have made it possible to assay hundreds of thousands of SNPs in a single reaction, allowing detailed GWAS. A typical GWAS relies on a large number of SNPs that are distributed evenly across the genome, using either a case–control or a cohort study design. Associations between the trait of interest and each SNP are tested using standard statistical methods. However, because of the large number of tests in a GWAS, criteria for statistical significance have to be modified. For example, in a GWAS with 1 million SNPs, the threshold of genome-wide statistical significance is $p < 5 \times 10^{-8}$.

Data on the molecular genetics of physical activity levels in humans are still scarce, although the first GWAS for activity level was published in 2009 (De Moor et al. 2009). The report included results from two cohort studies: 1,644 unrelated individuals from the Netherlands Twin Register and 978 subjects living in Omaha, Nebraska. None of the 1.6 million SNPs reached the commonly used threshold of genome-wide significance ($p = 5 \times 10^{-8}$), although SNPs in three genomic regions showed p-values less than 1×10^{-5}. The strongest associations were observed on chromosome 10q23.2 at the 3'-phosphoadenosine 5' phosphosulfate synthase 2 (*PAPSS2*) gene locus: The odds ratio (OR) for being an exerciser was 1.32 ($p = 3.81 \times 10^{-6}$) for the common T-allele of SNP rs10887741. The other two SNPs with $p < 1 \times 10^{-5}$ were rs12612420 ($p = 7.61 \times 10^{-6}$, OR = 1.43), located about 12 kilobases (kb) upstream of the first exon of the DNA polymerase-transactivated protein 6 (*DNAPTP6*) gene, and rs8097348 ($p = 6.68 \times 10^{-6}$, OR = 1.36), which is located about 236 kb upstream of chromosome 18 open reading frame 2 (*C18orf2*). The associations with previously reported physical activity **candidate genes** were explored as well. The strongest candidate gene association was detected with SNP rs12405556 ($p = 9.7 \times 10^{-4}$, OR = 1.24) at the leptin receptor (*LEPR*) gene locus.

The major advantage of the GWAS strategy is that it is not restricted by a priori hypotheses, as is the case in candidate gene studies. Moreover, with millions of measured and imputed SNPs, a GWAS covers the entire genome uniformly and has sufficient sensitivity to detect small to moderate gene effects of relatively common sequence variants if the sample size is large. A critical feature of genetic studies is replication: Findings of an individual study should be tested in other large cohorts with a similar phenotype and study design. If the associations are replicated, the case for the contribution of a gene and DNA sequence variant to the trait of interest becomes considerably stronger. Given that several large cohort studies with physical activity questionnaire data available have also recently completed GWAS SNP genotyping, we could have interesting new data with replication panels in the near future.

Individual Differences in Response to Regular Exercise

There are marked interindividual differences in the adaptation to exercise training. For example, in the HERITAGE Family Study, 742 healthy but sedentary participants followed an identical, well-controlled endurance training program for 20 weeks. Despite the identical training program, increases in $\dot{V}O_2max$ varied from no change to increases of more than 1 L/min (figure 24.9). This high degree of heterogeneity in responsiveness to a fully standardized exercise program in the HERITAGE Family Study was not accounted for by age, gender, or ethnic differences. A similar pattern of variation in training responses was observed for several other phenotypes, such as plasma high-density lipoprotein (HDL) cholesterol levels and submaximal exercise heart rate and blood pressure changes (Bouchard and Rankinen 2001). These data underline the notion that the effects of endurance training on cardiovascular and other relevant traits should be evaluated not only in terms of mean changes, but also in terms of response heterogeneity.

A number of questions come to mind as a result of observations such as those depicted in figure 24.9 and the others discussed previously. Are the high and low responses to regular exercise characterized by sig-

nificant familial aggregation; that is, are there families with mainly low responders and others in which all family members show significant improvements? Is individual variability a normal biological phenomenon reflecting genetic diversity? Can we identify SNPs, genes, and alleles that predict the ability to respond positively or adversely to regular exercise?

Genes and Responses to Exercise

We now turn our attention to the evidence for a role of specific gene and sequence variants in the range of responses to regular exercise. Blood pressure, lipids and lipoproteins, glucose and insulin, and cardiorespiratory endurance response phenotypes are discussed. For editorial considerations, studies reviewed from this point onward in this chapter are not referenced individually. However, the interested

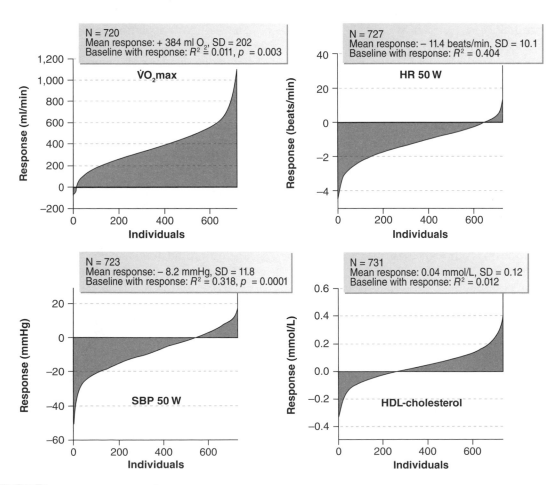

FIGURE 24.9 Heterogeneity of $\dot{V}O_2max$, submaximal exercise heart rate (HR 50 W), systolic blood pressure (SBP 50 W), and plasma HDL cholesterol training responses in the HERITAGE Family Study.

Reprinted, by permission, from C. Bouchard and T. Rankinen, 2001, "Individual differences in response to regular physical activity," *Medicine and Science in Sports and Exercise* 33(6): S446-451.

Major Areas of Interest Concerning the Genetics of Fitness and Adaptation to Exercise

- Familial aggregation and genetic contribution to human variation in the sedentary state
- Individual differences in the response to regular exercise
- Familial aggregation of the response variation to regular exercise
- Genes and alleles contributing to human variation in trainability

reader can find these references in the latest version of the Human Gene Map for Performance and Health-Related Fitness Phenotypes (Bray et al. 2009).

Another example of response heterogeneity relates to skeletal metabolism indicators and comes from the HERITAGE Family Study. The levels of nine enzymes involved in phosphocreatine metabolism, glycolysis, and oxidative metabolism were measured in muscle biopsies obtained before and after a 20-week endurance training program in 78 individuals from 19 nuclear families (Rico-Sanz et al. 2003). Exercise training induced statistically significant increases in all enzyme levels. Furthermore, all training responses showed significant familial aggregation: The between-family variance was 1.85 to 4.0 times greater than the variance within families.

Although exercise-related traits are mainly polygenic and multifactorial in nature, much can be learned from some monogenic disorders characterized by compromised exercise capacity or exercise intolerance. These disorders affect only a few individuals, but they provide interesting examples of genetic defects that have profound effects on the ability to perform physical activity, usually attributable to compromised energy metabolism. Although these genetic defects compromise exercise capacity, there is no evidence that overexpression of these genes leads to improved physical performance. However, it is important to understand the molecular mechanisms contributing to both ends of the distribution of cardiorespiratory endurance and its trainability. Table 24.3 lists some of the genes that have been associated with a decreased exercise capacity (Rankinen et al. 2004).

Genes and Blood Pressure Response to Regular Exercise

The 2007 update of the Human Gene Map for Performance and Health-Related Fitness Phenotypes included 17 genes from 23 studies that have been investigated in relation to exercise training–induced changes in hemodynamic phenotypes (Bray et al. 2009). Findings for 13 candidate genes (*AGTR1,* *AMPD1, APOE, BDKRB2, CHRM2, EDN1, FABP2, GNB3, HBB, KCNQ1, NFKB1, PPARA, TTN*) were based on a single study. However, with four candidate genes, the positive associations were reported in at least two studies. For example, in both the HERITAGE Family Study and the DNASCO Study cohorts, the angiotensinogen (*AGT*) Met235Thr polymorphism (in which threonine is substituted for methionine) was associated with endurance training–induced changes in diastolic blood pressure in men.

Similarly, an association between the angiotensin I converting enzyme (*ACE*) gene I/D (insertion or deletion of a sequence) polymorphism and training-induced left ventricular (LV) growth has been reported in two studies (figure 24.10) (Montgomery et al. 1997; Myerson et al. 2001). In 1997, Montgomery and coworkers reported that the *ACE* D-allele was associated with greater increases in LV mass and with septal and posterior wall thickness after 10 weeks of physical training in British Army recruits (figure 24.10*a*). A similar training paradigm was repeated a few years later, and the training-induced increase in LV mass was 2.7 times greater in the D/D genotype compared with the I/I **homozygotes** (figure 24.10*b*).

A third candidate gene with positive evidence of associations from multiple studies is endothelial nitric oxide synthase 3 (*NOS3*). In the HERITAGE Family Study, homozygotes for the glutamine allele at codon 298 (Glu298) had a reduction in submaximal exercise diastolic blood pressure that was more than three times greater than that of the homozygotes for the asparagine allele (Asp298Asp) after the training program. A similar pattern was evident with the systolic blood pressure and rate–pressure product training responses (Rankinen et al. 2000). In coronary artery disease patients, exercise training significantly improved acetylcholine-induced change in average peak velocity of coronary arteries. However, the training response was significantly blunted in the carriers of the *NOS3* –786C allele of a polymorphism located in the 5'-UTR of the *NOS3* gene compared with the patients who were homozygotes for the –786T-allele (Erbs et al. 2003).

TABLE 24.3   Genes Encoded by Nuclear and Mitochondrial DNA in Which Mutations Have Been Reported in Patients With Exercise Intolerance

Gene	OMIM number	Location
Nuclear DNA		
CPT2	255110	1p32
PGAM2	261670	7p13-p12
LDHA	150000	11p15.4
PYGM	232600	11q12-q13.2
PFKM	232800	12q13.3
ENO3	131370	17pter-p11
ACADVL	201475	17p13-p11
SGCA	600119	17q21
PHKA1	311870	Xq12-q13
PGK1	311800	Xq13
Mitochondrial DNA		
MTTL1	590050	3,230-3,304
MTND1	516000	3,307-4,262
MTTI	590045	4,263-4,331
MTTM	590065	4,402-4,469
MTTY	590100	5,826-5,891
MTCO1	516030	5,904-7,445
MTTS1	590080	7,445-7,516
MTTK	590060	8,295-8,364
MTCO3	516050	9,207-9,990
MTND4	516003	10,760-12,137
MTTL2	590055	12,266-12,336
MTTE	590025	14,674-14,742
MTCYB	516020	14,747-15,887

OMIM = Online Mendelian Inheritance in Man (www.ncbi.nlm.nih.gov/omim/).

Reprinted from Rankinen et al. 2004.

In the HERITAGE Family Study, a GWAS approach was applied to exercise training–induced changes in submaximal exercise (50 W) heart rate (HR 50). Following quality control procedures, 324,611 SNPs were available for analyses in 473 white adults from 99 families. The strongest single-SNP associations ($p = 8.1 \times 10^{-7}$) were detected with SNPs located on chromosome 2p25 in the tyrosine 3-monooxygenase/tryptophan 5-monooxygenase activation protein, theta polypeptide (*YWHAQ*) gene locus. In addition, a SNP in the RNA binding protein with multiple splicing (*RBPMS*) gene locus on chromosome 8p12 showed an association with $p = 4.0 \times 10^{-6}$. Furthermore, 37 additional SNPs were associated with the changes in HR 50 with p-values less than 8.0×10^{-5}. After removal of redundant

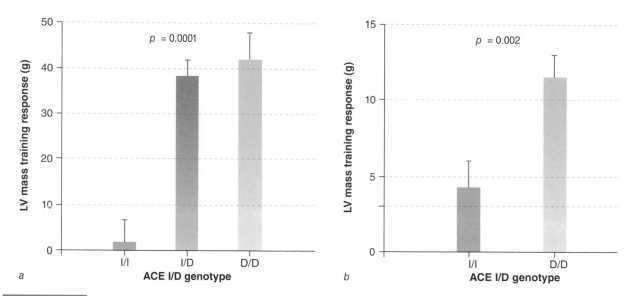

FIGURE 24.10 Panel *a* is modified from Montgomery and colleagues (1997) and depicts data from 140 healthy army recruits who participated in a 10-week basic training program. Panel *b* summarizes data from a replication study (Myerson et al. 2001) using a similar training program with 141 healthy recruits.

Reprinted, by permission, from T. Rankinen and C. Bouchard, 2005, Genes, genetic heterogeneity, and exercise phenotypes, In *Molecular and cellular exercise physiology*, edited by F.C. Mooren and K. Volker (Champaign, IL: Human Kinetics), 50.

SNPs, 30 markers were analyzed using a multivariable regression model with forward selection. In the final model, six SNPs each explained at least 3% of variance in ΔHR 50 (range from 3% to 6%), while another four SNPs contributed between 2% and 3% each (table 24.4). The full model of 10 SNPs explained 36% of the variance in HR 50 training response. To illustrate the combined contribution of

TABLE 24.4 HR 50 Training Response GWAS: Results of the Final Multivariate Regression Model

| | | | | | Regression model | | |
| | | | | | | Partial | |
SNP*	Chromosome	Map	Frequency	Gene	r^2	r^2 model	P-value
rs2979481	8	30,382,328	0.645	RBPMS	0.060	0.060	<.0001
rs6432018	2	9,639,347	0.524	YWHAQ	0.046	0.106	<.0001
rs2253206	2	208,100,223	0.522	CREB1	0.045	0.151	<.0001
rs1560488	4	90,444,858	0.791	GPRIN3	0.042	0.193	<.0001
rs10248479	7	115,395,591	0.876	TFEC	0.033	0.226	<.0001
rs857838	1	157,017,174	0.60	OR6N2	0.030	0.256	<.0001
rs909562	6	16,238,312	0.875	MYLIP	0.029	0.285	<.0001
rs4759659	12	129,403,241	0.538	PIWIL1	0.028	0.314	<.0001
rs2057368	14	54,373,759	0.804	GCH1	0.024	0.337	<.0001
rs4498613	9	95,545,460	0.808	PHF2 (60 kb)	0.022	0.359	.0001

*Most significant SNPs from the initial single-SNP analyses

Map = Nucleotide position on the chromosome

Frequency = Frequency of the major allele

From Rankinen et al. 2012.

the 10 SNPs with a partial R^2 >2%, a SNP summary score was constructed. Each SNP was recoded based on the number of alleles associated with a favorable HR 50 training response (homozygote = 2; heterozygote = 1; homozygote for unfavorable allele = 0), and the summary score was derived by summing up the 10 recoded SNPs. As shown in figure 24.11, subjects with a low summary score (nine or fewer favorable alleles) did not show any improvements in HR 50, while those who had a summary score of 16 or greater decreased HR 50 by more than 20 beats per minute.

Although a considerable number of studies have been published on the genetics of hemodynamic training responses, few of the associations have been replicated in multiple studies. The most convincing evidence so far is for an association between the *ACE* genotype and training-induced changes in LV mass, reported in two studies. The GWAS results on submaximal exercise heart rate training response just described represent a new and exciting approach to understanding the genetic architecture of cardiac adaptation to regular physical activity. However, additional studies are needed to confirm and permit full understanding of the physiological and clinical meaning of the GWAS findings.

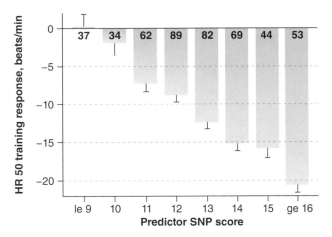

FIGURE 24.11 Submaximal exercise heart rate (HR 50) training responses (adjusted for age, sex, baseline body mass index, baseline HR 50) across SNP summary score categories in white HERITAGE Family Study subjects. See text for further explanation.

Reprinted from T. Rankinen et al. 2012, "The heritability of submaximal heart rate response to exercise training is accounted for by nine SNPs." *Journal of Applied Physiology*. In press. Used with permission.

Genes and Lipids and Lipoprotein Response to Regular Exercise

Although the first observations of the beneficial effects of regular physical activity on plasma lipid levels date back to the early 1970s, surprisingly few studies are available on the associations between

DNA sequence variation in candidate genes and training-induced changes in plasma lipids, lipoproteins, and apolipoproteins. In the HERITAGE Family Study, apolipoprotein E (*APOE*) genotypes were associated with training-induced changes in plasma low-density lipoprotein (LDL) and high-density lipoprotein (HDL) cholesterol and plasma triglyceride levels (Leon et al. 2004). In both black and white participants, regular exercise lowered the LDL cholesterol levels in the E2/E3 and E2/E4 genotypes but not in the E3/E3 homozygotes and the E3/E4 **heterozygotes**. The LDL cholesterol training response in the E4/E4 homozygotes showed an ethnic difference: Black participants showed a significant reduction, whereas in white participants the LDL cholesterol levels tended to increase. High-density lipoprotein and triglyceride training responses were associated with *APOE* genotypes only in white participants: The E2/E2, E2/E3, and E3/E3 genotypes showed more favorable changes in HDL and triglyceride levels than the E2/E4 and E4/E4 genotypes.

In a small cohort of elderly men, the *APOE* E2 allele carriers showed a greater increase in HDL cholesterol levels after a nine-month exercise training program than the E3 and E4 allele carriers (Hagberg et al. 1999). Other studies have suggested that there may be associations between HDL training responses and lipoprotein lipase and endothelial lipase genotypes.

Genes and Glucose and Insulin Response to Regular Exercise

Only a few candidate gene studies are available on exercise training–induced changes in glucose and insulin metabolism phenotypes. In obese type 2 diabetic Japanese women, carriers of the arginine allele at the Trp64Arg polymorphism in the β-3-adrenergic receptor (*ADRB3*) gene showed smaller reductions in fasting glucose and HbA1c (glycosylated hemoglobin) levels after a combined low-calorie diet and exercise training program compared with noncarriers. In healthy Japanese males, a three-month endurance training program resulted in a significant reduction in fasting plasma glucose levels in the Trp64Trp homozygotes and Trp64Arg heterozygotes but not in the Arg64Arg homozygotes. Another Japanese study reported suggestive associations between Pro-12Ala (alanine is substituted for proline) genotypes of the peroxisome proliferator-activated receptor γ (*PPARG*) gene and training-induced changes in fasting insulin and insulin resistance in healthy males. The Pro12Ala heterozygotes showed reductions in both trait values, whereas no changes were observed in the Pro12Pro homozygotes. In a small cohort of older hypertensive men, the *ACE* I/I homozygotes had the greatest improvements in insulin sensitivity and the greatest decline in the acute insulin response to glucose compared with those who were homozygous for the D-allele.

In the HERITAGE Family Study, a genome-wide **linkage** scan revealed a quantitative trait locus (QTL) on chromosome 7q31 for the fasting insulin response to endurance training in white families (Lakka et al. 2003). A single-nucleotide polymorphism (SNP) in the 5' untranslated region (UTR) of the leptin (*LEP*) gene, which is located within the QTL region, was not associated with the insulin training response. However, there was a significant interaction between the *LEP* genotype and a Lys109Leu polymorphism (lysine is replaced by leucine) of the leptin receptor (*LEPR*) gene (Lakka et al. 2004). Changes in fasting insulin levels did not differ among the *LEP* genotypes in those participants who were homozygotes for the *LEPR* Lys109 allele. However, among the *LEPR* Leu109 allele carriers, the *LEP* A/A homozygotes showed a significantly greater reduction in fasting insulin levels than the heterozygotes and the *LEP* G/G homozygotes (figure 24.12).

Again in the HERITAGE Family Study, global gene expression profiling was used to identify genes associated with insulin sensitivity training response (Teran-Garcia et al. 2005). Total RNA was extracted from vastus lateralis muscle biopsies from 16 sub-jects: Eight subjects were high responders for the gains in insulin sensitivity, and eight were age-, sex-, and body mass index (BMI)–matched nonresponders. A total of 47 transcripts were differentially expressed (at least 1.4- or minus 0.7-fold difference) at baseline, while another 361 transcripts showed differential expressions after 20 weeks of exercise training. Five genes *(SKI, FHL1, TTN, PDK4, CTBP1)* that exhibited at least a 50% difference in expression between high responders and nonresponders either at baseline or posttraining were selected for validation experiments. Association of the *FHL1* (four and a half LIM domains 1) gene (encoded on the X chromosome) with exercise training–induced changes in insulin metabolism phenotypes was further investigated by genotyping of three *FHL1* SNPs (Teran-Garcia et al. 2007) in the whole HERITAGE cohort of whites. SNP rs9018 was associated with disposition index and glucose disappearance training responses in white women, while in the white males, the same SNP showed a suggestive association with fasting insulin training response. Another SNP (rs2180062) was associated with fasting insulin, insulin sensitivity, disposition index, and glucose disappearance index responses in white males (Teran-Garcia et al. 2007).

Genes and Cardiorespiratory Endurance in Response to Regular Exercise

In addition to examining the role of genetic heterogeneity in the ability to improve the diabetes and cardiovascular disease risk factor profile in response to a fully standardized program, one can ask whether the same concept applies to the gains in cardiorespiratory endurance or exercise tolerance. A number of studies indicate that such is the case, and the genetic dissection of the trainability phenotype is under way.

In pairs of MZ twins, the $\dot{V}O_2$max response to standardized training programs showed six to nine times more variance between genotypes (between pairs of twins) than within genotypes (within pairs of twins) based on the findings of three independent studies (Bouchard et al. 1992). Thus, gains in absolute $\dot{V}O_2$max were much more heterogeneous between pairs of twins than within pairs of twins. The results of one such study are summarized in figure 24.13. The MZ twins exercised for 20 weeks using a standardized and demanding endurance training program (Bouchard et al. 1992). In the HERITAGE Family Study, the increase in $\dot{V}O_2$max in 481 individuals from 99 two-generation families of Caucasian descent showed 2.6 times more variance between families than within families, and the model-fitting analytical procedure yielded a maximal heritability

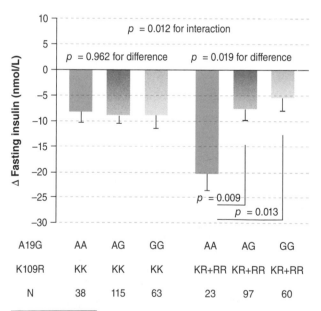

FIGURE 24.12 Exercise-induced changes in fasting insulin according to the leptin *(LEP)* A19G polymorphism and the leptin receptor *(LEPR)* K109R polymorphism.

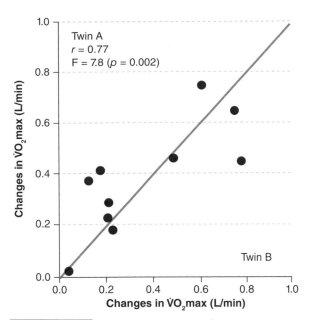

FIGURE 24.13 Training changes in $\dot{V}O_2$max among 10 pairs of MZ twins subjected to a standardized 20-week exercise training program.

Adapted from Prud'homme, Bouchard, et al. 1984.

estimate of 47% (Bouchard et al. 1999). Thus, the extraordinary heterogeneity observed for the gains in $\dot{V}O_2$ among adults is not random and is characterized by a strong familial aggregation (figure 24.14).

We are dealing here with extremely complex phenotypes, and the genetic architecture of their responsiveness to regular exercise is likely to depend on a large number of genes. Two favorite candidate genes have been the angiotensin I converting enzyme *(ACE)* and actinin, alpha 3 *(ACTN3)* genes. Several studies showed significant associations between the *ACE* or *ACTN3* genotypes and common indicators of fitness or physical performance. However, there are in both cases multiple discordant studies. Not surprisingly, most studies are statistically underpowered to detect small effect sizes, which is what one would expect in the relation between genetic variants and fitness traits. Thus, at this time, the data are still too fragmented and inconclusive to allow full evaluation of the role played by the *ACE* and *ACTN3* genes in variation in cardiorespiratory fitness and human trainability. The same conclusion can be reached about all the other candidate genes studied to date.

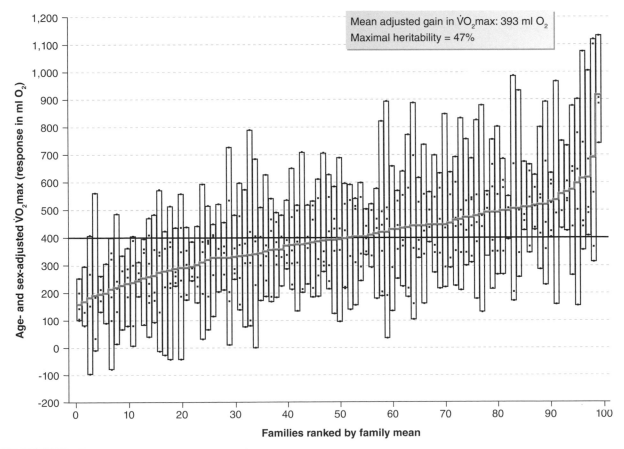

FIGURE 24.14 Familial aggregation of the $\dot{V}O_2$max changes in response to exercise training in the sample of whites of the HERITAGE Family Study.

Reprinted from C. Bouchard et al., 1999, "Familial aggregation of VO(2max) response to exercise training: results from the HERITAGE Family Study," *Journal of Applied Physiology* 87(3): 1003-1008. Used with permission.

A highly promising strategy, global skeletal muscle gene expression profiling combined with DNA sequence variation screening, was used to identify genes associated with $\dot{V}O_2$max training response status (Timmons et al. 2010). RNA expression profiling of pretraining skeletal muscle samples identified 29 gene transcripts that were strongly associated with $\dot{V}O_2$max training response in two independent exercise studies. Haplotype tagging SNPs in the 29 predictor genes were identified and genotyped in the HERITAGE Family Study. A multivariable regression analysis using the predictor gene SNPs and a set of SNPs from positional cloning (small genomic region) and candidate gene studies of the HERITAGE Family Study identified a set of 11 SNPs that explained about 22% of the variance in $\dot{V}O_2$max training response (table 24.5). Seven of the SNPs were from the RNA predictor gene set, and four were from prior HERITAGE projects (Timmons et al. 2010).

A full GWAS approach was applied to exercise training–induced changes in $\dot{V}O_2$max in the HERITAGE Family Study (Bouchard et al. 2011). In single-SNP analysis, altogether 39 SNPs were associated with $\dot{V}O_2$max training response at $p < 1.5 \times 10^{-4}$: Five SNPs showed associations at $p < 1 \times 10^{-5}$, while another 20 SNPs showed significance levels at $1.5 \times 10^{-5} < p < 9.3$

$\times 10^{-4}$. The strongest evidence of association ($p = 1.3 \times 10^{-6}$) was observed with a SNP located in the first intron of the acyl-CoA synthetase long-chain family member 1 (*ACSL1*) gene, located on chromosome 4q35. When all 39 SNPs were analyzed simultaneously in multivariate regression models, 21 SNPs were retained in the final model. Of these, nine SNPs explained at least 2% (range 2.2% to 7.0%) of the variance ($p < 0.0001$ for all), while seven markers contributed between 1% and 2% each. Collectively, these 16 SNPs explained 45% of the variance in $\dot{V}O_2$max training response, which is very close to the maximal heritability estimate of 47% reported previously in the HERITAGE Family Study. Finally, a predictor score was constructed using the 21 SNPs from the final regression model. Each SNP was recoded based on the number of high $\dot{V}O_2$max training response alleles: Low-response allele homozygote was assigned 0; heterozygote received 1; and homozygote for the high-response allele was assigned 2. The sum of the recoded SNPs was used as the predictor score. The score values ranged from 7 to 31. The difference in $\dot{V}O_2$max training response between those with the lowest (9 or fewer, N = 36, mean = +221 ml/min) and the highest (19 or more, N = 52, mean = +604 ml/min) was almost 400 ml/min (figure 24.15).

TABLE 24.5 Stepwise Regression Model for Standardized Residuals of Maximal Oxygen Consumption Training Response in the HERITAGE Family Study

Gene (SNP)	Identification method	Genomic location	Partial r^2	Model r^2	p-value
SVIL (rs6481619)	QTL	10p11.2	0.0411	0.0411	<.0001
SLC22A3 (rs2457571)	RNA predictor	6q26-q27	0.0307	0.0718	.0003
NRP2 (rs3770991)	QTL	2q33.3	0.0224	0.0942	.0017
TTN (rs10497520)	QTL	2q31	0.0204	0.1146	.0025
H19 (rs2251375)	RNA predictor	11p15.5	0.0268	0.1414	.0004
ID3 (rs11574)	RNA predictor	1p36.13-p36.12	0.02	0.1615	.0021
MIPEP (rs7324557)	QTL	13q12	0.0163	0.1778	.0051
CPVL (rs4257918)	RNA predictor	7p15-p14	0.0179	0.1957	.0031
DEPDC6 (rs7386139)	RNA predictor	8q24.12	0.0112	0.2069	.0185
BTAF1 (rs2792022)	RNA predictor	10q22-q23	0.0125	0.2194	.0122
DIS3L (rs1546570)	RNA predictor	15q22.31	0.0095	0.2289	.0279

Residual maximal oxygen consumption response values were adjusted for age, sex, baseline body weight, and baseline maximal oxygen consumption (n = 473). RNA determinations were made using U133 Plus 2.0 Affymetrix chips, normalized using MAS5.0, and tested for differential expression using significance analysis of microarrays (SAM). In total, ~1,000 probe sets were deemed differentially expressed (>1.5-fold change, <5% false discovery rate).

SNP = single-nucleotide polymorphism; QTL = quantitative trait locus.

Adapted from J.A. Timmons et al., 2010, "Using molecular classification to predict gains in maximal aerobic capacity following endurance exercise training in humans," *Journal of Applied Physiology* 108:1487-1496. Used with permission.

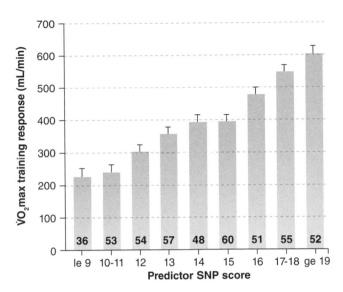

FIGURE 24.15 Age-, sex-, and baseline $\dot{V}O_2$max–adjusted $\dot{V}O_2$max training responses across nine GWAS predictor SNP score categories in HERITAGE whites. Number of subjects within each SNP score category is indicated inside each histogram bar.

Reprinted from C. Bouchard et al., 2011, "Genomic predictors of maximal oxygen uptake response to standardized exercise training programs," *Journal of Applied Physiology* 110: 1160-1170. Used with permission.

These last two studies point to a paradigm change in the strategies used to investigate the genomic and genetic basis of health-related fitness, performance, and trainability. Both studies provide strong support for the concept that it will soon become possible to identify not only who is likely to benefit most from being physically active, but also who is likely to exhibit an adverse response pattern.

Gene–Physical Activity Interactions on Obesity and Body Weight Regulation

Exercise genetic studies require a large sample size and a carefully monitored training program. Since such studies are very expensive and difficult to execute, some have relied on existing large observational, epidemiological cohorts as a less expensive alternative to intervention trials. Examples can be found in recent studies dealing with physical activity–*FTO* (fat mass and obesity associated) gene interactions on body weight and body composition. *FTO* was the first gene that was identified using high-density GWAS as a strong candidate for obesity-related phenotypes in Caucasians (Frayling et al. 2007). Minor alleles of obesity risk SNPs located in the first intron of the gene were associated with a 65% greater risk of obesity in the homozygote state: The homozygotes of the minor allele were 3 to 4 kg (6.6-8.8 lb) heavier,

and heterozygotes had 1 to 2 kg (2.2-4.4 lb) higher body weight than the common allele homozygotes. The population attributable risk (PAR) of *FTO* for obesity has been estimated to be as high as 20%. The initial finding has been replicated in several large cohorts, and the association holds both in children and in adults (Loos and Bouchard 2008).

Several observational studies showed that the *FTO* gene–related risk of obesity was modified by the physical activity level of the subjects (see, for instance, Andreasen et al. 2008; Rampersaud et al. 2008). The higher BMI level found in the *FTO* risk allele homozygotes was particularly evident in those who were sedentary, while the difference in BMI between the risk allele and common allele homozygotes was not significant among physically active individuals (Andreasen et al. 2008; Rampersaud et al. 2008). Given the cross-sectional nature of these studies, these findings could be interpreted in at least two ways: Either physical activity helps to prevent weight gain or accelerates weight loss in the risk allele homozygotes.

The exercise training–related weight loss hypothesis was formally tested in the HERITAGE Family Study (Rankinen et al. 2010). However, the results did not support the assumption that the risk allele homozygotes lose more weight with exercise. In fact, the data suggested that the risk allele homozygotes are resistant to training-induced changes in adiposity (figure 24.16). After 20 weeks of carefully supervised endurance training with 100% compliance, the *FTO* risk allele homozygotes did not lose fat mass, while the homozygotes for the non-risk allele showed a significant reduction in total adiposity. Thus, the experimental data would suggest that regular physical activity does not help the *FTO* risk allele carriers to lose weight. On the other hand, the weight gain prevention hypothesis is a very interesting one, but it remains to be tested in controlled clinical trials. The *FTO* example emphasizes the importance of testing genetic hypotheses related to the responsiveness to exercise training in appropriately designed studies. While observational studies can provide valuable insights as to how regular physical activity may interact with a genetic predisposition for a variety of health outcomes, they have limited value regarding the genetics of responsiveness to regular exercise.

Trait-Specific Response to Exercise

This chapter has provided ample evidence that the response to regular exercise varies considerably among individuals. One of the lessons from these studies is that there are people who do not enjoy

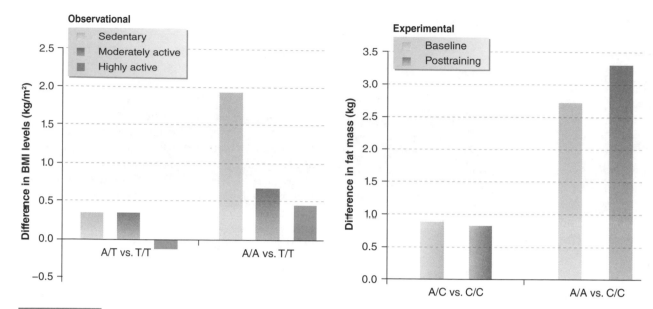

FIGURE 24.16 Different outcomes of *FTO* genotype-by-physical activity interactions on adiposity in observational and experimental studies. In a cross-sectional population study (left panel; see Andreasen et al. 2008 for details), the obesity risk allele of the *FTO* gene (rs9939609 A-allele) was associated with high body mass index in sedentary but not in physically active individuals. In a 20-week exercise training study (right panel; see Rankinen et al. 2010 for details), the difference in total adiposity observed in sedentary state (baseline) between the *FTO* risk allele (rs8050136 A-allele) and common allele (C) homozygotes became even greater after the training program because of smaller training-induced fat mass loss among the A/A homozygotes (−0.2 kg vs. −0.8 kg).

Part (a) copyright 2008 American Diabetes Association. From Diabetes®, Vol. 57, 2008; 95-101. Reprinted with permission from The American Diabetes Association. Part (b) data from Rankinen et al. 2010.

any improvement in a given risk factor or biological indicator of health or fitness as a result of exercise.

Fortunately, the heterogeneity in responsiveness to regular exercise is a trait-specific phenomenon. In other words, an individual who is a low responder for one phenotype may be an average or even a high responder for another trait. A way to illustrate this phenomenon is to calculate correlation coefficients between the training responses among various traits. A strong correlation would indicate a uniform response pattern across traits, whereas the lack of correlation would support the independence of the responses to regular exercise. Table 24.6 summarizes the correlation coefficients between $\dot{V}O_2$max and several cardiovascular disease risk factor training

responses in the HERITAGE Family Study cohort. The table clearly shows that endurance training–induced changes in risk factors do not correlate well with changes in cardiorespiratory fitness level. On the other hand, the correlations summarized in table 24.7 indicate that it is impossible to predict the effects of regular exercise on one risk factor by observing the training response of another risk factor. Thus far, we have no indication that there are people who are nonresponders across all the markers of risk for diseases.

These observations emphasize that even if a person cannot improve cardiorespiratory fitness with regular exercise, it is likely that he will achieve other significant health benefits from a physically active lifestyle. This has important public health implications.

Important Recent Research

- Individual differences exist in the response of risk factors to regular exercise.
- Risk factor responses are not correlated with the gains in $\dot{V}O_2$max.
- Risk factor response changes are poorly correlated among one another.
- There is no indication that there are individuals who are nonresponders across all fitness and risk factors.

Personalized Exercise Medicine

Physical activity, as a behavior, and cardiorespiratory fitness, as a state, are key traits that are already taken into account in health promotion and disease prevention strategies around the world. Even though regular exercise is globally perceived as a healthy behavior for everyone, considerable inter-individual differences are observed in the responses of accepted cardiometabolic risk factors to exercise programs, with a significant fraction of individuals experiencing no improvement for given traits and a minority developing adverse response patterns. These facts have not yet become part of the global discourse on physical activity, fitness, and health and have not been incorporated in public health policies. However, this is likely to change in the coming years.

TABLE 24.6 Correlation Coefficients Between Endurance Training–Induced Changes in Cardiorespiratory Fitness ($\dot{V}O_2$max) and Risk Factors in the HERITAGE Family Study

Response phenotype	Black participants Males	Black participants Females	White participants Males	White participants Females
Body weight	.20	.11	−.09	−.08
Fat mass	−.03	.05	−.19	−.18
Insulin	.05	.10	−.04	.03
Cholesterol	.13	−.05	.00	.01
Triglycerides	.05	−.01	.12	−.10
Systolic blood pressure	.22	.01	−.11	−.01

TABLE 24.7 Correlation Coefficients Between Risk Factor Training Responses in the HERITAGE Family Study Cohort

	Black participants FM	Ins	Chol	TG	SBP	White participants FM	Ins	Chol	TG	SBP
Females										
Body weight	.92	.24	.34	.23	−.06	.84	.17	.22	.06	.10
Fat mass		.16	.31	.23	.04		.14	.21	.09	.06
Insulin			.19	.07	−.04			−.02	.04	.11
Cholesterol				.25	−.06				.31	.16
Triglycerides					−.01					.14
Males										
Body weight	.84	.20	.29	.33	.24	.83	−.03	.13	.13	.06
Fat mass		.24	.14	.18	.16		.03	.07	.08	.04
Insulin			.23	.24	.19			.09	.03	−.05
Cholesterol				.33	.18				.31	−.11
Triglycerides					.15					.14

FM = fat mass; Ins = plasma insulin; Chol = plasma cholesterol; TG = plasma triglycerides; SBP = systolic blood pressure.

As biomedical and clinical research embraces molecular profiling, translational research is entering the promising era of "personalized medicine." Personalized preventive and therapeutic medicine promises the ability to target interventions and therapies to those who will most benefit and to find alternative treatments for those who are resistant to traditional therapies or are likely to develop adverse responses. The National Institutes of Health (NIH) has publicly committed to capitalizing on recent technological advances to fully explore this potential. One can predict that tailoring exercise programs, dietary recommendations, stress management, pharmacotherapies, and other interventions to individuals' genomic and epigenomic characteristics will become the cornerstone of personalized disease prevention recommendations.

The use of exercise genomic and epigenomic information will require a paradigm shift compared to current practice and public health policies. The genetic data summarized in the previous sections indicate that it will likely be possible soon to identify a priori who is expected to respond positively to an exercise program. This is admittedly just a beginning. Molecular predictor algorithms will become more powerful and reliable as research on this topic expands. In fact, the time is ripe for properly designed, very large studies in which regular exercise is used to increase cardiorespiratory fitness and improve the risk factor profile for cardiovascular disease, diabetes, and other morbidities. These would generate the strong data needed to develop and implement personalized exercise medicine programs based on reliable information about who will benefit the most, who will benefit the least, and who should consider other options because of the risk of maladaptive reactions. Personalized exercise medicine is our next destination.

Summary

The past decade has witnessed remarkable progress in genomics and human genetics. The availability of the DNA sequence of the human genome has changed our ability to study the genetic basis of complex multifactorial traits and to develop novel treatments for several chronic diseases. The recent advances in molecular genetics are starting to affect exercise science. The availability of powerful methods such as microarray technology to measure gene expression and targeting specific genes in knockout and transgenic animal models will greatly add to our research capability in the investigation of basic and applied exercise science issues.

Although the research on molecular genetics of physical activity, health-related fitness, and health-related outcomes is still in its infancy, we recognize that understanding the effects of DNA sequence variation on interindividual differences in responsiveness to acute exercise and regular exercise holds great promise. Such data not only would help to develop more concrete public health measures regarding the role of physical activity in the prevention and treatment of chronic diseases, but also would provide an opportunity to individualize preventive medicine.

Key Concepts

allele—Alternative form of a genetic locus; a single allele for each locus is inherited from each parent.

alternative splicing—Different ways of combining a gene's exons to generate alternative transcripts.

candidate gene—A gene suspected of being involved in a trait or a disease.

chromatin—Packaged DNA combined with histones in a cell nucleus.

epigenome—DNA methylation of cytosines and chemical modification of the histone proteins. Epigenetic events influence gene expression without a change in DNA sequence.

exon—An amino acid coding DNA sequence of a gene.

gene—Fundamental physical and functional unit of heredity. A gene is an ordered sequence of nucleotides located in a particular position on a particular chromosome that encodes a specific functional product (i.e., a protein or RNA molecule).

gene expression—Process by which a gene's coded information is converted into a functional product. Expressed genes include those that are transcribed into messenger RNA and then translated into protein and those that are transcribed into RNA but not translated into protein (e.g., transfer and ribosomal RNAs and small RNAs).

genome—All the DNA bases and sequences of a particular organism; the size of a genome is generally given as its total number of bases or base pairs.

genomic imprinting—Chemical modifications of DNA and histones that result in the silencing of a sequence or a gene cluster. Imprinting is typically established by parental origin.

heterozygote—Presence of two different alleles at a given genetic locus.

homozygote—Two identical alleles at a given genetic locus.

intron—DNA sequence that interrupts the amino acid coding sequence in a gene; an intron is transcribed into RNA but is spliced out of the message before it is translated into protein.

linkage—Proximity of two or more markers on a chromosome; the closer the markers, the lower the probability that they will be separated during DNA recombination at meiosis and hence the greater the probability that they will be inherited together.

promoter—A DNA site to which RNA polymerase will bind before initiating transcription.

single-nucleotide polymorphism (SNP)—DNA sequence variation that occurs when a single nucleotide (A, T, C, or G) in a genome sequence is altered.

transcription—Synthesis of an RNA copy from a sequence of DNA.

transcription factor—Protein that binds to the regulatory region of a gene and participates in the regulation of gene expression.

translation—Process by which the genetic code carried by messenger RNA directs the incorporation of amino acids and the synthesis of a protein.

Study Questions

1. How can we account for the fact that human tissues produce more proteins than there are genes?

2. Describe how differences in gene expression can influence a phenotype affected by regular exercise.

3. What are the major sources of human genetic variation?

4. How extensive is human genetic variation?

5. What is a quantitative trait locus?

6. How extensive are individual differences in response to regular exercise?

7. Is the *ACE* I/D polymorphism a good predictor of blood pressure response to regular exercise?

8. Describe how one can use the profile of expressed transcripts in skeletal muscle and DNA sequence variants to predict the response to a standardized exercise program.

9. What are the implications of the differences in the responsiveness to regular exercise for the promotion of physical activity?

10. Define the concept of personalized exercise medicine.

References

Andreasen, C.H., K.L. Stender-Petersen, M.S. Mogensen, S.S. Torekov, L. Wegner, G. Andersen, A.L. Nielsen, A. Albrechtsen, K. Borch-Johnsen, S.S. Rasmussen, J.O. Clausen, A. Sandbaek, T. Lauritzen, L. Hansen, T. Jørgensen, O. Pedersen, and T. Hansen. 2008. Low physical activity accentuates the effect of the FTO rs9939609 polymorphism on body fat accumulation. *Diabetes* 57:95-101.

Arden, N.K., and T.D. Spector. 1997. Genetic influences on muscle strength, lean body mass, and bone mineral density: A twin study. *Journal of Bone and Mineral Research* 12:2076-2081.

Bouchard, C., P. An, T. Rice, J.S. Skinner, J.H. Wilmore, J. Gagnon, L. Perusse, A.S. Leon, and D.C. Rao. 1999. Familial aggregation of VO$_2$max response to exercise training: Results from the HERITAGE Family Study. *Journal of Applied Physiology* 87:1003-1008.

Bouchard, C., E.W. Daw, T. Rice, L. Perusse, J. Gagnon, M.A. Province, A.S. Leon, D.C. Rao, J.S. Skinner, and J.H. Wilmore. 1998. Familial resemblance for VO$_2$max in the sedentary state: The HERITAGE Family Study. *Medicine and Science in Sports and Exercise* 30:252-258.

Bouchard, C., F.T. Dionne, J.A. Simoneau, and M.R. Boulay. 1992. Genetics of aerobic and anaerobic performances. *Exercise and Sport Sciences Reviews* 20:27-58.

Bouchard, C., L. Perusse, C. Leblanc, A. Tremblay, and G. Theriault. 1988. Inheritance of the amount and distribution of human body fat. *International Journal of Obesity* 12:205-215.

Bouchard, C., and T. Rankinen. 2001. Individual differences in response to regular physical activity. *Medicine and Science in Sports and Exercise* 33:S446-S451.

Bouchard, C., M.A. Sarzynski, T.K. Rice, W.E. Kraus, T.S. Church, Y.J. Sung, D.C. Rao, and T. Rankinen. 2011. Genomic predictors of the maximal O2 uptake response to standardized exercise training programs. *Journal of Applied Physiology* 110:1160-1170.

Bouchard, C., J.A. Simoneau, G. Lortie, M.R. Boulay, M. Marcotte, and M.C. Thibault. 1986. Genetic effects in human skeletal muscle fiber type distribution and enzyme activities. *Canadian Journal of Physiology and Pharmacology* 64:1245-1251.

Bray, M., J. Hagberg, L. Perusse, T. Rankinen, S. Roth, B. Wolfarth, and C. Bouchard. 2009. The Human Gene Map for Performance and Health-Related Fitness Phenotypes: The 2006-2007 update. *Medicine and Science in Sports and Exercise* 41:34-72.

De Moor, M.H.M., Y.-J. Liu, D.I. Boomsma, J. Li, J.J. Hamilton, J.-J. Hottenga, S. Levy, X.-G. Liu, Y.-F. Pei, D. Posthuma, R.R. Recker, P.F. Sullivan, L. Wang, G. Willemsen, H. Yan, E.J.C. De Geus, and H.-W. Deng. 2009. Genome-wide association study of exercise behavior in Dutch and American adults. *Medicine and Science in Sports and Exercise* 41:1887-1895.

Erbs, S., Y. Baither, A. Linke, V. Adams, Y. Shu, K. Lenk, S. Gielen, R. Dilz, G. Schuler, and R. Hambrecht. 2003. Promoter but not exon 7 polymorphism of endothelial nitric oxide synthase affects training-induced correction of endothelial dysfunction. *Arteriosclerosis, Thrombosis, and Vascular Biology* 23:1814-1819.

Frayling, T.M., N.J. Timpson, M.N. Weedon, E. Zeggini, R.M. Freathy, C.M. Lindgren, J.R. Perry, K.S. Elliott, H. Lango, N.W. Rayner, B. Shields, L.W. Harries, J.C. Barrett, S. Ellard, C.J. Groves, B. Knight, A.M. Patch, A.R. Ness, S. Ebrahim, D.A. Lawlor, S.M. Ring, Y. Ben-Shlomo, M.R. Jarvelin, U.

Sovio, A.J. Bennett, D. Melzer, L. Ferrucci, R.J. Loos, I. Barroso, N.J. Wareham, F. Karpe, K.R. Owen, L.R. Cardon, M. Walker, G.A. Hitman, C.N. Palmer, A.S. Doney, A.D. Morris, G.D. Smith, A.T. Hattersley, and M.I. McCarthy. 2007. A common variant in the FTO gene is associated with body mass index and predisposes to childhood and adult obesity. *Science* 316:889-894.

Hagberg, J.M., R.E. Ferrell, L.I. Katzel, D.R. Dengel, J.D. Sorkin, and A.P. Goldberg. 1999. Apolipoprotein E genotype and exercise training-induced increases in plasma high-density lipoprotein (HDL)- and HDL2-cholesterol levels in overweight men. *Metabolism* 48:943-945.

Lakka, T.A., T. Rankinen, S.J. Weisnagel, Y.C. Chagnon, H.M. Lakka, O. Ukkola, N. Boule, T. Rice, A.S. Leon, J.S. Skinner, J.H. Wilmore, D.C. Rao, R. Bergman, and C. Bouchard. 2004. Leptin and leptin receptor gene polymorphisms and changes in glucose homeostasis in response to regular exercise in nondiabetic individuals: The HERITAGE Family Study. *Diabetes* 53:1603-1609.

Lakka, T.A., T. Rankinen, S.J. Weisnagel, Y.C. Chagnon, T. Rice, A.S. Leon, J.S. Skinner, J.H. Wilmore, D.C. Rao, and C. Bouchard. 2003. A quantitative trait locus on 7q31 for the changes in plasma insulin in response to exercise training: The HERITAGE Family Study. *Diabetes* 52:1583-1587.

Leon, A.S., K. Togashi, T. Rankinen, J.P. Despres, D.C. Rao, J.S. Skinner, J.H. Wilmore, and C. Bouchard. 2004. Association of apolipoprotein E polymorphism with blood lipids and maximal oxygen uptake in the sedentary state and after exercise training in the HERITAGE Family Study. *Metabolism* 53:108-116.

Loos, R.J., and C. Bouchard. 2008. FTO: The first gene contributing to common forms of obesity. *Obesity Reviews* 9:246-250.

Modrek, B., and C. Lee. 2002. A genomic view of alternative splicing. *Nature Genetics* 30:13-19.

Montgomery, H.E., P. Clarkson, C.M. Dollery, K. Prasad, M.A. Losi, H. Hemingway, D. Statters, M. Jubb, M. Girvain, A. Varnava, M. World, J. Deanfield, P. Talmud, J.R. McEwan, W.J. McKenna, and S. Humphries. 1997. Association of angiotensin-converting enzyme gene I/D polymorphism with change in left ventricular mass in response to physical training. *Circulation* 96:741-747.

Myerson, S.G., H.E. Montgomery, M. Whittingham, M. Jubb, M.J. World, S.E. Humphries, and D.J. Pennell. 2001. Left ventricular hypertrophy with exercise and ACE gene insertion/deletion polymorphism: A randomized controlled trial with losartan. *Circulation* 103:226-230.

Rampersaud, E., B.D. Mitchell, T.I. Pollin, M. Fu, H. Shen, J.R. O'Connell, J.L. Ducharme, S. Hines, P. Sack, R. Naglieri, A.R. Shuldiner, and S. Snitker. 2008. Physical activity and the association of common FTO gene variants with body mass index and obesity. *Archives of Internal Medicine* 168:1791-1797. Erratum in *Archives of Internal Medicine* 169:453.

Rankinen, T., L. Perusse, R. Rauramaa, M.A. Rivera, B. Wolfarth, and C. Bouchard. 2004. The human gene map for performance and health-related fitness phenotypes:

The 2003 update. *Medicine and Science in Sports and Exercise* 36: 1451-1469.

Rankinen, T., T. Rice, L. Perusse, Y.C. Chagnon, J. Gagnon, A.S. Leon, J.S. Skinner, J.H. Wilmore, D.C. Rao, and C. Bouchard. 2000. NOS3 Glu298Asp genotype and blood pressure response to endurance training: The HERITAGE Family Study. *Hypertension* 36:885-889.

Rankinen, T., T. Rice, M. Teran-Garcia, D.C. Rao, and C. Bouchard. 2010. FTO genotype is associated with exercise training-induced changes in body composition. *Obesity* 18:322-326.

Rankinen, T., Y.J. Sung, M.A. Sarzynski, T. Rice, D.C. Rao, and C. Bouchard. 2012. The heritability of submaximal heart rate response to exercise training is accounted for by nine SNPs. *Journal of Applied Physiology.* In press.

Rice, T., E.W. Daw, J. Gagnon, C. Bouchard, A.S. Leon, J.S. Skinner, J.H. Wilmore, and D.C. Rao. 1997. Familial resemblance for body composition measures: The HERITAGE Family Study. *Obesity Research* 5:557-562.

Rico-Sanz, J., T. Rankinen, D.R. Joanisse, A.S. Leon, J.S. Skinner, W.H. Wilmore, D.C. Rao, and C. Bouchard. 2003. Familial resemblance for muscle phenotypes in the HERITAGE Family Study. *Medicine and Science in Sports and Exercise* 35:1360-1366. Corrigendum in Rankinen, T., C. Bouchard, and D.C. Rao. 2005. Familial resemblance for muscle phenotypes: The HERITAGE Family Study. *Medicine and Science in Sports and Exercise* 37:2017.

Roth, S.M., R.E. Ferrell, D.G. Peters, E.J. Metter, B.F. Hurley, and M.A. Rogers. 2002. Influence of age, sex, and strength training on human muscle gene expression determined by microarray. *Physiological Genomics* 10:181-190.

Simoneau, J.A., and C. Bouchard. 1995. Genetic determinism of fiber type proportion in human skeletal muscle. *FASEB Journal* 9:1091-1095.

Szathmary, E., F. Jordan, and C. Pal. 2001. Molecular biology and evolution. Can genes explain biological complexity? *Science* 292:1315-1316.

Teran-Garcia, M., T. Rankinen, R.A. Koza, D.C. Rao, and C. Bouchard. 2005. Endurance training-induced changes in insulin sensitivity and gene expression. *American Journal of Applied Physiology: Endocrinology and Metabolism* 288:E1168-E1178.

Teran-Garcia, M., T. Rankinen, T. Rice, A.S. Leon, D.C. Rao, J.S. Skinner, and C. Bouchard. 2007. Variations in the four and a half LIM domains 1 gene (FHL1) are associated with fasting insulin and insulin sensitivity responses to regular exercise. *Diabetologia* 50:1858-1866.

Timmons, J.A., S. Knudsen, T. Rankinen, L.G. Koch, M. Sarzynski, T. Jensen, P. Keller, C. Scheele, N.B. Vollaard, S. Nielsen, T. Akerström, O.A. Macdougald, E. Jansson, P.L. Greenhaff, M.A. Tarnopolsky, L.J. van Loon, B.K. Pedersen, C.J. Sundberg, C. Wahlestedt, S.L. Britton, and C. Bouchard. 2010. Using molecular classification to predict gains in maximal aerobic capacity following endurance exercise training in humans. *Journal of Applied Physiology* 108:1487-1496.

25

An Integrated View of Physical Activity, Fitness, and Health

William L. Haskell, PhD; Steven N. Blair, PED; and Claude Bouchard, PhD

©Photodisc

Regardless of a person's socioeconomic status, ethnicity, or culture, a common goal for most people is to live a healthy, enjoyable, peaceful, and productive life. As has been documented in a number of the chapters in this book, having a physically active lifestyle can contribute significantly to this goal. A major challenge for many people of all ages is building back into their otherwise sedentary lives sufficient activity to promote good health. Although regular activity is necessary, it is not sufficient in itself to ensure optimal health. Thus, a person's health-oriented physical activity plan needs to be integrated into a general wellness plan that includes a variety of other health behaviors, such as eating a healthy diet, reducing sedentary time such as sitting, avoiding tobacco, not abusing alcohol or drugs, obtaining adequate sleep, and managing stress. The activity plan will vary as life conditions change—being in school, changing jobs, moving from one community to another, getting married, having children, recovering from an illness, and retirement. As people move through the life span, their life goals tend to change, and so will some of the specific features of their activity plans. In this chapter, we present key challenges and issues that people encounter when attempting to integrate a health-oriented plan of physical activity into a technology-dominated environment that has reduced dramatically the need for human **energy expenditure**. We also present some strategies on how to successfully meet these challenges.

In most technologically advanced cultures where there has been a large decrease in the daily physical activity required of many adults to survive, it is generally understood that being sedentary and sitting throughout the day contribute to a variety of health problems. However, many people are unsuccessful in acting on this knowledge and remain sedentary or, at best, sporadically active. This inability to implement and maintain an effective activity plan can result from a number of countervailing factors that vary from time to time for a person as well as from person to person.

A number of factors appear to prevent the pursuit or maintenance of an active lifestyle. Factors frequently reported include lack of time, laziness, uncertainty about how to begin, existing illness or injury, not having the cognitive and behavioral skills to implement and sustain change in a complicated behavior, concerns about the risks of exercise, lack of access to convenient or safe areas to exercise, and confusion about the difference between being busy and being physically active. The challenge to the public and health professionals to overcome these barriers

has been only partly met. To achieve a much higher level of success will require more of the following:

- Research on human behavior change and the translation of that research into better programs that can be widely implemented
- Government policies that support an active lifestyle throughout the life span
- **Built environments** that make being physically active the default option
- Built environments that make physical activity convenient and safe wherever people live, work, and go to school
- Medical and health care systems that give priority to lifestyle change and disease prevention
- An educational system that teaches all students about the health benefits of activity, shows students how to initiate and sustain a successful activity program, helps them learn how to apply cognitive and behavioral strategies to be more active, and provides opportunities for participation in activities that contribute to a lifelong active lifestyle

To reverse the population trend toward a more sedentary lifestyle, individual behavior change needs to be supported by public policy that promotes health and physical education, preventive medical services, convenient access to exercise facilities and services, and construction of an activity-friendly built environment (National Physical Activity Plan 2010).

Physical Activity Versus Inactivity: Universal Value Versus Damaging Consequences

The failure to maintain a physically active lifestyle has major negative health consequences regardless of a person's health status, age, gender, race, or ethnicity. Yet for an increasing portion of the world's population, less and less physical activity is required to get where they need to go, to earn a living, to take care of their home and family, and to enjoy recreation. A multifactorial approach will be necessary to reverse this trend.

Universal Value of a Physically Active Lifestyle

The promotion of increased physical activity is an excellent public health intervention in that a physically active lifestyle has a positive impact in the prevention and treatment of a wide variety of chronic diseases, has unique and independent positive effects on physical and mental functioning and quality of life, acts synergistically with other behaviors to improve health, and contributes to the health and well-being of children and adults of all cultures and ethnicities throughout their life span.

This is not to say that physical activity is a panacea for maintaining good health, but rather that given the human constitution and biology, populations with a physically active lifestyle will have a health and wellness advantage over those who remain sedentary. The results from the INTERHEART study illustrate the population-based effects of a physically active lifestyle, in this case the prevention of myocardial infarction in people in 52 countries representing every inhabited continent on earth (Yusuf et al. 2004). This was a case–control study in which major behavioral and biological risk factors were determined for 15,152 men and women who had had a myocardial infarction and 14,820 age- and sex-matched control participants free of cardiovascular disease (CVD). Participants were considered "active" if they reported usually performing at least 4 h per week of moderate- or vigorous-intensity exercise or "inactive" if they were routinely less active than that. Regardless of other risk factors and independent of ethnicity and country of origin, fewer participants with myocardial infarction were classified as active compared with the matched controls. After adjustment for other risk factors, active persons had a significantly lower risk of myocardial infarction than inactive persons (odds ratio [OR] = 0.86; 95% confidence interval [CI] = 0.76-0.97), and the contribution of being inactive to having a myocardial infarction (population attributable risk [PAR]) was 12.2% ($p < .0001$). That a combination of lifestyle risk factors has a major impact on the risk of myocardial infarction in this highly diverse population is demonstrated in figure 25.1. When being active is added to nonsmoking and daily intake of fruits and vegetables, the OR for developing a myocardial infarction decreases to 0.21 (95% CI = 0.17-0.25).

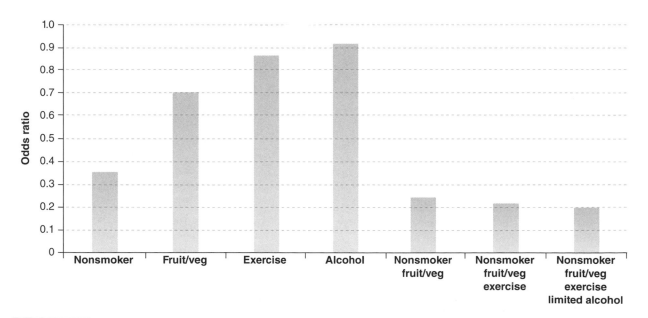

FIGURE 25.1 Odds ratios for nonsmokers versus current smokers, participants who consumed fruits and vegetables daily versus those with infrequent consumption, frequent exercisers (≥4 h/week) versus infrequent exercisers, and participants who consumed alcohol regularly (three or more times per week) versus those who consumed alcohol infrequently. Odds ratios are adjusted for age, country, and all other risk factors.

Reprinted from *The Lancet*, Vol. 364, S. Yusuf, S. Hawken, S. Ounpuu, T. Dans, A. Avzeum, F. Lanas, M. McQueen, A. Budaj, P. Pais, J. Varigos, and L. Lisheng, "Effect of potentially modifiable risk factors associated with myocardial infarction in 52 countries (the INTERHEART study) case-control study," pp 937-952, copyright 2004, with permission from Elsevier.

Physical activity promotion is an excellent public health approach because of its positive effects on various chronic medical conditions, physical and mental functions, and quality of life in children and adults regardless of culture or ethnicity throughout their life span.

Other studies have documented that people who are physically active, remain relatively lean, frequently eat fruits and vegetables (and less processed foods), don't smoke cigarettes, and drink limited amounts of alcohol have very favorable health outcomes. For example, in 84,129 nurses followed for 14 years, those who did not smoke, who performed ≥ 30 min of moderate- or vigorous-intensity exercise per day, and who ate a diet high in fruits and vegetables had a risk of heart attack or stroke that was 57% less (relative risk 0.43; 95% CI = 0.33-0.55) than that for women without these three characteristics (Stampfer et al. 2000). A separate analysis of data from this large group of nurses indicated that physical inactivity (<3.5 h/week) and excess weight (body mass index [BMI] ≥25.0) together could account for 31% of all premature deaths, 59% of deaths from CVD, and 21% of deaths from cancer among nonsmoking women (Hu et al. 2004). The combined effect of having normal weight (BMI = 18.5-24.9), not smoking (not a current smoker), consuming a moderate amount of alcohol (1-14 drinks per week), being physically active (moderate to high level), and having a higher cardiorespiratory fitness (top two-thirds) on all-cause mortality was examined in 38,110 men aged 20 to 84 years from the Aerobics Center Longitudinal Study (Byun et al. 2010). Compared with men with zero positive health factors, the multivariable-adjusted hazard ratios of all-cause mortality with one, two, three, four, and five positive health factors were 0.78, 0.61, 0.54, 0.43, and 0.39, respectively (p-value for trend < 0.001). These data strongly support the substantial value of integrating activity with other established health habits to prevent disease and promote health.

Challenge of Labor-Saving Technology

For some people, just performing their **required activities of daily living** provides sufficient physical activity to promote good health. For example, there still are farmers, construction workers, professional gardeners, and others who perform physical labor much of the day, as well as those who walk or ride bicycles an hour or more per day for transportation.

These individuals do not need to seek other activity for health outcomes. However, in technologically advanced societies, fewer and fewer people achieve an adequate level of physical activity just through the activities they typically perform for self- and home care, transportation, and occupation. Over a relatively short time—since the beginning of the industrial revolution circa 1800—a significant amount of the physical activity required for most people to survive has been eliminated and replaced by machines. The list of technological advances that reduce the daily energy expenditure of an increasing number of people throughout the world is large and continues to grow—electricity, steam and gas engines, telephones, automobiles, trains, elevators and escalators, tractors and cranes, television, computers, and the Internet.

Although few actual measurements have systematically documented the decline in energy expenditure attributable to all of these advances in technology, a number of measurements have been made of the energy cost associated with the physical activity required by various occupations. Also, several investigators have measured the physical activity or energy expenditure of people who live without the use of modern technologies. Montgomery (1978), who collected energy expenditure data on members of a primitive Indian tribe of hunters, gatherers, and farmers living in Peru, documented that their way of living required an average daily energy expenditure of about 60 and 44 kcal per kilogram of body weight per day for men and women, respectively. This is in comparison with about 34 and 32 kcal per kilogram of body weight per day for current-day healthy but sedentary middle-aged men and women, respectively (Simons-Morton et al. 2000).

Old Order Amish living in Ontario, Canada, refrain from driving automobiles and from using electrical appliances and other modern conveniences. Manual labor farming is their primary occupation. The amount of physical activity performed daily by 98 Old Order Amish men and women was assessed using standard questionnaires, pedometers, and a log for seven consecutive days (Bassett, Schneider, and Huntington 2004). The average number of steps per day for the Amish men was 18,425 and for women 14,196, and the men reported walking 12.0 h per week and the women 5.7 h per week. None of the men and only 9% of the women were obese (BMI >30), whereas 25% of the men and 26% of the women were overweight or obese (BMI >25) (see figure 25.2). By contrast, a majority of U.S. adults (20-74 years old) walk less than 2 to 3 h per week and accumulate less than 5,000 steps per day; approximately 65% have BMI >25; and 31% are considered

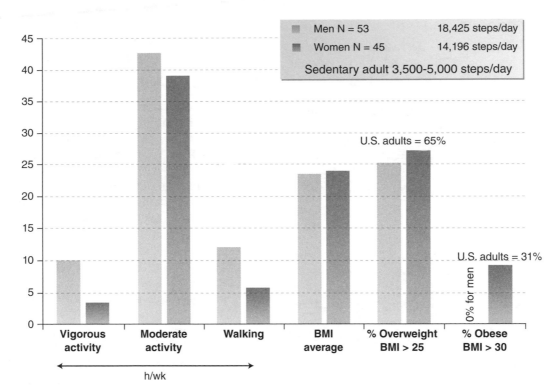

FIGURE 25.2 Old Order Amish who do not use modern technology on their farms are very physically active and exhibit much less obesity compared with men and women who use such technology in many aspects of their lives (see text for detailed explanation of data). The first three pairs of bars are for activities in hours per week. The bars for average BMI are in kg/m². and the last two sets of bars are expressed in percentage of overweight or obese individuals.

Data from Bassett, Schneider, and Huntington 2004 and Hedley et al. 2004.

obese (BMI ≥30) (Hedley et al. 2004). Similar trends were reported for boys and girls ages 9 to 13 years living in Old Order Amish or Old Order Mennonite communities in Canada as compared to similar-aged youth living in rural or urban communities (Esliger et al. 2010). Determining the physical activity patterns in populations who live a lifestyle somewhat similar to that of the general population in the past helps us understand the amount and nature of physical activity needed to prevent unhealthy weight gain and suffer less chronic disease.

Typical Occupational Activities

For most people, the opportunities to build increased physical activity back into their occupation are very limited. An increasing number of jobs require that for people to be most productive, they need to stay in one place, very frequently sitting, for much of the workday. Many office workers can now sit at their desks and through the use of advanced computer and communication technology perform most of their job tasks without even standing—they can communicate electronically with a coworker down the hall or a customer on the other side of the world. In the 1950s, most major office and manufacturing

buildings employed night patrolmen who walked the perimeter of the building for much of their shift. By the 1990s, most of these "patrolmen" were replaced by "watchmen" who, instead of walking to provide security, sat and watched a bank of video screens that displayed various locations in the building (one watchman can provide security to a greater area than one patrolman, thus increasing worker productivity). Several similar situations can be cited to demonstrate that individual worker productivity tends to be increased through new technologies that reduce workers' need to move about. Thus, for these types of jobs, there is an economic disincentive to increase opportunities for employees to be physically active during work hours. This situation is compounded when employees are required to work extended hours, which is a common practice. It is less expensive for companies to have fewer employees work longer hours than to hire additional employees, partly because of the cost of employee benefits other than salary, especially health insurance. As the use of technologies spreads to more occupations, especially in developing countries, there will be a further increase in the proportion of the population who become sedentary.

Typical Modes of Transportation

Daily physical activity has decreased substantially because of advancing technology relating to personal transportation. The rise in the use of the automobile during the 20th century significantly reduced average daily physical activity of many children and adults. Few programs have been successful in convincing people to substitute walking or bicycling for the use of their cars. Such a change would have a number of beneficial consequences in addition to the increased physical activity, including reductions in environmental pollution, traffic congestion, the need for new roads and parking structures, and—in some countries—reliance on foreign oil. The use of the automobile for transportation has been both facilitated and made necessary by the automobile-friendly design of many communities and by government subsidization of road and parking lot construction. These pro-automobile policies were partly a response to vigorous lobbying by automobile manufacturers and dealers, tire manufacturers, and the petroleum industry.

In many communities, safe sidewalks and bike paths are not available or do not lead from one functional destination to another (e.g., from a residential area to a shopping center, school, or office complex). The Institute of Medicine and Transportation Research Board of the U.S. National Academies released a report titled "Does the Built Environment Influence Physical Activity? Examining the Evidence" (Committee on Physical Activity, Health, Transportation, and Land Use 2005). The research on this topic is at an early stage, but preliminary results indicate associations between the built environment and physical activity. These findings need to be confirmed with prospective observations and ideally with experiments.

In 2010, the U.S. National Physical Activity Plan recommended that the following actions be taken to promote physical activity in the transportation, land use, and community design sector:

- Increase accountability of project planning and selection to ensure infrastructure supporting active transportation and other forms of physical activity.

- Prioritize resources and provide incentives to increase active transportation and other physical activity through community design, infrastructure projects, systems, policies, and initiatives.

- Integrate land-use, transportation, community design, and economic development planning with public health planning to increase active transportation and other physical activity.

- Increase connectivity and accessibility to essential community destinations to increase active transportation and other physical activity.

Typical Domestic Activities

As with the reduction in the need for physical activity on the job, technology has substantially reduced the amount of human energy required for domestic activities and self-care. Devices such as automatic clothes washers and dryers, dishwashers, self-propelled vacuum cleaners, and power lawn mowers make life much easier for many people but reduce their physical activity on most days (Lanningham-Foster, Nysse, and Levine 2003). Although the amount of energy saved by using these devices is quite small on any one day (<50 kcal/day), a reduction of only 25 kcal per day over the year would equal about 9,125 kcal, which is equivalent to the amount of energy in 1.2 kg (2.6 lb) of body mass (figure 25.3). To put this in context, the average weight gain for U.S. adults during the 1990s, when major increases in the prevalence of obesity were observed, was about 0.5 kg (1 lb) per year (U.S. Department of Health and Human Services [DHHS] 2004). A decline in energy expenditure of this magnitude contributes to the obesity epidemic

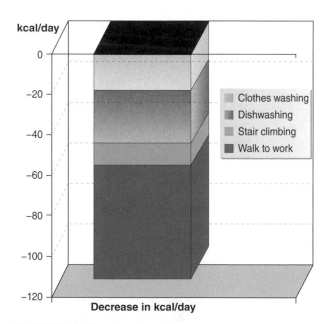

FIGURE 25.3 Participants (122 healthy adults) completed the indicated tasks with and without the aid of equipment or machines while the energy expenditure was measured. The distance walked or driven was 0.8 mi (1.28 km); the number of stairs climbed was 219; and the time spent washing dishes and clothes each day was 17.1 and 18.7 min, respectively.

Reprinted, by permission, from Macmillan Publishers Ltd: L. Lanningham-Foster, L.J. Nysse, and J.A. Levine, 2003, "Labor saved, calories lost: The energetic impact of domestic labor-saving devices," *Obesity Research* 11: 1178-1181, copyright 2003.

that has been observed in the United States over the past 25 years.

The automobile also has significantly reduced domestic-related physical activity by its frequent use for short trips for shopping, going to the cleaners and doing other errands, and taking children to school. It is highly unlikely that individuals will give up their dishwasher, clothes washer, or power lawn mower to increase their activity; but an improved system of safe sidewalks and bike paths throughout the community and financial disincentives, rather than incentives, to use the automobile might contribute to more physical activity to complete many short trips. Another example of small changes in activity having a substantial effect on energy expenditure is the observation that spontaneous activities and frequent "fidgeting" (can't sit still) can amount to a substantial number of calories expended every day. In one study, obese men and women were documented as sitting 164 min per day longer than lean individuals (Levine et al. 2005).

Typical Leisure Time

In the 1960s and 1970s, the prediction by experts regarding the impact of technology on U.S. society over the following 50 years was that many adults would spend much less time at their occupations and have a great deal of time to pursue "leisure" activities. This prediction has not been realized, with many adults working more than 40 h per week, spending more time commuting to work via automobile or mass transit, or working more than one job. Also, the increasing percentage of households in which both partners are gainfully employed has further increased the overall amount of time spent working, performing domestic chores, or commuting and has reduced time available for recreation or leisure activities.

Compounding the problem of reduced leisure time has been the widespread pursuit of sedentary leisure-time activities, especially watching television, video monitors, and computer screens. How most people spend their leisure time depends on interests, priorities, and opportunities. It is in the area of leisure time that most people have the greatest opportunity to build physical activity back into their lives. There is a paradox regarding leisure time and amount of physical activity. Surveys show that the most frequently given reason for not being more physically active is lack of time, but this is in a population that reports at least several hours of TV watching each day. The problem is one of assigning priorities to leisure-time activities, not an absolute lack of leisure time itself. This issue appears especially important for children. A national survey in 2009 of television and video viewing and computer use showed that 2- to 5-year-old boys and girls devoted more than 32 h a week, and those ages 6 to 11 years more than 28 h per week, to entertainment media—and about 75% of this time was spent watching television (NielsenWire 2009) (table 25.1). With every passing year, new technologies are providing more and more options for sedentary activities among children and youth using computers, video players, and other electronic devices. This trend carries many detrimental consequences, as illustrated by the newly uncovered relationships between the amount of sedentary time or sitting time in a normal day and the increased risk of morbidities and premature death (Katzmarzyk et al. 2009; Warren et al. 2010). (See chapter 4 for additional information on this topic.)

Developing and Implementing Physical Activity Plans

Developing and implementing strategies that will help generally inactive people to become more physically active need to be a high priority for government agencies and private organizations. There are no simple solutions to reversing the downward trend in activity for many people, but research and experience have provided a number of good leads on what needs to be done.

The National Physical Activity Plan was implemented in 2010 in an attempt to get Americans more physically active (National Physical Activity Plan

TABLE 25.1 Average Weekly Viewing of TV and Peripheral Devices by U.S. Children

Age		Average hours: minutes per week				
	Total	TV	DVR	DVD	VCR	Game console
2-5 years	>32 h	24:51	1:29	4:33	0:45	1:12
6-11 years	>28 h	22:09	0:59	2:28	0:18	2:23

Data from Nielsen Wire, "TV Viewing Among Kids at an Eight-Year High," October 28, 2009.

2010). A number of other countries already have similar plans. The U.S. plan is composed of recommendations that are organized into eight societal sectors:

- Business and Industry
- Education
- Health Care
- Mass Media
- Parks, Recreation, Fitness, and Sports
- Public Health
- Transportation, Land Use, and Community Design
- Volunteer and Non-Profit

For each sector, the plan presents strategies aimed at promoting physical activity. For each strategy, specific tactics are outlined that communities, organizations and agencies, and individuals can use to address the strategy. The scope of the National Physical Activity Plan is indicative of the magnitude of the challenge that we are facing when attempting to reengineer physical activity back into the life of the citizenry.

Developing Good Physical Activity Knowledge, Attitudes, and Skills

Given the tendency for current society to engineer a great deal of physical activity out of most people's lives, there is a public health need to develop system approaches to reintroduce activity so that most people can achieve active lifestyles throughout their life span. Because of the complexity of physical activity as a behavior and the multitude of factors that influence the process in which a person becomes and remains physically active, there is no one set formula for achieving this goal. However, adequate and accurate information, a positive attitude, and good skills are valuable tools in achieving an active lifestyle.

Knowledge

For most people, understanding why an active lifestyle is important for their health and what options they have to become and stay active is a good first step to an active lifestyle. Ideally, they acquire this knowledge during their youth from parents as well as teachers, but it is never too late to obtain such information from teachers, health care providers, exercise specialists, books, videos, and the Internet. The more relevant or personalized this information is, the more likely it will be retained and acted upon.

As in other areas of health behavior, just having the facts about what needs to be done and how to do it is not sufficient to ensure that the behavior will be performed.

The primary reason many people are physically active is not that they want to prevent disease or maintain their physical independence. Some are active because they enjoy participating in sports and other active recreational activities. The goal of many hikers or backpackers is to enjoy the outdoors; promoting good health is a secondary issue for them. Other people, primarily for environmental or cost reasons, walk or use bicycles as their major means of transportation. For these people, it is important to encourage and possibly reward their behavior and find ways to support their active lifestyle by making the activity as convenient as possible. Also, there still is a segment of the population who obtain adequate physical activity during their occupational or domestic activities. These people need to know whether their "required" activity meets the dose of activity that promotes good health and how to supplement this required activity if necessary.

Attitudes

One goal of any physical activity intervention should be to promote positive attitudes about the experience and seek ways to maintain or even enhance these attitudes over time. Although positive attitudes don't automatically lead to the desired behavior, negative attitudes, especially when derived or reinforced by negative experiences associated with activity, are a major deterrent to achieving an active lifestyle. A variety of factors can go into the formation of attitudes about physical activity, including past personal experiences with exercise; attitudes about exercise on the part of parents, teachers, physicians, mentors, and other role models; and the importance of the potential benefits of being active or the negative consequences of being inactive. People who have been sedentary for a number of years often view themselves as inactive and have a difficult time viewing themselves as ever being physically active or fit. This negative attitude about activity frequently is made worse when the person is obese or has a chronic illness that contributes to a decrease in his functional capacity. A change in attitude may occur only when the person has actually been successful in increasing activity rather than as a precursor to an increase in activity.

Skills

A major reason many people continue to be physically active during leisure time is the enjoyment it provides. For activities that require a special skill, such as most sports, having a certain level of skill significantly adds to the enjoyment of participation. Beginning to develop such skills during childhood is thought to facilitate participation as an adult. Thus, including programs for "lifetime activities" in schools for all students may be a way to facilitate sport participation in adults as well as youth. Programs like these include activities such as tennis, volleyball, soccer, basketball, cycling, jogging and running, weight training, and dance. The focus is not on competition or winning but on attaining basic skills, understanding the health and performance benefits, and enjoying participation. This type of program is in sharp contrast to a majority of school programs that focus on competitive football, basketball, and baseball for only the highly skilled and motivated students. For some basic activities that can be a major component of a health-oriented activity program, no or few special skills are needed. The best example is walking, which for many adults is, or should be, a major component of their activity plan. We do not have extensive data showing that teaching "lifetime activities" in schools will increase activity levels, but it seems clear that highly competitive sports are not the answer to mass sedentary behavior. It would be hard to emphasize competitive sport more than it has been emphasized in the United States over the past several decades. For instance, consider the number of pages of the typical daily newspaper that are devoted to sport coverage in contrast to international events. Yet this interest in competitive sport has not resulted in an active society.

Two Approaches to Physical Activity Plans

As part of a comprehensive program designed to promote good health, all individuals should have a physical activity plan. Ideally, this plan should be individualized to meet each person's current needs (goals), health status, physical capacity, interests, skills, schedule, and resources, at the same time taking into account current required daily activity. The plan should include activities to be proscribed because of the risk of injury or illness, as well as activities prescribed for their health benefits. In its simplest form, this plan is no more than the current public health recommendations of "moderate-intensity activity for at least 150 minutes per week." However, although good advice, this very general guideline is primarily directed at the average sedentary middle-aged person and does not address any number of specific issues that are deterrents to maintaining

an active lifestyle. Many inactive individuals would benefit from a more detailed and personalized plan.

■ **Although current U.S. physical activity guidelines (moderate-intensity activity for ≥150 min per week, vigorous-intensity activity for ≥75 min per week, or a combination of moderate- and vigorous-intensity activities [with 1 min of vigorous-intensity equal to 2 min of moderate-intensity activity]) are valuable for informing individuals on what they need to do, a personalized activity plan should consider goals; current activity; health status; and family, work, and social obligations.**

Class Approach

The traditional approach many have used to build more activity into their lives is to join an exercise facility or exercise class. This approach can be the major part of a successful plan for some people because they engage in activity in a safe place with proper facilities (exercise equipment, showers), good exercise instruction and leadership, and a positive social environment. Some classes are designed for people with special interests or needs and can provide a very personalized program or meet special needs, for example the medical supervision provided in many classes for people with CVD, diabetes, or chronic pulmonary disease.

Among the limitations to this approach are the logistics of time and location. For many busy people with work and family responsibilities, it is very difficult to visit an exercise facility or attend a class several times per week. It is not necessarily the time spent in class but the time required to commute to a facility that may be a major deterrent. For people with limited incomes, the financial costs to join a facility or class may make participation difficult or impossible. Also, participating in a typical exercise class two times per week may not fully replace the decrease in energy expenditure experienced by many people with otherwise very sedentary lifestyles.

For example, think of an administrative assistant who, with access to improved computer and communication technology, reduces the amount of time walking around the office building by 5 min per hour (40 min per 8-h day or 200 min per five-day workweek) and drives a car to work rather than taking a bus, which required 20 min more walking per workday (20 × 5 = 100 min/week). He needs to replace the energy he used to expend walking versus sitting for 300 min or 5 h per week. If the net decrease in energy expenditure between walking and sitting at a desk or driving averages 2.0 **metabolic equivalents (METs)** (3.5 METs walking – 1.5 METs sitting/driving = 2.0 METs), then he needs to replace about 600 MET-minutes of activity per week (300 min at 2.0 METs). To replace this amount of physical activity during two 40-min exercise classes per week, it would be necessary to exercise at an average intensity of approximately 7.5 METs (vigorous or hard intensity). At three 40-min exercise sessions per week, the average intensity would need to be approximately 5.0 METs. This example illustrates the fact that energy expenditure of daily activity is an important determinant of the weight gain epidemic that has been observed around the world.

Active Living Concept

A second approach to building physical activity back into one's life that has received much more attention of late focuses on making or taking opportunities throughout the day to be active. So, instead of going to a class or a gym to work out several times a week, people make time to be active while they go on with daily life. Although such recommendations, in a somewhat isolated way, already had been made by many people and organizations (e.g., take the stairs, park in the back of the lot), starting in around 1990, this approach has begun to be formalized, discussed, evaluated, debated, and promoted. In Canada, this idea of integrating physical activity into daily life to create an active lifestyle was labeled as **active living** (Fitness Canada 1991). Although active living has been given a variety of meanings and definitions, it was defined by Fitness Canada (1991) as "a way of life in which physical activity is valued and integrated into daily life" (p. 4) or by Makosky (1994) as "a way of life in which individuals make useful, pleasurable and satisfying physical activities as an integral part of their daily routine" (p. 272).

A key to success is to help sedentary individuals learn to apply cognitive and behavioral strategies such as self-monitoring, goal setting, and problem solving to increase physical activity (Blair et al. 2001). Some individuals may use these approaches to increase their lifestyle physical activities; others may use them to adopt and maintain more traditional forms of physical activity such as walking, jogging, or sport. Behavioral intervention approaches have been found to be comparable to more traditional and structured approaches in helping sedentary persons to become and stay more physically active up for up to 24 months (Dunn et al. 1999; Wilcox et al. 2008).

This concept, be it called active living, lifestyle activity, or by some other name, is in part supported by the research that has evaluated the health and performance benefits of shorter versus longer bouts of activity, but the two ideas are not the same. Most of the experimental studies evaluating short bouts of activity have used bouts in the 8- to 10-min range, yielding few data on the value of discrete bouts of moderate-intensity exercise of 5 min or less. Short bouts of vigorous exercise, such as stair climbing, spread throughout the day have produced improvements in health-related fitness. For example, in a study of young sedentary women, those who performed stair climbing five days per week for seven weeks (climbed 199 steps in 135 s, one time per day in week 1 progressing to six times per day in weeks 6 and 7) had a reduction in heart rate and blood lactate concentration during submaximal exercise and an increase in high-density lipoprotein (HDL) cholesterol compared with women who remained sedentary (Boreham, Wallace, and Nevill 2000). Some epidemiological studies relating habitual physical activity to chronic disease or all-cause mortality have collected data supporting the concept that activity accumulated throughout the day in bouts of various lengths and intensities confers health benefits (Lee, Sesso, and Paffenbarger 2000).

Overcoming Barriers to Achieving a Physically Active Lifestyle

As noted earlier in this chapter, there are a number of barriers to initiating and especially maintaining a physically active lifestyle.

Barriers

Frequently, barriers relate to the need for sedentary people to make time to build activity back into their daily life. Often people feel too busy with work, family, commuting, and social engagements to take time to be active. As individuals age, a major barrier to being physically active is the concern about the health or injury risks of increased activity, especially if the person has led a sedentary life or has one or more medical conditions.

Yet even some very busy women who have a full-time job and a family to care for make time to be physically active—their answer is that their health and appearance are a priority and that they have good organizational and time management skills. It is a major challenge for health professionals to help people give regular activity a sufficiently high priority to overcome time demands and other barriers. Somehow the negative health consequences of being inactive, as well as the benefits of being active, need to be made salient to the person as a first step in an activity plan.

Strategies for Increasing Physical Activity

Strategies for helping people achieve an active lifestyle work differently for different people. For example, some people are "loners" and prefer to exercise on their own, whereas others benefit from having access to a highly social environment. Some employees may find it very difficult to make time for a 20-min walk alone during lunch but will readily commit to walking and talking with a coworker on a regular basis. Appointments to go for a walk or perform other exercise with a friend, colleague, or family member should carry the same obligation as any other type of appointment. For people who benefit from positive social interactions, well-led exercise classes of numerous types can facilitate frequent and long-term participation.

Those who believe they are too busy to make time for physical activity need to closely review their schedules and look for even small opportunities to begin to build activity back—climbing the stairs instead of using the escalator, taking a 10-min walk during at least one coffee break each workday, spending some of the extra time between plane connections walking in the terminal. For some people, performing these short bouts of activity is the first step in building more activity into their lives and thinking of themselves as active rather than sedentary.

For some people, interaction with an authority figure or role model may help make an active lifestyle a higher priority. The results of numerous surveys have indicated that if a person's physician or other health care provider makes a strong recommendation regarding an active lifestyle and monitors to even a small extent, the likelihood that the person will become or stay active is increased. People who are respected for how they lead their lives can be positive role models for people of all ages in developing an active lifestyle; this seems to be true throughout the life span.

Older individuals who are afraid that physical activity may be harmful because of their age or physical condition should receive specific personalized instructions from their health care provider or staff regarding their activity plan, including an opportunity to get questions answered and a schedule for follow-up. There are many options even for people with quite low exercise capacities attributable to injury, illness, or age to perform health-promoting physical activity. At the very low end of the activity–

fitness range, a person should at least get out of a bed or chair and move about periodically throughout the day to help prevent thrombosis, orthostatic hypotension, insulin resistance, muscle weakness, and loss of balance. To provide additional assurance for at-risk patients that activity can be increased with relative safety, supervised exercise classes or individual instruction by a well-trained exercise leader should be considered.

One procedure that appears to help many people improve adherence to their activity plan is "self-monitoring" (Dunn et al. 1999). This procedure can take different forms but basically consists of developing a logging or tracking system that allows individuals to compare their activity performance over time with their activity plan. For example, a person could record her activity plan on a calendar or electronic organizer for the next three months (e.g., walk briskly 30 min at noon Monday, Wednesday, and Friday and in the morning on Saturday and Sunday) and then check off or document each time she performs that activity. This process provides a visual history of adherence to the plan and a frequent reminder when an activity session is missed.

The recent rapid development of electronic technologies has provided many new strategies to help individuals become and stay more physically active. A wide variety of accelerometer-based monitors can track physical activity (especially steps) and provide immediate feedback to the individual, or the data can be viewed on a mobile phone or uploaded to a personalized website. Investigators interested in understanding how these technologies can be used to promote physical activity are evaluating numerous approaches using wireless communication and customized software that work with both mobile phones and social networking applications such as Twitter and Facebook. These studies are still in their early stages, and there is no evidence as yet regarding what approaches work best for which audiences.

Integrating Physical Activity With Other Aspects of Life

Much of how we spend our daily lives is based on a number of trade-offs among those things we have to accomplish daily to survive and meet our obligations to others and those things we choose to do during our free or leisure time. In attempting to build more physical activity into existing busy schedules, we must recognize that some balance needs to be achieved among these various interests and time commitments but that physical activity needs to be given sufficient priority to ensure a physically active lifestyle. All too often, physical activity gets dropped from the

Riding bicycles to run errands is one way to integrate physical activity into daily life.

schedule because of the immediate need to get other things done, with the notion that the activity will get done later. Is it more important to go to a movie, watch a television show, read a book, or have coffee with a friend than to go for a 30-min walk, bicycle ride, or swim? People have to make this decision individually, but they should do so fully aware of the long-term consequences of being physically inactive.

Need to Modify the Activity Plan as Life Evolves

As people progress through life, their activity plan should be considered a dynamic and not a static tool for helping to maintain good health. Although the overall plan of maintaining an active lifestyle remains the same, some specific features of the plan will need to change in response to changes in the person's age, living environment, physical fitness, and health.

Frequently it is during major life transition periods, such as leaving high school or college, getting a full-time job, changing jobs, moving to a new home, getting married, having children, and retirement, that a number of health-related behaviors, including physical activity, encounter pressure to change. With each of these life changes, a person takes on new responsibilities, the daily schedule changes, and frequently the opportunities for activity decrease. When people are young and in school, the activity plan should include not only the activity performed at recess, in physical education classes, or as part of

sport teams, but also non-school-related activity such as supervised and unsupervised play or recreation, household chores, and personal transportation. Once out of high school, people typically have less structured time for activity, and the plan will substantially depend on college or work obligations. When people who have been in college take a full-time job, frequently moving to a new community and meeting new friends, the activity plan will be influenced by the need to identify new space or facilities, find time for activity, and find new persons to exercise with.

> The basic components of an effective activity plan remain relatively constant over the life span, but its implementation will change with age, other life commitments (school, employment, family), health status, and personal goals.

With more and more people living into their 70s, 80s, and 90s, becoming or remaining physically active takes on increased public health importance. In addition to helping to reduce the incidence or severity of various chronic diseases, frequent activity is critical to maintain physical functioning, independent living, and adequate quality of life. For example, in older men and women who show early indicators of loss of mobility, a six-month program of endurance and strength training significantly improved gait speed and performance on a physical functioning test battery (LIFE Study Investigators et al. 2006). Many people who reach their late 60s or older realize that if they had known they were going to live so long they would have taken better care of themselves. Of particular importance is the need for older persons to retain sufficient muscle strength, endurance, balance, and flexibility to independently perform a wide range of activities of daily living. A comprehensive retirement plan (or nonretirement plan if a person continues to work) should include an activity plan that considers the special needs and goals of the older person. This activity plan should focus on maintaining the capacity to independently climb stairs, get out of a bathtub, dress and undress, walk several blocks without stopping, carry groceries home from the market, make the bed, wash dishes and clothes, and perform other self-care and household chores. In this regard, the results of numerous research projects, along with increasing experience from community-based exercise programs for older persons, have documented the importance of a personal physician's endorsement of an activity plan, personalized instruction on how to conduct an effective and safe plan, and convenient and safe facilities in which to carry out the plan.

Integration of Physical Activity With Other Health Behaviors for Optimal Health

Whereas physical activity has a number of unique and independent effects on a person's health, it also interacts in a variety of important ways with other health behaviors or conditions. To obtain the greatest benefit from a physically active lifestyle, a person should consider how activity interacts with such behaviors as sleep, stress management, and nutrition. For example, if the activity being performed to enhance health significantly interferes with a person's ability to get adequate sleep, then health status and quality of life will be negatively affected. Similarly, physical activity and other strategies for managing stress will interact. If the activity is viewed by the person as another burden on limited time and adds to feelings of stress or anxiety, the activity plan should be revised or other adjustments made to reduce the person's overall time burden.

Because food is the source of all the energy and nutrients required for physical activity, eating and being physically active should be closely linked for optimal health. The most obvious link is the relationships among the number of calories consumed, the number of calories expended during physical activity, and body mass, especially adiposity. To achieve optimal long-term health, calories consumed need to be balanced against calories expended at an optimal body mass—a BMI between 18.5 and 25 is generally recommended. Over weeks or months, if the calories expended during activity significantly increase, then calorie intake will need to be increased to maintain a specific body weight, and the opposite will be true if a sustained decrease in activity occurs. Maintaining this calorie balance at an optimal body weight is not an easy task for many people but appears to be easier to achieve at higher levels of activity (Hill et al. 2003). At lower levels of activity where total daily energy expenditure is less than 30 kcal/kg, calorie intake can easily exceed calorie expenditure, resulting in an increase in body weight. Also, to maintain body weight with a sedentary lifestyle and low calorie intake, very careful food selection is needed to obtain the diversity of micronutrients required for optimal health.

> Moderate-intensity physical activity has numerous health benefits that occur independently of other health behaviors, but such activity appears to be even more effective when well integrated with other health-promoting actions.

Maximizing Benefits and Minimizing Risks

Physical activity can aggravate existing disease or cause a variety of injuries, as well as provide substantial health and performance benefits. (See chapter 18 for a more detailed discussion of risks associated with increased activity.) The activity plan should be individualized to minimize these risks while maximizing benefits as defined by the primary goals of the participant. For generally healthy older men and women, performing up to 150 min per week of moderate-intensity activity such as walking is associated with a very low risk of injury (LIFE Study Investigators et al. 2006). To obtain this balance of benefits versus risks, the plan should be individualized based on age, past medical history, current health status, other health-related behaviors, and physical fitness. Attention to risk should increase if the person has cardiorespiratory, metabolic, or musculoskeletal disease; has had joint or bone injuries; is obese; is at high risk of CVD because of smoking, hypertension, or dyslipidemia; or has a very low exercise capacity attributable to age or disability. Other factors that increase the risk of exercise-induced injury include shoes that don't provide appropriate support for walking, jogging, or running; slick or uneven exercise surfaces that contribute to falls (especially in older persons); and high environmental temperatures and humidity or very low temperatures, which can cause heat or cold injuries. For a comprehensive discussion of how to develop a personalized activity plan that considers all of these issues, refer to *ACSM's Guidelines for Exercise Testing and Prescription* (American College of Sports Medicine 2009) or the *Physical Activity Guidelines Advisory Committee Report* (Part G, Section 10) (Physical Activity Guidelines Advisory Committee [PAGAC] 2008).

> Because physical activity can aggravate existing disease or cause injury as well as provide health benefits, people should consider personal characteristics (illness, prior injury), types of activity (high intensity or impact), and environmental conditions (slippery surfaces, high or low temperature) that can substantially increase risk when they design a personalized activity plan.

Research Questions and Issues

This book has cited scientific documentation that a physically active lifestyle leads to a higher state of health and wellness across the life span. The results of this wide range of research support current recommendations by a number of government agencies, health organizations, and medical and exercise associations. However, several important questions need to be answered through scientific investigation to provide a more comprehensive understanding of the biological, clinical, and behavioral issues involved in maintaining an active lifestyle throughout life. Following is a summary of key research issues that should be investigated.

Most of the data supporting a favorable relationship between higher levels of physical activity or physical fitness and morbidity or mortality attributable to coronary heart disease, stroke, type 2 diabetes, and various cancers are from observational studies that are limited in their ability to demonstrate causality. Well-designed randomized clinical trials (RCTs)

are needed in persons at increased risk for specific clinical events to test whether an increase in activity significantly reduces specific clinical events.

A low rate of clinical events such as heart attacks, strokes, and type 2 diabetes, if attributable to higher levels of physical activity, must be mediated by biological changes. For example, it appears that the reduced risk of coronary artery disease may be attributable to improvements in a number of biological processes including lipid metabolism, blood pressure, inflammation, fibrinolysis, insulin-mediated glucose uptake, and coronary artery dilation. However, which of these or other biological changes are necessary to reduce coronary heart disease clinical events needs to be established, and the situation is similar for other clinical outcomes. Knowing what biological changes are necessary to improve clinical status will be valuable in establishing who is most likely to receive benefit for a given outcome and to more accurately develop the dose component of the activity plan.

A small body of research has demonstrated that the tissues and organs challenged by regular exercise adapt by activating the transcription of genes in key pathways and attenuating the transcription of others. Different types of exercise modalities are associated with different profiles of changes in gene expression and thus protein synthesis. Further research of this type is warranted because it has great potential for establishing how exercise affects basic biological functions and how adaptation to exercise occurs at the molecular level.

Several issues regarding the dose of activity required for specific health outcomes still need to be resolved using innovative scientific methods. For example, experimental data are needed on what health outcomes will be produced through an accumulation of 30 to 60 min of activity throughout the day in bouts of just several minutes' duration each, as well as on the specific benefits derived from just two long bouts of activity per week, like 90-min hikes on each day of the weekend.

The role of physical activity in preventing unhealthy weight gain, contributing to weight loss, and preventing weight regain after weight loss is not well understood and requires substantial scientific investigation. Controversy surrounds how much activity is needed to prevent undesired weight gain or significantly contribute to weight loss. Also, the relative benefits of endurance versus resistance exercise in the management of body mass and composition are not well understood. Anecdotes and nonscientific observations suggest that some individuals appear to be quite sedentary yet maintain weight over time.

Others are relatively active but gain weight. These individual differences are probably attributable to genetic and familial factors, but specific information on these points is needed. To more effectively conduct research on physical activity and obesity, we need measurement tools that accurately assess energy intake and energy expenditure in real-life situations over extended periods of time.

Studies have demonstrated considerable interindividual differences in the response to a change in **physical activity level**, a substantial fraction of which is attributable to genetic variation. Research on this topic in human populations exposed to highly standardized exercise programs needs to continue to reveal the true basis of the heterogeneity in responsiveness to regular exercise and to progressively move the field toward individualized exercise recommendations and personalized preventive medicine. Similarly, studies using genomics and other "-omics" technologies, including epigenomics, are necessary to elucidate why regular exercise–induced changes in metabolic risk factors do not correlate with improvements in fitness.

An area that has received only limited attention thus far is the potential deleterious effects of sedentary time and sitting time in particular. Comprehensive research programs are urgently needed to enable understanding of the exact circumstances under which sedentary time exerts its deleterious effects. An important issue for future investigation is whether the potential harmful effects of sedentary time are offset by maintaining a reasonable level of fitness or by engaging regularly in some forms of physical activity that meet the current guidelines.

One of the most challenging issues facing exercise and health professionals is how to help a large proportion of the population achieve a physically active lifestyle throughout the life span. Innovative research is needed to evaluate theory-based activity interventions that are targeted at specific subsets of the population. Priority needs to be given to persons in medically underserved populations, including those with low incomes and limited formal education and ethnic minorities. Much additional research is needed on the effect of the built environment on physical activity patterns. We also need to evaluate how environmental changes accompanied by behavioral and educational interventions might interact to affect activity levels. As noted earlier, we now have the National Physical Activity Plan (2010), which emphasizes strategies and tactics that need to be implemented in eight identified sectors. This approach is comprehensive and the task is massive, but the plan is one that we must now adopt and implement.

Summary

The primary objective for having an active lifestyle throughout the life span is to contribute to a healthy, enjoyable, productive, and long life. It is possible for some minority of people to be quite sedentary throughout most of their life and still have success and satisfaction, remain reasonably free of major diseases, and avoid weight gain during a long life. In the future, we may be able to identify the genetic and biological profile that makes this situation possible; but for the present, we do not have the knowledge or procedures to determine who these people might be. Thus, the public health goal should be to facilitate an active lifestyle throughout the life span for the entire population by means of the following:

- Improved education to increase the public's knowledge about why physical activity is important and how to go about developing a physically active lifestyle
- Changes in the built environment that make safe activity available to all
- Adoption of policies that encourage activity in all aspects of life (occupation, education, transportation, retirement)
- A health care system that aggressively promotes disease prevention, health promotion, and quality of life by means of improved health behaviors, including a physically active lifestyle

Key Concepts

active living—Way of life in which physical activity is valued and integrated into daily life.

built environment—Environment as modified by the construction of any structure or place including, but not limited to, homes, commercial buildings, schools, churches, roads, sidewalks, parks, and recreation centers in the context of its potential effects on daily physical activity.

energy expenditure—Sum of three factors:
- Resting energy expenditure to maintain basic body functions (approximately 65% of total energy requirements)
- Processing of food eaten throughout the day, which includes digestion, absorption, transport, and disposition of nutrients (about 10% of total energy requirements)
- Nonresting energy expenditure, primarily in the form of physical activity (about 25% of total energy requirements)

metabolic equivalent (MET)—For definition, see page 19.

physical activity level—For definition, see page 212.

required activities of daily living—Activities required for a person to survive and lead a productive life.

Study Questions

1. Briefly describe three major changes in the United States that have significantly contributed to the reduction in "daily required physical activity" for a large proportion of the population.

2. What information would you want to have before helping a 43-year-old healthy but very sedentary woman, who is a senior administrative assistant in a large corporation and lives with her husband and two teenage daughters, develop her physical activity plan?

3. Contrast the advantages and disadvantages of a class-based activity program and an "active living" approach for a 55-year-old business executive who frequently travels to other cities.

4. Provide four examples of how the built environment has reduced the opportunities for many people to perform physical activity throughout the day.

5. In helping to design a physical activity plan for an obese (BMI = 36) 17-year-old girl, what information would you want to know about her activity history, medical status, school activities, family arrangements, and social activities?

6. Describe the role that positive or negative attitudes about physical activity have in relation to maintaining an active lifestyle throughout the life span. What conditions or experiences lead to negative attitudes? Propose strategies to improve these attitudes.

7. Briefly discuss three topics in the area of physical activity and health that require more research, and suggest specific research projects that would help address these issues.

References

American College of Sports Medicine. 2009. *ACSM's guidelines for exercise testing and prescription.* 8th ed. Philadelphia: Lippincott Williams & Wilkins.

Bassett, D.R., P.L. Schneider, and G.E. Huntington. 2004. Physical activity in an Old Order Amish community. *Medicine and Science in Sports and Exercise* 36:79-85.

Blair, S.N., A.L. Dunn, B.H. Marcus, R.A. Carpenter, and P. Jaret. 2001. *Active living every day—20 weeks to lifelong vitality.* Champaign, IL: Human Kinetics.

Boreham, C.A., F.M. Wallace, and A. Nevill. 2000. Training effects of accumulated daily stair-climbing exercise in previously sedentary young women. *Preventive Medicine* 30:277-281.

Byun, W., J.C. Sieverdes, X. Sui, S.P. Hooker, C.D. Lee, T.S. Church, and S.N. Blair. 2010. Effect of positive health factors and all-cause mortality in men. *Medicine and Science in Sports and Exercise* 42:1632-1638.

Committee on Physical Activity, Health, Transportation, and Land Use, Transportation Research Board and Institute of Medicine. 2005. *Does the built environment influence physical activity? Examining the evidence.* Special Report 282. Washington, DC: National Academy of Sciences. http://onlinepubs.trb.org/onlinepubs/sr/sr282.pdf. Accessed December 3, 2010.

Dunn, A.L., B.H. Marcus, J.B. Kampert, M.E. Garcia, H.W. Kohl III, and S.N. Blair. 1999. Comparison of lifestyle and structured interventions to increase physical activity and cardiorespiratory fitness: A randomized trial. *Journal of the American Medical Association* 281:327-334.

Esliger, D.W., M.S. Tremblay, J.L. Copeland, J.D. Barnes, G.E. Huntington, and D.R. Bassett Jr. 2010. Physical activity profile of Old Order Amish, Mennonite, and contemporary children. *Medicine and Science in Sports and Exercise* 42:296-303.

Fitness Canada. 1991. *Active living: A conceptual overview.* Ottawa: Government of Canada.

Hedley, A.A., C.L. Ogden, C.L. Johnson, M.D. Carroll, L.R. Curtin, and K.M. Fegal. 2004. Prevalence of overweight and obesity among US children, adolescents, and adults; 1999-2002. *Journal of the American Medical Association* 292:2847-2850.

Hill, J.O., H.R. Wyatt, G.W. Reed, and J.C. Peters. 2003. Obesity and the environment: Where do we go from here? *Science* 299:853-855.

Hu, F.B., W.C. Willett, T. Li, M.J. Stampfer, G.A. Colditz, and J.E. Manson. 2004. Adiposity compared to physical activity in predicting mortality among women. *New England Journal of Medicine* 351:2694-2703.

Katzmarzyk, P.T., T.S. Church, C.L. Craig, and C. Bouchard. 2009. Sitting time and mortality from all causes, cardiovascular disease, and cancer. *Medicine and Science in Sports and Exercise* 41:998-1005.

Lanningham-Foster, L., L.J. Nysse, and J.A. Levine. 2003. Labor saved, calories lost: The energetic impact of domestic labor-saving devices. *Obesity Research* 11:1178-1181.

Lee, I-M., H.D. Sesso, and R.S. Paffenbarger. 2000. Physical activity and coronary heart disease risk in men: Does the duration of exercise episodes predict risk? *Circulation* 102:981-986.

Levine, J.A., L.M. Lanningham-Foster, S.K. McCrady, A.C. Krizan, L.R. Olson, P.H. Kane, M.D. Jensen, and M.M. Clark. 2005. Interindividual variation in posture allocation: Possible role in human obesity. *Science* 307:584-586.

LIFE Study Investigators, M. Pahor, S.N. Blair, M. Espeland, R. Fielding, T.M. Gill, J.M. Guralnik, E.C. Hadley, A.C. King, S.B. Kritchevsky, C. Maraldi, M.E. Miller, A.B. Newman, W.J. Rejeski, S. Romashkan, and S. Studenski. 2006. Effects of a physical activity intervention on measures of physical performance: Results of the lifestyle interventions and independence for Elders Pilot (LIFE-P) study. *Journals of Gerontology: Series A, Biological Sciences and Medical Sciences* 61:1157-1165.

Makosky, L. 1994. The active living concept. In *Toward active living: Proceedings of the International Conference on Physical Activity, Fitness and Health,* ed. H.A. Quinney, L. Gauvin, and A.E. Wall (pp. 272-276). Champaign, IL: Human Kinetics.

Montgomery, E. 1978. Toward representative energy expenditure data: The Machiguenga Study. *Federation Proceedings* 37:61-64.

National Physical Activity Plan. www.physicalactivityplan.org/NationalPhysicalActivityPlan.pdf. Accessed December 3, 2010.

NielsenWire. http://blog.nielsen.com/nielsenwire/media_entertainment/tv-viewing-among-kids-at-an-eight-year-high/. Accessed December 3, 2010.

Physical Activity Guidelines Advisory Committee. 2008. *Physical Activity Guidelines Advisory Committee report, 2008.* Washington, DC: U.S. Department of Health and Human Services. www.health.gov/paguidelines/committeereport.aspx. Accessed December 4, 2010.

Simons-Morton, D.G., P. Hogan, A.L. Dunn, L. Pruitt, A.C. King, B.D. Levine, and S.T. Miller. 2000. Characteristics of inactive primary care patients: Baseline data from the Activity Counseling Trial. *Preventive Medicine* 31:513-521.

Stampfer, M.J., F.B. Hu, J.E. Manson, E.B. Rimm, and W.C. Willett. 2000. Primary prevention of coronary heart disease in women through diet and exercise. *New England Journal of Medicine* 343:16-22.

U.S. Department of Health and Human Services. 2004. *Advance data from vital and health statistics: 347, Mean body weight, height and body mass index, United States 1960-2002.* Washington, DC: U.S. Department of Health and Human Services.

Warren, T.Y., V. Barry, S.P. Hooker, X. Sui, T.S. Church, and S.N. Blair. 2010. Sedentary behaviors increase risk of cardiovascular disease mortality in men. *Medicine and Science in Sports Exercise* 42:879-885.

Wilcox, S., M. Dowda, L.C. Leviton, J. Bartlett-Prescott, T. Bazzarre, K. Campbell-Voytal, R.A. Carpenter, C.M. Castro, D. Dowdy, A.L. Dunn, S.F. Griffin, M. Guerra, A.C. King, M.G. Ory, C. Rheaume, J. Tobnick, and S. Wegley. 2008. Active for life: Final results from the translation of two physical activity programs. *American Journal of Preventive Medicine* 35:340-351.

Yusuf, S., S. Hawken, S. Ounpuu, T. Dans, A. Avzeum, F. Lanas, M. McQueen, A. Budaj, P. Pais, J. Varigos, and L. Lisheng. 2004. Effect of potentially modifiable risk factors associated with myocardial infarction in 52 countries (the INTERHEART study): Case-control study. *Lancet* 364:937-952.

Index

Note: The italicized *f* and *t* following page numbers refer to figures and tables, respectively.

About the Contributors

Loretta DiPietro, PhD, MPH, is a professor in and chair of the Department of Exercise Science at the George Washington University School of Public Health and Health Services. Dr. DiPietro's background includes field work and laboratory- and clinic-based studies of physical activity, and she has received research funding from NIA in this area. DiPietro is a fellow of the American College of Sports Medicine and a recipient of the R. Tait McKenzie Prize (AAPHERD), and she served as the epidemic intelligence service officer for the Centers for Disease Control and Prevention from 1990 to 1991.

Kirk I. Erickson, PhD, is an assistant professor in the Department of Psychology and a member of the Center for the Neural Basis of Cognition at the University of Pittsburgh. Dr. Erickson, an expert in the field of cognitive neuroscience and aging, has focused on changes in brain and cognition that occur in old age and the factors that prevent or reverse senescence. He has found that participating in consistent routines of aerobic exercise, maintaining a healthy diet, and being intellectually engaged can both prevent and reverse cognitive and brain decay. Dr. Erickson and his colleagues have designed and published results of a randomized trial assessing the effect of exercise on neurocognitive function using functional magnetic resonance imaging and have demonstrated that the brain remains plastic and modifiable into old age. His research attempts to bridge molecular neuroscience with human neuroscience with the hope of developing a translational model of age-related cognitive impairment.

Peter A. Farrell, PhD, is a professor in the Department of Exercise and Sport Science and holds an adjunct appointment in the physiology department at East Carolina University. Dr. Farrell has studied hormonal adaptations to both acute and chronic physical activity since 1980. He has authored more than 90 peer-reviewed publications and chapters with a special emphasis on how physical activity alters insulin secretion, insulin action, and hormonal regulation of muscle protein synthesis. Dr. Farrell conducted some of the original research on the role of endor-

phins during exercise. He has been supported by the National Institutes of Health for his work centering on the regulation of skeletal muscle protein synthesis after resistance exercise. He is a leader in the promotion of regular exercise for people with type 1 and type 2 diabetes mellitus. Farrell is a member of both the American College of Sports Medicine and the American Physiological Society and is a fellow in the National Academy of Kinesiology. He has also been the winner of a Citation Award from the American College of Sports Medicine (2004).

Laurie J. Goodyear, PhD, is the senior investigator and section head at the Joslin Diabetes Center in Boston and an associate professor of medicine at Harvard Medical School. Dr. Goodyear has an MS degree in exercise physiology and a PhD in cell biology. She has authored and coauthored numerous publications and obtained several NIH grants. Dr. Goodyear is a member of the American Physiological Society, the American Diabetes Association, and the American College of Sports Medicine. She is a recipient of several awards, including the Mary K. Iacocca Fellowship while she was at the Joslin Research Laboratory (1992); the New Investigator Award from the American College of Sports Medicine (1993); the Career Development Award from the American Diabetes Association (1993); and the Alumna of the Year Award from the School of Public Health at the University of South Carolina (2002)

Howard J. Green, PhD, is a distinguished professor emeritus in the Department of Kinesiology at the University of Waterloo in Ontario. Dr. Green has more than 45 years of teaching and research experience in muscle physiology and has published more than 260 research articles, chapters, and reviews. He has done extensive research on muscle, which has addressed exercise and training, altitude, and disease (e.g., COPD, CHF), including investigation of the cellular basis of weakness and fatigue that accompany those disease states. Dr. Green is a member of the Canadian Society for Exercise Physiology and the American College of Sports Medicine. Among other awards, he has received the 1987 Outstanding Research Award from the Canadian Association of Sport Sciences, the 2001 Award of Excellence in Research from the University of Waterloo, and the 2002 Citation Award (Research) from the American College of Sports Medicine.

Marc Hamilton, PhD, is a professor at the Pennington Biomedical Research Center in Baton Rouge, Louisiana. His doctoral degree is in exercise physiology, and he has received the distinguished alumni award from the Department of Exercise Science in the School of Public Health at the University of South Carolina. He currently directs the Inactivity Physiology Laboratory, which focuses on identifying the underlying molecular triggers impairing health during physical inactivity and on developing effective pragmatic lifestyle solutions for people living in a sedentary society. Dr. Hamilton's laboratory conducted some of the original research identifying a causal basis for explaining why sedentary behaviors are unhealthy independent of exercise. The laboratory also performed work to establish biochemical and molecular reasons why maintaining low-intensity physical activity throughout the entire day is important for clinically relevant outcomes. Through Dr. Hamilton's frequent speaking invitations and collaborative contributions, he has been one of the leaders for this young and interdisciplinary field of inactivity physiology focusing on the science to combat the problems of "too much sitting" (sedentary behaviors).

Adrianne E. Hardman, MSc, PhD, is professor emeritus of human exercise metabolism at the School of Sport, Exercise and Health Sciences at the University of Loughborough, United Kingdom. She is known internationally for research on the acute effects of exercise bouts on lipoprotein metabolism—particularly in the postprandial period. She was a member of the Scientific Advisory Board for the 1992 United Kingdom National Fitness Survey; a speaker on issues of fractionalization (intermittent versus continuous exercise) at the 2000 Consensus Symposium organized by Health Canada and the Centers for Disease Control and Prevention; and the recipient of the first research grants funded by the British Heart Foundation for studies on the acute effects of exercise. She has received many invitations to speak on this topic at scientific meetings in the United Kingdom, Europe, and beyond. Professor Hardman is a member of the American College of Sports Medicine and an honorary fellow of the British Association of Sport and Exercise Sciences. She is coauthor, with Dr. David Stensel, of a U.K. textbook titled *Physical Activity and Health: The Evidence Explained*, now in its second edition.

Jennifer M. Hootman, PhD, ATC, FACSM, FNATA, is an epidemiologist in the Arthritis Program at the Centers for Disease Control and Prevention in Atlanta. She trained clinically in sports medicine, practiced as a certified athletic trainer for 10 years, and taught anatomy and joint evaluation at the university level. Dr. Hootman's doctoral work in epidemiology focused on injury and physical activity epidemiology methods, and she completed a postdoctoral fellowship in arthritis epidemiology at the Centers for Disease Control and Prevention. Dr. Hootman serves as an associate editor for the *Journal of Athletic Training* and is coeditor of the *Journal of Physical Activity and Health*. She is a fellow of the American College of Sports Medicine and the National Athletic Trainers' Association. She is chair of the Association of Rheumatology Health Professionals Research Committee. She has won many awards, including the Student Research Award from the Southeast American College of Sports Medicine in 1998; the Student Travel Award from the American Public Health Association in 2000; the Journal of Athletic Training, Clint Thompson Award for Outstanding Non-Research Manuscript in 2004; the Journal of Athletic Training, Kenneth L. Knight Award for Outstanding Research Manuscript in 2008; and Notable Poster Awards from the American College of Rheumatology in 2008 and 2009.

Edward T. Howley, PhD, is professor emeritus in the Department of Kinesiology, Recreation and Sport Studies at the University of Tennessee. Dr. Howley is the coauthor of an exercise physiology textbook and coeditor of a fitness testing and prescription textbook. He taught exercise physiology for more than 30 years, during which time he received numerous awards for outstanding teaching. Dr. Howley has served as president of American College of Sports Medicine and as a member of the Science Board of the President's Council on Physical Fitness and Sports. In 2007-2008 he served on the Physical Activity Guidelines Advisory Committee. He currently serves as editor-in-chief of *ACSM's Health & Fitness Journal*. In 2007 he was recognized for his contributions with the ACSM Citation Award.

Ian Janssen, PhD, is an associate professor and Canada research chair in physical activity and obesity in the School of Kinesiology and Health Studies and the Department of Community Health and Epidemiology at Queen's University in Ontario, Canada. Dr. Janssen has written more than 100 peer-reviewed publications in the area of obesity, physical activity, and cardiovascular disease. He has expertise and training in exercise physiology and physical activity epidemiology. Awards include the Governor General's Academic Gold Medal Award from Queen's University (2002), an Early Researcher Award from the province of Ontario (2006), a New Investigator Award from the Canadian Institutes of Health Research (2007), and the Canadian Society for Exercise Physiology Young Investigator Award (2007).

Peter T. Katzmarzyk, PhD, FACSM, is a professor and the associate executive director for Population Science at the Pennington Biomedical Research Center in Baton Rouge, Louisiana. He also holds the Louisiana Public Facilities Authority Endowed Chair in Nutrition. Dr. Katzmarzyk received his PhD in exercise science from Michigan State University, and he completed a postdoctoral fellowship at Laval University. Dr. Katzmarzyk has published his research findings in more than 220 scholarly journals and books and regularly participates in the scientific meetings of several national and international organizations. A member of the Canadian Society for Exercise Physiology and fellow of the American College of Sports Medicine, Dr. Katzmarzyk has won the American College of Sports Medicine New Investigator Award (2003) and the Canadian Society for Exercise Physiology Young Investigator Award (2002) and was recognized as a Queen's National Scholar, Queen's University (2002-2007).

Michael J. LaMonte, PhD, MPH, is an epidemiologist in the Department of Social and Preventive Medicine, School of Public Health and Health Professions, University at Buffalo (New York), where he teaches courses on epidemiological methods, physical activity, and cardiovascular disease and conducts research in these areas. He formerly worked in the epidemiology division at the Cooper Institute and also held a position as director of the exercise testing laboratory and research program at LDS Hospital (Cardiology Division), University of Utah, in Salt Lake City. He completed a postdoctoral fellowship in physical activity epidemiology at the University of South Carolina. He is a fellow of the American College of Sports Medicine and member of the American Heart Association.

I-Min Lee, MBBS, MPH, ScD, is trained in medicine and epidemiology. She currently serves as associate professor of medicine at Harvard Medical School and associate professor of epidemiology at the Harvard School of Public Health. Her research interests focus on the role of physical activity in promoting health and enhancing longevity and on women's health. She has authored more than 230 scientific publications, most of which relate to physical activity. Dr.

Lee is widely recognized for her expertise in physical activity and health. She has been invited to serve on several expert panels and to speak at numerous national and international meetings. She was selected to serve on the Physical Activity Guidelines Advisory Committee, which provided the scientific basis for the 2008 U.S. Department of Health and Human Services *Physical Activity Guidelines for Americans,* the first comprehensive physical activity guidelines issued by the U.S. federal government. She is coeditor of a textbook, *Epidemiologic Methods in Physical Activity Studies,* published by Oxford University Press in 2009.

Neil McCartney, PhD, is a professor and Dean of the Faculty of Applied Health Sciences at Brock University in St. Catharines, Ontario. He was formerly a professor in the Department of Kinesiology at McMaster University, where he was also the director of the McMaster University Centre for Health Promotion and Rehabilitation as well as the director of the McMaster University Cardiac Rehabilitation and Seniors' Exercise and Wellness programs. Dr. McCartney is internationally recognized for his work on strength training in seniors and other special populations such as those with coronary artery disease, neuromuscular disorders, and spinal cord injury. His research has demonstrated the safety and efficacy of resistance training in people with heart disease and the effectiveness of long-term resistance training in healthy seniors and has shown that resistance training can promote positive changes in the muscle phenotype of individuals with neuromuscular disorders such as limb girdle and fascioscapulohumeral muscular dystrophy.

Neville Owen, PhD, is head of the Behavioural Epidemiology Laboratory at the Baker IDI Heart and Diabetes Institute in Melbourne, Australia, and professor of health behaviour at the University of Queensland. His research aims to inform the primary prevention of diabetes, heart disease, and cancer and deals with the environmental, social, and personal-level determinants of behavioral risk factors—primarily lack of physical activity and sedentary behaviors (television viewing, sitting in automobiles, desk- and screen-bound work). His program includes descriptive and analytic epidemiology studies, analyses of environmental determinants, and trials of broad-reach interventions. He has been a contributing author to national and international policy documents on physical activity and chronic disease prevention.

Russell R. Pate, PhD, is a professor in the Department of Exercise Science in the Arnold School of Public Health at the University of South Carolina. He has taught exercise science for more than 25 years. He is a nationally recognized expert on physical activity and physical fitness. He was the lead author of the landmark CDC-ACSM position statement on physical activity and public health, published in the *Journal of the American Medical Association*. He served on the advisory committees for the 2005 *Dietary Guidelines for Americans* and the 2008 *Physical Activity Guidelines for Americans*. Currently he chairs the Coordinating Committee for the National Physical Activity Plan and serves on the Committee on Childhood Obesity Prevention of the Institute of Medicine. Dr. Pate is a member of the American College of Sports Medicine and previously served as its president. He has been the recipient of Citation Awards from the American College of Sports Medicine and the Alliance Scholar Award from the American Alliance for Health, Physical Education, Recreation and Dance.

Stuart M. Phillips, PhD, is a professor in the Department of Kinesiology at McMaster University in Ontario. Dr. Phillips received his BSc in biochemistry, his MSc in nutritional biochemistry, and his PhD in physiology (exercise). He has authored more than 100 peer-reviewed publications in human exercise physiology with a focus on skeletal muscle and nutritional physiology. A member of the Canadian Society for Exercise Physiology and a fellow of the College of Sports Medicine, Dr. Phillips has delivered more than 100 scientific and public presentations and has worked with agencies such as the International Olympic Committee in developing nutritional guidelines for athletes.

John S. Raglin, PhD, is a professor and director of graduate studies in the Department of Kinesiology at Indiana University in Bloomington. He is a fellow of the American College of Sports Medicine, the American Psychological Association, and the American Academy of Kinesiology and Physical Activity. Dr. Raglin's research focuses on how psychological and physiological variables interact to influence sport performance and exercise behavior.

Tuomo Rankinen, PhD, is an associate professor in the Human Genomics Laboratory at Pennington Biomedical Research Center in Louisiana. Dr. Rankinen has 15 years of research experience on gene–physical activity interactions on health-related outcomes. He has authored more than 200 original scientific and professional publications, most of them on the genetics of responsiveness to exercise training. He is also the lead author of the Human Gene Map for Performance and Health-Related Fitness Phenotypes review, an annual review that has been published in Medicine and Science in Sports and Exercise since

2001. Rankinen has been the project director of the HERITAGE Family Study and has received independent research funding from the National Institutes of Health and the National Heart, Lung, and Blood Institute. He is a member of the American College of Sports Medicine (fellow), the American Physiological Society, the American Society of Human Genetics, and the American Heart Association. He is the recipient of the International Olympic Committee Sport Science Award for the HERITAGE Family Study (1999) and the American College of Sports Medicine New Investigator Award (2001).

Robert Ross, PhD, FACSM, received his doctoral degree in exercise physiology from the Université de Montréal in 1992. He is currently a professor in the School of Kinesiology and Health Studies and the School of Medicine, Endocrinology and Metabolism at Queen's University in Ontario. His research focuses on the characterization of obesity and the development of lifestyle-based strategies designed to prevent and reduce obesity and related comorbid conditions. Dr. Ross is recognized internationally as a leader in the area of obesity, physical activity, and metabolism and has published more than 150 manuscripts and book chapters on these and related subjects. He was recently awarded a research chair at Queen's University. He is a past president of the Canadian Society for Exercise Physiology, is a fellow of the American College of Sports Medicine, and is currently the director of the Center for Obesity Research and Education at Queen's.

Thomas Rowland, MD, is a pediatric cardiologist. He has extensive research experience in exercise in children. He served as director of the pediatric exercise physiology laboratory at the Baystate Medical Center in Springfield, Massachusetts; president of the North American Society for Pediatric Exercise Medicine; editor of *Pediatric Exercise Science*; and director of pediatric cardiology at Baystate Medical Center from 1975 to 2005. He published the textbook *Children's Exercise Physiology* (2005), providing state-of-the-art information in the field. Dr. Rowland is a member of the North American Society for Pediatric Exercise Medicine and the American College of Sports Medicine. He was the winner of the Honor Award from the New England chapter of the American College of Sports Medicine in 1988.

Roy J. Shephard, MB, BS, MD (London), PhD, DPE, DLL, is a professor emeritus of applied physiology at the University of Toronto. For some 60 years he has written, conducted research, and taught at the university level in all areas of exercise physiology. Dr. Shephard has won Honour Awards and served as president of both the American College of Sport Medicine and the Canadian Society of Exercise Physiology. He has also received honorary doctorates from Ghent University, Université de Montréal, University of Toronto, and the Université de Québec à Trois Rivières. Dr. Shephard has written more than 100 books and about 1,800 papers on many aspects of exercise physiology. He has worked as a professor of applied physiology; director of the School of Physical Education and Health; and professor of preventive medicine, Faculty of Medicine, University of Toronto.

Mark S. Tremblay, PhD, has a bachelor of commerce degree in Sports Administration and a bachelor of Physical and Health Education degree from Laurentian University. His graduate training was from the University of Toronto, where he obtained his MSc and PhD from the Department of Community Health, Faculty of Medicine with a specialty in exercise science. Dr. Tremblay is the director of Healthy Active Living and Obesity Research (HALO) at the Children's Hospital of Eastern Ontario Research Institute and professor of pediatrics in the Faculty of Medicine, University of Ottawa. He is a fellow of the American College of Sports Medicine, chief scientific officer of Active Healthy Kids Canada, chair of the ParticipACTION Research Work Group, chair of the Canadian Physical Activity Guidelines Project, and former Dean of Kinesiology at the University of Saskatchewan. Dr. Tremblay has published more than 120 papers and book chapters in the areas of childhood obesity, physical activity measurement, exercise physiology, exercise endocrinology, and health surveillance.

Willem van Mechelen, MD, PhD, FACSM, FECSS, is a professor of occupational and sports medicine in the Department of Public and Occupational Health at the VU University Medical Center in Amsterdam. Dr. van Mechelen has practical and research experience in occupational medicine, work and health, work-related musculoskeletal disorders, sports medicine, physical fitness and physical activity epidemiology, sport injury epidemiology, and applied exercise physiology. He is a fellow of the American College of Sports Medicine and the European College of Sport Sciences. Dr. van Mechelen is or has been an editorial board member of 10 sports medicine journals, a board member of the Dutch College of Sports Medicine, president of the Dutch Society for Human Movement Sciences, and a board member of the Dutch Epidemiological Society. He is chairman of the WHO Europe Network on Health-Enhancing Physical Activity and chairman of the EU Contact Group Sport. He is a recipient of the 2010 ACSM Citation Award.

Esther M.F. van Sluijs, PhD, is an investigator scientist at the Medical Research Council Epidemiology Unit at the Institute of Metabolic Science in Cambridge, United Kingdom. Dr. van Sluijs received her MSc in Human Movement Sciences and her PhD in Public Health and Epidemiology with a focus on physical activity promotion in general practice at the VU University in Amsterdam, The Netherlands. She is a member of the EMGO Institute/Department of Public and Occupational Health at the VU University Medical Centre in Amsterdam.

Evert A.L.M. Verhagen, PhD, FECSS, is a senior researcher of the Department of Public and Occupational Health at the VU University Medical Centre in Amsterdam. He is a board-certified epidemiologist and human movement scientist. He leads a modest group of 10 researchers, who conduct research in the area of sports medicine (ranging from injury prevention to prescribed exercises for patients with type 2 diabetes mellitus). He is a member of a great number of national or international committees, is on the editorial board of the *Journal of Science and Medicine in Sports,* and is associate editor of the *British Journal of Sports Medicine.*

Gregory S. Wilson, PED, FACSM, is chair and professor in the Department of Exercise and Sport Science at the University of Evansville. Dr. Wilson has cowritten 12 chapters for various edited textbooks dealing with topics in exercise and sport psychology, and he served as editor for two additional textbooks. He has also published numerous articles on a wide variety of topics ranging from the effects of anxiety and overtraining on sport performance to the effects of coping strategies on physical health in college students. Dr. Wilson is a fellow of the American College of Sports Medicine and a member of the American Alliance for Health, Physical Education, Recreation and Dance. He has been the recipient of both the Exemplary Teacher of the Year Award and the Dean's Teaching Award from the University of Evansville.

About the Editors

Claude Bouchard, PhD, is the director of the Human Genomics Laboratory at Pennington Biomedical Research Center, a campus of the Louisiana State University System, where he also holds the John W. Barton Sr. chair in genetics and nutrition. He was director of the Physical Activity Sciences Laboratory at Laval University, Quebec City, Canada, for over 20 years. Dr. Bouchard holds a BPed from Laval University, an MSc in exercise physiology from the University of Oregon at Eugene, and a PhD in population genetics from the University of Texas at Austin.

For four decades, his research has dealt with the role of physical activity, and the lack thereof, on physiology, metabolism, and indicators of health, taking into account genetic uniqueness. He has performed research on the contributions of gene sequence variation and the benefits to be expected from regular activity in terms of changes in cardiovascular and diabetes risk factors.

Dr. Bouchard has served as program leader for four consensus conferences and symposia pertaining to various aspects of physical activity and health. He has published more than 1,000 scientific papers and has edited several books and monographs dealing with physical activity and health.

Dr. Bouchard is the recipient of the Willendorf Award from the International Association for the Study of Obesity, the Sandoz Award from the Canadian Atherosclerosis Society, the Albert Creff Award of the National Academy of Medicine of France, and four honoris causa doctorates (Katholieke Universiteit Leuven, University of South Carolina, University of Guelph, and Brock University). He is a foreign member of the Royal Academy of Medicine of Belgium and a member of the Order of Canada.

Dr. Bouchard is former president of the Canadian Society for Applied Physiology, the North American Association for the Study of Obesity, and the International Association for the Study of Obesity. He is a fellow of the American College of Sports Medicine, the American Heart Association, the American Society of Nutrition, and the American Association for the Advancement of Science.

Steven N. Blair, PhD, is a professor at the Arnold School of Public Health at the University of South Carolina in Columbia. His research focuses on the associations between lifestyle and health, with a specific emphasis on exercise, physical fitness, body composition, and chronic disease. As one of the most highly cited exercise scientists currently active in research, Dr. Blair has published more than 550 articles, chapters, and books in scientific and professional literature. He also was the senior scientific editor for the *U.S. Surgeon General's Report on Physical Activity and Health*.

Dr. Blair has received numerous awards, including the Honor Award from the American College of Sports Medicine, Population Science Award from the American Heart Association, U.S. Surgeon General's Medallion, Folksam Epidemiology Prize from the Karolinska Institute of Stockholm, and a MERIT award from the National Institutes of Health. He also has received honorary doctoral degrees from universities in the United States, Belgium, and England.

Dr. Blair is a fellow of the American College of Epidemiology, Society of Behavioral Medicine, American College of Sports Medicine, American Heart Association, and American Academy of Kinesiology and Physical Education. He was also elected to membership in the American Epidemiological Society. He was the first president of the National Coalition for Promoting Physical Activity and is a past president of the American College of Sports Medicine and the American Academy of Kinesiology and Physical Education.

William L. Haskell, PhD, is emeritus professor of medicine in the Stanford Prevention Research Center and the Division of Cardiovascular Medicine, Stanford School of Medicine. He holds an honorary MD degree from Linkoping University in Sweden.

For more than 40 years, his research has investigated the relationships between physical activity and health. He has been involved at the national and international levels in the development of physical activity and fitness guidelines and recommendations for physical activity in health promotion and disease prevention.

Dr. Haskell has served as principal investigator on major NIH-funded research projects demonstrating the health benefits of physical activity. For the past 17 years, he has been a member of the planning committee and faculty for the CDC-sponsored research course on physical activity and public health. From 1968 to 1970, he was program director for the President's Council on Physical Fitness and Sports. He also served as chair of the Physical Activity Guidelines Advisory Committee for the U.S. Department of Health and Human Services, which documented the scientific basis for the 2008 Physical Activity Guidelines for Americans. From 2008 to 2010 he was a scientific advisor to the World Health Organization for the development of *Global Recommendations on Physical Activity for Health* (2010) and to the United Kingdom Health Ministries for the development of physical activity and sedentary behavior guidelines for the home countries. Currently he is chair of the International Review Panel for the Evaluation of Exercise and Sports Sciences in the Nordic Countries.

He is past president of the American College of Sports Medicine and founder and past president of the American College of Sports Medicine Foundation. He was a fellow with the Exercise and Rehabilitation Council, American Heart Association, and American Association of Cardiovascular and Pulmonary Rehabilitation.

*You'll find other outstanding
physical activity and health
promotion resources at*

www.HumanKinetics.com

In the U.S. call

1-800-747-4457

Australia	08 8372 0999
Canada	1-800-465-7301
Europe	+44 (0) 113 255 5665
New Zealand	0800 222 062

 HUMAN KINETICS
The Information Leader in Physical Activity & Health
P.O. Box 5076 • Champaign, IL 61825-5076 USA